Terrestrial Ecosystems and Biodiversity

The Handbook of Natural Resources, Second Edition

Series Editor:
Yeqiao Wang

Volume 1
Terrestrial Ecosystems and Biodiversity

Volume 2
Landscape and Land Capacity

Volume 3
Wetlands and Habitats

Volume 4
Fresh Water and Watersheds

Volume 5
Coastal and Marine Environments

Volume 6
Atmosphere and Climate

Terrestrial Ecosystems and Biodiversity

Edited by
Yeqiao Wang

CRC Press
Taylor & Francis Group
Boca Raton London New York

CRC Press is an imprint of the
Taylor & Francis Group, an **informa** business

CRC Press
Taylor & Francis Group
6000 Broken Sound Parkway NW, Suite 300
Boca Raton, FL 33487-2742

First issued in paperback 2022

ISBN 13: 978-1-03-247442-7 (pbk)
ISBN 13: 978-1-138-33391-8 (hbk)

DOI: 10.1201/9780429445651

Library of Congress Cataloging-in-Publication Data

Names: Wang, Yeqiao, editor.
Title: Handbook of natural resources / edited by Yeqiao Wang.
Other titles: Encyclopedia of natural resources.
Description: Second edition. | Boca Raton: CRC Press, [2020] | Revised edition of: Encyclopedia of natural resources. [2014]. | Includes bibliographical references and index. | Contents: volume 1. Ecosystems and biodiversity — volume 2. Landscape and land capacity — volume 3. Wetland and habitats — volume 4. Fresh water and watersheds — volume 5. Coastal and marine environments — volume 6. Atmosphere and climate. |
Summary: "This volume covers topical areas of terrestrial ecosystems, their biodiversity, services, and ecosystem management. Organized for ease of reference, the handbook provides fundamental information on terrestrial systems and a complete overview on the impacts of climate change on natural vegetation and forests. New to this edition are discussions on decision support systems, biodiversity conservation, gross and net primary production, soil microbiology, and land surface phenology. The book demonstrates the key processes, methods, and models used through several practical case studies from around the world" — Provided by publisher.
Identifiers: LCCN 2019051202 | ISBN 9781138333918 (volume 1 ; hardback) | ISBN 9780429445651 (volume 1 ; ebook)
Subjects: LCSH: Natural resources. | Land use. | Climatic changes.
Classification: LCC HC85 .E493 2020 | DDC 333.95—dc23
LC record available at https://lccn.loc.gov/2019051202

Visit the Taylor & Francis Web site at
http://www.taylorandfrancis.com

and the CRC Press Web site at
http://www.crcpress.com

Contents

Preface ... ix
About The Handbook of Natural Resources ... xi
Acknowledgments ... xiii
Aims and Scope ... xv
Editor ... xvii
Contributors ... xix

SECTION I Biodiversity and Conservation

1 Assessment of Community-Based Activities to Reduce Human–Elephant Conflict in Nepal ... 3
Bill Buffum, Thomas P. Husband, and Sandeep Shrestha

2 Biodiversity: Agriculture ... 17
Deborah Stinner

3 Biodiversity: Climate Change ... 23
Lisa Freudenberger and Jan Hanning Sommer

4 Biodiversity: Conservation ... 35
Nigel E. Stork

5 Biodiversity: Habitat Suitability ... 45
Betty J. Kreakie

6 Biodiversity: Tropical Agroforestry ... 55
Thomas P. Husband

7 Biodiversity: Values ... 65
David Ehrenfeld

8 Biomes ... 75
Paul Reich and Hari Eswaran

9 Decision-Making and Monitoring Strategies in Natural Resource Management and Conservation ... 81
Brian D. Gerber and Perry J. Williams

10 Diversity: Species 87
Jeffrey D. Corbin and Brittany L. Oakes

11 Exotic and Invasive Species 93
Jan Pergl and Petr Pyšek

12 Keystone and Indicator Species 105
Barry R. Noon

13 Population Genetics 113
Michael T. Clegg

14 Prioritizing Conservation Sites for Biodiversity Protection in Uganda....... 119
Samuel Ayebare, Grace Nangendo, and Simon Nampindo

15 Weeds: Seed Banks and Seed Dormancy 129
Lynn Fandrich

SECTION II Ecosystem Type, Function, and Service

16 Community Forestry: Sustainability and Equity Issues 135
Bill Buffum

17 Ecosystem Services: Evaluation 143
James Boyd

18 Ecosystem Services: Land Systems Approach 151
Richard Aspinall

19 Ecosystem Services: Pollinators and Pollination 157
Victoria A. Wojcik

20 Ecosystems: Diversity 165
L. M. Chu

21 Ecosystems: Forest Nutrient Cycling 177
Neil W. Foster and Jagtar S. Bhatti

22 Ecosystems: Functions and Services 183
Robert Costanza

23 Ecosystems: Soil Animal Functioning 191
Alan J. Franzluebbers

24 Ecotone 197
Finn C. Pillsbury and Debra P. C. Peters

25 Environmental Goods and Services: Economic and Non-Economic Methods for Valuing 205
Daniel E. Campbell

26 Forests: Temperate Evergreen and Deciduous 213
Lindsay M. Dreiss and John C. Volin

27 Forests: Tropical Rain 227
Richard T. Corlett

28 Savannas and Grasslands 235
Niti B. Mishra and Kenneth R. Young

SECTION III Ecological Processes

29 Agriculture: Organic 251
John P. Reganold and Jonathan M. Wachter

30 Crops and the Atmosphere: Trace Gas Exchanges 261
Jürgen Kreuzwieser and Heinz Rennenberg

31 Desertification 267
Uriel N. Safriel

32 Ecological and Evolutionary Processes 281
Roland C. de Gouvenain and Gopalasamy Reuben Clements

33 Ecology: Functions, Patterns, and Evolution 289
Fernando Valladares

34 Land Plants: Origin and Evolution 293
Elizabeth R. Waters

35 Leaves: Elevated CO_2 Levels 299
G. Brett Runion, Seth G. Pritchard, and Stephen A. Prior

36 Net Ecosystem Production (NEP) 305
Mark E. Harmon

37 Nitrogen Deposition and *Deyeuxia angustifolia* Encroachment in Alpine Tundra 315
Shengwei Zong and Kunwei Wang

38 Nitrogen Fixation: Biological 321
Robert H. Burris

39 Ozone: Crop Yield and Quality Effects 325
Håkan Pleijel

40 Global Plant Breeding 331
Manjit S. Kang

SECTION IV Ecosystem Monitoring

41 Altitudinal Belts: Global Mountains, Patterns, and Mechanisms 339
Baiping Zhang and Fang Zhao

42 Biodegradation and Bioremediation 349
Owen P. Ward and Ajay Singh

43 Decision Support: Monitoring, Reporting, and Forecasting Ecological Conditions of Appalachian Trail 359
Yeqiao Wang

44 Estimation of Net Primary Production and Its Spatial and Temporal Variations in Northeast China: Remote Sensing 365
Xiaoyi Guo, Jianjun Zhao, and Hongyan Zhang

45 Optical Properties of Reflected Light from Leaves 383
Zhongqiu Sun

46 Remote Sensing of CO_2 Emissions from Wildfires 393
Meng Guo

Index 403

Preface

Terrestrial Ecosystems and Biodiversity is the first volume of *The Handbook of Natural Resources, Second Edition (HNR)*. This volume consists of 46 chapters authored by 66 contributors from 11 countries. The contents are organized in four sections: *Biodiversity and Conservation* (15 chapters); *Ecosystem Type, Function and Service* (13 chapters); *Ecological Processes* (12 chapters); and *Ecosystem Monitoring* (6 chapters).

A variety of ecosystems exist in a given place on the planet Earth. The summation of the Earth's ecosystems forms the biosphere. The dynamics of terrestrial ecosystem are influenced by natural and anthropogenic factors from local to global scales. Ecological complexity and stability are maintained by a diversity of interacting species or populations that have profound effects on the structure and function of ecosystems. Biodiversity is the diversity of life from genes to species to ecosystems. Biodiversity, productivity, and nutrient cycling are all connected and change through time. Biodiversity conservation is essential in protection of organisms and their habitats because of the increasing threats from environmental and climate change. Climate and biodiversity are interacting in multiple and complex ways. The impact of environmental and climate change is becoming so severe that a large proportion of the world's species are threatened with extinction. Overexploitation of natural resources can degrade the ecological and evolutionary processes that sustain these ecosystems. Proper ecosystem management and monitoring are essential to maintain the planet as a provider of different ecosystem services. Evaluation of ecosystem services involves biophysical analysis, ecological conditions, and changes in those conditions, resulting from environmental management, stresses, and protection.

With the challenges and concerns, the 46 chapters in this volume cover topics in *Biodiversity and Conservation*, including population genetics, diversity of species, exotic and invasive species, seed banks and seed dormancy, keystone and indicator species, biomes, climate change and biodiversity, biodiversity conservation, biodiversity value, habitat suitability, as well as monitoring strategies and prioritization for management and conservation planning; in *Ecosystem Type, Function and Service*, including sustainability and equity issues of community forestry, evaluation of ecosystem services, land system approach in ecosystem services, pollinators and pollination, ecosystem diversity, forest nutrient cycling, ecosystem function and services, soil animal functioning, ecotone, valuing economic and non-economic environmental goods and services, temperate evergreen and deciduous forests, tropical rain forests, savannas and grasslands; in *Ecological Processes*, including organic agriculture, trace gas-exchanges between crops and atmosphere, desertification, ecological and evolutionary processes, functions, patterns and evolution of ecology, origin and evolution of land plants, net ecosystem production, nitrogen deposition, biological fixation, ozone's crop yield and quality effects, and global plant breeding; and in *Ecosystem Monitoring*, including patterns and mechanisms of altitudinal belts of global mountains, biodegradation and bioremediation, decision support in monitoring, reporting and forecasting ecological conditions, estimation of net primary production and spatial and temporal variation, remote sensing of CO_2 emission from wildfire, and optical properties of reflected light from leaves.

The chapters provide updated knowledge and information in general environmental and natural science education and serve as a value-added collection of references for scientific research and management practices.

Yeqiao Wang
University of Rhode Island

About The Handbook of Natural Resources

With unprecedented attentions to the changing environment on the planet Earth, one of the central focuses is about the availability and sustainability of natural resources and the native biodiversity. It is critical to gain a full understanding about the consequences of the changing natural resources to the degradation of ecological integrity and the sustainability of life. Natural resources represent such a broad scope of complex and challenging topics.

The Handbook of Natural Resources, Second Edition (HNR), is a restructured and retitled book series based on the 2014 publication of the *Encyclopedia of Natural Resources (ENR).* The ENR was reviewed favorably in February 2015 by CHOICE and commented as *highly recommended for lower-division undergraduates through professionals and practitioners.* This HNR is a continuation of the theme reference with restructured sectional design and extended topical coverage. The chapters included in the HNR provide authoritative references under the systematic relevance to the subject of the volumes. The case studies presented in the chapters cover diversified examples from local to global scales, and from addressing fundamental science questions to the needs in management practices.

The Handbook of Natural Resources consists of six volumes with 241 chapters organized by topical sections as summarized below.

Volume 1. Terrestrial Ecosystems and Biodiversity
Section I. Biodiversity and Conservation (15 Chapters)
Section II. Ecosystem Type, Function and Service (13 Chapters)
Section III. Ecological Processes (12 Chapters)
Section IV. Ecosystem Monitoring (6 Chapters)

Volume 2. Landscape and Land Capacity
Section I. Landscape Composition, Configuration and Change (10 Chapters)
Section II. Genetic Resource and Land Capability (13 Chapters)
Section III. Soil (15 Chapters)
Section IV. Landscape Change and Ecological Security (11 Chapters)

Volume 3. Wetlands and Habitats
Section I. Riparian Zone and Management (13 Chapters)
Section II. Wetland Ecosystem (8 Chapters)
Section III. Wetland Assessment and Monitoring (9 Chapters)

Volume 4. Fresh Water and Watersheds
Section I. Fresh Water and Hydrology (16 Chapters)
Section II. Water Management (16 Chapters)
Section III. Water and Watershed Monitoring (8 Chapters)

Volume 5. Coastal and Marine Environments
Section I. Terrestrial Coastal Environment (14 Chapters)
Section II. Marine Environment (13 Chapters)
Section III. Coastal Change and Monitoring (9 Chapters)

Volume 6. Atmosphere and Climate
Section I. Atmosphere (16 Chapters)
Section II. Weather and Climate (16 Chapters)
Section III. Climate Change (8 Chapters)

With the challenges and uncertainties ahead, I hope that the collective wisdom, the improved science, technology, and awareness and willingness of the people could lead us toward the right direction and decision in governance of natural resources and make responsible collaborative efforts in balancing the equilibrium between societal demands and the capacity of natural resources base. I hope that this *Handbook of Natural Resources* series can help facilitate the understanding about the consequences of changing resource base to the ecological integrity and the sustainability of life on the planet Earth.

Yeqiao Wang
University of Rhode Island

Acknowledgments

I am honored to have this opportunity and privilege to work on *The Handbook of Natural Resources, Second Edition (HNR)*. It would be impossible to complete such a task without the tremendous amount of support from so many individuals and groups during the process. First and foremost, I thank the 342 contributors from 28 countries around the world, including Australia, Austria, Brazil, China, Cameroon, Canada, Czech Republic, Finland, France, Germany, Hungary, India, Israel, Japan, Nepal, New Zealand, Norway, Puerto Rico, Spain, Sweden, Switzerland, Syria, Turkey, Uganda, the United Kingdom, the United States, Uzbekistan, and Venezuela. Their expertise, insights, dedication, hard work, and professionalism ensure the quality of this important publication. I wish to express my gratitude in particular to those contributors who authored chapters for this HNR and those who provided revisions from their original articles published in the *Encyclopedia of Natural Resources*.

The preparation for the development of this HNR started in 2017. I appreciate the visionary initiation of the restructure idea and the guidance throughout the preparation of this HNR from Irma Shagla Britton, Senior Editor for Environmental Sciences and Engineering of the Taylor & Francis Group/CRC Press. I appreciate the professional assistance and support from Claudia Kisielewicz and Rebecca Pringle of the Taylor & Francis Group/CRC Press, which are vital toward the completion of this important reference series.

The inspiration for working on this reference series came from my over 30 years of research and teaching experiences in different stages of my professional career. I am grateful for the opportunities to work with many top-notch scholars, colleagues, staff members, administrators, and enthusiastic students, domestic and international, throughout the time. Many of my former graduate students are among and/or becoming world-class scholars, scientists, educators, resource managers, and administrators, and they are playing leadership roles in scientific exploration and in management practice. I appreciate their dedication toward the advancement of science and technology for governing the precious natural resources. I am thankful for their contributions in HNR chapters.

As always, the most special appreciation is due to my wife and daughters for their love, patience, understanding, and encouragement during the preparation of this publication. I wish my late parents, who were past professors of soil ecology and of climatology from the School of Geographical Sciences, Northeast Normal University, could see this set of publications.

Yeqiao Wang
University of Rhode Island

Aims and Scope

Land, water, and air are the most precious natural resources that sustain life and civilization. Maintenance of clean air and water and preservation of land resources and native biological diversity are among the challenges that we are facing for the sustainability and well-being of all on the planet Earth. Natural and anthropogenic forces have affected constantly land, water, and air resources through interactive processes such as shifting climate patterns, disturbing hydrological regimes, and alternating landscape configurations and compositions. Improvements in understanding of the complexity of land, water, and air systems and their interactions with human activities and disturbances represent priorities in scientific research, technology development, education programs, and administrative actions for conservation and management of natural resources.

The chapters of *The Handbook of Natural Resources, Second Edition (HNR)*, are authored by world-class scientists and scholars. The theme topics of the chapters reflect the state-of-the-art science and technology, and management practices and understanding. The chapters are written at the level that allows a broad scope of audience to understand. The graphical and photographic support and list of references provide the helpful information for extended understanding.

Public and private libraries, educational and research institutions, scientists, scholars, resource managers, and graduate and undergraduate students will be the primary audience of this set of reference series. The full set of the HNR and individual volumes and chapters can be used as the references in general environmental science and natural science courses at different levels and disciplines, such as biology, geography, Earth system science, environmental and life sciences, ecology, and natural resources science. The chapters can be a value-added collection of references for scientific research and management practices.

Editor

Yeqiao Wang, PhD, is a professor at the Department of Natural Resources Science, College of the Environment and Life Sciences, University of Rhode Island. He earned his BS from the Northeast Normal University in 1982 and his MS degree from the Chinese Academy of Sciences in 1987. He earned the MS and PhD degrees in natural resources management and engineering from the University of Connecticut in 1992 and 1995, respectively. From 1995 to 1999, he held the position of assistant professor in the Department of Geography and the Department of Anthropology, University of Illinois at Chicago. He has been on the faculty of the University of Rhode Island since 1999. Among his awards and recognitions, Dr. Wang was a recipient of the prestigious Presidential Early Career Award for Scientists and Engineers (PECASE) in 2000 by former U.S. President William J. Clinton, for his outstanding research in the area of land cover and land use in the Greater Chicago area in connection with the Chicago Wilderness Program.

Dr. Wang's specialties and research interests are in terrestrial remote sensing and the applications in natural resources analysis and mapping. One of his primary interests is the land change science, which includes performing repeated inventories of landscape dynamics and land-use and land-cover change from space, developing scientific understanding and models necessary to simulate the processes taking place, evaluating consequences of observed and predicted changes, and understanding the consequences of change on environmental goods and services and management of natural resources. His research and scholarships are aimed to provide scientific foundations in understanding of the sustainability, vulnerability and resilience of land and water systems, and the management and governance of their uses. His study areas include various regions in the United States, East and West Africa, and China.

Dr. Wang published over 170 refereed articles, edited *Remote Sensing of Coastal Environments* and *Remote Sensing of Protected Lands*, published by CRC Press in 2009 and 2011, respectively. He served as the editor-in-chief for the *Encyclopedia of Natural Resources* published by CRC Press in 2014, which was the first edition of *The Handbook of Natural Resources.*

Contributors

Richard Aspinall
James Hutton Institute
Aberdeen, Scotland

Samuel Ayebare
Uganda Country Program
Wildlife Conservation Society
Bronx, New York

Jagtar S. Bhatti
Canadian Forest Service
Northern Forestry Centre
Edmonton, Alberta, Canada

James Boyd
Resources for the Future
Washington, District of Columbia

Bill Buffum
Department of Natural Resources Science
University of Rhode Island
Kingston, Rhode Island

Robert H. Burris
Department of Biochemistry
College of Agriculture and Life Sciences
University of Wisconsin
Madison, Wisconsin

Daniel E. Campbell
Atlantic Ecology Division (AED)
National Health and Environmental Effects Research Laboratory (NHEERL)
U.S. Environmental Protection Agency (EPA)
Narragansett, Rhode Island

L. M. Chu
School of Life Sciences
Chinese University of Hong Kong
Hong Kong, China

Michael T. Clegg
Department of Ecology and Evolutionary Biology
University of California, Irvine
Irvine, California

Gopalasamy Reuben Clements
School of Science and Technology
Sunway University
Petaling Jaya, Selangor, Malaysia

Jeffrey D. Corbin
Department of Biological Sciences
Union College
Schenectady, New York

Richard T. Corlett
Xishuangbanna Tropical Botanical Garden
Chinese Academy of Sciences
Yunnan, China

Robert Costanza
Institute for Sustainable Solutions
Portland State University
Portland, Oregon

Roland C. de Gouvenain
Department of Biology
Rhode Island College
Providence, Rhode Island

Lindsay M. Dreiss
Department of Natural Resources and the Environment
University of Connecticut
Storrs, Connecticut

David Ehrenfeld
Department of Ecology, Evolution, and Natural Resources
Rutgers University
New Brunswick, New Jersey

Hari Eswaran
National Resources Conservation Service
U.S. Department of Agriculture (USDA-NRCS)
Washington, District of Columbia

Lynn Fandrich
Oregon State University
Corvallis, Oregon

Neil W. Foster
Canadian Forest Service, Ontario
Sault Ste. Marie, Ontario, Canada

Alan J. Franzluebbers
Agricultural Research Service
U.S. Department of Agriculture (USDA-ARS)
Watkinsville, Georgia

Lisa Freudenberger
Center for Development Research (ZEF)
University of Bonn
Bonn, Germany

Brian D. Gerber
Department of Natural Resource Science
University of Rhode Island
Kingston, Rhode Island

Meng Guo
School of Geographical Sciences
Northeast Normal University
Changchun, China

Xiaoyi Guo
Key Laboratory of Geographical Processes and Ecological Security in Changbai Mountains
Ministry of Education, School of Geographical Sciences
Northeast Normal University
Changchun, China

Mark E. Harmon
Oregon State University
Corvallis, Oregon

Thomas P. Husband
Department of Natural Resources Science
University of Rhode Island
Kingston, Rhode Island

Manjit S. Kang
Department of Plant Pathology
Kansas State University
Manhattan, Kansas

Betty J. Kreakie
Atlantic Ecology Division
Office of Research and Development
U.S. Environmental Protection Agency (EPA)
Narragansett, Rhode Island

Jürgen Kreuzwieser
Institute of Forest Botany and Tree Physiology
University of Freiburg
Freiburg, Germany

Niti B. Mishra
University of Texas at Austin
Austin, Texas

Simon Nampindo
Uganda Country Program
Wildlife Conservation Society
Bronx, New York

Grace Nangendo
Uganda Country Program
Wildlife Conservation Society
Bronx, New York

Barry R. Noon
Department of Fish, Wildlife, and Conservation Biology
Colorado State University
Fort Collins, Colorado

Brittany L. Oakes
Department of Biological Sciences
Union College
Schenectady, New York

Jan Pergl
Department of Invasion Ecology
Institute of Botany
Academy of Sciences of the Czech Republic (CAS)
Průhonice, Czech Republic

Debra P. C. Peters
Jornada Experimental Range
Agricultural Research Service
U.S. Department of Agriculture (USDA-ARS)
Las Cruces, New Mexico

Finn C. Pillsbury
Jornada Experimental Range
Agricultural Research Service
U.S. Department of Agriculture (USDA-ARS)
Las Cruces, New Mexico

Håkan Pleijel
Applied Environmental Science
Göteborg University
Göteborg, Sweden

Seth G. Pritchard
Department of Biology
College of Charleston
Charleston, South Carolina

Stephen A. Prior
National Soil Dynamics Laboratory
Agricultural Research Service
U.S. Department of Agriculture (USDA-ARS)
Auburn, Alabama

Petr Pyšek
Department of Invasion Ecology
Institute of Botany
Academy of Sciences of the Czech Republic (CAS)
Průhonice, Czech Republic

John P. Reganold
Department of Crop and Soil Sciences
Washington State University
Pullman, Washington

Paul Reich
National Resources Conservation Service
U.S. Department of Agriculture (USDA-NRCS)
Washington, District of Columbia

Heinz Rennenberg
Intitute of Forest Botany and Tree Physiology
University of Freiburg
Freiburg, Germany

G. Brett Runion
National Soil Dynamics Laboratory
Agricultural Research Service
U.S. Department of Agriculture (USDA-ARS)
Auburn, Alabama

Uriel N. Safriel
Center of Environmental Conventions
Jacob Blaustein Institutes for Desert Research
Department of Ecology, Evolution and Behavior
Hebrew University of Jerusalem
Jerusalem, Israel

Sandeep Shrestha
Department of Environmental Science and Engineering
Kathmandu University
Dhulikhel, Nepal

Ajay Singh
Department of Biology
University of Waterloo
Waterloo, Ontario, Canada

Jan Hanning Sommer
Center for Development Research (ZEF)
University of Bonn
Bonn, Germany

Deborah Stinner
Ohio Agriculture Research and Development Center
The Ohio State University
Wooster, Ohio

Nigel E. Stork
Environment Futures Centre
Griffith School of Environment
Griffith University
Brisbane, Queensland, Australia

Zhongqiu Sun
Key Laboratory of Geographical Processes and Ecological Security in Changbai Mountains
Ministry of Education
School of Geographical Sciences
Northeast Normal University
Changchun, China

Fernando Valladares
Centro de Ciencias Medioambientales
Madrid, Spain

John C. Volin
Department of Natural Resources and the Environment
University of Connecticut
Storrs, Connecticut

Jonathan M. Wachter
Department of Crop and Soil Sciences
Washington State University
Pullman, Washington

Kunwei Wang
School of Geographical Sciences
Northeast Normal University
Changchun, China

Yeqiao Wang
Department of Natural Resources Science
University of Rhode Island
Kingston, Rhode Island

Owen P. Ward
Department of Biology
University of Waterloo
Waterloo, Ontario, Canada

Elizabeth R. Waters
Department of Biology
San Diego State University
San Diego, California

Perry J. Williams
Department of Natural Resources and Environmental Science
University of Nevada
Reno, Nevada

Victoria A. Wojcik
Pollinator Partnership
San Francisco, California

Kenneth R. Young
University of Texas at Austin
Austin, Texas

Baiping Zhang
State Key Lab for Resources and Environment Information System
Institute of Geographic Sciences and Natural Resources Research
Chinese Academy of Sciences
Beijing, China

Hongyan Zhang
Key Laboratory of Geographical Processes and Ecological Security in Changbai Mountains
Ministry of Education, School of Geographical Sciences
Northeast Normal University
Changchun, China

Fang Zhao
State Key Lab for Resources and Environment Information System
Institute of Geographic Sciences and Natural Resources Research
Chinese Academy of Sciences
Beijing, China

Jianjun Zhao
Key Laboratory of Geographical Processes and Ecological Security in Changbai Mountains
Ministry of Education
and
Urban Remote Sensing Application Innovation Center
School of Geographical Sciences
Northeast Normal University
Changchun, China

Shengwei Zong
Key Laboratory of Geographical Processes and Ecological Security in Changbai Mountains
Ministry of Education
School of Geographical Sciences
Northeast Normal University
Changchun, China

I

Biodiversity and Conservation

1

Assessment of Community-Based Activities to Reduce Human–Elephant Conflict in Nepal

Introduction 3
Methods 4
Site Selection • Statistical Analysis
Results 7
Major Problems in the Villages • Community Forestry • Individual Activities to Reduce Human–Elephant Conflict • Community-Based Activities to Reduce Human–Elephant Conflict • Attribute Ratings in the Study Villages • Correlations between Attributes
Discussion 13
Were Communities with Active CFs More Successful at Implementing HEC Activities than Communities without CFs? • What Other Factors Made Some Communities More Successful than Others in Implementing Community-Based HEC Programs? • How Can the National Park Further Strengthen the Capability of Communities to Implement Community-Based HEC Activities?
References 15

Bill Buffum and
Thomas P. Husband
University of Rhode Island

Sandeep Shrestha
Kathmandu University

Introduction

Human–elephant conflict (HEC) is a growing problem in Nepal [1,2]. The Asian elephant (*Elephas maximus*) is the most dangerous wildlife species in Nepal, causing more than 40% of the conflicts with humans and 70% of human casualties [3]. More than 290 incidents of elephant damage were reported in the buffer zones of Chitwan National Park and Parsa Wildlife Reserve between 2008 and 2012, including 21 deaths [4]. Most of the incidents occurred in the winter months [2]. The major cause for the increasing HEC is the increasing fragmentation of remaining forests coupled with increasing elephant populations due to their migration from India [5].

Community-based approaches have been recommended to address HEC in Nepal [6] and India [7]. Several studies have assessed elephant damage in Nepal in terms of human casualties, crop damage and economic loss, and recommended mitigation measures to reduce human–wildlife conflict in Nepal [4,8,9]. However, no study to date has examined why some communities in Nepal are better able to implement community-based HEC programs than others. Our study assessed the implementation of community-based HEC activities in the buffer zone of Chitwan National Park. We focused on the construction of solar electric fencing, the major community-based HEC approach in our study sites, but also assessed a number of other community-based HEC activities. Solar electric fencing is considered to be the most cost-effective approach to reducing damage from wild elephants [1].

We were particularly interested to learn if communities involved in the management of community forests (CFs) were better able to collaborate in community-based HEC activities than communities that were not managing a CF. Governments around the world are increasingly devolving authority for forest management to the local level through community forestry programs [10–12]. This approach has been particularly widespread in Nepal, where more than 13,000 CF user groups in Nepal are managing 25% of the total area of national forests [13]. CFs are recognized as providing a range of economic and social benefits to the participating communities [10–12,14]. An analysis of CFs in Nepal found evidence of improved governance and concluded that CFs "have much pertinent experience to offer in village level planning and decision making", p. 43 [15]. Another study reported that CF user groups had effectively negotiated with external agencies for development projects in their villages such as rural electrification [16]. Therefore, we expected that communities managing CFs would possess the organizational skills to implement more effective community-based HEC programs.

Our study addressed three research questions:

1. Are communities with active CFs more successful at implementing HEC activities than communities without CFs?
2. What other factors make some communities more successful than others in implementing community-based HEC programs?
3. What can the government do to strengthen the capability of communities to implement community-based HEC activities?

Methods

Site Selection

We implemented the study in the buffer zone of Chitwan National Park in southern Nepal near the border with India. The area of the national park is 932 km^2 with an elevation range of 150–950 m above sea level [17]. The climate of the area is classified as sub-tropical. We selected 17 villages in the buffer zone of the national park based on the findings of a previous study that assessed the number and types of HEC incidents in the buffer zone between 2008 and 2012 [4]. We selected 15 villages with the highest number of HEC incidents in five Village Development Committees (VDC) and included two villages in Kumjoj VDC that had fewer HEC incidents but were located in the same general area (Table 1.1).

We visited each village during January and February 2015 and conducted focus group sessions with the local residents to discuss how they are addressing conflict with elephants. Each focus group session lasted 2–4 hours and involved an average of nine participants. After each session, the villagers took us on a tour of their HEC sites, which provided another opportunity for us to ask questions. We also collected secondary data from office records and libraries of the Chitwan National Park and the Department of National Parks and Wildlife Conservation.

Prior to visiting the first village, we developed a list of 15 attributes to quantify in each village. The first attribute was the total number of HEC incidents between 2008 and 2012 according to Pant and Hockings (2013). The other 14 attributes included socio-economic status, land tenure security, experience in implementing community-based HEC activities, experience with community forestry, and several attributes related to the impact and status of the HEC activities. We developed criteria to rank each of our new attributes on either a 1–3 or a 1–5 scale, and compiled an open-ended questionnaire to elicit the information we required to rate each attribute. This type of attribute ranking has been used in other studies of community-based activities [18,19]. We field-tested the rating system in the first three villages and made some adjustments before visiting the remaining villages (Table 1.2).

TABLE 1.1 Information about Study Villages, Including the Number of Participants in the Focus Groups, and the Number of HEC Incidents between 2008 and 2012

Village Development Committee	Ward No.	Total Households	Number of Focus Group Participants[a]	Number of HEC Incidents[b] Property	Crop	Human Casualty	All Types
Ayodhyapuri	1	200	10	7	2	0	9
Ayodhyapuri	4	50	9	3	0	0	3
Ayodhyapuri	5	100	12	6	5	0	11
Ayodhyapuri	6	428	8	10	3	0	13
Ayodhyapuri	7	650	7	5	4	0	9
Ayodhyapuri	8	250	10	10	28	1	39
Ayodhyapuri	9	1,000	11	8	5	4	17
Bagauda	5	100	8	5	11	0	16
Bagauda	6	200	9	6	3	0	9
Gardi	1	800	16	2	7	1	10
Gardi	4	700	8	6	2	1	9
Kalyanpur	6	190	9	5	11	0	16
Kalyanpur	9	300	6	2	4	0	6
Kumroj	2	250	8	2	0	0	2
Kumroj	3	2,000	8	3	0	0	3
Thori	3	90	6	2	1	1	4
Thori	5	250	7	3	3	0	6

[a] In the current study.
[b] Based on Pant and Hockings (2013).

TABLE 1.2 Ranking Criteria for Village Attributes

Rank	Ranking Criteria
	HEC Incidents between 2008 and 2012
	Total number of HEC incidents according to Pant and Hockings (2013)
	HEC Problem Rating
1	Community considers conflict with elephants to be their worst problem
2	Community considers conflict with elephants to be their second worst problem
3	Community considers conflict with elephants to be their third worst problem
	Socio-economic Status
1	Almost all residents are subsistence farmers or agricultural laborers, living in traditional houses
2	Most residents are subsistence farmers or agricultural laborers, living in traditional houses
3	Approximately half of residents are commercial farmers living in improved houses with concrete walls or tin roofs
4	Most residents are commercial farmers living in improved houses with concrete walls or tin roofs
5	Almost all residents are commercial farmers living in improved houses with concrete walls or tin roofs
	Land Tenure Security
1	Almost all residents are landless
2	Most residents are landless
3	Approximately half residents are landless
4	Most residents are not landless
5	Almost no residents are landless

(Continued)

TABLE 1.2 (*Continued*) Ranking Criteria for Village Attributes

Rank	Ranking Criteria
	Capability of Community Intuitions to Implement HEC Activities
1	Community lacks organizational capability to implement HEC activities
2	Community has limited organizational capability to implement HEC activities
3	Community has average organizational capability to implement HEC activities
4	Community has good organizational capability to implement HEC activities
5	Community has excellent organizational capability to implement HEC activities
	Status of Community Forest or Community Grassland
1	No CF or CG
2	Existing CF or CG that is not yet registered
3	Existing CF or CG that is registered
	Experience with Community Forest or Community Grassland
1	No experience with CF or CG
2	CF or CG not very active and members report poor management
3	CF or CG is moderately active with average management
4	CF or CG active with good management, and supports some village development activities
5	CF or CG is very active, well managed, and supports many village development activities
	Community Forest Support for HEC Activities
1	No CF or CG
2	Existing CF or CG but no support for HEC activities
3	Existing CF or CG that supports HEC activities
	Scale of HEC Program
1	Community has not implemented community-based HEC activities
2	Community implemented one major community-based HEC activity
3	Community implemented two major community-based HEC activities
4	Community implemented three major community-based HEC activities
5	Community implemented three major and two minor community-based HEC activities
	Impact of HEC Activities during First Year
1	No reduction of damage from wildlife
2	Minor reduction of damage from wildlife
3	Moderate reduction of damage from wildlife
4	High reduction of damage from wildlife
5	Very high reduction of damage from wildlife
	Current Impact of Fencing on Rhinos
1	No reduction of damage from wildlife
2	Minor reduction of damage from wildlife
3	Moderate reduction of damage from wildlife
4	High reduction of damage from wildlife
5	Very high reduction of damage from wildlife
	Current Impact of Fencing on Elephants
1	No reduction of damage from wildlife
2	Minor reduction of damage from wildlife
3	Moderate reduction of damage from wildlife
4	High reduction of damage from wildlife
5	Very high reduction of damage from wildlife

(*Continued*)

TABLE 1.2 (*Continued*) Ranking Criteria for Village Attributes

Rank	Ranking Criteria
	Current Status of Fencing
1	Most of the fencing damaged and/or electricity not provided
2	Many areas of fence damaged and/or electricity not available
3	Some portions of the fence functioning, other portions damaged, or electricity not available
4	Some openings in fence, electricity generally available
5	Fencing functioning well with electricity
	Resistance to Flooding of Fencing
1	Flooding has caused extreme damage to fencing
2	Flooding has caused major damage to fencing
3	Flooding has caused moderate but repairable damage to fencing
4	Flooding has caused only minor damage to fencing
5	Flooding has not caused any damage to fencing
	Motivation to Implement Additional HEC Activities
1	Community has lost all motivation to implement HEC activities
2	Community has limited motivation to implement HEC activities
3	Community has average motivation to implement HEC activities

Statistical Analysis

We analyzed the data using IBM SPSS Statistics (v.25, Armonk, NYL IBM Corp). We classified the number of HEC incidents as interval data, and the ranking data for all other attributes as ordinal data. We tested for significant correlations using Kendal's Tau, which is considered the most appropriate nonparametric test when the sample size is small and there are many tied ranks [20]. We used one-tailed tests when we had predicted positive correlations and two-tailed tests when we did not have a specific prediction.

Results

Major Problems in the Villages

Eleven of the 17 focus groups considered wild elephants to be the most serious problems in their village, and all of the focus groups reported that the number of HEC incidents had increased during the previous 2 years. The most frequent damage was eating and trampling agricultural crops, which occurred in all of the 17 villages studied. The crops most often damaged by elephants were rice, wheat, and maize. Elephants also damaged houses or grain storage structures in all 17 villages. The elephants generally came into the villages between July and December. Humans had been injured or killed in four of the study villages during the previous 2 years. Single elephants did the most damage, although sometimes herds of elephants entered the villages.

The villagers offered different explanations for the increasing problems with elephants. Many believed that the national park was not large enough to support the increasing population of elephants and that a limited food supply in the national park was causing the elephants to search for food in the neighboring villages. Others attributed the problem to the release of older domesticated elephants that were previously used for tourism purposes in the national park. These elephants were unafraid of humans and accustomed to being fed by them. Furthermore, they tended to be larger than wild elephants.

Four of the 17 villages reported that flooding from the Riyo River was an even greater problem than elephant conflict. Flooding had repeatedly damaged their agricultural land and homes, and made it harder for them to implement activities to prevent elephants from entering their villages. Two villages reported that their most serious problem was landlessness or lack of drinking water supply (Table 1.3).

TABLE 1.3 Three Most Serious Problems in Each Study Site

Ward	Worst Problem	Second Problem	Third Problem
Ayodhyapuri-1	Elephant	Flooding	Chital
Ayodhyapuri-4	Elephant	Flooding	Rhino
Ayodhyapuri-5	Elephant	Rhino	Chital
Ayodhyapuri-6	Flooding	Elephant	Wild boar
Ayodhyapuri-7	Landlessness	Tiger	Elephant
Ayodhyapuri-8	Elephant	Chital	Tiger
Ayodhyapuri-9	Elephant	Tiger	Rhino
Bagauda-5	Flooding	Elephant	Rhino
Bagauda-6	Flooding	Elephant	Rhino
Gardi-1	Elephant	Flooding	Rhino
Gardi-4	Elephant	Flooding	Rhino
Kalyanpur-6	Elephant	Flooding	Rhino
Kalyanpur-9	Elephant	Flooding	Rhino
Kumroj-2	Elephant	Rhino	Chital
Kumroj-3	Elephant	Rhino	Chital
Thori-3	Flooding	Elephant	Chital
Thori-5	Water supply	Elephant	Flooding

Other wildlife species reported as serious problems in some villages included Bengal tiger (*Panthera tigris tigris*), Indian rhinoceros (*Rhinoceros unicornis*), chital (*Axis axis*), and Indian boar (*Sus scrofa cristatus*), all of which live in the national park but frequently enter the villages (Table 1.3).

Community Forestry

The community forestry program of Nepal allows villages to request the Forest Department to give them legal rights to manage a block of national forest near the village. The village must establish a CF management group that will coordinate the management of the CF and ensure that the forest is managed sustainability. The CF management group must also maintain accurate financial records and ensure that CF products are distributed equitably. We found that 10 of the 17 study villages were participating in the community forestry program, and one was managing a community grassland (CG). Most of the CFs were not yet formally registered, which limited their management options. Nevertheless, most of the CFs and the CG appeared to be functioning well and providing valuable services to the communities.

The CF members were generally able to purchase timber or thatch grass once per year after making a nominal payment to the CF. The payments were deposited in the CF fund, which was managed by the CF management committee and used to hire forest guards to patrol the forest. CF funds were also used to support village development activities, including building gabion structures for flood control, supporting local schools, building community irrigation systems, improving access roads to the community, and providing loans to individual CF members for activities such as creating fish ponds or installing biogas systems. Some CFs produced bamboo for flood prevention activities or provided free fuelwood to community members for events such as funerals and marriages. In four of the study sites, CFs provided valuable support to community-based HEC programs, including repair of the solar electric fences by the CF guards, and provision of CF timber to replace fence posts that were damaged by elephants.

Individual Activities to Reduce Human–Elephant Conflict

Individuals in the villages implemented a number of activities to reduce elephant damage. The most common individual activity was shouting or making noise by striking tins to drive the elephants away. A few individuals constructed private watchtowers near their agriculture fields, or planted timber trees in their fields, which was less profitable than producing food crops but less likely to be damaged by elephants. However, these individual approaches had much less impact than community-based activities that are implemented on a much larger scale.

Community-Based Activities to Reduce Human–Elephant Conflict

The communities implemented nine types of community-based activities to reduce human–elephant conflict. We classified these as primary activities, which involve construction and require significant inputs from the community, and secondary activities, which involve fewer community inputs (Table 1.4).

Solar Electric Fencing: The major HEC activity was construction of solar electric fences between the national park and the villages. Because electricity was generally not available in the area, this activity required installation of batteries charged by solar panels. Most of the villages could not afford to implement this activity on their own, but fortunately, the national park provided technical and financial assistance. Each participating village established a "functional group" (FG) to manage the activity composed of eight to ten members. The national park provided wires, solar batteries, solar panels, other necessary items, and, in some cases, timber from the national park for fence posts. The villagers provided labor to construct the solar fencing, which included carrying the poles to the site, placing and fixing poles, and attaching the fencing wire. The FGs assigned one or more guards to monitor elephant damage to the fence. The guards were also responsible for turning on the electric current each evening and turning it off in the morning. If any major damage to the fence took place, the guard would inform the FG, which organized the villagers to repair it. Most communities paid the guards in cash or with food grains. In some communities, the households took turns as guards on a volunteer basis.

The solar electric fencing was highly effective at deterring elephants during the first year. However, the elephants quickly learned that they could avoid electric shocks by throwing trees at the fence and breaking the wires. The elephants also learned that they could knock down the fences by pushing against the posts. Therefore, the elephants started coming back into many of the villages during the second year,

TABLE 1.4 Community-Based Activities to Human–Elephant Conflict in the 17 Study Sites

	Primary Activities			Secondary Activities					
VDC and Ward	Solar Electric Fence	Dam	Watchtower	Siren	Elephant Guard	Torchlight	Chili Powder	Tractor Patrol	Community Burning
Ayodhyapuri-1	X		X	X		X			
Ayodhyapuri-4	X		X						
Ayodhyapuri-5	X		X						
Ayodhyapuri-6	X								
Ayodhyapuri-7	X								
Ayodhyapuri-8	X		X			X			
Ayodhyapuri-9	X	X	X						
Bagauda-5	X		X						
Bagauda-6	X		X		X				
Gardi-1	X	X	X		X		X	X	X
Gardi-4	X						X		X
Kalyanpur-6	X	X	X		X				
Kalyanpur-9	X		X						
Kumroj-2									
Kumroj-3	X								
Thori-3	X		X						
Thori-5	X								

X, implemented in village.

despite the efforts of the community to maintain the fence. The fencing was much more effective at controlling rhinos, which generally avoided the fences after getting shocked once. Flooding of the Riyo River has been an even greater problem for many communities. Flooding repeatedly swept many poles and wires of fenced areas, so the communities found it hard to cope with these two independent problems (elephant and flooding damage) simultaneously. In some villages, major repair work on the fences was required three times per year.

Dams: Several communities constructed earthen dams and built their solar fences on top of them. These earthen dams reduced the risk of flood damage to the fences, and made it harder for elephants to damage the electric fencing. In some cases, the villagers constructed the dams by hand. In other cases, the national park provided an excavator for the dam construction. In both cases, the community established a FG to coordinate the construction. One person from every household was required to provide free labor. In some cases, villagers owning land near the dam constructed fishponds in the excavated areas, which provided an additional source of food and income.

The combination of dams and electric fencing was much more effective at reducing elephant damage. However, in some cases the elephants started damaging the fences by throwing trees on the wires from the bottom of the dams. They also damaged the dams by digging into the soil. However, the dams were still considered to be very effective, despite the significant amount of labor required to construct and maintain them.

Community Watchtower: The third activity we categorized as a "primary activity" was constructing community watchtowers. The participating villages established FGs to coordinate the construction of the towers. The national park provided the timber, which was harvested inside the park and transported to the village by tractors. One member of each household participated in harvesting the timber and constructing the watchtower. Some villages hired contractors to construct the towers. During June–December, when elephants were likely to enter the village, groups of villagers took turns sleeping in the towers. When an elephant entered the village, the villagers tried to deter them by making loud noise, lighting fires, or shining torchlights at the elephants. In some cases, they phoned the army and asked them to come drive the elephants away. This approach was successful at some sites, but elephants attacked and damaged some of the towers.

The villagers also experimented with a number of minor activities that we classified as "secondary activities."

Siren: One village tried using a siren that had been provided by the national park. The siren was stored in a household at the center of the village. When an elephant approached, the siren was used to frighten the elephant and to let the other villagers know that the elephant is present. This approach was effective in some cases, but did not always deter the elephant.

Elephant Guard: The national park helped three villages to experiment with a domesticated elephant guard. The national park provided two domesticated elephants to each village for 45 days. The communities were responsible for providing food to the elephant and an elephant driver. The presence of these large domesticated elephants deterred wild elephants from entering the village. This activity was effective, but the domesticated elephants were only available for one season.

Torchlight: The national park provided large torchlights to two villages. A team of three younger people in each community was selected to operate the torchlights. The community alerted the team if an elephant was seen in the village. Then, the team focused the torchlights on the elephant from three directions, and other villagers shouted or burned firewood. In one village, this activity was effective. However, after 1 year, the battery died and the village could not afford to replace it. In the other village, the elephants became aggressive when the torchlights were focused on it, and the community abandoned this approach.

Chili Powder: In two villages, the national park donated 500 sacks of chili powder, which the villagers mixed with cow dung. They placed the sacks along the electric fence and in different locations in the village. They burned the mixture at night when an elephant entered the ward. However, the villagers themselves developed eye problems from the smoke when winds shifted, and the fumes made some elephants become more aggressive. After trying this method five times with limited success, the communities abandoned it.

Tractor Patrol: In one village, the national park provided a tractor and a torch light for patrolling during the crop-harvesting season. A group of six young people were involved in tractor patrolling on a rotational basis. When an elephant entered the village, they blew the horn of the tractor and focused the torchlight on the elephant. However, this method was not effective.

Community Burning: In two villages where residents had been killed by elephants during the previous year, the community collected firewood and stored it in multiple locations in their agricultural fields. The national park also provided some wood, which the villagers transported to their villages on bullock carts. When elephants entered the villages, groups of villagers lit fires in different locations. This activity was moderately successful, but some elephants continued to enter the villages.

Attribute Ratings in the Study Villages

Based on the focus group sessions and our discussions with community members during the field visits, we rated the 15 community attributes for the 17 villages (Tables 1.5A and 1.5B).

Correlations between Attributes

We predicted that many of the 15 attributes would be positively correlated. Our statistical tests of these predictions (one-tailed tests) found a number of significant correlations (Tables 1.6A and 1.6B). We also tested for correlations (two-tailed) between attributes for which we did not predict positive correlations. None of these test results was significant except for community capability with problem rating (Kendal's Tau = 0.543, $p < 0.016$, $N = 17$) and community ability with flood resistance (Kendal's Tau = 0.612, $p < 0.006$, $N = 16$).

TABLE 1.5A Rating of Village Attributes for the Study Sites

Site	HEC Incidents[a]	HEC Problem Rating	Socio-economic Status	Land Tenure	Capability	CF Status	CF Experience	CF HEC Support
Ayodhyapuri-1	9	3	2	4	3	2	3	3
Ayodhyapuri-4	3	3	2	4	4	1	1	1
Ayodhyapuri-5	11	3	2	4	4	1	1	1
Ayodhyapuri-6	13	2	3	4	2	2	3	2
Ayodhyapuri-7	9	1	2	3	4	2	3	2
Ayodhyapuri-8	39	3	3	4	5	2	3	2
Ayodhyapuri-9	17	3	2	4	4	2	2	2
Bagauda-5	16	2	2	4	4	1	1	1
Bagauda-6	9	2	3	3	3	1	1	1
Gardi-1	10	3	2	4	4	1	1	1
Gardi-4	9	3	2	2	4	1	1	1
Kalyanpur-6	16	3	2	4	5	2	3	2
Kalyanpur-9	6	3	2	4	5	2	4	3
Kumroj-2	2	3	3	4	5	3	5	1
Kumroj-3	3	3	4	4	5	3	5	3
Thori-3	4	2	2	4	3	3	4	3
Thori-5	6	2	2	4	2	2	3	1

[a] Based on Pant and Hockings (2013).

TABLE 1.5B Rating of Village Attributes for the Study Sites

Site	Scale of HEC Program	Impact after First Year	Current Impact Elephant Damage	Current Impact Rhino Damage	Current Status of Fencing	Flood Resistance of Fencing	Current Motivation
Ayodhyapuri-1	3	5	1	3	2	2	3
Ayodhyapuri-4	3	3	1	3	2	2	3
Ayodhyapuri-5	3	4	1	3	2	2	3
Ayodhyapuri-6	2	3	2	3	2	2	2
Ayodhyapuri-7	2	5	3	3	3	4	2
Ayodhyapuri-8	3	5	2	3	3	4	4
Ayodhyapuri-9	4	5	2	4	4	4	4
Bagauda-5	3	5	2	3	3	3	3
Bagauda-6	4	5	3	3	3	2	4
Gardi-1	5	5	1	5	2	4	3
Gardi-4	2	5	1	4	3	4	3
Kalyanpur-6	5	5	2	5	5	2	4
Kalyanpur-9	3	5	3	4	4	5	4
Kumroj-2	1	1	1	1	[a]	[a]	3
Kumroj-3	2	2	2	4	4	5	3
Thori-3	3	5	2	2	3	2	3
Thori-5	2	2	1	2	2	2	2

[a] Kumroj-2 did not construct solar electric fence, so we did not provide a rating for these attributes.

TABLE 1.6A Significant Correlations between Attributes for Which We Predicted Positive Correlations (Kendal's Tau, One-Tailed)

	Elephant Incidents	Problem Rating	Socio-economic	Land Tenure	Community Capability	CF Status	CF Experience
Elephant incidents	1.000	ns					ns
Problem rating	ns	1.000					
Socio-economics			1.000				
Land tenure				1.000	ns		
Community capability					1.000	ns	
CF status						1.000	0.920**
CF experience						0.920**	1.000
CF HEC support						0.648**	0.654**
Scale of HEC program	0.393*	ns	ns	ns	ns	ns	ns
First-year impact	0.429*	ns			ns	ns	0.576**
Impact on elephants	Ns	ns			ns	ns	ns
Impact on rhinos	ns	0.427*			0.378*	ns	0.453*
Electric fence status	Ns	ns			0.580**	ns	ns
Flood resistance of fencing							
Current motivation	Ns	0.471*			0.469*	ns	ns

* Significant at the 0.05 level (one-tailed).
** Significant at the 0.01 level (one-tailed).
ns = not significant. $N = 17$ for all tests, except those involving electric fence status or flood resistance for which $N = 16$.

TABLE 1.6B Significant Correlations between Attributes for Which We Predicted Positive Correlations (Kendal's Tau, One-Tailed)

	CF HEC Support	Scale of HEC Program	First-Year Impact	Impact on Elephants	Impact on Rhinos	Electric Fence Status	Flood Resistance of Fencing	Current Motivation
Elephant incidents	ns	0.393*	0.429*	ns	ns	ns		ns
Problem rating	ns		Ns	ns	0.427*	ns		0.471*
Socio-economics		ns						
Land tenure		ns						
Community capability	Ns	ns	ns	ns	0.378*	0.580**		0.469*
CF status	0.648**	ns	ns	ns	ns	ns		ns
CF experience	0.654**	ns	ns	ns	ns	ns		ns
CF HEC support	1.000	ns	ns	0.426*	ns	0.388*		ns
Scale of HEC program	Ns	1.000	0.576**	ns	0.453*	ns	ns	0.592**
First-year impact	Ns	0.576**	1.000	0.377*	0.383*	ns		0.434*
Impact on elephants	0.426*	ns	0.377*	1.000	ns	0.560**	ns	ns
Impact on rhinos	Ns	0.453*	0.383*	ns	1.000	0.468**	0.497*	0.375*
Electric fence status	0.388*	ns	ns	0.560**	0.468**	1.000	0.488*	0.555**
Flood resistance of fencing		ns		ns	0.497*	0.488*	1.000	ns
Current motivation	Ns	0.592**	0.434*	ns	0.375*	0.555**	ns	1.000

* Significant at the 0.05 level (One-tailed).

** Significant at the 0.01 level (One-tailed).

ns = not significant. $N = 17$ for all tests, except those involving electric fence status or flood resistance for which $N = 16$.

Discussion

Were Communities with Active CFs More Successful at Implementing HEC Activities than Communities without CFs?

We assessed three attributes related to community forestry: the legal status of the CF, the range of activities implemented by the CF, and the amount of support that the CF contributed to HEC programs. As expected, these three attributes were positively correlated. We also detected a number of positive correlations between the CF attributes and the effectiveness of the HEC programs. CF experience was positively correlated with the scale of the HEC program, the impact of the HEC programs during the first year, and the current impact on rhino damage. CF contributions to HEC programs were positively correlated with the current impact of HEC programs on elephant damage and with the current status of the electric fences.

We expected that villages with successful CFs would have greater motivation to implement additional HEC activities, but we did not detect any correlations between these attributes. We believe that this was largely due to the extensive flooding in the area, which repeatedly damaged the electric fencing in many of the villages and was significantly correlated to the current status of the fencing. We learned that HEC programs in some villages had initially very successful, but had failed by the time of our study. We originally assumed that the success of the HEC activities would be linked to the effort and capability of the communities, but the flooding turned out to be a confounding factor. Our findings agreed with research conducted 1 year after our study, which reported that only 26% of the electric fences in the buffer zone of the Chitwan National Park were operating effectively due to severe flood damage [21].

Another factor was the small sample size of our study. If we had been able to assess a greater number of villages, we may have detected a greater number of significant correlations between our CF attributes and the impact of the HEC programs. Nevertheless, we did not detect any significant negative correlations between these attributes. Therefore, despite the limitations of our study, we believe that our results support the hypothesis that experience with community forestry supports the implementation of more effective community-based HEC programs. Our findings agree with other studies that concluded that CFs helped communities develop improved governance and village-level planning, including negotiating with external agencies for development projects in their villages [15,16].

Our findings on community capability also support our belief that strong community institutions have an important impact on a village's ability to manage community-based programs. We did not expect correlations between our community capability attribute and our CF attributes, because we knew that some very capable and motivated communities did not have any suitable land for establishing a CF, even though they would have liked to create one. However, our community capability attribute was positively correlated with impact on rhino damage, status of electric fencing, and community motivation to implement additional activities.

What Other Factors Made Some Communities More Successful than Others in Implementing Community-Based HEC Programs?

We expected that communities with lower socio-economic status and less land tenure security would be less able to implement HEC programs than wealthier communities. However, we did not detect any significant correlations between these two attributes and any other attributes. We were encouraged by this finding and believe that the financial support provided by the national park has made it possible for both rich and poor communities to participate in community-based HEC activities.

We assessed the current impact of the HEC programs with two attributes: impact on elephant damage and impact on rhino damage. As expected, these two attributes were closely correlated, even though many communities reported that their HEC programs had been much more successful in reducing damage by rhinos. In addition, both attributes were correlated with the impact of the program during the first year, and with the current status of the fence, as we expected. We also expected that these two attributes would be correlated with community capability, but this was only the case for the impact on rhino damage.

We expected that the motivation of communities to implement additional HEC activities would be correlated with community capability, the scale of the HEC program, the impact during the first year, the status of the electric fence, and the impact on elephant and rhino damage. We detected significant correlations for all of these attributes, except for impact on elephant damage. We believe these correlations would have been even stronger without the confounding factor of the flooding.

How Can the National Park Further Strengthen the Capability of Communities to Implement Community-Based HEC Activities?

Based on our findings, we offer the following recommendations to the national park: (a) seek ways to speed up the process of CF registration, which would help communities develop the institutional capability to implement community-based HEC programs and also provide an additional source of local support for the HEC programs; (b) encourage communities in flood-prone areas to build solar fences on top of dams, and provide financial support to the communities for dam construction; and (c) provide more intensive training to the community FGs on repair and maintenance of the solar batteries and associated wiring, and encourage the FGs to maintain a supply of wire, iron nails, and poles, so immediate repairs can be done after elephant damage.

References

1. Sapkota, S., A. Aryal, S.R. Baral, and M.W. Hayward, Economic analysis of electric fencing for mitigating human-wildlife conflict in Nepal. *BIOONE*, 2014. **5**(7): pp. 237–243.
2. Neupane, D., R.L. Johnson, and T.S. Risch, How do land-use practices affect human-elephant conflict in Nepal? *BIOONE*, 2017. **2017**. doi:10.2981/wlb.00313.
3. Bajimaya, S., Managing human-wildlife conflict in Nepal, in *Biodiversity Conservation in Nepal: A Success Story*, K.P. Acharya, M. Dhakal, Editors. 2012, Department of National Parks and Wildlife Conservation: Kathmandu, Nepal.
4. Pant, G. and M. Hockings, *Understanding the Nature and Extent of Human-Elephant Conflict in Central Nepal*. 2013, The University of Queensland: Central, Nepal.
5. Pradhan, N.M., A.C. Williams, and M. Dhakal, Current status of Asian elephants in Nepal. *Gajah*, 2011. **35**: pp. 87–92.
6. Cordingley, M., Participatory development in Nepal: challenges and opportunities for conservation in managing human elephant conflict. *Gajah*, 2008. **29**: pp. 41–44.
7. Zimmermann, A., T.E. Davies, N. Hazarika, and S. Wilson, Community-based human-elephant conflict management in Assam. *Gajah*, 2009. **30**: pp. 34–40.
8. Shrestha, R., A case study on human-wildlife conflict in Nepal (With particular reference to human-elephant conflict in Eastern and Western Terai regions). 2007: World Wildlife Fund (WWF).
9. Yadav, B., C. Dutta, M.K. Chalise, and C. Williams, Human-Asian wild elephant (Elephas maximus) conflicts and its socio-economic consequences in and around the protected areas of Central Terai, Nepal. *Banko Janakari*, 2014. **24**(1): pp. 47–54.
10. Carter, J., Recent Experience in Collaborative Management - CIFOR Occasional Paper No. 43. 2005, Centre for International Forestry Research: Bogor.
11. Lawrence, A., Beyond the second generation: towards adaptiveness in participatory forest management. *Veterinary Science, Nutrition and Natural Resources*, 2007. **2**(28): pp. 1–15.
12. Menzies, N.K., *'Global Gleanings' Lessons from Six Studies of Community Based Forest Management, Report for Ford Foundation's Environment and Development Affinity Group*. 2002, Ford Foundation: Berkley.
13. Pokharel, B.K. and D. Paudel, Impacts of armed conflicts on community forest user groups in Nepal: can community forestry survive and contribute to peace building at local level? *European Tropical Forest Research Network*, 2005. **2005**(43/44): pp. 83–86.
14. Buffum, B., Why is there no tragedy in these commons? An analysis of forest user groups and forest policy in Bhutan. *Sustainability*, 2012. **4**(7): pp. 1448–1465.
15. Carter, J., B. Pokharel, and R.R. Parajuli, *Two Decades of Community Forestry in Nepal: What Have We Learned?* 2011, Nepal Swiss Community Forestry Project: Kathmandu, Nepal.
16. Schreckenberg, K. and C. Luttrell, Participatory forest management: a route to poverty reduction? *International Forestry Review*, 2009. **11**(2): pp. 221–238.
17. DNPWC, *The Elephant Conservation Action Plan of Nepal*. 2008, Department of National Parks and Wildlife Conservation: Kathmandu, Nepal.
18. Buffum, B., A. Lawrence, and K.J. Temphel, Equity in community forests in Bhutan. *International Forestry Review*, 2010. **12**(3): pp. 187–199.
19. Malla, Y., H.R. Neupane, and P.J. Branney, Why aren't poor people benefiting more from community forestry? *Journal of Forest and Livelihood*, 2003. **3**(1): pp. 78–90.
20. Field, A.P., *Discovering Statistics Using SPSS*. 2nd ed. 2005, Sage Publications: London. p. 779.
21. Banikoi, H., S. Thapa, N. Bhattarai, and R.C. Kandel, Mitigating human-wildlife conflict in Nepal: a case study of fences around Chitwan National Park. *ICIMOD Working Paper 2017/14*. 2017, International Centre for Integrated Mountain Development: Kathmandu, Nepal.

2

Biodiversity: Agriculture

Deborah Stinner
The Ohio State University

Introduction 17
Agriculture's Impact on Wild Biodiversity Globally 17
Structure and Function of Agroecosystems and Biodiversity 18
Offsite Impacts of Agricultural Activities 19
Loss of Diversity within Agricultural Species 19
Conclusion 19
Biodiversity as a Principle of Agroecosystem Management
References 21

Introduction

No other human activity has greater impact on the Earth's biodiversity than agriculture. From its origins some 12,000 years ago, the goal of agriculturists has been to enhance production of desired species over competing species. Expansion of human agricultural activity around the globe historically has resulted in significant impacts on global biodiversity in four major ways: 1) loss of wild biodiversity and species shifts resulting from conversion of native ecosystems by agroecosystems; 2) influence of agroecosystems' structure and function on agrobiodiversity; 3) offsite impacts of agricultural practices; and 4) loss of genetic diversity among and within agricultural species.[1] Although agriculture and biodiversity often are inversely related, biodiversity enhancement can be a key organizing principle in sustainable agroecosystems.[2–4]

Agriculture's Impact on Wild Biodiversity Globally

Historically, the earliest subsistence farmers and pastoralists had low population densities and limited technology and their small-scale patchworks of fields, pastures, and home gardens had little net effect on global biodiversity. In some ecosystems, agricultural activity may have actually increased biodiversity because more diverse habitats and ecotones were created—a pattern that may still exist in some areas.[2] However, as surplus agricultural production allowed human populations to increase and with the development of civilizations, the impacts of agriculture on wild biodiversity increased, even to the point that biodiversity loss may have contributed to the decline of some ancient civilizations. Since 1650, there has been at least a 600% increase in the worldwide deforestation of native ecosystems for agriculture and wood extraction that have resulted in radical changes to wild biodiversity globally.

Wild biodiversity is more threatened now than at any time since the extinction of the dinosaurs, with nearly 24% of all mammals, 12% of birds, and almost 14% of plants threatened with extinction.[2] If current trends continue, it is estimated that at least 25% of the Earth's species could become extinct or drastically reduced by the middle of this century.[2] Conversion of natural ecosystems to agroecosystems is a primary cause of these alarming trends.[2,4,5] At least 28% of the Earth's land area currently is devoted to agriculture to some degree.[1] Intensive agriculture dominates 10% of the earth's total land area and is part of the landscape mosaic on another 17%, while extensive grazing covers an additional 10–20%.[2] Nearly half of the global temperate broadleaf and mixed-forest and tropical and subtropical dry and monsoon broadleaf forest ecosystems are converted to agricultural use (45.8% and 43.4%, respectively).[1,2] However, agriculture's greatest impact has been on grassland ecosystems, including temperate grasslands, savannas, and shrublands (34.2%); flooded grasslands and savannas (20.2%); and montane grasslands and shrublands (9.8%).[1] Combined, 64.2% of the Earth's grassland ecosystems have been converted to agriculture, primarily for production of cereal grasses—maize, rice and wheat.[1,2] In the past 20 years, net expansion of agricultural land has claimed approximately 130,000 km^2/yr globally, mostly at the expense of forest and grassland ecosystems, but also from wetlands and deserts.[1]

The native ecosystems that agriculture has replaced typically had high biodiversity. A hectare of tropical rain forest may contain over 100 species of trees and at least 10–30 animal species for every plant species, leading to estimates of 200,000 or more total species.[6] In contrast, the world's agroecosystems are dominated by some 12 species of grains, 23 vegetable crops, and about 35 fruit and nut crop species.[1] Furthermore, conversion of native ecosystems to agriculture causes dramatic shifts in ecosystem structure and function that affect ecosystem processes above and below ground including energy flow, nutrient cycling, water cycling, food web dynamics, and biodiversity at all trophic levels.[7] The amount of wild biodiversity loss depends on the degree of fragmentation of the native landscape. Whereas some species require vast continuous areas of native habitat, many can survive as long as the appropriate size and number of patches with connecting corridors of native habitat are left intact and provided that barriers to species movement—such as road and irrigation networks—are limited. However, when conversion leads to critical levels of native landscape fragmentation, chain reactions of biodiversity loss have been observed as interdependent species lose the resources they need to survive.[2] Loss of wild biodiversity at this level leads to loss of numerous ecosystem benefits that are essential to agriculture, e.g., 1) drought and flood mitigation; 2) soil erosion control and soil quality regeneration; 3) pollination of crops and natural vegetation; 4) nutrient cycling; and 5) control of most agricultural pests.[5]

Structure and Function of Agroecosystems and Biodiversity

The structure and function of agroecosystems are largely determined by local context, including interaction of ecological conditions (including bio-, geo-, and chemical) with social factors, including farmers' economic needs, cultural and spiritual values, and social structure and technology. Two types of agrobiodiversity have been defined:[8] planned biodiversity is the specific crops and/or livestock that are planted and managed and associated biodiversity is nonagricultural species that find the environment created by the production system compatible (e.g., weeds, insect and disease pests, predators and parasites of pest organisms, and symbiotic and mutualistic species).[8] Planned and associated biodiversity can enhance stability and predictability of agroecosystems.[5] Traditional forms of agriculture—such as home gardens and shade coffee farms in the New and Old World tropics[4,6,8] and traditional Amish dairy farms in North America[9]—have a complex and diverse spatial and vertical structure and high planned and associated biodiversity. For example, traditional neotropical agroforestry systems commonly contain over 100 annual and perennial plant species per field.[4] Traditional agroecosystems create landscape patterns of small-scale diverse patches with many edges, habitat patches, and corridors for wild biodiversity.

In contrast to traditional agroecosystems, the vertical and horizontal structure of modern industrial agroecosystems is simplified into monocultures on a large scale that create landscape patterns of widespread extreme genetic uniformity with few edges, habitat patches, and corridors for dispersal.[2] For example, in the United States 60–70% of the total soybean area is planted with two to three varieties, 72% of the potato area with four varieties, and 53% of the cotton area with three varieties.[6] The structure and function of industrial livestock agriculture impose similar negative impacts on biodiversity worldwide.[10] Livestock operations for all major species—particularly swine, poultry, beef, and dairy—are becoming increasingly concentrated, with feed produced in monocultures and brought to the animals in feedlots. Even in more extensive grazing operations, although good management can increase plant biodiversity,[3,10] these systems replace native forests and/or grasslands that once supported highly diverse complexes of coadapted plants and migratory grazing and browsing ungulates and their predators.[10]

Offsite Impacts of Agricultural Activities

The third major way that agriculture impacts global biodiversity is through the direct and indirect offsite effects of the various managements used to maintain their structure and function. Growing annual species in large monocultures goes against ecological forces of plant community succession; therefore, a great deal of intervention is required to maintain high levels of production. Fertilizers applied to maximize production of crop plants create favorable habitat for other plant species that are adapted to nutrient-enriched conditions, including alien invasive species. Tillage, herbicides, and genetic engineering may prevent competition between crop plants and annual and perennial weeds. Widespread monocultures of nutrient enriched plants create an easily exploited resource for insect pests and disease organisms. Insecticides, fungicides, and genetic engineering may protect crops from these competitors. Furthermore, conventional cropping agroecosystems are notoriously leaky (i.e., the sheer volume of external inputs being applied in combination with soil disturbance and decreased soil quality often exceeds the capacity of the agroecosystem to absorb and process the inputs).[1] Concentrated livestock agriculture also can be a major source of chemical and biological pollution. As a result of these many factors, sediment, excess fertilizer, manure, and pesticides run off into streams and down into groundwater. Hydrological alterations to land and natural streams in combination with chemical and biological pollution cause considerable reductions in aquatic biodiversity that can extend throughout whole watershed systems.[2] The Hypoxia in the Gulf of Mexico is a dead zone that covers 18,000 km^2 where aquatic biodiversity has been drastically reduced by impacts of agriculture in the Mississippi River watershed.[2] A new concern regarding offsite impacts of modern agriculture on biodiversity is the genetic pollution that can result as genetically modified crops expand worldwide.[1] Possible transfer of genes for resistance to weeds, insects, fungi, and viruses could overwhelm wild populations and communities.[2]

Loss of Diversity within Agricultural Species

Of 7000 crop species, less than 2% are currently important, only 30 of which provide an estimated 90% of the world's calorie intake—with wheat, rice, and maize alone providing more than half of plant-derived calories.[1] Some 30–40 animal species have been used extensively for agriculture worldwide, but fewer than 14 account for over 90% of global livestock production, whereas some 30% of international domesticated breeds are threatened with extinction.[1] There are additional trends of decreased varietal and land-race diversity within crop species as more farmers adopt modern high-yielding varieties. [1,4] These alarming trends have prompted government policy recommendations whose purpose is to 1) ensure that current agricultural genetic diversity in plants is preserved in seed banks and plant and germplasm collections (ex situ) or as growing crops (in situ), particularly wild relatives of major crops and livestock breeds in their centers of origin; and 2) ensure that wild crop and livestock relatives are conserved in carefully identified natural systems1.[5]

Conclusion

Biodiversity as a Principle of Agroecosystem Management

Although industrial agriculture is generally inversely related to biodiversity, there are promising examples of alternative agroecosystems that protect and enhance biodiversity and are also highly productive.[2–4] Some of these include: 1) organic agriculture; 2) sustainable agriculture; 3) permaculture;

4) natural system agriculture; 5) holistic management; and 6) ecoagriculture.[2,3] These models are based on ecological principles and the assumption that biodiversity can contribute significantly to sustainable agricultural production. Within ecoagriculture, the following strategies are proposed to protect and enhance wild biodiversity: 1) create biodiversity reserves that also benefit local farming communities; 2) develop habitat networks in nonfarmed areas; 3) reduce (or reverse) conversion of wild lands to agriculture by increasing farm productivity; 4) minimize agricultural pollution; 5) modify management of soil, water, and vegetation resources; and 6) modify farming systems to mimic natural ecosystems.[2] Examples of specific management practices that sustain or enhance biodiversity include: 1) hedgerows; 2) dykes with wild herbage; 3) polyculture; 4) agroforestry; 5) rotation with legumes; 6) dead and living mulches; 7) strip crops, ribbon cropping, and alley cropping; 8) minimum tillage, no tillage, and ridge tillage; 9) mosaic landscape porosity; 10) organic farming; 11) biological pest control and integrated pest management; 12) plant resistance; and 13) germplasm diversity.[11] New research, particularly if conducted in a participatory mode with farmers, should lead to many more ways to protect and enhance biodiversity, including: rotational grazing of high-diversity grasslands for dairy and beef cattle production, timber and pulp production systems that use perennial plants, high-diversity mixtures of single annual crops and/or rotational diversity, and precision agriculture that closely matches small-scale soil conditions with optimal crop genotypes.[5] Although specific management practices are helpful, also needed are whole-farm planning approaches and decision-making processes that encompass farmers' values and economic needs, in addition to environmental concerns for biodiversity[2] (e.g., holistic management).[3] More research and education and policies that encourage farmers and consumers to appreciate the ecological, economic, and quality-of-life values of biodiversity are needed to thwart current threats to global biodiversity and agrobiodiversity, and to address the many challenges and opportunities of global sustainable food security.[4]

FIGURE 2.1 Ecoagricultural strategies in practice.

References

1. Wood, S.; Sebastian, K.; Scherr, S.J. *Pilot Analysis of Global Ecosystems: Agroecosystems*; International Food Policy Research Institute and World Resources Institute: Washington, DC, 2000.
2. McNeely, J.A.; Scherr, S.J. *Ecoagriculture: Strategies to Feed the World and Save Wild Biodiversity*; Island Press: Washington, 2003.
3. Stinner, D.H.; Stinner, B.R.; Martsolf, E. Biodiversity as an organizing principle in agroecosystem management: Case studies of holistic resource management practitioners in the USA. Agric. Ecosys. Environ. **1997**, *62* (2), 199–213.
4. http://www.wri.org/wri/sustag/lab-01.html (accessed February 2003).
5. http://www.cast-science.org/bid.pdf (assessed December 2002).
6. Altieri, M.A. The ecological role of biodiversity in agroecosystems. Agric. Ecosys. Environ. **1999**, *74* (1–3), 19–31.
7. Swift, M.J.; Anderson, J.M. Biodiversity and Ecosystem Function in Agricultural Systems. In *Biodiversity and Ecosystem Function*; Schuleze, E.D., Mooney, H.A., Eds.; Springer-Verlag: New York, 1993; 15–41.
8. Vandermeer, J. Biodiversity loss in and around agroecosystems. In *Biodiversity and Human Health*; Grifo, F., Rosenthal, J., Eds.; Island Press: Washington, DC, 1997; 111–127.
9. Moore, R.H.; Stinner, D.H.; Kline, D.; Kline, E. Honoring creation and tending the garden: Amish views of biodiversity. In *Cultural and Spiritual Values of Biodiversity*; Posey, D.A., Ed.; United Nations Environment Programme, Intermediate Technology Publications: London, 1999; 305–309.
10. Blackburn, H.W.; de Haan, C. Livestock and biodiversity. In *Biodiversity in Agroecosystems*; Collins, W.W., Qualset, C.O., Eds.; CRC: New York, 1999; 85–99.
11. Paoletti, M.G.; Pimental, D.; Stinner, B.R.; Stinner, D. Agroecosystem biodiversity: Matching production and conservation biology. In *Biotic Diversity in Agroecosystems*; Paoletti, M.G., Pimental, D., Eds.; Elsevier: New York, 1992; 3–23.

3

Biodiversity: Climate Change

Lisa Freudenberger
and Jan Hanning
Sommer
University of Bonn

Introduction 23
The Impact of Climate Change on Biodiversity 24
Impact on Individuals, Populations, and Habitats • Impacts on the Interactions between Species and the Community • Impact on Ecosystems and Ecosystem Services • System Resilience to Climate Change
Methodological and Conceptual Approaches to Assess the Impacts of Climate Change 26
Experimental Approaches • Modeling Approaches and Scenario Development • Vulnerability Indices and Relative Assessments • Complexities of Climate Change Impacts on Biodiversity and Non-Knowledge Management
Climate Change Mitigation and Adaptation 28
Politics of Climate Change and Biodiversity 29
Conclusion 29
Acknowledgments 30
References 30
Bibliography 33

Introduction

Biodiversity, the diversity of life on Earth and its evolution, is closely linked to climate conditions, dynamics, and climate history.[1–3] From the poles to the tropics, very strong climate gradients exist that in turn influence the distribution of species and the patterns of biodiversity.[4,5] Temperature and precipitation are the main factors in this regard. There is strong empirical evidence that broad-scale gradients of species richness can be best predicted by global climatic conditions in combination with topographical parameters such as altitude and topographical diversity.[6–8] The habitat of each species is delimited, amongst others, by its physiological level of tolerance to deal with climatic extremes, such as drought or frost. Moreover, climate affects the characteristics of soils such as the availability of nutrients and the water cycle. Finally, not only are temperature and precipitation important but also the length and intensity of solar radiation. The sun represents the global source of energy and hence facilitates primary productivity, the production of plant biomass by photosynthesis, and influences metabolism rates in general. The cycling of energy within an ecosystem is positively correlated with its complexity and functionality and thus with its species richness, genetic diversity, and biomass productivity. This is leading to higher energy dissipation causing a positive feedback loop.[9,10]

Climate change means a mid- to long-term change in one or more climate parameters, such as an increase of temperature, or changes in the magnitude and frequency of extreme events such as droughts and heavy rainfall events. It is to a large extent driven by a rise of the concentration of greenhouse gasses in the atmosphere, such as carbon dioxide.[11] Carbon dioxide, on the other hand, also has a fertilizing effect as it supports plant growth and biomass production.[12]

Climate and biodiversity are interdependent, as biodiversity also influences the climate on the micro- to mesoscale. Biodiversity facilitates the provisioning of ecosystem services including climatically relevant functions such as temperature regulation and water cycling. It has been shown, for example, that higher amounts of biomass have a cooling effect (e.g., the climate in cities can be improved by parks) and dead wood in forests is leading to the stabilization of microclimatic conditions.[13] As a result, climatic conditions are not only affecting the distribution and evolution of biodiversity but also vice versa.[14,15]

The Impact of Climate Change on Biodiversity

Climate change is affecting biodiversity at all its levels and at different spatial and temporal scales. It has a potential impact not only on the individual and species level but also on the ecosystem level affecting ecosystem functionality and services. For a systematic classification, see Geyer et al.[16]

Impact on Individuals, Populations, and Habitats

Probably the most studied impacts of climate change can be found on the individual and population levels of animals and plants. These include changes in their behavior and physiology comprising their metabolism, growth rates, life cycling, immune functions, and also their population dynamics such as changes in gene pools, dispersal abilities, population growth rates, or sex ratios.[17] Climate change is also often having an impact on species distributions through changes in habitat quality. This may result in the spatial shift of species distributions if new suitable areas are available and if the species is able to reach them. However, in many cases it is more likely that species will not be able to adapt, and populations or even whole species may become extinct as conditions turn unsuitable.[18] For example, corals are suffering from increasing water temperatures (see Table 3.1).

Species in some ecosystem types are more vulnerable than others as, e.g., at mountain tops where no cooler refuges are available when temperature rise drives species uphill.[19] But also other types of physical boundaries such as roads, dams, or fragmented habitats in general may restrict species to escape unsuitable conditions.[20]Therefore, habitat connectivity and landscape permeability are critical to mitigate those negative effects. Fragmentation may also cause the loss of genetic diversity, as species populations are becoming increasingly isolated and genetic exchange is delimited. This leads to a decrease in genetic or intraspecific diversity.[21] Reduced levels of intraspecific diversity are also relevant for the adaptive capacity of species under changing environmental conditions since higher genetic diversity

TABLE 3.1 Coral Bleaching

Coral reefs are described as the rainforests of the sea as they represent the most diverse marine habitat. They are, however, very vulnerable to climate change since water temperature increase is causing large-scale diebacks of coral reefs through a process called coral bleaching. Corals live in symbiosis with algae (zooxanthellae) that grow in corals, give them their specific color, and provide them with nutrients. When the temperature increases, zooxanthellae start to produce toxins leading to the release of the algae from the coral and the loss of the corals' color. Corals are able to survive a limited time without nutrients from the symbiotic algae but a general increase of water temperature and temperature fluctuations during the last years coupled with other stressing factors such as overfishing, water pollution, acidification, and mechanical destruction has already caused the loss of large proportions of coral reefs. About 19% of the original coral reefs have already been lost and about 35% are threatened. Predictions estimate that about 50% of the world's human population will live along coasts by 2015.[23] Altogether, these stressors are likely to lead to ever increasing pressures on coral reefs and the loss of essential ecosystem services they provide.

increases the probability that some individuals are able to survive and reproduce under the changing conditions.[22] Lower genetic diversity on the other hand is likely to negatively affect the persistence of species under changing conditions.

Impacts on the Interactions between Species and the Community

Climate change impacts species interactions and community structure in many different ways. Often, this is a consequence of temporal or spatial decoupling between the interacting species, resulting in changes or loss of tropical relationships and other interactions.[24] A popular example is the temporal decoupling between the earlier flowering times of plants well before the awakening from hibernation of bees as their pollinators.[25] At the same time, new interactions may become apparent under climate change, e.g., through the spread of invasive species.[26] "Winners" of climate change may extend their distribution ranges into areas where there are fewer competitors or predators and can reach high densities. All these changes may lead to general alterations in species composition and community structure having an impact in turn on other biotic and abiotic factors of whole ecosystems.

Impact on Ecosystems and Ecosystem Services

Climate change has a direct impact on abiotic conditions such as humidity, evaporation, wind patterns, microclimatic conditions, snow loads, water bodies, and the condition of soils. Impacts on ecosystem structure, processes, and dynamics include changes in photosynthetic activities, and the frequency and intensity of disturbances, e.g., from floods or fire events, nutrient cycling, and succession[27] Finally, not only species are likely to shift their distribution under altered climatic conditions but also ecosystem types.[28,29] The melting of the permafrost soil leading to a spatial increase of the Taiga vegetation or the drying of the Mediterranean region are just two examples of climate change impacts on ecosystem-type distributions.

Changes in ecosystem functions, processes, and distributions will also affect ecosystem services.[30] In some areas of the world, extreme events such as droughts and floods will have tremendous effects on the availability of food, fiber, timber, and water. Climate change will also influence several supporting services, such as water cycling, primary production, and nutrient cycling as well as different regulating services like pest regulation, seed dispersal, and water purification. Finally, climate change is having a general impact on the appearance of ecosystems and therefore on their spiritual, religious, or aesthetical value. Altogether, these impacts on ecosystems may directly or indirectly influence human society and feed conflicts about scarce resources, food, and clean water.

System Resilience to Climate Change

Climate change impacts are not uniformly distributed and different biodiversity elements will suffer from climate change to varying degrees. The vulnerability of biodiversity, e.g., an ecosystem or a certain species, is a function of its exposure and sensitivity toward climate change as well as its ability to adapt to these changes.[11,31] This means that an element of biodiversity that has a high exposure towards climate change can still be less vulnerable than another element with low exposure as long as it is either less sensitive or has the ability to adapt to these changes, e.g., through the spatial shift of a species distribution. This is connected to the related concept of resilience. Traditionally, resilience has been described as the ability of a system or a species to return to its original state after disturbance. However, recent trends in ecological studies have highlighted the dynamic behavior of ecosystems, therefore giving rise to a different definition where resilience means merely the ability of a system to return to a stable state that could be different from its original one.[32] According to this concept, systems can have different stable states and it is possible that an ecosystem can transform to another without losing its functionality. The potential nonlinear impact of climate change on ecosystems and biodiversity has also been described

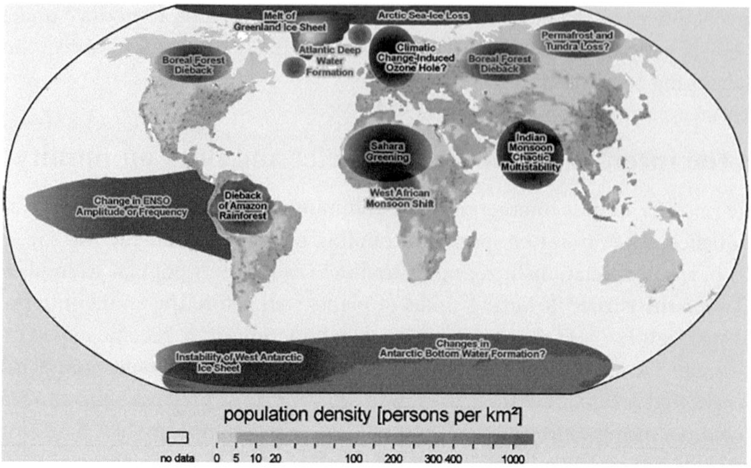

FIGURE 3.1 Map of global tipping elements.
Source: Adapted from Schellnhuber & Held,[55] Lenton et al.,[33] by permission of Oxford University Press.

TABLE 3.2 Tipping Points

Tipping points describe points in time when a system does not develop any more in a way that can be predicted by previous linear relationships but reaches a point of no return beyond which a trend has either significantly changed its pace or its direction. Tipping elements in the climate system describe global subsystems that are likely to shift to a completely different state if climate change continues and tipping points are the corresponding points in time (see Figure 3.1).[33] These elements have been self-stabilizing over a long period of time, and may most likely not recover once they are destroyed. Climate change impacts become much more difficult to predict after tipping points have been reached or complex feedback mechanisms are getting more important.

with the concept of tipping points (see Figure 3.1 and Table 3.2). These are "points of no return" beyond which it is impossible to stop or even reverse the negative impacts from climate change.[33] Different definitions for all of these concepts can be found in literature and a discussion on the relation between them and their differentiation is still ongoing.

Methodological and Conceptual Approaches to Assess the Impacts of Climate Change

The impacts of climate change on biodiversity are often difficult to assess since predictions of the future are difficult to verify with recent or historical data, especially if the expected future conditions have never existed in the past. Gaining evidence by conducting field experiments is very time consuming and may provide results only at a very late temporal stage, and lab experiments are not able to tackle the response of biodiversity and ecosystems in their full complexity. Additionally, climate change–induced impacts are often overshadowed and camouflaged by other stressors such as landscape fragmentation, land degradation, landuse intensification, pollution, deforestation, or over-exploitation in general. Nevertheless, different methodologies have been developed to assess the impacts of climate change on biodiversity.[34–36]

Experimental Approaches

Classical empirical research is designed as a systematic experiment that can take place either in the field or within the laboratory. Species are exposed to different environmental conditions, e.g., higher temperature or lower precipitation rates, alone or in combination to test how climate change may affect species and ecosystems. While lab experiments have the advantage of enabling the control of external factors, field experiments capture a higher level of complexity. Different studies have been conducted monitoring, e.g., growth rates of plants under increased temperature and higher carbon dioxide levels.[37] Empirical evidence is of high importance to understand the underlying processes of climate change and biodiversity interactions. However, their practical value often remains limited as they are usually focused on a relatively narrow research question, a specific location, a particular ecosystem type, and a specific set of species. Nevertheless, empirical research is providing the real world data necessary to extrapolate and estimate the future impact of climate change.

Modeling Approaches and Scenario Development

The impacts of climate change on biodiversity have often been assessed through modeling approaches. Modeling approaches can be categorized as either statistical or mechanistic, although hybrid forms of both also exist. Statistical models use empirical data to find correlations between, e.g., the number of species and recent climatic conditions that may be used to assess the impact of future climate conditions on species richness. For example, biodiversity parameters for an area of interest can be compared to those of other areas in the world that today already have climate conditions similar to the expected future ones of the area of interest (space for time substitution approach).[38,39] Mechanistic or process-based models incorporate generally known relationships with the aim to reproduce the processes relevant for the ecosystem or species studied. Common examples for the application of mechanistic models are vulnerability assessments of ecosystems under climate change scenarios.[28] Models provide valuable information for research, nature conservation, and decision making but are often associated with a high degree of uncertainty. Climate change scenarios are also often combined with different specifications to assess, e.g., the impact of different policy options.[40,41] As climate change is modeled based on different emission scenarios, modeling outputs can be highly dependent on the socio-economic assumptions being made under the specific scenario. Furthermore, there are different uncertainties that are resulting from complex climate–biodiversity–society interactions.[42,43] When interpreting the results of model outputs, it is therefore important to take all model restrictions into consideration and to understand the results as one out of many potential futures.

Vulnerability Indices and Relative Assessments

While modeling approaches usually rely to some extent on quantitative empirical data from the past, there are other approaches that are utilizing heuristics to assess potential future impacts of climate change. These assessments are often only relative measurements or indices and can only be interpreted when comparing different species, ecosystems, or areas. A common concept to assess the impacts of climate change on biodiversity is, e.g., the vulnerability framework.[31] Different indicators or expert knowledge are used to assess the sensitivity, adaptive capacity, and exposure to climate change and are combined in a vulnerability index. These relative assessments have the advantage of combining various different proxies often with different units of measurement while being easy to interpret.[44] However, they do not provide true quantitative data that can be directly validated by field measurements or interpreted on its own without comparing the index value to different areas or times.

Complexities of Climate Change Impacts on Biodiversity and Non-Knowledge Management

Scenario development and modeling are very important tools to deal with the uncertainties and knowledge gaps of climate change and its impacts. However, it should be noted that dealing with ecosystems and climate change means dealing with uncertainties and non-knowledge.[45]

The precautionary concept that is promoted by the United Nations Convention on Biodiversity (CBD), which was signed at the Rio Earth Summit in 1992, becomes particularly relevant in terms of biodiversity conservation under climate change.[46] It endorses a proactive policy approach, where future risks are minimized and actions are not necessarily prevented by knowledge gaps. The impacts of climate change on biodiversity are characterized by a high level of complexity, feedback mechanisms, and non-linear dependencies. It is difficult to capture this complexity through simple statistical relationships and to assess future impacts of climate change on biodiversity. Different concepts to describe these complex relationships have been introduced, all of which deal with the high level of uncertainty. Although we can develop many different models and scenarios for future climate change, each of them corresponds only to one of many possible futures and it is very unlikely that the real future will exactly match a model that is being developed today. Nevertheless, we can use these models or even non-quantitative scenarios and heuristics to identify trends and paint pictures of how the future might look.

Climate Change Mitigation and Adaptation

Conservation biology used to be a relatively isolated political field motivated mainly by ethical or even religious obligations to protect nature. In the past 10 years, a more anthropocentric perspective has advanced where biodiversity is not only appreciated for its inherent value but also for the services that are provided to human society. From this perspective, biodiversity conservation is indispensable for the provision of ecosystem services that humanity is depending on, such as the production of food or the filtration of water.[47] Another development has been the consideration of the interconnections between climate change and biodiversity. It has been realized by both, the biodiversity conservation as well as the climate change community, that synergies as well as conflicts between both policy fields exist that need to be dealt with.[48–51] There are two categories, mitigation and adaptation, that conceptualize how nature conservation can contribute to the management of climate change impacts. While mitigation corresponds to actions undertaken to alleviate future climate change, adaptation measures are aiming at enabling biodiversity to persist under climate change. Mitigation approaches mainly refer to strategies targeting the reduction of greenhouse gases in the atmosphere through either the minimization of emissions, e.g., the protection of carbon sources, or the maximization of greenhouse gas uptake, e.g., the protection or restoration of carbon sinks. On the other hand, adaptation measurements assume that at least some intensity of climate change will take place and that biodiversity can only be protected effectively if adaptive strategies are employed. Despite the straightforward definition of mitigation and adaptation, both are often not isolated but interconnected. In many cases, adaptation measures are only applicable if climate change is not exceeding critical thresholds and, vice versa, ecosystems are more likely to contribute effectively to climate change mitigation if they are in a healthy condition.

During the last few years, a new concept has become increasingly important, namely, ecosystem-based adaptation and mitigation.[52,53] Ecosystem-based adaptation refers to the utilization and strengthening of natural functions of ecosystems to make societies less vulnerable to the negative impacts of climate change. One example for ecosystem-based adaptation would be the abandonment of alluvial areas for construction in order to maintain space for flood protection, but other examples also exist where natural solutions can provide more effective and efficient options than technological advances for adaptation to climate change. Ecosystem-based mitigation follows the same approach but with the aim of increasing the capacity of ecosystems to store and capture carbon, e.g., through the conservation of natural carbon sinks and sources. Ecological engineering, on the other hand, focuses on actions coupling both technological and ecological solutions to benefit society.[54] These often relate to the restoration or enhancement of natural ecosystem functions or the use of these functions within technological solutions. One of the reasons that these two concepts, ecosystem-based actions and ecological engineering, have gained in relevance is their potential to explicitly combine aspects of climate change policy with biodiversity conservation. Table 3.3 gives examples for mitigation and adaptation strategies for all three options. While nature conservation and climate change policy have often

TABLE 3.3 Examples for Strategies Aiming at the Mitigation of Climate Change and Adaptation toward Its Impacts

	Mitigation of climate change through higher carbon sequestration	**Adaptation** to increased frequency of floods through climate change
Technological solution	Use of carbon capture and storage technology (CCS)	Building of artificial dams for flood protection
Ecological engineering	Rewetting of peatlands to increase carbon capture	Restoration and control of alluvial areas for flood protection
Ecosystem based	Protection of peatlands to maintain natural carbon capture functionality	Abandoning alluvial areas for natural flood protection

acted independently from each other, it has been realized that collaborative approaches often provide synergies and beneficial effects for both policy areas. Furthermore, it turned out that pure technological solutions aiming at the mitigation and adaptation to climate change often entail substantial costs and efforts. Ecosystem-based actions and ecological engineering have the potential to provide cost-effective strategies to cope with future climate change.

Politics of Climate Change and Biodiversity

Different international initiatives exist to tackle the challenge of climate change and its impact on biodiversity. While both the United Nations Framework Convention on Climate Change (UN-FCCC) and the Convention on Biological Diversity (CBD) were founded at the Earth Summit 1992 in Rio de Janeíro, they remained relatively isolated in the following 20 years. It is only recently that a better integration of climate change policy and biodiversity conservation goals is getting higher priority on the political agenda. The UN-REDD (Reducing Emissions from Deforestation and Forest Degradation) program, which is part of the Clean Development Mechanism of the Kyoto Protocol[11] and currently probably the most powerful instrument for international climate change mitigation, aims at the reduction of greenhouse gases through the creation of financial incentives paid for the mitigation of deforestation and forest degradation. To avoid potential negative effects on biodiversity and society, it has been developed towards the REDD+ mechanism, with a stronger focus on multiple benefits including biodiversity conservation. Climate change has also been officially included as a main crosscutting issue of the CBD since 2004 after the establishment of an Ad Hoc Technical Expert Group (AHTEG) on climate change and biodiversity in 2001. During the last few years, the Rio Conventions Pavilion has been founded at the 10th Conference of Parties of the CBD in Nagoya (Japan, 2010) as a platform aiming at the identification of synergies between the different conventions that were established in 1992 in Rio. In 2012, as a pendant to the Intergovernmental Panel on Climate Change, a new Intergovernmental Platform on Biodiversity and Ecosystems Services (IPBES) has been established, an independent intergovernmental body to strengthen the science policy interface in the field of biodiversity and ecosystem services. In consideration of the various interactions between climate change and biodiversity, a collaboration and coordination between both policy areas is likely to become increasingly necessary.

Conclusion

Climate change and biodiversity represent both very complex issues, and many answers to the question of how they are interacting are still elusive. However, there is strong evidence that climate change will have negative impacts on biodiversity, ecosystem service supply, and thereby on human society. The paradigm of natural sciences to base research on empirical data can only be partly fulfilled in the context of climate change research, as research questions include predictions about possible future developments not experienced in the past. Nevertheless, the expected pace and magnitude of climate change suggests that prompt and flexible actions are needed to prevent unbearable and irreversible negative impacts. Both climate change mitigation and adaptation strategies need to be implemented, incorporating concepts of conservationists and climate researchers in order to gain the best possible achievements.

Acknowledgments

The authors thank the Center for Development Research, University of Bonn, the Nees Institute for the Biodiversity of Plants, University of Bonn, the Centre for Econics and Ecosystem Services, and the University for Sustainable Development Eberswalde for their support in conducting research on climate change and biodiversity and for the development of this entry.

References

1. O'Brien, E. Water-energy dynamics, climate, and prediction of woody plant species richness: An interim general model. J. Biogeogr. **1998**, *25* (2), 379–398.
2. Hawkins, B.A.; Field, R.; Cornell, H.V.; Currie, D.J.; Guégan, J.F.; Kaufman, D.M.; Kerr, J.T.; Mittelbach, G.G.; Oberdorff, T.; O'Brien, E.M.; Porter, E.E.; Turner, J.R.G. Energy, water and broad-scale geographic patterns of species richness. Ecology **2003**, *84* (12), 3105–3117.
3. Currie, D.J.; Mittelbach, G.G.; Cornell, H.V.; Field, R.; Guégan, J.F.; Hawkins, B.A.; Kaufman, D.M.; Kerr, J.T.; Oberdorff, T.; O'Brien, E.; Turner, J.R.G. Predictions and tests of climate-based hypotheses of broad-scale variation in taxonomic richness. Ecol Lett. **2004**, *7* (12), 1121–1134.
4. Barthlott, W.; Mutke, J.; Rafiqpoor, M.D.; Kier, G.; Kreft, H. Global centres of vascular plant diversity. Nova Acta Leopoldina, *92*, 1–83.
5. Kier, G.; Mutke, J.; Dinerstein, E.; Ricketts, T.H.; Küper, W.; Kreft, H.; Barthlott, W. Global patterns of plant diversity and floristic knowledge. J. Biogeogr. **2005**, *32* (7), 1107–1116.
6. Field, R.; Hawkins, B.A.; Cornell, H.V.; Currie, D.J.; Diniz-Filho, J.A.F.; Guégan, J.F.; Kaufman, D.M.; Kerr, J.T.; Mittelbach, G.G.; Oberdorff, T.; O'Brien, E.M.; Turner, J.R.G. Spatial species-richness gradients across scales: A meta-analysis. J. Biogeogr. **2009**, *36* (1), 132–147.
7. Sommer, J.H.; Kreft, H.; Kier, G.; Jetz, W.; Mutke, J.; Barthlott, W. Projected impacts of climate change on regional capacities for global plant species richness. Proc. R. Soc. B Biol. Sci. **2010**, *277* (1692), 2271–2280.
8. Kreft, H.; Jetz, W. Global patterns and determinants of vascular plant diversity. Proc. Natl. Acad. Sci. **2007**, *104* (14), 5925–5930.
9. Fath, B.D.; Jørgensen, S.E.; Patten, B.C.; Straskraba, M. Ecosystem growth and development. Biosystems **2004**, *77* (1–3), 213–228.
10. Jørgensen, S.E.; Patten, B.C.; Straškraba, M. Ecosystems emerging. Ecol. Model. **2000**, *126* (2–3), 249–284.
11. IPCC. *Climate Change 2007: Impacts, Adaptation and Vulnerability,* Working Group II Contribution to the Fourth Assessment Report of the Intergovernmental Panel on Climate Change; IPCC Secretariat: Geneva, 2008.
12. Ainsworth, E.A.; Rogers, A. The response of photosynthesis and stomatal conductance to rising [CO_2]: Mechanisms and environmental interactions. Plant Cell Environ. **2007**, *30* (3), 258–270.
13. Norris, C.; Hobson, P.; Ibisch, P.L. Microclimate and vegetation function as indicators of forest thermodynamic efficiency. J. Appl. Ecol. **2012**, *49* (3), 562–570.
14. Battersby, S. Call in the clouds. New Sci. **2013**, *218* (2923), 32–35.
15. Bonan, G.B. Forests and climate change: Forcings, feedbacks, and the climate benefits of forests. Science **2008**, *320* (5882), 1444–1449.
16. Geyer, J.; Kiefer, I.; Kreft, S.; Chavez, V; Salafsky, N.; Jeltsch, F.; Ibisch, P.L. Classification of climate-change-induced stresses on biological diversity. Conserv. Biol. **2011**, *25* (4), 708–715.

17. Walther, G.-R.; Post, E.; Convey, P.; Menzel, A.; Parmesan, C.; Beebee, T.J.C.; Fromentin, J.-M.; Guldberg, O.H.; Bairlein, F. Ecological responses to recent climate change. Nature **2002**, *416* (6879), 389–395.
18. Thomas, C.D.; Cameron, A.; Green, R.E.; Bakkenes, M.; Beaumont, L.J.; Collingham, Y.C.; Erasmus, B.F.N.; de Siqueira, M.F.; Grainger, A.; Hannah, L.; Hughes, L.; Huntley, B.; van Jaarsveld, A.S.; Midgley, G.F.; Miles, L.; Huerta, M.A.O.; Peterson, A.T.; Phillips, O.L.; Williams, S.E. Extinction risk from climate change. Nature **2004**, *427* (6970), 145–148.
19. Thuiller, W. Climate change threats to plant diversity in Europe. Proc. Natl. Acad. Sci. **2005**, *102* (23), 8245–8250.
20. Opdam, P.; Wascher, D. Climate change meets habitat fragmentation: linking landscape and biogeographical scale levels in research and conservation. Biol. Conserv. **2004**, *117* (3), 285–297.
21. Keller, I.; Largiader, C.R. Recent habitat fragmentation caused by major roads leads to reduction of gene flow and loss of genetic variability in ground beetles. Proc. R. Soc. B Biol. Sci. **2003**, *270* (1513), 417–423.
22. Jump, A.S.; Marchant, R.; Peñuelas, J. Environmental change and the option value of genetic diversity. Trends Plant Sci. **2009**, *14* (1), 51–58.
23. Wilkinson, C., Ed. *Status of Coral Reefs of the World: 2008;* Townsville, Australia, 2008.
24. Winder, M.; Schindler, D.E. Climate change uncouples trophic itneractions in an aquatic ecosystem. Ecology **2004**, *85* (8), 2100–2106.
25. Kudo, G.; Ida, T.Y. Early onset of spring increases the phenological mismatch between plants and pollinators. Ecology **2013**, *94* (10), 2311–2320.
26. Mainka, S.A.; Howard, G.W. Climate change and invasive species: Double jeopardy. Integr. Zoolog. **2010**, *5* (2), 102–111.
27. Littell, J.S.; Oneil, E.E.; McKenzie, D.; Hicke, J.A.; Lutz, J.A.; Norheim, R.A.; Elsner, M.M. Forest ecosystems, disturbance, and climatic change in Washington State, USA. Climatic Change **2010**, *102* (1–2), 129–158.
28. Scholze, M.; Knorr, W.; Arnell, N.W.; Prentice, I.C. A climate-change risk analysis for world ecosystems. Proc. Natl. Acad. Sci. **2006**, *103* (35), 13116–13120.
29. Gonzalez, P.; Neilson, R.P.; Lenihan, J.M.; Drapek, R.J. Global patterns in the vulnerability of ecosystems to vegetation shifts due to climate change. Global Ecol. Biogeogr. **2010**, *19* (6), 755–768.
30. Mooney, H.; Larigauderie, A.; Cesario, M.; Elmquist, T.; Guldberg, O.H.; Lavorel, S.; Mace, G.M.; Palmer, M.; Scholes, R.; Yahara, T. Biodiversity, climate change, and ecosystem services. Curr. Opin. Environ. Sustainability **2009**, *1* (1), 46–54.
31. Schröter, D. Ecosystem service supply and vulnerability to global change in Europe. Science **2005**, *310* (5752), 1333–1337.
32. Walker, B.H.; Gunderson, L.H.; Kinzig, A.P.; Folke, C.; Carpenter, S.; Schultz, L. A handful of heuristics and some propositions for understanding resilience in social-ecological systems. Ecol. Soc. **2006**, *11* (1), 13.
33. Lenton, T.M.; Held, H.; Kriegler, E.; Hall, J.W.; Lucht, W.; Rahmstorf, S.; Schellnhuber, H.J. Tipping elements in the Earth's climate system. Proc. Natl. Acad. Sci. U.S.A. **2008**, *105* (6), 1786–1793.
34. Guisan, A.; Thuiller, W. Predicting species distribution: offering more than simple habitat models. Ecol Lett. **2005**, *8* (9), 993–1009.
35. Araujo, M.B.; Rahbek, C. How does climate change affect biodiversity? Science **2006**, *313* (5792), 1396–1397.
36. Algar, A.C.; Kharouba, H.M.; Young, E.R.; Kerr, J.T. Predicting the future of species diversity: macroecological theory, climate change, and direct tests of alternative forecasting methods. Ecography **2009**, *32* (1), 22–33.

37. Ghannoum, O.; Phillips, N.G.; Conroy, J.P.; Smith, R.A.; Attard, R.D.; Woodfield, R.; Logan, B.A.; Lewis, J.D.; Tissue, D.T. Exposure to preindustrial, current and future atmospheric CO_2 and temperature differentially affects growth and photosynthesis in *Eucalyptus*. Global Change Biol. **2010**, *16* (1), 303–319.
38. Hijmans, R.J.; Graham, C.H. The ability of climate envelope models to predict the effect of climate change on species distributions. Global Change Biol. **2006**, *12* (12), 2272–2281.
39. Dolgener, N.; Freudenberger, L.; Schneeweiss, N.; Ibisch, P.L.; Tiedemann, R. Projecting current and potential future distribution of the Fire-bellied toad *Bombina bombina* under climate change in north-eastern Germany. Reg. Environ. Change **2013**, doi:10.1007/s10113-013-0468-9.
40. Moss, R.H.; Edmonds, J.A.; Hibbard, K.A.; Manning, M.R.; Rose, S.K.; van Vuuren, D.P.; Carter, T.R.; Emori, S.; Kainuma, M.; Kram, T.; Meehl, G.A.; Mitchell, J.F.B.; Nakicenovic, N.; Riahi, K.; Smith, S.J.; Stouffer, R.J.; Thomson, A.M.; Weyant, J.P.; Wilbanks, T.J. The next generation of scenarios for climate change research and assessment. Nature **2010**, *463* (7282), 747–756.
41. Parry, M.L.; Rosenzweig, C.; Iglesias, A.; Livermored, M.; Fischere, G. Effects of climate change on global food production under SRES emissions and socio-economic scenarios. Global Environ. Change **2004**, *14* (1), 53–67.
42. Hansen, A.J.; Neilson, R.P.; Dale, V.H.; Flather, C.H.; Iverson, L.R.; Currie, D.J.; Shafer, S.; Cook, R.; Bartlein, PJ. Global change in forests: Responses of species, communities, and biomes. BioScience **2001**, *51* (9), 765.
43. Freudenberger, L.; Schluck, M.; Hobson, P.; Sommer, H.; Cramer, W.; Barthlott, W.; Ibisch, P.L. A view on global patterns and interlinkages of biodiversity and human development. In *Interdependence of Biodiversity and Development Under Global Change;* Ibisch, P.L., Vega, A.E., Herrmann, T.M., Eds.; Secretariat of the Convention on Biological Diversity: Montreal, Quebec, Canada, 2010; Vol. 54, pp. 37–56.
44. Saltelli, A.; Nardo, M.; Saisana, M.; Tarantola, S. Composite indicators–The controversy and the way forward. In *Statistics, Knowledge and Policy: Key Indicators to Inform Decision Making;* OECD, Ed., 2005, pp. 359–373.
45. Ibisch, P.L.; Hobson, P. Blindspots and sustainability under global change: Non-knowledge illiteracy as a key challenge to a knowledge society. In *Global Change Management: Knowledge Gaps, Blindspots and Unknowables,* 1st Ed.; Ibisch, P.L., Geiger, L., Cybulla, F., Eds.; Nomos: Baden-Baden, 2012; vol. 1, pp. 15–54.
46. CBD, Ed. Preamble of the convention on biological diversity. In *Handbook of the Convention on Biological Diversity*; Earthscan Publications: London, Sterling, VA, 1992.
47. *Reid, W.V. Ecosystems and Human Well-Being: General Synthesis: A Report of the Millennium Ecosystem Assessment; Island Press: Washington, DC, 2005.*
48. Heller, N.E.; Zavaleta, E.S. Biodiversity management in the face of climate change: A review of 22 years of recommendations. Biol. Conserv. **2009**, *142* (1), 14–32.
49. Poiani, K.A.; Goldman, R.L.; Hobson, J.; Hoekstra, J.M.; Nelson, K.S. Redesigning biodiversity conservation projects for climate change: examples from the field. Biodivers. Conserv. **2011**, *20* (1), 185–201.
50. Freudenberger, L.; Hobson, P.R.; Schluck, M.; Ibisch, P.L. A global map of the functionality of terrestrial ecosystems. Ecol. Complexity **2012**, *12*, 13–22.
51. Freudenberger, L.; Hobson, P.; Schluck, M.; Kreft, S.; Vohland, K.; Sommer, H.; Reichle, S.; Nowicki, C.; Barthlott, W.; Ibisch, P.L. Nature conservation: Prioritysetting needs a global change. Biodivers. Conserv. **2013**, 22 (5), 1255–1281.
52. Munang, R.; Thiaw, I.; Alverson, K.; Mumba, M.; Liu, J.; Rivington, M. Climate change and Ecosystem-based Adaptation: A new pragmatic approach to buffering climate change impacts. Curr. Opin. Environ. Sustainability **2013**, *5* (1), 67–71.
53. *Doswald, N.; Osti, M. Ecosystem-Based Approaches to Adaptation and Mitigation - Good Practice Examples and Lessons Learned in Europe; BfN: Bonn, 2011.*

54. Mitsch, W.J.; Jørgensen, S.E. Ecological engineering: A field whose time has come. Ecol. Eng. **2003**, *20* (5), 363–377.
55. Schellnhuber, H.J.; Held, H. How fragile is the Earth system. In *Managing the Earth: The Eleventh Linacre Lectures*; Briden, J.C., Downing, T.E., Eds.; Oxford University Press: Oxford, UK, 2002; 5–34.

Bibliography

Blair, T.; Schellnhuber, H.J.; Cramer, W.P., Eds. *Avoiding Dangerous Climate Change*; Cambridge University Press: Cambridge, England, 2006.

Hilty, J.A.; Chester, C.C.; Cross, M.S., Eds. *Climate and Conservation: Landscape and Seascape Science, Planning, and Action*; Island Press: Washington, DC, 2012.

Kapos, V.; Campbell, A. *Review of the Literature on the Links between Biodiversity and Climate Change: Impacts, Adaptation, and Mitigation*; Secretariat of the Convention on Biological Diversity: Montreal, Quebec, Canada, 2009.

Lovejoy, T.E.; Hannah, L.J., Eds. *Climate Change and Biodiversity*; Yale University Press: New Haven, 2005.

Maes, F.; Cliquet, A.; Du Plessis, W.; McLeod-Kilmurray, H. Biodiversity and climate change. *Linkages at International, National and Local Levels* (The Iucn Academy of Environmental Law); Edward Elgar Publishing Ltd, 2013.

Pearson, R. Driven to Extinction: The Impact of Climate Change on Biodiversity; Natural History Museum: London, UK, 2011.

UNEP-CMC, Ed. *Carbon and Biodiversity: A Demonstration Atlas*; UNEP-WCMC: Cambridge, UK, 2008.

4

Biodiversity: Conservation

Introduction 35
The Dimensions of Biodiversity: How Much Do We Know? 35
Phyletic and Genetic Diversity • Species Diversity: How Many Species Are There and How Many Have Been Described? • Where Is Biodiversity Found?
Conservation 39
Protected Areas • Threats to Biodiversity and the Environment • Conservation Biology • Global Action
Extinction Rates 42
Conclusion 42
References 42
Bibliography 43

Nigel E. Stork
Griffith University

Introduction

The term "biodiversity," the contraction of "biological diversity," came into common usage following the signing of the Convention on Biological Diversity at the United Nations Conference on Environment (commonly known as the "Earth Summit") in Rio de Janeiro in 1992. This Convention was signed by 193 nations (ratified by 168) and is one of the most widely recognized international treaties. Prior to the Convention, biodiversity was a rapidly developing concept among scientists. It was first featured in a book edited by Wilson and Peters [1] and signified a rising concern for the world's fauna and flora. This also followed several decades of concern for the world's environment as the conservation movement also began to grow.

But what is biodiversity? The Convention defines biodiversity as "the variability among living organisms from all sources including terrestrial, marine and other aquatic ecosystems and ecological complexes of which they are a part; this includes diversity within species, between species and ecosystems." However, biodiversity means different things to different people and is more complex than this simple definition. It encompasses not just hierarchies of taxonomic and ecological scales and other more or less independent scales such as geographic scales and scaling in body size of organisms. A search on Google for the definition of biodiversity comes up with an interesting word cloud (Figure 4.1).

The Dimensions of Biodiversity: How Much Do We Know?

Phyletic and Genetic Diversity

Classical taxonomic systems had traditionally grouped most organisms into two Kingdoms, Animalia and Plantae. In these systems, a wide array of micro-organisms had been loosely grouped together in a third Kingdom, Protista. Recognizing this group was a "grab bag," in the last few decades, other Kingdoms have been recognized, and current classifications now replace Protista with separate

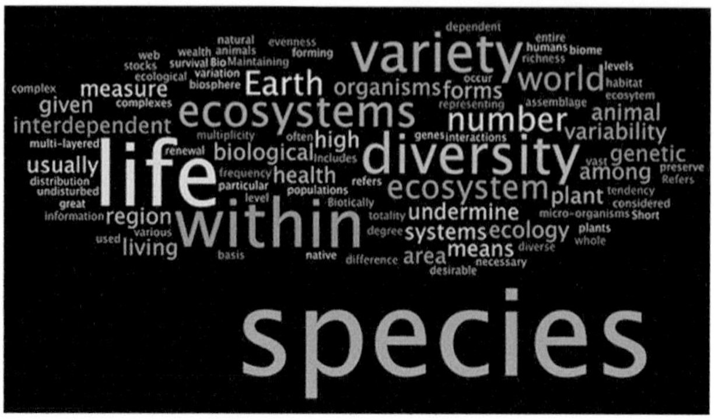

FIGURE 4.1 (See color insert.) This word cloud results from a Google search for the definition of biodiversity. (http://thepimmgroup.org/1022/defining-biodiversity-toward-a-consensus/.)

Kingdoms of Bacteria, Protozoa, Chromista, and Fungi.[2,3] Alternative classifications, also based on ribosomal (r)RNA sequences, group all life into three Domains: Bacteria, Archaea, and Eucarya, which are above Kingdoms in the taxonomic classification system[4] (Figure 4.2). In this classification where the lines represent genetic distance, it is clear that all the multicellular groups are relatively closely related genetically compared to some of the former Protista. It also shows that animals, plants, and fungi share a great deal of genetic material.

Darwin observed in his *Origin of Species* that "systematists are far from being pleased in finding variability in important characters," and yet it is this genetic component that is the basis for evolution, and without it, evolution and adaptation cannot take place. From the origins of first life on Earth, genetic variation, adaptation, and evolution have led to the enormous diversity of life that we see today and in past extinct life-forms. The distribution of genetic variation within populations of individuals can also tell us a lot about migration between populations, past processes of colonization, and the effects of habitat disruption and loss. How variation is distributed within individuals of a population can also tell us about breeding structures (inbreeding versus outbreeding), reproductive success of different individuals, and a whole range of other patterns. Such genetic variation also leads to morphological variation in

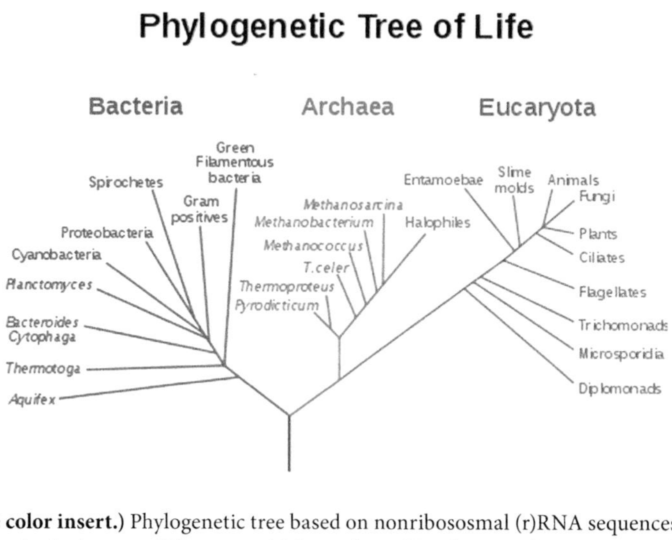

FIGURE 4.2 (See color insert.) Phylogenetic tree based on nonribososmal (r)RNA sequences groups all life into three Domains Bacteria, Archaea, and Eucarya, which are above Kingdoms in the taxonomic classification system. **Source:** Adapted from Woese, et al.[4]

individuals—the kind of variation we see everyday when we look at ourselves as humans and other animals. In addition, the environment an individual lives in can also have an impact and lead to additional environmental variation. For example, the amount of food a larva of a Rhinoceros beetle has available to it will determine not just the size of the adult beetle but also the shape. Larger males will have larger horns and hence may be more competitive when courting females.

Species Diversity: How Many Species Are There and How Many Have Been Described?

Taxonomists have been naming and describing species since the Swedish biologist Carolus Linnaeus (later known as Carlvon Linné) created the universal system of naming species (his 10th edition of the Systema Natura published in 1758 is generally considered the start of modern taxonomy).[5] However, determining just how many species have been described is not a simple task as taxonomists often disagree on what constitutes a species because individuals can vary so much. The proportions of known synonyms range from over 80% and 90% in some alga genera, 7% to 80% (32% overall) for insects, 33% to 88% for groups of seed plants, 38% for world molluscs, 50% for marine fish, and 40% for all marine species.[6] Considering past rates of synonymy, it is probable that overall, at least 20% of the currently estimated described species are yet to be discovered synonyms. There are only a few groups of organisms where it is likely that all or virtually all species on Earth have been described, such as the birds and large mammals. For others, such as the insects, nematodes, and fungi, the task is still far from complete. Estimates suggest that 1.5–1.8 million species have been named and described so far, and about two-thirds of these have been cataloged in global databases.

The famous English evolutionary biologist J.B.S. Haldane was asked in the 1930s what he could deduce about "the Creator" from his creation and he replied "an inordinate fondness for beetles." Not surprisingly, about 400,000 species of beetles have been described (about 25% of all named species), and with other insects and other arthropods, this brings up a total of around 1 million species. But just how many species there are including the undescribed ones remains a mystery and is difficult to estimate In the past, some have suggested that there might be as few as 2 million species, whereas others have suggested as high as 100 million species. Three different methods now suggest that estimates of around 5 ± 3 million species are more likely. Two of these use the new and extensive (but as yet incomplete) catalogs of described species to provide estimates for all species. The first method used extrapolation of the rate of discovery of selected higher taxa to predict 8.7 million (±1.3 million SE) eukaryotic species globally, of which 25% (2.2 million ± 0.18 million SE) were marine.[7] The second method modeled global species richness based on the rates of species description and predicted that 0.3 million marine and 1.7 million terrestrial species may exist.[8] A third model used uncertainty analysis of the specificity of arthropods on tropical trees to suggest there may be 6.1–7.8 million terrestrial arthropod species)[9,10] This latter method showed that the key variable for global insect species richness is the level of host specificity of insects to individual tree species. It also has 90% confidence intervals of 3.6–13.7 million species, suggesting that higher global estimates are extremely unlikely. The finding that marine organisms comprise only about 25% of all species on Earth is surprising, given that first, oceans cover 70% of the surface and second, that at the higher taxonomic level of Phylum, most phyla are found in the marine environment with the terrestrial fauna being less rich. However, the marine environment has few barriers to dispersal, whereas land and freshwater habitats are highly fragmented. The largest group in terms of species, after the arthropods, is the plants with an estimated 400,000 species, and most of these are terrestrial. The explosion of diversity that arose through the evolution of the flowering plants and the associated insects that live and feed on them may account for the apparent disproportionate proportions of these two groups from the total.

The International Census of Marine Microbes (ICoMM) estimates that 3.6×10^{30} microbial cells account for more than 90% of the total biomass in the oceans, and there is the possibility that viral particles may be up to 100 times greater than this number. Archaea, one of the two prokaryotic Domains of life, dominate oceanic mid-waters. For prokaryotes, normal species definitions do not apply, and the term phylotype is used. Great progress has been made in recent years through the use of genomics to characterize the diversity of phylotypes in different habitats, but achieving global phylotype estimates for all prokaryotes seems to be some way off.

Where Is Biodiversity Found?

In the terrestrial environment, it is an almost universal principle across all taxa that the diversity of species is highest close to the equator and decreases toward the poles. At the same time, diversity is highest in lowlands and decreases with altitude. Similarly, areas with higher rainfall have the highest diversity at a given latitude or altitude; hence, tropical rainforests are renowned for their high diversity. Consequently, countries with large areas of tropical rain forest feature highly in the list of those countries with the largest numbers of species of plants, birds, and mammals (Table 4.1). There is some evidence that in tropical forests there is a midaltitude peak in species richness, but this result may well be spurious because of sampling problems. Table 4.2 shows the diversity of plant species found in small areas of different forests around the world where typically researchers have measured and identified to species, every tree stem above 1 cm diameter. The highest diversity of plants is found in lowland tropical rain forest plots, particularly those with high annual rainfall and a less distinct seasonal pattern such as the plots in Lambir Hills (Malaysia) and Yasuni (Ecuador). With this high plant richness is an associated high insect and other arthropod species richness. For the marine realm, patterns of species richness do not always follow the same diversity patterns. Here, the diversity of organisms associated with coral reefs gives a high species richness in the tropics for some groups, but the deep-sea welling's of nutrients in higher latitudes result in the reversal of the normally low diversity in lower latitudes for some groups that are dependent on these nutrients.

TABLE 4.1 World Ranking of Mega-Biodiversity Countries as Assessed by Estimates of the Numbers of Species of Mammals, Birds, and Flowering Plants Found in These Countries

Country	Mammals	Birds	Flowering Plants
Mexico	450	1026	25,000
Indonesia	436	1531	27,500
Zaire	415	1,096	11,000
Brazil	394	1,635	55,000
China	394	1,244	30,000
Colombia	359	1,695	50,000
Peru	344	1,678	17,121
India	316	1,219	15,000
Venezuela	305	1,296	20,000
Ecuador	302	1,559	18,250
Cameroon	297	874	8,000
Malaysia	286	736	15,000
Australia	252	751	15,000
South Africa	247	790	23,000
Panama	218	926	9,000
Papua New Guinea	214	708	10,000
Vietnam	213	761	7,000
Costa Rica	205	850	11,000
Philippines	153	556	8,000
Madagascar	105	253	9,000

Source: Adapted from Paine & Status.[11]

TABLE 4.2 The Number of Individuals and Number of Species Richness of Trees in a Selected List of Surveyed 50- and 25-hectare Plots around the World

CTFS Plot Location	Latitude	Plot Size (ha)	No. spp	No. Trees
Wind River, USA	45.8	25.6	25	31,162
Wabikon Lake Forest, USA	45.6	25.6	36	57,388
Haliburton Forest, Canada	45.3	13.53	30	46,339
Changbaishan, China	42.4	25	52	38,902
Smithsonian Cons. Biol. Inst. USA	38.9	25.6	65	41,031
Yosemite National Park, USA	37.8	25.6	23	34,458
Gutianshan, China	29.3	24	159	140,700
Xishuangbanna, China	21.6	20	468	95,834
Huai Kha Khaeng, Thailand	15.6	50	251	72,500
Mo Singto, Thailand	14.4	30.5	262	no data
Mudumalai, India	11.6	50	72	25,500
Barro Colorado Island, Panama	9.2	50	299	208,400
Khao Chong, Thailand	7.5	24	593	121,500
Sinharaja, Sri Lanka	6.4	25	204	193,400
Korup, Cameroon	5.1	50	494	329,000
Lambir, Malaysia	4.2	52	1,182	359,600
Pasoh, Malaysia	3.0	50	814	335,400
Yasuni, Ecuador	-0.7	50	1,114	151,300

Source: Data from the Center for Forest Science, Smithsonian Institution.

Abbreviations: CTFS, The Center for Tropical Forest Science; ha, Hectare; Smithsonian Cons. Biol. Inst., Smithsonian Conservation Biology Institute; Spp., Species.

Note: These data are for all trees on the plots that are greater than 10 cm in diameter (as measured on tree trunks at around 1.5 meters above ground).

There has been some debate about whether within the tropical forests there are more species in the forest canopy than in the ground layer (including soil), with the focus largely being on arthropods. For this reason, some have called the forest canopy "the last biological frontier." However, evidence from a long-term study using equal sampling of the forest and canopy layers shows that of the 1500 beetle species collected, roughly 25% are canopy specialists, 25% are ground specialists, and 50% are shared between these strata)[12] Beetles are probably a good indicator for other arthropods because they comprise 40% of described arthropods and they are biologically very diverse. So, although the canopy of trees is a very important stratum for biodiversity, because that is where most photosynthesis, flowering, and fruiting take place, it seems that the soil and the ground layer are equally important for biodiversity as that is where the roots of trees are found, where seeds develop into young plants, and where decomposition takes place.

The 17th-century poet Jonathan Swift wrote "So nat'ralists observe, a flea hath smaller fleas that on him prey, and these have smaller fleas that bite 'em, and so proceed ad infinitum." This observation that larger organisms have smaller parasitic ones is true about life. Some groups of parasites can be very host specific, such as lice, whereas others, such as fleas, are less host specific and may be found on a wide range of host animals. In general, the larger, more abundant, and more geographically widespread species of organisms, such as the mammals or large fish, have the richest parasitic load. The obvious conclusion from Swift's observation is that most species should be parasites, but this has yet to be statistically tested.

Conservation

Conservation, the protection of animals, plants, and other organisms and their habitats probably first appeared in Assyria in 700 BC where game reserves were set aside for royal hunting. In the United Kingdom, some of the best-protected and most species-rich forests, such as Windsor Great Park and

Epping Forest, were protected to conserve deer for hunting by royalty from the 15th and 16th centuries. This had followed a sharp decline in the area of forests, particularly oak, as large areas had been cleared for timber to build warships in the 1500s. This kind of conservation for sustainable use—in this case of deer—is different from what some conservationists would consider conservation—which is to protect species. The so-called conservation movement is often more concerned with the latter kind of conservation.

Protected Areas

Across the world, an average of 10–15% of the terrestrial environment is protected in some way or other.[13] In contrast, there is far less marine protection, with only 1.2% protected.[14] Much of the oceans are considered international waters with very limited international environmental protection.

The International Union for Conservation of Nature (IUCN), originally called the International Union for the Preservation of Nature (IUPN) and for a short time also called the World Conservation Union, is the world's oldest and largest global environmental organization. Its mission is "to influence, encourage and assist societies throughout the world to conserve the integrity and diversity of nature and to ensure that any use of natural resources is equitable and ecologically sustainable." IUCN is the largest global environmental network with more than 1000 government and non-governmental member organizations, and almost 11,000 volunteer scientists in more than 160 countries. Its program of work is discussed and agreed every 4 years at IUCN's World Conservation Congress. IUCN recognizes six categories of protected management areas: Ia) Strict Nature Reserve and Ib) Wilderness Area, II) National Park, III) Natural Monument or Feature, IV) Habitat/Species Management Area, V) Protected Landscape/Seascape, and VI) Protected Area with sustainable use of natural resources. Each of these gives different levels of protection or sustainable use or both.

Threats to Biodiversity and the Environment

In the last century, there has been a very rapid increase in the world's human population and as a result, increasing demands on the environment for food, water, and other commodities. Efforts to improve agricultural productivity have increased the levels of pest organisms, and as a response chemicals have been developed to manage weeds, fungi, and pest insects and other animals. In the 1960s, concerns were first raised about environmental pollution through the use of toxic chemicals as insecticides and herbicides. Much of the concern was about health risks to humans and other animals but also because of the harm to the environment. Rachel Carson's *Silent Spring*[15] was at the vanguard of this movement. At the same time, the industrial revolution resulted in problems of industrial waste and chemical pollution of water systems. Other sometimes more insidious forms of environmental pollution include acid rain caused by the release of nitrous oxide and sulfur dioxide from the burning of fossil fuels and subsequently deposited through rain as nitric acid and sulfuric acid during rainfall events. Often, the acid rain, which causes die-back and often death of trees, was affecting areas of forests away from the area of initial production and pollution.

Perhaps as much as 70–80% of all terrestrial biodiversity is found in tropical forests. Conservation biologists talk about the importance of old growth or primary forests, particularly tropical rainforests as these contain the greatest diversity of life. The term "virgin" forests is nowadays largely ignored as virtually all forests have been impacted directly or indirectly by man in some way. It was in 1979 that Norman Myers identified the loss of tropical forests as a serious environmental issue.[16] At the time, he suggested that shifting cultivators, provided with new road access, was the major cause of forest loss. Now, selective logging of forest for timber or clear felling to provide new areas for agriculture are seen as the major threats. Subsequently, many suggested that the loss of species associated with cleared primary forest might cause a manmade extinction crisis. However, the large-scale losses have not been observed as yet perhaps due to targeted conservation efforts and the fact that in many places logged forests have remained as secondary forests and that the extinction of many species threatened by habitat loss has been delayed, thus forming an extinction debt that may occur sometime in the future. As of 2010, about 50% of the world's tropical forests fell into the broad secondary-forest category.[17] Although many species that are normally found in primary forests appear to be able to survive in some kinds of secondary forests, particularly selectively logged forests, it is clear that for others there is no substitute for primary forests.[18]

Climate change is already having effects on biodiversity and, given the Intergovernmental Panel on Climate Change (IPCC) scenarios for increasing levels of carbon dioxide (CO_2) in the atmosphere, rising temperatures, and changing rainfall patterns, both the distribution and the existence of many species will be affected.[19] Modeling of the impact of rising temperatures on biodiversity suggests that species that are restricted in distribution to mountaintops will move up the mountains and many will disappear, particularly in the tropics. A 5°C rise in temperature could mean a roughly 1000-meter rise in elevation for species if they are to continue in the same climate envelope. For lowland species, a similar temperature increase could mean they would need to move many thousands of kilometers poleward if there are no higher elevations available. In the marine environment, increasing CO_2 levels will increase acidification of seas, and this could lead to the demise of many corals. Rising temperatures, changing rainfall patterns, and ocean acidification are likely to act synergistically with other threats such as hunting, overharvesting, and habitat loss to increase biodiversity loss and extinctions. Precisely how they may act together and what this may mean for biodiversity is still largely unknown.

Conservation Biology

The science concerned with the management of biodiversity is usually referred to as conservation biology. The management and conservation of biodiversity draw not only on the ecology of species but also on a range of disciplines including animal behavior, chemistry, genetics, geography, demography, medicine, political science, and philosophy. It is a truly multidisciplinary field. Strategies for conservation will often combine theoretical disciplines such as population biology with landscape management options such as establishing protected areas of one kind or another. Others will include the removal and protection of endangered species with reintroductions back into the wild, once viable populations have been created. Other mechanisms will involve the removal or management of threats such as hunting, overfishing, or overharvesting and the reduction or removal of pest animals and weeds. Other activities include controlling the movement or activities of people that are likely to be detrimental to the protected species or protected areas. Such actions usually require government legislation and policy development in addition to financial resources and dedicated staff.

Global Action

A number of approaches have been adopted to tackle the problem of conservation of biodiversity, and these differ in the way that the problems and goals have been defined. Conservation International, for example, adopted an idea first identified by Norman Myers[20] and targeted the protection of biodiversity through the identification of global biodiversity hotspots. They identify these areas as being places where there are more than 1500 endemic (not found anywhere else) species and 70% habitat change or loss. So far, they have identified 36 biodiversity hotspots, which they claim cover 2.3% of the terrestrial land surface and yet include 50% and 40% of the world's plants and vertebrates, respectively.[21] BirdLife International focuses, as the name suggests, on the world's birds and because this is the best known of all organisms, their Important Bird Areas is based on the occurrence of key bird species that are vulnerable to global extinction or whose populations are otherwise irreplaceable.

Extinction Rates

Many scientists believe that the extent and rate of environmental change is so great that the Earth is heading toward a catastrophic "Holocene" mass extinction event like the five mass extinctions that occurred in the geological past and separately wiped out 65–95% of the species that existed at the time. For one of the best-known groups, birds, it appears that extinctions are already at between 50 to 100 times the so-called "background rate" and could be a 1000 times that rate in the next 50 to 100 years.[22] The background rate is a very rubbery figure of how many species extinctions per year there may have been per million existing species—for the last 600 million years, some have suggested that this might be around one extinction per year per million species.[23,24] IUCN has worked with biologists around the world to develop a list of extinct, threatened, and endangered species called the Red List of Threatened Species. In 2012, the list of species that are recognized as having gone extinct in the last 200 years stood at about 1500 species of plants and animals. The status of about 45,000–60,000 species is monitored, particularly birds, mammals, and other vertebrates, and so many species of other groups lack close observation. Of course, recognizing when a species has disappeared, particularly for species that are difficult to observe, is difficult, and so many more species may have gone extinct without anyone recognizing the fact. At the same time, some species, which were thought to have gone extinct are miraculously rediscovered. One example of this is the Lord Howe Island stick insect, which had disappeared soon after the introduction of the black rat on this small (16 km^2) island in 1918. It was rediscovered on a very small island off the coast of Lord Howe Island in 2001, and a plan to reintroduce it was put in place alongside a plan to eradicate the black rat.

A large proportion of the species that are known to have gone extinct in the last 200 years are from islands and are species that had a geographically very small range. Their extinction was caused predominantly by habitat loss, by hunting, or by invasive organisms. There is strong evidence that now most extinctions are of perhaps more widespread species in continental areas, and their demise is due to habitat loss and other threats.

Conclusion

The term "biodiversity" arose in the 1980s when concern for our environment was increasing and led eventually to global agreements to increase the conservation of animals and plants. In spite of these agreements, environmental degradation and climate change are likely to continue to impact severely on the world's biota. In coming decades, extinction rates will rise, and many iconic species as well as many largely unknown species will disappear.

References

1. Wilson, E.O.; Peters, F.M. *Biodiversity.* National Academy Press: Washington, D.C., 1988; 326–332 pp.
2. Cavalier-Smith, T. Only six kingdoms of life. Proc. R. Soc. Lond. B **2004**, *271* (1545), 1251–1262.
3. Simpson, A.G.B.; Andrew, R.J. The real 'kingdoms' of eukaryotes. Curr. Biol. **2004**, *14* (17), R165–R167.
4. Woese, C.R.; Kandler, O.; Wheelis, M.L. Towards a natural system of organisms: Proposal for the domain Archaea, Bacteria, and Eucarya (Euryarchaeota/Crenarchaeota/kingdom/evolution). PNAS **1990**, *87*, 4576–4579.
5. Linnaeus, C. (in Latin). Systema naturae per regna trianaturae: Secundum classes, ordines, genera, species, cum char-acteribus, differentiis, synonymis, locis (10th edition ed.). Stockholm: Laurentius Salvius. 1758.
6. Costello, M.J.; May, R.M.; Stork, N. E. Can we name Earth's species before they go extinct? Science. **2013**, *339*, 413–416, DOI:10.1126/science.1230318

7. Mora, C.; Tittensor, D.P.; Adl, S.; Simpson, A.G.B.; Worm, B. How many species are there on Earth and in the Ocean? PLoSBiol. **2011**, *9* (8), e1001127, doi:10.1371/journal. pbio.1001127.
8. Costello, M.J.; Wilson, S.P.; Houlding, B. Predicting total global species richness using rates of species description and estimates of taxonomic effort. Syst. Biol. **2012**, *61*, 871–883.
9. Hamilton, A.J.; Basset, Y.; Benke, K.K.; Grimbacher, P.S.; Miller, S.E.; Novotný, V.; Samuelson, G.A.; Stork, N.E.; Weiblen, G.D.; Yen, J.D.L. Quantifying uncertainty of tropical arthropod species richness. Amer. Nat. **2010**, *176*, 90–95.
10. Hamilton, A.J.; Basset, Y.; Benke, K.K.; Grimbacher, P.S.; Miller, S.E.; Novotný, V.; Samuelson, G.A.; Stork, N.E.; Weiblen,G.D.; Yen, J.D.L. Correction. Amer. Nat. **2011**, *177*, 544–545.
11. Stork, N.E.; Grimbacher, P.S. Beetle assemblages from an Australian tropical rainforest show that the canopy and the ground strata contribute equally to biodiversity. Proc. Roy. Soc. Series B. **2006**, *273*, 1965–1975.
12. Soutullo, A. Extent of the Global Network of Terrestrial Protected Areas. Cons. Biol. **2010**, *24*, 362–363.
13. Global Ocean Protection: *Present Status and Future Possibilities*: IUCN, Gland. 2010; 96 pp.
14. Carson, R. *Silent Spring*: Houghton, Mifflin, Boston, USA.1962.
15. *Myers, N. The Sinking Ark: A New Look at the Problem of Disappearing Species: Oxford, U.K., Pergamon, 1979; 307 pp.*
16. Chazdon, R.L.; Peres, C.A.; Dent, D; Sheil, D; Lugo, A.E; Lamb, D; Stork, N.E; Miller, S. The potential for species conservation in tropical secondary forests. Conserv. Biol. **2009**, *23*, 1406–1417
17. Gibson, L.; Lee, T.M.; Koh, L.P.; Brook, B.W.; Gardner, T.A.; Barlow, J.; Peres, C.A.; Bradshaw, C.J.A.; Laurance, W.F.; Lovejoy, T.E.; Sodhi. N.S. Primary forests are irreplaceable for sustaining tropical biodiversity. Nature **2011**,*478*, 378–381, doi:10.1038/nature10425.
18. IPCC Fourth Assessment Report: *Climate Change*: 2007; 104 pp.
19. Myers, N. Threatened biotas: 'hotspots' in tropical forests. Environmentalist **1988**, *8*, 1–20.
20. Mittermeier, R.A.; Turner, W.R.; Larsen, F.W.; Brooks, T.M.; Gascon, C. Global biodiversity Conservation: The Critical role of Hotspots. In *Biodiversity Hotspots. Distribution and Protection of Conservation Areas*. Frank, E.Z., and Jan, C.H, Eds.; Springer: Heidelberg 2011; 3–20 pp.
21. Pimm, S.L.; Raven, P.H. Extinction by numbers. Nature **2000**, *403*, 843–845.
22. May, R.M.; Lawton, J.H.; Stork, N.E. Assessing extinction rates. In *Extinction Rates*: John, H., Robert, M., Eds.;Oxford University Press: Oxford, 1995; 1–24 pp.
23. Pimm, S.K.; Russell, G.J.; Gittleman, J.L.; Brooks, T.M. The future of biodiversity. Science **1995**, *269*, 347–349.
24. Giam, A.; Brook, B.W.; Gregory, S.D.; Koh, L.P.; Gibson, L.; Stork, N.E.; Fordham, D.A.; Bradshaw, C.J.A. A global assessment of threats to biodiversity across continents and islands. Nature (in review).

Bibliography

Lindenmayer, D.; Burgman, M. *Practical Conservation Biology;* CSIRO Publishing: Collingwood, Australia, 2005; 433–445 pp.

Norse, E.A., Crowder, L.B. (Eds.) *Marine Conservation Biology*; The Science of Maintaining the Sea's Biodiversity,Island Press: ISBN 978-1-55963-662-9, 2005.

Primack, R.B. *Essentials of Conservation Biology*, 2nd Ed.; Sunderland, MA: Sinauer, 1998.

5

Biodiversity: Habitat Suitability

Introduction 45
Species Fitness and Habitat Quality • Habitat Suitability and Niche Theory • The "n-Dimensional Hypervolume"
Habitat Suitability Modeling 47
Data Requirements • Limitations due to Data and Approach • Limitations due to Biological Assumptions • A Brief Introduction to Mechanistic Approaches
Applications of Habitat Suitability Models 49
Applications for Habitat Change • Applications for Invasive Species Management • Applications for Endangered Species Management
Conclusion 50
Acknowledgments 50
References 51

Betty J. Kreakie
U.S. Environmental Protection Agency (EPA)

Introduction

Habitat suitability is the capacity of a habitat to support a particular species.[1] To fully understand habitat suitability and how it is used in applied research, it is necessary to understand concepts of fitness, habitat quality, and basic niche theory. The remaining portion of the introduction will cover these concepts. The subsequent section discusses how these theories are used to build spatial models of species distribution. The final main section explores some of the areas of conservation and land management that use these types of habitat suitability models.

Species Fitness and Habitat Quality

In general, the fitness of a population is the number of births per generation (or birth rate) minus the number of deaths per generation (death rate).[2] Furthermore, a population is in suitable habitat only if the environment can support the population (e.g., the population sustains itself and is able to maintain its population size). The fitness of the population must, therefore, be greater than (birth rate greater than death rate) or equal to zero. If we were to look at a species existing in a habitat of optimal quality, we would expect to observe two possible fitness responses.[3] First, if the population is at equilibrium—perhaps this population has existed for several generations in this optimal habitat—the birth rate would be equal to the death rate. This population is said to be at carrying capacity, the largest population size the habitat can support with given levels of resources. Alternatively, if the population has not reached carrying capacity and the birth rate is higher than the death rate, the population size would be increasing. Species, however, do not exist only in optimal habitats (for reasons we will discuss later).

Species can be found in a range of habitats of varying quality, which would result in varying levels of fitness.[4] To define habitat suitability, we need to quantify the relationship between fitness and the habitat of a particular area.

A logical assumption might be that a species would only exist in habitats where it could maximize its fitness. Natural selection preferentially selects populations that contribute the largest number of offspring to the next generation. Accordingly, we might expect to see populations existing only in optimal habitats because those populations are being preferentially selected. However, existence in an optimal habitat is not always the case.[5] The possible reasons for living in a range of habitat are nearly infinite. However, numerous reasons are due to the dynamic nature of our world; most characteristics of the environment are not static and therefore populations are unlikely to reach equilibrium at a given location.[6] Climates change, as does land cover. Often, populations are limited in dispersal capabilities and cannot move or migrate to follow these changes in geographic space.[7] If they are able to migrate, it is not certain if this new area will provide their optimal habitat or even be sufficient to support the population. Additionally, they might not be able to compete with an established competitor population in a new location. Despite any changes in climate or land cover, the established population might still be comfortable in their habitat range and feel no pressure to acquiesce the habitat to the displaced population.

Habitat Suitability and Niche Theory

We cannot talk about defining habitat suitability without at least a cursory discussion of the niche theory.[8] The niche theory has a long history and a complex literature, which is well beyond the scope of this text. However, for a thorough review of the niche theory history and a modern synthesis, see Chase and Leibold.[9] For our purposes of defining habitat suitability, the broad Chase and Leibold definition of "niche" is appropriate:

> *"[T]he joint description of the environmental conditions that allow a species to satisfy its minimum requirements so that birth rate of a local population is equal to or greater than its death rate along with the set of per capita effects of that species on these environmental conditions."*

As you can see, the first part of the broad niche definition is nearly the definition of habitat suitability presented earlier. In fact, through history the "per capita effects of that species on these environmental conditions" portion of the niche definition was often ignored or omitted. Some will refer to this more concise or limited definition of niche as the Grinnellian Niche, based on the work of Joseph Grinnell.[10]

The "n-Dimensional Hypervolume"

G. Evelyn Hutchinson was perhaps the first, and definitely the most famous, to propose defining a niche by what he called the "n-dimensional hypervolume".[11] His concept of defining the niche by its "n-dimensional hypervolume" is still essentially the core of habitat suitability quantification. The concept is as follows. For a given species, elements of the environment are measured at each known species occurrence location. In other words, each time that species is observed, environmental characteristics are measured at that location. The list of environmental characteristics depends on the type of species (i.e., if the species is a plant, terrestrial animal, fish, fungus, etc.), the availability or ability to measure a particular variable, and the researchers' assumptions about the species. Typically, the measured parameters might include temperature, precipitation, elevation, pH (soil or water), habitat type, and so forth. From these parameters, researchers would be able to define the niche as the multidimensional space where that species "lives." The niche is not defined as latitude and longitude extents on a landscape but instead in terms of physiological tolerances to environmental parameters. It might even be possible to assume that where (in that multidimensional space) we observe the highest population numbers is where the habitat is optimal. Reciprocally, areas with low or diminishing population numbers might be classified as sub-optimal (or fringe) habitats. This method of defining a niche's n-dimensional hypervolume is also a means to explore the range of habitat suitability.[12]

Habitat Suitability Modeling

Besides pure ecological curiosity, understanding the range extent and spatial distribution of a species' suitable habitat is often necessary for effective land management and species conservation (the following section will concentrate on exploring some specific applications). Based only on observational data, our ability to understand species' habitat suitability throughout an entire range is limited. These limitations are mainly due to sampling. To understand comprehensively the behavior of species habitat suitability throughout the entire range, we would need to sample the entire range. Additionally, we would need to sample areas that are not habitats. This allows for predictions across the entire landscape and not just in areas of known occurrence. Furthermore, we would need to conduct this sampling of habitats and non-habitats throughout the year and over many years. This would allow us to understand how the species moves across the landscape throughout the year and how this spatial distribution changes between years. In others words, we would need to sample everywhere all of the time. Even with high sample intensity, not observing your species during a survey does not necessarily indicate a true absence. Some species are cryptic; they may be present in a location but maybe simply undetected by the researcher. The compounded effects of these sampling limitations necessitate habitat modeling. There are numerous statistical approaches used to carry out this type of research (see Guisan and Zimmermann[13] for a review).

The previous discussion of habitat suitability involved relating environmental parameters to a species' fitness. Habitat suitability modeling takes the previous definition subtly in a new direction. Habitat modeling incorporates environmental information from sampled habitats and nonhabitats and makes predictions about the likelihood of species occurrence in an unsampled location.[12,14] The more similar a novel unsampled location is to the optimal in the species' habitat multidimensional space (again returning to the concept "n-dimensional hyperspace"), we would predict a higher probability of occurrence.[8] This, of course, is contingent on such factors as the ability of the species to reach this unsampled location. This type of modeling and subtle variations of it has been called many names: species distribution modeling, habitat suitability index, niche modeling, ecological niche factor analysis, climate envelopment modeling, and numerous others. However, they are all generally based on the same simple concept; there is some means to predict the likelihood of occurrence (or predicted suitability) in an unsampled location based on what we know about the habitat at a known occurrence. Figure 5.1 provides an illustration of typical habitat suitability results.[15]

Data Requirements

Habitat suitability modeling is employed because we are simply not capable of sampling the entire range of a species. Even though habitat suitability modeling approaches are employed to compensate for insufficient observational data, there are still considerable data requirements. We discussed how most modeling approaches require information about both habitats and non-habitats; in other words, descriptions of where the species was observed as well as where it was not observed. As there are large amounts of presence-only data available (e.g., herbarium records, bird banding), modeling approaches have been developed to utilize this type of presence-only data.[16,18] For this reason and others, a model approach should be selected based on type of data available. As mentioned in the preceding text, some species may be more cryptic than others; therefore it is important to understand the detection probability of the particular species.[19,20] It is also important to understand any sampling bias in the data. For example, if a bird species was only sampled in the summer, it is unlikely that sampling data can be used to make predictions about winter locations. Or if one area was sampled with a much higher frequency than any other portion of the range, this will skew the overall occurrence predictions. It is likely that predictions

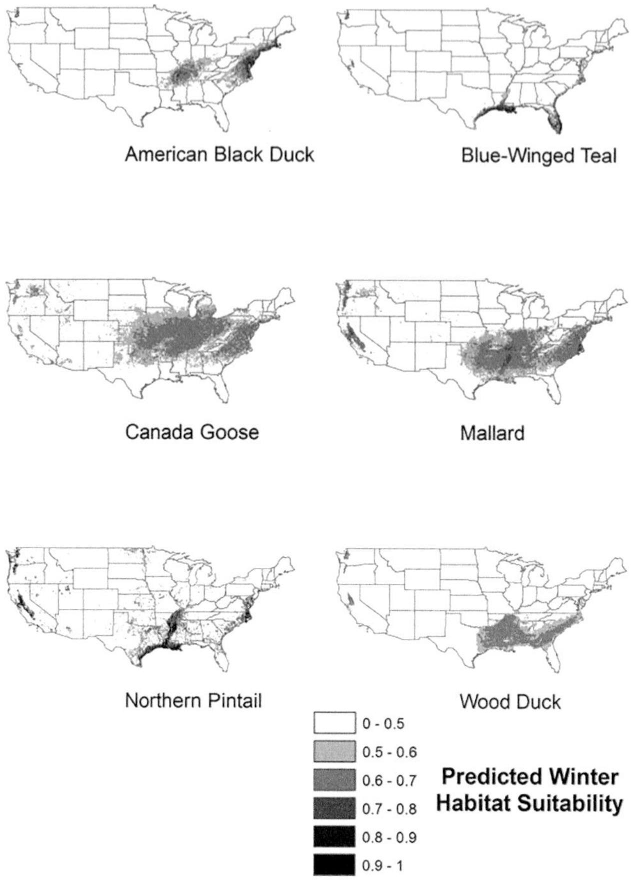

FIGURE 5.1 Map of winter-predicted suitability for six waterfowl species in the continental United States. Predictions were created using MaxEnt[16] with data from the U.S. Geological Survey's Banding Bird Laboratory 1990–1999 data.[17] Temperature, precipitation, elevation, and water table depth were the predicted variables used to model the predicted distribution of these species' habitat.[15]
Source: Adapted from Kreakie et al.[15]

will be high for areas that are similar to intensively sampled locations (regardless of true relationship to optimal habitat) unless the sample bias is statistically corrected.

Limitations due to Data and Approach

The predictions from habitat suitability modeling must be interpreted carefully with a clear understanding of the limitations associated with the data and modeling method.[21,22] Species data that are spatially sparse or temporally sporadic may impact the uncertainty we have when interpreting the results.[22–24] In addition to understanding the limitations of the species data, the same level of scrutiny is needed for the environmental parameters. Variables that are included and omitted might lead to drastically different predictions of species occurrence probability. Often it is possible to create predictions that seem to perform very well; yet in reality, these results are meaningless. Due to issues of overfitting the model predictions to the sample data, the results may be irrelevant to species. To alleviate some concerns about model selection and performance, the general reader of the research results does not need to have an in-depth understanding of the mathematics behind all modeling approaches and methods of performance evaluation. Researchers should demonstrate to readers that they recognize these issues and have taken steps to alleviate some of the ramifications.

Limitations due to Biological Assumptions

The shortcomings of habitat suitability modeling are more than methodological. In using habitat models, we make assumptions about the biology of a species that should be considered when interpreting the results.[25] The model predictions are based on the assumption that the species can successfully disperse to all locations. This might not be the case. There may be barriers in the landscapes that the species simply cannot cross (e.g., roads or rivers). It might simply be too great a distance for the species to travel. Model predictions are based on sightings of a species, which we assume is living in a suitable habitat and is static in time.[26] However, it is possible for a species to be found in an area where it has reduced fitness (what ecologists call sink habitat).[27,28] Therefore, model predictions could be based on a habitat to which the species is actually maladapted. Also, in general, model predictions are based on a single species, when in reality all species live in communities where they interact with different species.[29,30]

These shortcomings of habitat suitability modeling do not mean that we should abandon this area of research. The following section will present several areas of research where these approaches are very useful. However, understanding the shortcomings will help to increase the effectiveness of their use.[31]

A Brief Introduction to Mechanistic Approaches

Our discussion of habitat suitability modeling has focused entirely on methods that are considered correlative approaches. These methods attempt to elucidate a relationship between species location data and environmental parameters. However, there is an area of research that makes an attempt to define habitat range from a completely different angle. Mechanistic models estimate occurrence likelihood based on resource requirements and physiological constraints of a species.[32,33] This approach requires a detailed understanding of the physiological tolerances of a species, which may require extensive laboratory work. Once a researcher determines these thresholds, they determine the spatial distribution of the appropriate habitat. Kearney et al. have significantly advanced this research avenue with their detailed work on reptiles and amphibians.[34,35]

Applications of Habitat Suitability Models

There are numerous uses for the predictions created from habitat suitability modeling, and these results have been applied in a wide range of disciplines. The following provides a concise summary of a few key examples in ecology. In addition, here are some further references for more insight into the applications of habitat suitability modeling.[36–39]

Applications for Habitat Change

One of the main uses of habitat suitability modeling is to predict how species may respond to changes in habitat.[40] Researchers gain information about the multidimensional habitat space in which the species is currently living and assume the species will continue to be found in the same multidimensional space in the future. Then data is gathered about how the landscape is predicted to look in the future. These future landscapes are used to predict the spatial distribution of the species' future habitat. This type of future scenario modeling has been used to predict how species may respond to converting a habitat from one type to another. For example, Broton et al. calculated the predicted effect of different irrigation scenarios on bird habitat.[41] Using this type of scenario modeling, Sánchez-Cordero forecasted how endemic mammal species of Mexico might respond to deforestation.[42] This approach is commonly used to predict new species range maps for various climate change scenarios. Early work in this area by Iverson et al. made predictions about different tree species in eastern United States.[43,44] Iverson's work illustrates how some species might expand their range limits while others may potentially contract.

Applications for Invasive Species Management

Habitat suitability modeling can be a useful tool in managing invasive species. Once researchers understand the native habitat of a potentially invasive species, they can use the information to predict which areas of the introduced habitat might be at a higher risk of invasion.[45,46] This work helps establish areas that might need to be monitored or regulated more closely.[47] For example, Ward used this approach to make predictions about the potential distribution of six ant species in New Zealand, finding a propensity for lowlands and coastal areas.[48] This also included information about the time duration of the invasion, illustrating a positive correlation between the time spent in the introduced habitat and its range. Zambrano et al. used these methods to make predictions about fish (carp and tilapia) invasions in stream reaches.[49] This study illustrated that each species presents different patterns of risk across the landscape. The majority of work assumes that the habitat requirements do not shift between native and introduced habitats, which may not always be true.[50]

Applications for Endangered Species Management

Habitat suitability has a wide range of applications for the management of endangered species. For species in the United States that are officially listed as endangered or threatened, legislation mandates that key habitats be identified. To accomplish this goal, the U.S. Fish and Wildlife Service (USFWS) developed methods to calculate an area's habitat suitability index (see Schamberger et al.[51] for more detail on methods and application). Additionally, endangered species management may require introduction into novel areas. These introductions could be from captivity- bred population (e.g., zoo breeding programs) or from struggling populations that are in isolated pockets of habitats. Habitat suitability modeling is used to identify areas that might optimize results. This application is really an attempt to identify areas that will maximize fitness (i.e., maximizing birth rate while minimizing the death rate). Understanding an area's potential reproductive fitness for a species is critical for this work. Often, the number of individuals introduced is low, and potential stochastic effects can be high.

Conclusion

Ecologists have taken the principles of niche theory and successfully developed tools for conservation biology and land management. By understanding where on the landscape a given species is presently flourishing (i.e., where it has the highest fitness), we can make inferences about how a species may respond in unsampled areas and novel habitat or to scenarios of climate or land change. This area of research continues to improve and expand. Remotely sensed environmental data is increasing the spatial extent of the modeling endeavors.[52,53] As a result, regional and global distribution modeling are becoming more accessible. Also, large biological databases make it possible for researchers to share their survey data and receive rangewide information about their target species without sampling the entire range.[54] The Global Biodiversity Information Facility (GBIF) collects and manages biodiversity data and provides it freely to researchers via the internet (http://www.gbif.org). Improved technology and methods will expand the spatial extent of study areas, allow for finer resolution predictions, and reduce our uncertainty in results. These improvements should enhance our ability to effectively mitigate ecological problems.

Acknowledgments

I would like to thank all the reviewers for their useful comments and suggestions. I would also like to thank Dwaine Kreakie for all his diligent editorial work. The U.S. EPA's Office of Research and Development (ORD) tracking number for this entry is ORD-003218.

References

1. Bernstein, C.; Krebs, J.R.; Kacelnik, A. *Distribution of Birds Amongst Habitats: Theory and Relevance to Conservation. Bird Population Studies: Relevance to Conservation and Management*; Oxford University Press: Oxford, UK, 1991; 317–345.
2. Pianka, E.R. *Evolutionary Ecology*, 7th Ed. eBook; Eric R.P. Ed.; 2011.
3. Futuyma, D.J.; Moreno, G. The evolution of ecological specialization. Annu. Rev. Ecol. Syst. **1988**, 207–233.
4. Johnson, M.D. Measuring habitat quality: A review. Condor **2007**, *109*, 489–504.
5. Terborgh, J. Habitat selection in Amazonian birds. In *Habitat Selection in Birds*; Cody, M., Ed.; Academic Press: New York, 1985; 311–338.
6. Franklin, J. Moving beyond static species distribution models in support of conservation biogeography. Diversity and Distributions **2010**, *16*, 321–330.
7. Araujo, M.B.; Guisan, A. Five (or so) challenges for species distribution modelling. J. Biogeogr. **2006**, *33*, 1677–1688.
8. Hirzel, A.H.; Le Lay, G. Habitat suitability modelling and niche theory. J. Appl. Ecol. **2008**, *45*, 1372–1381.
9. Chase, J.M.; Leibold, M.A. *Ecological Niches: Linking Classical and Contemporary Approaches*; University of Chicago Press: Chicago, IL, 2003; 212 p.
10. Grinnell, J. The niche-relationships of the California Thrasher. The Auk. **1917**, *34*, 427–433.
11. Hutchinson, G.E. Cold Spring Harbor Symposium on Quantitative Biology. Concluding Remarks **1957**, *22*, 415–427.
12. Pulliam, H.R. On the relationship between niche and distribution. Ecol. Lett. **2000**, *3*, 349–361.
13. Guisan, A.; Zimmermann, N.E. Predictive habitat distribution models in ecology. Ecol. Model. **2000**, *135*, 147–186.
14. Augustin, N.; Mugglestone, M.; Buckland, S. An autologistic model for the spatial distribution of wildlife. J.Appl. Ecol. **1996**, *33*, 339–347.
15. Kreakie, B.J.; Fan, Y.; Keitt, T.H. Enhanced migratory waterfowl distribution modeling by inclusion of depth to water table data. PloS One **2012**, *7*, e30142.
16. Phillips, S.J.; Dudík, M. Modeling of species distributions with Maxent. New extensions and a comprehensive evaluation. Ecography **2008**, *31*, 161–175.
17. Buckley, P.; Francis, C.M.; Blancher, P.; DeSante, D.F.; Robbins, C.S.; Smith, G.; Cannell, P. The North American Bird Banding Program: Into the 21st Century (El Programa de Anillaje de Aves de Norte América: Hacia el Siglo 21). J. Field Ornithol. **1998**, *69* (4), 511–529.
18. Hirzel, A.H.; Helfer, V.; Metral, F. Assessing habitat-suitability models with a virtual species. Ecol. Model. **2001**, *145*, 111–121.
19. Royle, J.A.; Nichols, J.D.; Kéry, M. Modelling occurrence and abundance of species when detection is imperfect. Oikos **2005**, *110*, 353–359.
20. Mackenzie, D.I. Modeling the probability of resource use: The effect of, and dealing with, detecting a species imperfectly. J. Wildlife Manag. **2006**, *70*, 367–374.
21. Hirzel, A.; Guisan, A. Which is the optimal sampling strategy for habitat suitability modelling. Ecol. Model. **2002**, *157*, 331–341.
22. Barry, S.; Elith, J. Error and uncertainty in habitat models. J. Appl. Ecol. **2006**, *43*, 413–423.
23. Stockwell, D.R.B.; Peterson, A.T. Effects of sample size on accuracy of species distribution models. Ecol. Model. **2002**, *148*, 1–13.
24. Wisz, M.S.; Hijmans, R.; Li, J.; Peterson, A.T.; Graham, C.; Guisan, A. Effects of sample size on the performance of species distribution models. Divers. Distributions **2008**, *14*, 763–773.
25. Cassini, M.H. Ecological principles of species distribution models: The habitat matching rule. J. Biogeogr. **2011**, *38*, 2057–2065.

26. Guisan, A.; Theurillat, J.P.; Deil, U.; Loidi, J. Equilibrium modeling of alpine plant distribution: How far can we go?, Vegetation and climate. A selection of contributions presented at the 42nd Symposium of the International Association of Vegetation Science, Bilbao, Spain, 26–30 July 1999. Gebrüder Borntraeger Verlagsbuchhandlung 2000, 353–384.
27. Dias, P.C. Sources and sinks in population biology. Trends Ecol. Evol. **1996**, *11*, 326–330.
28. Hanski, I.; Hanski, I.A. *Metapopulation Ecology*; Oxford University Press: Oxford, UK, 1999; 313 p.
29. Austin, M.P. Spatial prediction of species distribution: An interface between ecological theory and statistical modelling. Ecol. Model. **2002**, *157*, 101–118.
30. Elith, J.; Leathwick, J.R. Species distribution models: Ecological explanation and prediction across space and time. Annu. Rev. Ecol. Evol. Syst. **2009**, *40*, 677–697.
31. Hernandez, P.A.; Graham, C.H.; Master, L.L.; Albert, D.L. The effect of sample size and species characteristics on performance of different species distribution modeling methods. Ecography **2006**, *29*, 773–785.
32. Kearney, M.; Porter, W. Mechanistic niche modelling: Combining physiological and spatial data to predict species' ranges. Ecol. Lett. **2009**, *12*, 334–350.
33. Guisan, A.; Thuiller, W. Predicting species distribution: Offering more than simple habitat models. Ecol. Lett. **2005**, *8*, 993–1009.
34. Kearney, M.; Porter, W.P. Mapping the fundamental niche: Physiology, climate, and the distribution of a nocturnal lizard. Ecology **2004**, *85*, 3119–3131.
35. Kearney, M.; Phillips, B.L.; Tracy, C.R.; Christian, K.A.; Betts, G.; Porter, W.P. Modelling species distributions without using species distributions: the cane toad in Australia under current and future climates. Ecography **2008**, *31*, 423–434.
36. Brooks, T.M.; Mittermeier, R.A.; da Fonseca, G.A.B.; Gerlach, J.; Hoffmann, M.; Lamoreux, J.F.; Mittermeier, C.G.; Pilgrim, J.D.; Rodrigues, A.S.L. Global biodiversity conservation priorities. Science **2006**, *313*, 58–61.
37. García, A. Using ecological niche modelling to identify diversity hotspots for the herpetofauna of Pacific lowlands and adjacent interior valleys of Mexico. Biol. Conserv. **2006**, *130*, 25–46.
38. Naidoo, R.; Balmford, A.; Costanza, R.; Fisher, B.; Green, R.E.; Lehner, B.; Malcolm, T.; Ricketts, T.H. Global mapping of ecosystem services and conservation priorities. Proc. Nat. Acad. Sci. **2008**, *105*, 9495–9500.
39. Costa, G.C.; Nogueira, C.; Machado, R.B.; Colli, G.R. Sampling bias and the use of ecological niche modeling in conservation planning: A field evaluation in a biodiversity hotspot. Biodivers. Conserv. **2010**, *19*, 883–899.
40. Pearson, R.G.; Dawson, T.P. Predicting the impacts of climate change on the distribution of species: Are bioclimate envelope models useful? Global Ecol. Biogeogr. **2003**, *12*, 361–371.
41. Brotons, L.; Mañosa, S.; Estrada, J. Modelling the effects of irrigation schemes on the distribution of steppe birds in Mediterranean farmland. Biodivers. Conserv. **2004**, *13*, 1039–1058.
42. Sánchez-Cordero, V.; Illoldi-Rangel, P.; Linaje, M.; Sarkar, S.; Peterson, A.T. Deforestation and extant distributions of Mexican endemic mammals. Biol. Conserv. **2005**, *126*, 465–473.
43. Iverson, L.R.; Prasad, A.M. Predicting abundance of 80 tree species following climate change in the eastern United States. Ecol. Monogr. **1998**, *68*, 465–485.
44. Iverson, L.R.; Prasad, A.; Schwartz, M.W. Modeling potential future individual tree-species distributions in the eastern United States under a climate change scenario: A case study with *Pinus virginiana*. Ecol. Model. **1999**, *115*, 77–93.
45. Peterson, A.T. Predicting the geography of species' invasions via ecological niche modeling. Q. Rev. Biol. **2003**, *78*, 419–433.
46. Thuiller, W.; Richardson, D.M.; PYŠEK, P.; Midgley, G.F.; Hughes, G.O; Rouget, M. Niche-based modelling as a tool for predicting the risk of alien plant invasions at a global scale. Global Change Biol. **2005**, *11*, 2234–2250.

47. Herborg, L.M.; Jerde, C.L.; Lodge, D.M.; Ruiz, G.M.; MacIsaac, H.J. Predicting invasion risk using measures of introduction effort and environmental niche models. Ecol. Appl. **2007**, *17*, 663–674.
48. Ward, D.F. Modelling the potential geographic distribution of invasive ant species in New Zealand. Biol. Invasions **2007**, *9*, 723–735.
49. Zambrano, L.; Martínez-Meyer, E.; Menezes, N.; Peterson, A.T. Invasive potential of common carp (*Cyprinus carpio*) and Nile tilapia (*Oreochromis niloticus*) in American freshwater systems. Can. J. Fisheries Aqua. Sci. **2006**, *63*, 1903–1910.
50. Broennimann, O.; Treier, U.A.; Müller-Schärer, H.; Thuiller, W.; Peterson, A.; Guisan, A. Evidence of climatic niche shift during biological invasion. Ecol. Lett. **2007**, *10*, 701–709.
51. Schamberger, M.; Farmer, A.H.; Terrell, T.J. Habitat Suitability Index (HSI) Models: Introduction. 1982, U.S. Fish and Wildlife Service Special Report FWS-OBS-82-10.
52. Kerr, J.T.; Ostrovsky, M. From space to species: Ecological applications for remote sensing. Trends Ecol. Evol. **2003**, *18*, 299–305.
53. Bradley, B.A.; Fleishman, E. Can remote sensing of land cover improve species distribution modelling? J. Biogeogr. **2008**, *35*, 1158–1159.
54. Corsi, F.; de Leeuw, J.; Skidmore, A. *Modeling Species Distribution with GIS. Research Techniques in Animal Ecology*; Columbia University Press: New York, **2000**; 389–434.

6

Biodiversity: Tropical Agroforestry

Introduction 55
Agroforestry 55
Agroforestry Defined • The Value of Agroforestry
Biodiversity 57
Biodiversity Defined • The Value of Biodiversity
Trends in the World Forests 58
Agroforestry and Biodiversity Preservation 59
Studies of Agroforestry and Biodiversity 60
Planned Biodiversity 60
Certifications and Biodiversity 60
Conclusion 61
Acknowledgments 61
References 61

Thomas P. Husband
University of Rhode Island

Introduction

The goal of this entry is to introduce the reader to basic concepts of agroforestry and how they relate to landscape ecology and the enhancement of biodiversity in the tropics. In recent years, the term *biodiversity* has become a household word throughout much of the world. According to the Food and Agriculture Organization of the United Nations (FAO), "'Biological diversity' encompasses the variety of existing life forms, the ecological roles they perform and the genetic diversity they contain. In forest ecosystems, biological diversity allows species to adapt continuously to dynamically evolving environmental conditions, to maintain the potential for tree breeding and improvement (to meet human needs for goods and services and changing end-use requirements), and to support their ecosystem functions."[1] The role of agroforestry in maintaining or enhancing biodiversity, particularly in the tropics, is the primary focus of this encyclopedia entry.

Agroforestry

Agroforestry Defined

According to Young,[2] "Agroforestry is a collective name for land-use systems in which woody perennials (e.g., trees, shrubs, etc.) are grown in association with herbaceous plants (e.g., crops, pastures) and/or livestock in a spatial arrangement, a rotation or both, and in which there are both ecological and economic interactions between the tree and non-tree components of the system." Gascon et al.[3] simply defined agroforestry as "the practice of integrating trees and other large woody perennials on farms and throughout the agricultural landscape." Agroforestry practices include[2] trees on cropland, plantation crop combinations, multistory-tree gardens, hedgerow intercropping, boundary planting, trees on erosion-control structures, windbreaks and shelterbelts, biomass transfer, trees on rangeland or pastures, plantation crops with pastures, live fences, fodder banks, woodlots with multipurpose management, and reclamation forestry leading to multiple use. Agroforestry involves mixing trees or shrubs or both with non-woody crops and sometimes livestock for the purpose of crop diversification, improved profitability, and improved environmental stewardship. Agroforestry may involve the introduction of long-term crops (trees and shrubs) into open agricultural landscapes, or conversely, the introduction of agricultural husbandry of a wide range of crops and livestock into existing woodlands. Agroforestry crops include wood fiber; both native and exotic tree species; non-wood crops from trees; traditional agricultural crops such as horticultural crops, field crops, forages, special crops, and livestock; and non-timber forest crops. Agroforestry practices involve growing woody perennials, usually trees, along with more than one farm crop growing on the same land area at the same time, which can cause both positive and negative competition between crops for light, water, and nutrients. Successful implementation of agroforestry practices requires good planning and management.

The Value of Agroforestry

According to Garret,[4] "When properly designed and integrated, agroforestry can protect crops and improve crop yields, shelter livestock, reduce animal stress while improving weight gain, and enhance resource stewardship and land conservation. From creating consumer markets and reducing nonpoint source pollution to protecting waterfowl habitat and developing new farm crops, agroforestry is moving to the forefront in meeting the needs of landowners and natural resource professionals while keeping the family farm economically viable and the environment in which we live healthy." Gold and Garret[5] point out that, in the tropics, agroforestry has been long practiced and widely accepted by farmers as an alternative to traditional slash-and-burn agriculture and to conventional agriculture where slopes are steep and on marginal land. As a result, ecologically based agroforestry has helped to restore and maintain biodiversity, stabilize the ecology of farms and their watersheds, help sustain the production of basic human needs, and create marketplace opportunities for millions of rural poor.[6–8]

Much important advancement in tropical agroforestry has been made through the work of scientists from various institutions around the world. Among those institutions is the World Agroforestry Centre (ICRAF), an independent not-for-profit research organization. ICRAF is part of an alliance of the Consultative Group on International Agricultural Research (CGIAR) centers that are "dedicated to generating and applying the best available knowledge to stimulate agricultural growth, raise farmers' incomes, and protect the environment." ICRAF's vision is a rural transformation in the developing world as smallholder households strategically increase their use of trees in agricultural landscapes to improve their food security, nutrition, income, health, shelter, energy resources, and environmental sustainability. The Centre's mission is to generate science-based knowledge about the diverse roles that trees play in agricultural landscapes, and use its research to advance policies and practices that benefit the poor and the environment.[9] ICRAF lists the various benefits of agroforestry, including the following:

- Enriching the asset base of poor households with farm-grown trees.
- Enhancing soil fertility and livestock productivity on farms.
- Linking poor households to markets for high-value fruits, oils, cash crops, and medicines.
- Balancing improved productivity with the sustainable management of natural resources.
- Maintaining or enhancing the supply of environmental services in agricultural landscapes for water, soil health, carbon sequestration, and biodiversity.

Agroforestry is increasingly recognized as a valuable strategy with the potential to diversify production of agricultural and forest products for greater social, economic, and environmental benefits.[3] Relative to agroforestry, ICRAF has outlined seven key challenges inter-related to the Millennium Development Goals (MDG) and the Water, Energy, Health, Agriculture, and Biodiversity (WEHAB) initiative.[6,10]

They include the following: i) Help eradicate hunger through basic, proper food production systems in disadvantaged areas based on agroforestry methods; ii) lift more rural poor from poverty through market-driven, locally led tree cultivation systems that generate income and build assets; iii) advance health and nutrition of the rural poor through agroforestry systems; iv) conserve biodiversity through integrated conservation development solutions based on agroforestry technologies, innovative institutions, and better policies; v) protect watershed services through agroforestry-based solutions that enable the poor to be rewarded for their provision of these services; vi) assist the rural poor to better adapt to climate change, and to benefit from emerging carbon markets, through tree cultivation; and vii) build human and institutional capacity in agroforestry research and development. Agroforestry has had a significant role in increasing productivity of landscapes for the benefit of human populations, thus contributing to many of these goals.

Many of the MDGs focus on environmental stability and the integration of the principles of sustainability into policies and programs to reverse the loss of environmental services. According to Garrity,[6] the environmental services that are of greatest relevance are watershed protection, biodiversity conservation, and climate change mitigation and adaptation. A major emerging trend in agroforestry research and development is in the area of enhancing environmental services while meeting farmers' needs for food and income. The latter is critical, as about 90% of the biodiversity resources in the tropics are located in human dominated or working landscapes.[6] The intensification of agroforestry systems can reduce the exploitation of protected areas,[11,12] increase biodiversity in agricultural landscapes, and actually increase the species and within-species diversity of trees in farming systems.[4,6,13]

Biodiversity

Biodiversity Defined

Biodiversity comprises the complexity of all life forms. The Convention on Biological Diversity defines its area of concern as "... the variability among living organisms from all sources, including, *inter alia*, terrestrial, marine and other aquatic ecosystems and the ecological complexes of which they are part; this includes diversity within species, between species, and of ecosystems."[14] Myers[15] states, "Biodiversity embraces the totality of life forms, from the planetary spectrum of species to subunits of species (races, populations) together with ecosystems and their ecological processes."

Sometimes thought of simply as the number of species in a given area, the concept of biodiversity is much more complex. In addition to a count of species, or *species richness*, it includes elements of *species diversity, genetic diversity, functional diversity*, and *ecological diversity*. Species diversity includes both the number of species in an area and the measure of their relative abundance or spatial distribution. For example, in Table 6.1, each of the two hypothetical communities has three species with a total of 100 individuals. The species richness is the same; however, as calculated by the Shannon Index,[16] the species diversity is higher in Community 2. As with most calculations of species diversity, the Shannon Index takes into account evenness, or a measure of the relative proportions represented by each species. Community 2 has a higher degree of evenness, and thus higher species diversity.

TABLE 6.1 The Species Diversity of Two Different Hypothetical Communities

Species	Community 1	Community 2
A	90	34
B	5	33
C	5	33
Total Individuals	100	100
Species Diversity (H')	0.39	1.09

Genetic diversity is now being considered more frequently as a critical element of biodiversity.[17] This includes various measures of the variety of genetic material within a species or a population. Low *genetic diversity* can lead to short-, medium-, and long-term problems, such as inbreeding depression that shifts mean phenotypes in a direction that causes a reduction in fitness. *Functional diversity*, another important component, includes the biological and chemical processes such as energy flow and matter recycling needed for the survival of species, communities, and ecosystems.[18] Functional diversity also comprises the many ecological interactions among species (e.g., competition, predation, parasitism, mutualism, etc.), as well as the natural disturbances (e.g., fire, rain events, etc.) that many species and communities require. Ecological diversity considers the variety of terrestrial and aquatic ecosystems that are found in an area of the Earth.[18] Together, these four components comprise biodiversity.

The Value of Biodiversity

The values gained from biodiversity are associated with different scales and the individual components outlined above, including landscapes, ecosystems, species, populations, genes, and various ecosystem functions. Usually, the value of biodiversity must be judged on the basis of the scale or component critical to the situation; thus no single measure of biodiversity value can be employed universally. Generally, four different types of values can be recognized with regard to biodiversity.[19,20] First, biodiversity could be considered to have its own value, often called its *intrinsic* value. Cultural, social, aesthetic, and ethical benefits are all components of intrinsic value. Second is the *utilitarian* or *material goods* value of biodiversity, which includes all of the subsistence and commercial benefits of species or their genes by various sectors of society.[15] Third, biodiversity may have *serependic* value, that is, the belief that parts of an ecosystem may have future, but as of yet unknown, values to humans. Last, biodiversity has a *functional* or *ecosystems services* value, that is, it contributes to ecosystem life support functions and the preservation of ecological structure and integrity. At some level, all of these types of values may be considered ethnocentric and depend upon the context of the human culture in which they are viewed. In general, biodiversity is considered valuable to humans because it provides us with natural resources (e.g., food, water, wood paper products, energy, medicines), natural services (e.g., air and water purification, soil fertility, waste disposal, pest control), and aesthetic pleasure (e.g., wildlife, national parks and nature reserves, natural scenic beauty). Tropical ecosystems, where most agroforestry systems are employed, are particularly recognized for their significant biodiversity because they contain over half of the terrestrial species on our planet.[21]

Trends in the World Forests

One important role of agroforestry with respect to environmental services is that of ameliorating the downward trends in the world's forests. Based upon 2010 statistics,[1] the total forest in the world was estimated to be slightly more than 4 billion hectares or about 31% of the total land area of the Earth (Figure 6.1). Tropical forests make up about 47% of this total.[1] According to the FAO,[1] deforestation continues at an alarming rate, mostly the result of conversion to agricultural land. In recent years, this loss has been slowed significantly around the world by the planting of new forest trees, landscape restoration, and natural expansion.[1] Overall, net global forest loss from 2000 to 2010 was 13 million hectares per year, down from 16 million hectares per year for the period 1990–2000.[22] The greatest losses from 2000 to 2010 were incurred in South America where over 3.9 million hectares were lost each year, followed by Africa, which lost about 3.4 million hectares per year (Figure 6.2). Clearly, the forests of the tropics have been the most impacted in the world.

For a variety of reasons, trees are being planted at increasing rates around the world. Even so, forest plantations comprise only about 4–5% of the total forest area.[1] Of all forests, less than 12% are designated as having conservation of biodiversity as their primary function.[1] Although deforestation, including conversion to agricultural land, continues at an alarmingly high rate, agroforestry likely can play an important role in lessening the impact of this trend.

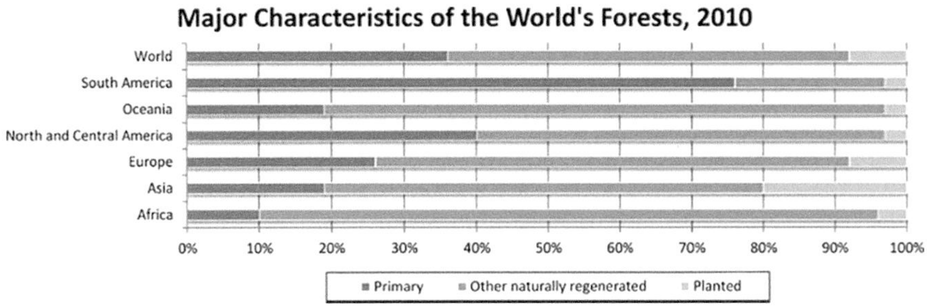

FIGURE 6.1 Major characteristics of the world's forests, 2010.
Source: From http://www.fao.org/forestry/26026-0f6e61f1755a602cea05a15f9a1e7c450.jpg, accessed October 2013; FAO Global Forest Resource Assessment 2010. FAO Forestry Paper No. 163. http://www.fao.org/docrep/013/i1757e/i1757e.pdf, accessed October 2013.[1]

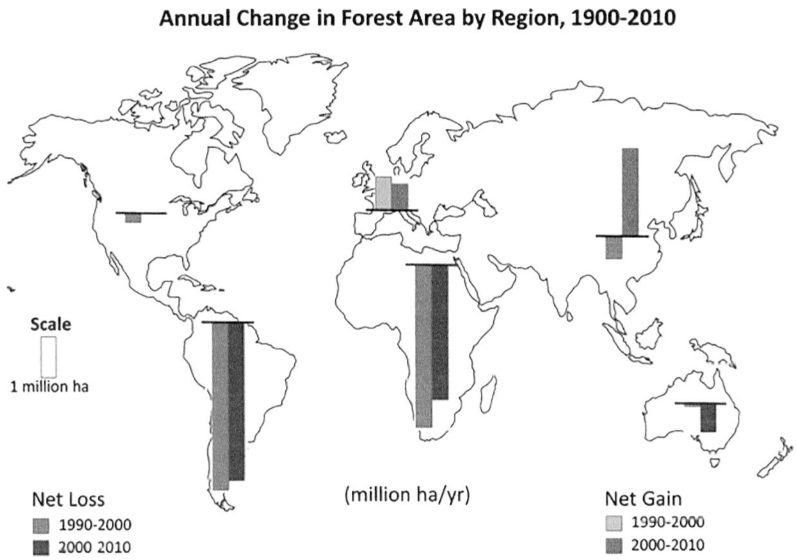

FIGURE 6.2 Changes in forest area by region of the world, 1990 2010.
Source: From FAO Global Forest Resource Assessment 2010. FAO Forestry Paper No. 163. http://www.fao.org/docrep/013/i1757e/ i1757e.pdf, accessed October 2013.[1]

Agroforestry and Biodiversity Preservation

The two major approaches to protecting biodiversity, particularly in forested environments, are i) creating protected parks and ii) establishing agroforestry areas. Parks have proven to be an effective way of protecting biodiversity, although land-use pressures compromise many parks, and their management is often underfunded.[23] In addition, parks are often subject to practices such as illegal poaching and overuse by visitors. Additionally, in some countries, it is politically unpopular to "lock up" needed resources in protected parks. An alternative approach is to use agroforestry to enhance the biodiversity of existing habitats. Agroforestry also may be a useful tool for reducing the impact of deforestation and edge effects on tropical parks and biodiversity corridors that link protected habitats.[24,25] At the same time, agroforestry can produce sustainable harvests of food crops and wood and perhaps contribute to the mitigation of greenhouse gases that contribute to climate change.[6,26]

Studies of Agroforestry and Biodiversity

The study of the relationship between agroforestry and biodiversity has steadily advanced in the past decade.[24] The impacts of agroforestry on biodiversity have been studied in coffee,[24,25,27–36] cacao,[37,38] tea,[3,39] banana and plantain,[36,40] rubber,[41] and several other crops. Among the agroforestry crops, biodiversity of coffee-dominated landscapes has perhaps been the most studied. Studies in traditional shaded coffee plantations have shown that agroecosystems can serve as a refuge for biodiversity.[28–30] However, most of the initial work on coffee and biodiversity was conducted on birds, insects, and arthropods.[38,42–54]

The relationship between agroforestry systems and the natural biodiversity of an area has been found to be very complicated. Much more research is needed to acquire an understanding of the ability of agroforestry systems to support the natural ecosystem structure and function. To date, some generalities have been supported: complex agroforestry systems are better than monocrop systems; shade coffee is better than sun coffee; and systems using native plants tend to be generally more biologically diverse. As landscape changes in the world seem to be inevitable, scientists and managers alike must examine whether ecosystem management strategies, such as those employing agroforestry, can limit ecological losses while satisfying economic and societal gains. This examination of tradeoffs has been conducted for several agroforestry crops in a number of countries, but there is still far too little known about the human dimension of land-use change and the consequences for ecosystem functioning and biodiversity retention.[55]

Planned Biodiversity

Biodiversity in the tropics has benefited through the implementation of the so-called planned biodiversity. Planned biodiversity refers to the crops and livestock introduced and sustained in an agroecosystem by a farmer, whereas unplanned biodiversity includes all other organisms (e.g., soil flora and fauna, herbivores, carnivores, decomposers) that exist in or colonize the agroecosystem. A number of agroforestry practices appear to promote planned biodiversity, namely, home gardens, annual systems, and perennial systems.[56] Perhaps the most notable is the practice of creating home gardens.[13] Home gardens, as defined by Perfecto and Vandermeer [56] are "land-use systems involving multipurpose trees and shrubs in association with annual and perennial agricultural crops as well as livestock (usually small animals) that are intensively managed by family labor within the compound of individual houses." Practiced in the humid tropics for centuries,[57] home garden systems are often very diverse, have several strata in the canopy, and provide multiple ecological and agricultural functions.[58–63] Although most of the research in home gardens is qualitative and descriptive,[56] more recent research that has employed cluster and other multivariate quantitative analyses[60,62,64,65] supports earlier descriptive work in evaluating the role of home gardens in the conservation of biodiversity. Kumar and Nair[66] documented that most studies of home gardens, whether descriptive or quantitative and experimental, demonstrate that these agroforestry systems have a high planned biodiversity.

Certifications and Biodiversity

Certifications associated with biodiversity have had an increasingly role in promoting biodiversity in tropical agroforestry systems. For example, research on avifauna has led to the formulation of certification criteria by the Smithsonian Migratory Bird Center for the so-called Certified Bird Friendly® farms.[67] Although the criteria are well thought out and represent a significant step toward growing agroforestry crops while supporting biodiversity, the number of farms certified in the program is relatively low. For example, only 28 coffee farms or co-operatives are now certified, representing about 1,402 individual growers in 7 countries. The total production of certified farms is only slightly over 9.6 million pounds of coffee per year, with the total land area equal to 7,332 hectares. To give some perspective to these numbers, there were 133,308,000 bags of coffee produced in 2010 by over 26 million people in some 70 countries.[68] If a figure of 60 kilograms per bag is assumed (bag weights vary somewhat from country to country), this equates to approximately 8 billion kilograms of coffee grown in the world. The total area on which coffee is grown peaked in 2000 at 10,719,591 and was last estimated in 2009 to be about 9,841,317 hectares worldwide.[69] Thus, Certified Bird Friendly® farms represent, relative to the pounds of coffee grown worldwide and the area on which it is grown, only 0.05% and 0.07%, respectively.

Other programs that are contributing to promoting biodiversity in agroforestry landscapes include Rainforest Alliance and UTZ certifications.[70–72] These programs offer third-party certification, verification, and validation for about 100 different crops, including coffee, tea, bananas, chocolate, and even roses. For example, to qualify for Rainforest Alliance certification, farms must meet the standards of the Sustainable Agriculture Network.[73] In 2003, various large- and medium-sized companies started selling Rainforest Alliance Certified coffee in North America, Europe, and Asia. By 2006, global retail sales of Rainforest Alliance Certified coffee, bananas, and chocolate exceeded U.S. $1 billion with more than 10,000 certified farms in 15 countries. In subsequent years, certification has grown steadily and expanded to many other crops such as flowers, pineapple, and tea. These types of certifications appear to be making significant contributions to sustainability and the preservation of biodiversity in tropical countries around the world.

Conclusion

Agroforestry systems in the tropics sometimes resemble natural rain forests, particularly in terms of their structure. Through the incorporation of trees on farmlands and bringing degraded lands under agroforestry systems, many areas now support the ecological functions and biodiversity as do more natural tropical forests. Agroforestry has been thought of as a promising biodiversity-friendly land-use strategy that conserves a significant proportion of tropical rain forest diversity while providing significant economic returns.[56] There exists a profound need for more quantitative information on the potential trade-offs between biodiversity loss and agricultural intensification in the tropics.

Acknowledgments

I wish to thank CIRAD (the French research center working with developing countries to tackle international agricultural and development issues) for providing me with the facilities and technical support necessary to complete this task. I also wish to thank the University of Rhode Island for supporting this work through a sabbatical leave. I especially wish to thank Dr. Philippe Vaast for reviewing the manuscript and offering constructive criticism.

References

1. FAO. Plant genetic resources: their conservation in situ for human use. Prepared in collaboration with the United Nations Educational, Scientific and Cultural Organization (UNESCO); UNEP and IUCN: Rome, 1989.
2. Young, A. *Agroforestry For Soil Conservation*; ICRAF Science and Practice of Agroforestry: BPCC Wheatons Ltd, Exeter, UK. 1986.
3. Gascon, C.; de Fonseca, G.A.B.; Sechrest, W.; Billmark, K.A.; Sanderson, J. Chapter 1: Biodiversity conservation in deforested and fragmented tropical landscapes: an overview. In *Agroforestry and Biodiversity Conservation in Tropical Landscapes*; Schroth, G., de Fonseca, G. A.B., Harvey, C.A., Gascon, C., Vasconcelos, H.L., Izac, A-M.N., Eds.; Island Press: Washington, D.C., 2004; 15–32.
4. Garrett, H.E., Ed.; *North American Agroforestry: An Integrated Science and Practice*; 2nd Ed. Agronomy Society of America: Madison, Wisconsin, USA, 2009.

5. Gold, M.A.; Garrett, H.E. Agroforestry nomenclature, concepts, and practices. In *North American Agroforestry: An Integrated Science and Practice*; 2nd Ed.; Garrett, H.E., Ed.; Agronomy Society of America: Madison, Wisconsin, USA, 2009; 45–55.
6. Garrity, D.P. Agroforestry and the achievement of the Millennium Development Goals. Agrofor. Syst. **2004**, *61*, 5–17.
7. Russell, D.; Franzel, S. Trees of prosperity: Agroforestry, markets and the African smallholder. Agrofor. Syst. **2004**, *61*, 345–355.
8. Nair, P.K.R.; Allen, S.C.; Bannister, M.E. Agroforestry today: An analysis of the 750 presentations to the 1st World Congress of Agroforestry 2004. J. For. **2005**, *103*, 417–421.
9. http://www.worldagroforestry.org/about_us. (accessed September 2011).
10. Jose, S. Agroforestry for ecosystem services and environmental benefits: an overview. Agrofor. Syst. **2009**, *76*, 1–10.
11. Murniati, G.; Garrity D.P.; Gintings, A.N. The contribution of agroforestry systems to reducing farmers' dependence on the resources of adjacent national parks: a case study from Sumatra Indonesia. Agrofor. Syst. **2001**, *52*, 171–184.
12. Garrity, D.P.; Amoroso, V.B., Koffa, S.; Catacutan, D.; Buenavista, G.; Fay, P.; Dar, W.D. Landcare on the poverty-protection interface in an Asian watershed. In *Integrated Natural Resource Management: Linking Productivity, the Environment, and Development.* Campbell, B.M., Sayer, J.A., Eds.; CABI Publishing: Cambridge, MA, USA. 2003, 195–210 pp.
13. Nair, P.K.R. *An Introduction to Agroforestry.* Springer: Dordrecht, the Netherlands, 1993.
14. Heywood, V.H.; Bates, I. Introduction. In *Global Biodiversity Assessment*; Heywood, V.H., Watson, R.T., Eds.; Cambridge University Press: Cambridge, England, UNEP, 1995;1–21 pp.
15. Myers, N. Environmental services of biodiversity. Proc. Natl. Acad. Sci. **1996**, *93*, 2764–2769.
16. Weaver, W.; Shannon, C.E. *The Mathematical Theory of Communication*; University of Illinois: Urbana, Illinois, 1949.
17. Atta-Krah, K.; Kindt, R.; Skilton, J.N.; Amaral, W. Managing biological and genetic diversity in tropical agroforestry. Agrofor. Syst. **2004**, *61*, 183–194.
18. *Miller, G.T.; Spoolman, S. Living in the Environment; 18th Ed., Cengage: Stamford, CT, USA, 2012.*
19. Nunes, P.A.D.; van der Bergh, J.C.J.M. Economic valuation of biodiversity: sense or nonsense? Ecol. Econ. **2001**, *39*, 203–222.
20. Swift, M.J.; Izac, A.-M.N.; van Noordwijk, M. Biodiversity and ecosystem services in agricultural landscapes—are we asking the right questions? Agric. Ecosyst. Environ. **2004**, *104* (1), 113–134.
21. Wiens, J.J.; Graham, C.H.; Moen, D.S.; Smith, S.A.; Reeder, T.W. Evolutionary and ecological causes of the latitudinal diversity gradient in hylid frogs: treefrog trees unearth the roots of high tropical diversity. Am. Nat. **2006**, *168*, 579–596.
22. Food and Agriculture Organization of the United Nations. *State of the World's Forests*; Rome, Italy, 2011.
23. Bruner, A.G.; Gullison, R.E.; Rice, R.E.; Fonseca, G.A. Effectiveness of parks in protecting tropical biodiversity. Science **2001**, *291*, 125–128.
24. Husband, T.P.; Abedon, D.H.; Paton, P.W.C. Surveying Mammal Biodiversity in Coffee-Dominated Landscapes in Costa Rica (2005–2007). Abstract of presentation at the Second International Symposium for the Multistrata Agroforestry Systems with Perennial Crops: Making Ecosystem Services Count for Farmers, Consumers, and the Environment: Turrialba, Costa Rica, Sept. 17–21,2007.
25. Husband, T.P.; Abedon, D.H.; Donelan, E.; Paton, P.W.C. Do in Coffee-Dominated Landscapes Support Mammal Biodiversity: Costa Rica 2005–2008. Abstract of presentation at the Second World Agroforestry Congress: Nairobi, Kenya, Aug. 23–29, 2009.
26. Albrecht, A.; Kandji, S.T. Review: Carbon sequestration in tropical agroforestry systems. Agriculture, Ecosyst. Environ. **2003**, 99, 15–27.

27. Philpott, S.M.; Arendt, W.J.; Armbrecht, I.; Bichier, P.; Diestch, T.V.; Gordon, C.; Greenberg, R.; Perfecto, I.; Reynoso-Santos, R.; Soto-Pinto, L.; Tejeda-Cruz, C.; Williams-Linera, G.; Valenzuela, J.; Zolotoff; J. M. Biodiversity Loss in Latin American Coffee Landscapes: Review of the Evidence on Ants, Birds, and Trees. Conserv. Biol.**2008** 22 (5), 1093–1105.
28. Escalante, E.E. Coffee and agroforestry in Venezuela. Agrofor. Today **1995**, 7 (3–4), 5–7.
29. Perfecto, I.; Snelling, R. Biodiversity and the transformation of a tropical agroecosystem: ants in coffee plantations. Ecological Applications **1995**, 5, 1084–1097.
30. Perfecto, I.; Rice, R.A.; Greenberg, R.; Van der Voort, M.E. Shade coffee: a disappearing refuge for biodiversity. BioSci. **1996**, *46*, 598–608.
31. Perfecto I.; Mas, A.; Dietsch, T.; Vandermeer, J. Conservation of biodiversity in coffee agroecosystems: a tri-taxa comparison in southern Mexico. Biodiver. Conserv. **2003**, *12*, 1239–1252.
32. Moguel, P.; Toledo, V.M. Biodiversity conservation in traditional coffee systems in Mexico. Conserv. Biol. **1999**, *12* (1), 1–11.
33. Klein, A.-M.; Steffan-Dewenter, I.; Buchori, D.; Tscharntke, T. Effects of land-use intensity in tropical agroforestry systems on coffee flower-visiting and trap-nesting bees and wasps. Conserv. Biol. **2002**, *16* (4), 1003–1014.
34. Mas, A. H.; Dietsch, T. V. An index of management intensity for coffee agroecosystems to evaluate butterfly species richness. Ecol. Appl. **2003**, *13*, 1491–1501.
35. McNeely, J.A. Nature vs. nurture: managing relationships between forests, agroforestry and wild biodiversity. Agrofor. Syst. **2004**, *61–62* (1–3), 155–165.
36. Bhagwat, S.A.; Willis, K.J.; Birks, H.J.B.; Whittaker, R.J. Agroforestry: a refuge for tropical biodiversity? Trends Ecol. Evol. **2008**, *23* (5), 261–267.
37. Bentley, J.W.; Boa, E.; Stonehouse, J. Neighbor trees: shade, intercropping, and cacao in Ecuador. Hum. Ecol. **2004**, *32* (2), 241–270.
38. Asare, R. A review on cocoa agroforestry as a means for biodiversity conservation. Paper presented at World Cocoa Foundation Partnership Conference, Brussels, May 2006.
39. van Biervliet, O.; Wisniewski, K.; Daniels, J.; Vonesh, J. R. Effects of tea plantations on stream invertebrates in a global biodiversity hotspot in Africa. Biotropica 2009, *41* (4), 469–475.
40. Harvey, C.A.; Gonzalez, J.; Somarriba, E. Dung beetle and terrestrial mammal diversity in forests, indigenous agroforestry systems and plantain monocultures in Talamanca, Costa Rica. Biodivers. Conserv. 2006, *15* (2), 555–585.
41. Penot, E. Prospects for conservation of biodiversity within productive rubber agroforests in Indonesia (CIRAD-Tera, Montpellier (France)), Conference International Workshop on Management of secondary and logged-over forests in Indonesia, CIFOR Bogor, Indonesia, Nov. 17–19, 1997.
42. Wunderle, J.; Latta S. Avian abundance in sun and shade coffee plantations and remnant pine forest in the Cordillera Central, Dominican Republic. Ornitologia Neotrop. 1996, 7, 19–34.
43. Estrada, A.; Coates-Estrada, S.; Merrit, D., Jr. Anthropogenic landscape changes and avian diversity at Los Tuxtlas, Mexico. Biodiver. Conserv. 1997, *6*, 19–43.
44. Greenberg, R.; Bichier P.; Cruz Angon A.; Reitsma R. Bird populations in shade and sun coffee plantations in Central Guatemala. Conserv. Biol. 1997a, 11, 448–459.
45. Greenberg, R.; Bichier P.; Sterling J. Bird populations in rustic and planted shade coffee plantations of Eastern Chiapas, Mexico. Biotropica 1997b, *29*, 501–514.
46. Calvo, L.; Blake, J. Bird diversity and abundance on two different shade coffee plantations in Guatemala. Birds Conserv. Int. 1998, 8, 297–308.
47. Wunderle, J.M., Jr. Avian distribution in Dominican shade coffee plantations: area and habitat relationships. J. Field Ornithol. **1999**, *70*, 58–70.
48. Johnson, M.D. Effects of shade tree species and crop structure on the arthropod and bird communities in a Jamaican coffee plantation. Biotropica **2000**, *32*, 133–145.
49. Nestel, D.; Dickschen, F. The foraging kinetics of ground ant communities in different coffee agroecosystems. Oecologia **1990**, *84*, 58–63.

50. Ibarra-Núñez, G.; Garcia Ballinas, J.A.; Moreno-Próspero, M.A. La comunidad de artrópodos de dos cafetales con diferentes prácticas agrícolas: el caso de los Hymenópteros Resúmenes, XXVIII Congreso Nacional de Entomología, Sociedad Mexicana de Entomología;. Universidad de las Américas: Cholula, Puebla, Mexico, 1993.
51. Perfecto, I.; Vandermeer, J.H. Understanding biodiversity loss in agroecosystems: reduction of ant diversity resulting from transformation of the coffee ecosystem in Costa Rica. Entomology. Trends Agric. Sci. **1994**, *2*, 7–13.
52. Perfecto, I.; Vandermeer, J.H. The quality of the agroecological matrix in a tropical montane landscape: ants in coffee plantations in southern Mexico. Conserv. Biol. **2002**, *16*, 174–182.
53. Perfecto, I.; Vandermeer J.H.; Hanson, P.; Cartín V. Arthropod biodiversity loss and the transformation of a tropical agro-ecosystem. Biodivers. Conserv. **1997**, *6*, 935–945.
54. Ibarra-Núñez, G.; Garcia-Ballinas, J.A. Diversidad de tres familias de arañas tejedoras (Araneae: Araneidae, Tetragnathidae, Theridiiae) en cafetales del soconusco, Chiapas, Mexico. Folia Entomol. Mexicana **1998**, *102*, –20.
55. Steffen-Dewenter, I.; Kessler, M.; Barkmann, J.; Bos, M.M.; Buchori, D. Tradeoffs between income, biodiversity, and ecosystem functioning during tropical rainforest conversion and agroforestry intensification. Proc. Natl. Acad. Sci. USA. **2007**, *104*, 4973–4978.
56. Perfecto, I.; Vandermeer, J. Biodiversity Conservation in Tropical Agroecosystems. Ann. N. Y. Acad. Sci. **2008**, *1134*, 173–200.
57. Montagnini, F. Homegardens of Mesoamerica: biodiversity, food security, and nutrient management. In *Tropical Homegardens: A Time-Tested Example of Sustainable Agroforestry*. Kumar, B.M., Nair, P.K.R., Eds.; Springer: the Netherlands, 2006; 1–23.
58. Fernandes, E.C.M.; Nair, P.K.R. An evaluation of the structure and function of tropical homegardens. Agric. Syst. **1986**, *21*, 279–310.
59. Coomes, O.T.; Ban, N. Cultivated plant species diversity in home gardens of an Amazonian peasant village in northeastern Peru. Econ. Bot. **2004**, *58*, 420–434.
60. Kehlenbeck, K.; Maass, B.L. Crop diversity and classification of homegardens in Central Sulawesi, Indonesia. Agro. Syst. **2004**, *63*, 53–62.
61. Major, J.; Clement C.R.; DiTommaso, A. Influence of market orientation on food plant diversity of farms located on Amazonian Dark Earth in the region of Manaus, Amazonas, Brazil. Econ. Bot. **2005**, *59*, 77–86.
62. Peyre, A.; Guidal, A.; Wiersum, K.F.; Bongers, F. Dynamics of homegarden structure and function in Karala, India. Agrofor. Syst. **2006**, 66, 101–115.
63. Pandey, C.B.; Rai, R.B.; Singh, L.; Singh, A.K. Homegardens of Andaman and Nicobar, Indian. Agric. Syst. **2007**, *92*, 1–22.
64. Blanckaert, I.; Swennen, R.L.; Paredes Flores, M.; Rosas Lopez, R.; Lira Saade, R. Floristic composition, plant uses and management practices in homegardens of San Rafael Coxcatlan, Valley of Tehuacan-Cuicatlan, Mexico. J. Arid Environ. **2004**, *57*, 39–62.
65. Albuquerque, U.P.; Andrade, L.H.C.; Caballero, J. Structure and floristics of homegardens in Northeastern Brazil. J. Arid Environ. **2005**, *62*, 491–506.
66. *Kumar, B.M.; Nair, P.K.R. Tropical Homegardens: A Time-Tested Example of Sustainable Agroforestry; Springer: the Netherlands, 2006.*
67. http://nationalzoo.si.edu/scbi/migratorybirds/coffee/farmer. cfm. (accessed September 2011).
68. http://www.ico.org/prices/po.htm. (accessed September 2011).
69. http://faostat.fao.org/. (accessed October 2011).
70. Perfecto, I.; Vandermeer, J.; Mas A.; Soto Pinto, L. Biodiversity, yield, and shade coffee certification. Ecol. Econ. **2005**, *54*, 435–446.
71. http://www.rainforest-alliance.org/agriculture/certification. (accessed 5 October 2011).
72. http://www.utzcertified.org/en/aboututzcertified. (accessed October 2011).
73. http://sanstandards.org/sitio/. (accessed October 2011).

7

Biodiversity: Values

Introduction 65
Intrinsic Values 66
Instrumental Values 68
The Values Battleground 70
Conflict Resolution 71
References 72
Bibliography 73

David Ehrenfeld
Rutgers University

Introduction

The idea that biodiversity has values predates the word "biodiversity," which only came into common use as a contraction of "biological diversity" in the 1980s. Biodiversity can be defined as the variety of life-forms in a given place, with "life-forms" ranging from genotypes to species and other taxonomic categories to species assemblages to entire ecosystems. Historically, values have been accorded to "wilderness," "natural resources," "species diversity," "genetic diversity," "community and ecosystem diversity," and the more general term "nature," which subsumes them all. Each of these and the other terms related to biodiversity has had its own set of values associated with it, but these values can be broadly divided into two categories, *intrinsic* and *instrumental.*

Intrinsic values are those believed to be inherent in biodiversity itself, rather than in any usefulness to humans that the biodiversity may confer. Thus, species considered useless for human purposes may yet be accorded intrinsic values. The values ascribed to wilderness and the values invoked by the animal rights movement are frequently intrinsic.

Instrumental values, conversely, are associated with components of biodiversity or entire natural systems that have a known, assumed, or suspected usefulness in the lives of people. The values assigned to natural resources and to biodiversity as a source of human health and ecosystem stability are usually instrumental. Since the growth of the physical environmental sciences and ecology, these instrumental values have gained steadily in importance, often overshadowing the older intrinsic values.

Any aspect of biodiversity may have both intrinsic and instrumental values associated with it. There is considerable overlap, and the lines separating the two systems can sometimes be blurred or missing.

In the modern world, the subject of biodiversity's values is frequently raised as a justification for environmental protection or conservation. When this happens, there can be sharp polarization between those who value the preservation or restoration of biodiversity primarily because it serves specific needs of humanity, including ecosystem health, and those who justify conservation because it maintains the order of nature without reference to human needs. Nevertheless, there is not always a contradiction between instrumental and intrinsic values, and a coexistence of the two positions can occur.

Intrinsic Values

The study of biodiversity in the West owes much to the early influence of Aristotle, whose *Generation of Animals* inspired both Islamic and western Christian scientists. In other parts of the world, there have also been long-established traditions of knowing and valuing biodiversity. A comprehensive and systematic discussion of the place and value of nature in each of the world's major religions and regions (including Shamanism, Shintoism, Taoism, Confucianism, Zoroastrianism, Buddhism, Hinduism, Judaism, Christianity, Islam, and others) can be found in the book *Religion and the Order of Nature,* by Seyyed Hossein Nasr.[1]

In 9th-century Baghdad, the biologist, Islamic philosopher, and grandson of an African slave Abu Uthman Amr ibn Bahr al-Kinani al-Fuqaimi al-Basri, known as al-Jahiz, published his multivolume work, the *Book of Animals,* some of which survives. al-Jahiz, who was the first to provide a scientific classification of biodiversity (animals), described approximately 350 animal types, and, 1000 years before Darwin and Wallace, gave a concise explanation of evolution by means of natural selection. To al-Jahiz, the value of biodiversity was not primarily in the usefulness of any of its particular elements but in demonstrating the wonders of God's creation: "For this purpose the small and slight carries as much weight as the great and vast."[2] In other words, God's imprint on every species or ecological system is what establishes its value—the value is inherent and not related to the particular properties of one species or another. On the other hand, al-Jahiz does see a special instrumental value to biodiversity because he believes that contemplation of it will lead people to a better appreciation of God.

Nearly 900 years later, in a different age, a different part of the world, and a different culture, the pioneering botanist John Ray, often described as the father of English natural history, found similar values in biodiversity. Ray, an Anglican priest, began the modern plant classification system that was later continued and developed by Linnaeus, and was the first to give a biological definition of the term "species."[3] Ray's book, *The Wisdom of God Manifested in the Works of the Creation* (1691), was enormously popular and was published in numerous editions, continuing long after his death in 1705. In this book, he considered biodiversity to include species of plants, animals, and microbes, as well as anatomical structures and physiological systems.[4]

To Ray, the value of biodiversity was clear, and strikingly similar to that of al-Jahiz. From the start, he makes plain what that value is: "If the Number of Creatures be so exceeding great, how great, nay immense, must need be the Power and Wisdom of him who formed them all!" Unlike al-Jahiz, however, Ray takes great pains to distance himself from a utilitarian valuing of biodiversity: "How can Man give Thanks and Praise to God for the Use of his Limbs and Senses, and [for] those his good Creatures which serve for his Sustenance, when he cannot be sure they were made in any Respect for him; nay, when 'tis as likely they were not, and that he doth abuse them to serve Ends for which they were never intended?" The value of biodiversity, says Ray, is entirely to God and for God. Quoting the book of Proverbs, he wrote: "God made all things for himself.... He hath no Need of them: he hath no use of them. No, he made them for the manifestation of his Power, Wisdom, and Goodness, and that he might receive from the Creatures that were able to take Notice thereof, his Tribute of Praise."

Later in his book, Ray qualifies this argument somewhat by cataloging a long list of practical instrumental values of plant and animal species to humans, including values that have yet to be discovered and the value of intellectual stimulation. He notes that even apparently worthless or noxious creatures (especially insects) may provide sustenance to other species that we do value. Nevertheless, the ultimate value of biodiversity that Ray always returns to is the intrinsic value relating to God.[4]

The celebration of nature for its own sake was established in 18th-century North America by the naturalists and artists Mark Catesby and John James Audubon, and by the father-and-son team of John and William Bartram. John Bartram, who traveled primarily in Virginia, Pennsylvania, New York, and the southeastern states, collected seeds and live specimens of every species of new plant he could find, bringing them back to his garden in Philadelphia [still preserved as a 46-acre (19-ha) U.S. National Historic Landmark]. A Quaker farmer with a rudimentary education, John corresponded with many of the leading European botanists, including Linnaeus, who referred to him as the greatest natural botanist in the world. Some Bartram plant specimens remain in the collection of the British Museum (Natural History).[4]

William Bartram is primarily known for his travels in the southeastern United States, especially Florida. His book, commonly titled *Bartram's Travels* (1791), gained him international fame.[5,6] Coleridge's poem *Xanadu* was inspired by the book's descriptions of Florida springs and sinkholes. William's written descriptions and paintings of plants, insects, birds, reptiles, and fish were widely circulated, as were the specimens he sent back to his father in Philadelphia. John and William's most celebrated discovery, in 1765, was a small, flowering tree found by the Altamaha River in Georgia, which they named *Franklinia alatamaha*, after John's friend Benjamin Franklin. Soon extinct in the wild, *Franklinia* survives in cultivation as descendants of specimens in John Bartram's garden.

The Bartrams, especially John, were practical people, always looking for new species, areas with fertile soils, and mineral deposits that might be put to use. Yet they also devoted enormous effort to finding and describing species (like *Franklinia*) that seemed of no practical use whatsoever, whose value was, as far as they knew, entirely intrinsic; and their enthusiasm at discovering these species remains vivid to this day.

The 19th-century Prussian explorer and naturalist Alexander von Humboldt carried out daring explorations, especially in South America, founded the science of physical geography (the Humboldt Current, etc.), also established new frontiers in the study of linguistics and education, and described many new species of animals and plants, some of which were named after him. von Humboldt's discoveries of species and ecosystem processes reflected his delight in nature for its own sake, irrespective of utilitarian values. In his celebrated work *Views of Nature*, translated into English in 1850, Humboldt wrote: "When the active spirit of man is directed to the investigation of nature, or when in imagination he scans the vast fields of organic creation, among the varied emotions excited in his mind there is none more profound or vivid than that awakened by the universal profusion of life."[7] von Humboldt's feelings were mirrored in the writings of the many 18th-, 19th-, and 20th-century naturalist explorers who spread out around the globe searching for and collecting new species, many of which ended up in great private and museum collections.

From the days of Ray to the present, the theme of the intrinsic value of biodiversity has reappeared many times in various ways. Henry David Thoreau, writing in *Walden; or, Life in the Woods* (1854), described and celebrated the diversity of life in his native Massachusetts. He often distanced himself from instrumental values, as, for example, when he extols vegetarianism: "I have no doubt that it is part of the destiny of the human race, in its gradual improvement, to leave off eating animals," and he continues, "All nature is your congratulation.... The greatest gains and values are farthest from being appreciated."

In consideration of the intrinsic values of biodiversity, the concept of animal rights should be mentioned, because in this case, the value of animals, including domesticated animals, is completely disconnected from their utility to humans. This formulation of intrinsic worth has, historically, led to contentious debates with utilitarians, even to ridicule, as in Samuel Butler's satiric chapter on the "Rights of Vegetables," in his novel *Erewhon* (1872). More recently, in support of intrinsic rights, the law scholar Christopher D. Stone advanced the argument that trees should have legal standing in U.S. courts, in his 1972 book *Should Trees Have Standing? Toward Legal Rights for Natural Objects*.[8] This notion was echoed in the same year by U.S. Supreme Court Justice William O. Douglas, who wrote, "the voice of the inanimate (non-human) object should not be stilled" in a famous legal dissent. He was referring to groves of trees, alpine meadows, aquatic insects, otters, elk, and other parts of biodiversity, stating that they had a right to be represented in court by those humans who had an "intimate relation" to them.[9]

The English ecologist Charles S. Elton wrote: "The first [reason for conservation], which is not usually put first, is really religious. There are some millions of people in the world who think that animals have a right to exist and be left alone, or at any rate that they should not be persecuted or made extinct as species. Some people will believe this even when it is quite dangerous to themselves." Although he was well aware of its instrumental values, the first great value of biodiversity, Elton asserted, was embodied in "the right relation between man and living things."[10]

Biologist David Ehrenfeld has pointed out that it is easy to exaggerate the utilitarian values of biodiversity: "When everything is called a resource, the word loses all meaning"; and he gave examples, including contemporary ecological debates over ecosystem diversity and stability. Concluding a discussion of intrinsic values, he cited the biblical account of Noah's ark. Noah is described as taking at least one male and female of each animal species on board the ark, presumably including the worthless and noxious ones. Ehrenfeld called this strategy the "Noah Principle": species should be conserved (valued) simply "because they exist." In his discussion of the Noah Principle, he cited contemporary arguments for the conservation of pathogens and vermin—even the smallpox virus.[11]

Instrumental Values

Arguments for the instrumental values of biodiversity are also long established. An oft-quoted passage from the Hebrew Bible (Deuteronomy 20:19,20) states: "When in your war against a city you have to besiege it a long time in order to capture it, you must not destroy its trees, wielding the axe against them. You may eat of them, but you must not cut them down. Are trees of the field human to withdraw before you into the besieged city? Only trees that you know do not yield food may be destroyed."[12] This passage forms the principal basis for the Jewish commandment of *bal tashhit* (do not destroy), an ancient series of Jewish environmental regulations, which were rabbinically extended to a wide range of transgressions, including the overgrazing of the countryside, unjustified killing of animals or feeding them toxic foods, sport hunting, extinction of species, destruction of cultivated varieties of plants, air and water pollution, overconsumption of anything, and the waste of resources.[13,14] The original source of the commandment and much of the Jewish system of biodiversity values that followed were certainly utilitarian or instrumental in nature. There are, however, some partial exceptions. The general Jewish prohibition against cruelty to animals, *ts'ar baalei hayyim* (pain of living things) is based on the following passage in Deuteronomy 22:6: "If, along the road, you chance on a bird's nest ... with fledglings or eggs and the mother sitting over [them], do not take the mother together with her young. Let the mother go, and take only the young, in order that you may fare well and have a long life."[12] This value seems largely intrinsic, although a modern interpretation would include the ecological utility of leaving the female to reproduce in subsequent seasons. Whether this is part of the biblical intent is not clear.

The English naturalist Gilbert White, whose book *The Natural History and Antiquities of Selbourne* (1789) inspired study and appreciation of the intrinsic value of nature around the world, was, nevertheless, one of the first to quantify a species' value. In his Letter XXVI to Daines Barrington, White, in an early example of ecological economics, carefully calculated the value, per pound, of the common rush, *Juncus effusus*. Rushes, when dried and dipped in bacon fat or other grease, made an inexpensive substitute for candles, giving "a good clear light.... Thus a poor family will enjoy 5 1/2 hours of comfortable light for a farthing." White estimated that one-and-a-half pounds of dried rushes would last a family as much as a year.[15]

The modern drive to assign instrumental values to biodiversity dates to 1864, when the American linguist, lawyer, legislator, writer, and a founder of environmental science George Perkins Marsh, following in the intellectual footsteps of von Humboldt, published his book *Man and Nature; or Physical Geography as Modified by Human Action.* Marsh, who traveled in Europe, the Middle East, and North Africa, and who read easily and widely in 20 languages, was the first to provide a truly comprehensive description of many of the instrumental values of biodiversity. He demonstrated that forests promote even- stream flow, mitigate flooding, protect against avalanches, stabilize climate, prevent soil erosion, and halt the movement of sand dunes, besides providing valuable raw materials. He showed the importance of birds in limiting outbreaks of destructive insects and in sowing seeds. He documented in great detail the economic benefits and liabilities of exotic plant species, the impact of earthworms on soil porosity and fertility, the usefulness of insects in destroying dead and decaying animal and vegetable matter, and the importance of micro-organisms in the formation of deposits of lime and in the process of fermentation. Above all, he sounded a very modern warning that we destroy biodiversity at our own peril: "If man is destined to inhabit the Earth much longer... he will learn to put a wiser estimate on the [value of the] works of creation."[16]

Marsh's warnings about the critical values to people of biodiversity have been echoed and amplified in the 20th and 21st centuries. The terms most associated with these values are "natural capital" and "ecosystem services," both of which indicate their instrumental nature. In a major paper entitled "Biodiversity loss and its impact on humanity," Bradley Cardinale and 16 other authors noted that "ecosystem services are the suite of benefits that ecosystems provide to humanity." Those pertaining to the role of biodiversity include food, wood, fresh water, climate regulation, pollination, and pest and disease control. The authors state, "There is now unequivocal evidence that biodiversity loss reduces the efficiency by which ecological communities capture biologically essential resources, produce biomass, [and] decompose and recycle biologically essential nutrients.... There is mounting evidence that biodiversity increases the stability of ecosystem functions through time.... Diverse communities are more productive because they contain key species that have a large influence on productivity, and differences in functional traits among organisms increase total resource capture." The authors note that "there are (a small number of) instances where increased biodiversity may be deleterious," and give examples, cautioning "against making sweeping statements that biodiversity always brings benefits to society."[17]

In a section called "Valuing biodiversity," the paper continues: "Economists have developed a wide array of tools to estimate the value of natural and managed ecosystems and the market and non-marketed services that they provide. Although there are good estimates of society's willingness to pay for a number of non-marketed ecosystem services, we still know little about the marginal value of biodiversity (that is, value associated with changes in the variation of genes, species and functional traits) in the production of those services."[17] The value of natural capital has been further explored in a related paper by Ehrlich, Kareiva, and Daily, who describe software-based models that "help decision makers by quantifying and mapping the generation, distribution and economic value of ecosystem services under alternative scenarios."[18]

The idea that the maintenance of and exposure to biodiversity, from microbes to ecosystems, improve human health and well-being has gained many adherents. This improvement comes about in a number of ways. First, many medicinals are obtained directly or indirectly from biological species. Examples are statin drugs from *Aspergillus*, *Pleurotus*, and *Monascus* fungi, an antidiabetic from gila monster lizard (*Heloderma suspectum*) saliva, a powerful painkiller from the venom of marine cone snails (*Conus*), the anticancer drug taxol from the Pacific yew (*Taxus brevifolia*), and the potent antimalarial artemisinin from wormwood (*Artemisia annua*).[19]

Second, mounting evidence has shown that childhood exposure to diverse species of bacteria and fungi, and probably to intestinal parasites, protects against the development of asthma and allergies. Children who grow up in environments, such as traditional farms, that provide a wide range of microbial exposure develop less asthma and allergy than children in more sanitary environments. Similarly, it has been proposed that exposure to certain bacteria may prevent or treat particular cancers. For example, the bacterium *Helicobacter pylori*, although causing gastric ulcers and gastric cancer, may reduce the risk of esophageal cancer. The diversity of the microbiome, especially that of the skin and internal organs, is increasingly linked to human health.[20–24]

There is a third, more sweeping association between biodiversity and health. In the words of the author and childhood development expert Richard Louv, "A growing body of research links our mental, physical, and spiritual health directly to our association with nature." Louv has coined the term "nature deficit disorder" to describe a constellation of maladies that afflict the increasing number of children who spend most of their time indoors, out of contact with the natural world. For many children in industrial nations, their relationship with nature is intellectual, not experiential. The consequences, it is claimed, include attention deficit disorder, obesity, depression, and eventually hypertension and heart disease.[25] Similar maladies afflict adults. The modern popularization of this idea can be traced to biologist E. O. Wilson's use, beginning in 1984, of the older term "biophilia," in which he claimed that human beings have a deep, evolved relationship with other species, which often confers a mutual survival value.[26,27] Even earlier, the botanist Hugh Iltis ascribed a similar importance to biodiversity.[28]

The Values Battleground

The proponents of the intrinsic values of biodiversity have at times struggled with those who favor instrumental values. One of the first and most long-lasting examples of this conflict was the controversy, beginning in 1908, about whether to dam California's Tuolumne River in Yosemite National Park's Hetch Hetchy Valley to create a reservoir for the City of San Francisco. Gifford Pinchot, early advocate of the conservation of natural resources, and eventually the first Chief Forester of the U.S. Forest Service, was strongly in favor of the dam. He wrote: "The first principle of conservation is development, the use of the natural resources now existing on this continent for the benefit of the people who live here now." On the other side was John Muir, naturalist, writer, inventor, and founder of the Sierra Club. Muir, whose writing had earlier prompted the creation of Yosemite Park, wrote: "God has cared for these trees, saved them from drought, disease, avalanches, and a thousand ... leveling tempests and floods; but He cannot save them from fools." Both men appealed to their friend President Theodore Roosevelt, but the issue was not decided until after Roosevelt had left office. In 1914, the U.S. Congress legislated the damming of the Hetch Hetchy Valley. John Muir died several months later.[29] In 1987, the U.S. Secretary of the Interior proposed restoration of the valley, and a year later, the U.S. Bureau of Reclamation concurred. The Sierra Club immediately endorsed breaching the dam. The battle continues, with fierce arguments on both sides. In a 2012 editorial entitled "Hetch Hetchy's Past and Future," *The New York Times* again supported breaching the dam: "Ninety-nine years ago, this newspaper argued against the construction of the dam. We lost, but the much bigger loss was nature's."[30]

Since Hetch-Hetchy, there have been countless struggles in nearly every country on Earth between those who seek to develop natural resources for some practical purpose and those who wish to preserve the species and ecosystems that would be lost to development. In virtually all of these cases, the arguments against or for the preservation of biodiversity echo the sorts of values expressed by Pin- chot and Muir.

Much of the fight about intrinsic versus instrumental values centers on the attempt by some economists to quantify the worth to humanity of biodiversity. A United Nations Environmental Program initiative, "The Economics of Ecosystems and Biodiversity," puts the annual value of ecosystems services at $5 trillion, with the loss of forests worldwide amounting to a loss of ecosystems services of $2–4.5 trillion a year. Ecological economist Robert Costanza argues that we need to understand the economic value of biodiversity and should create new institutions for managing natural capital if we are to promote the public good. "We've used up all the frontier," says Costanza. "People are recognizing this, but our institutions haven't caught up." Ignoring the value of natural capital has serious economic consequences, Costanza and others point out.[31]

In opposition, biologist Douglas J. McCauley claims that although putting a value on their ecosystem services can occasionally be useful in protecting specific parts of biodiversity, "We will make more progress in the long run by appealing to people's hearts rather than to their wallets. If we oversell the message that ecosystems are important because they provide services, we will have effectively sold out on nature." McCauley provides a detailed list of reasons why he believes that placing a dollar value on nature is a poor idea.[32]

Some scientists have stated that putting a dollar value on nature ignores both the very approximate and incomplete nature of the economic calculations involved and the vast extent of our ignorance about ecological systems. It has been pointed out that when we give a specific value to the ecosystem services of biodiversity, any destructive exploitation of that biodiversity that yields higher-dollar profits will appear to outweigh the value of the biodiversity itself, justifying its destruction. Moreover, it is claimed, an emphasis on the quantification of instrumental values leads, by analogy, to the problematic notion that the parts of the world with the highest number of species, mostly tropical, are more deserving of conservation funding than less biodiverse regions, which are mostly in temperate or colder latitudes. Wealth of species is surely important, it is noted, but there are more ways to value the biodiversity of a region than by counting its species.[33]

Conflict Resolution

There has been controversy over the values of biodiversity, but there has also been effective coexistence and integration. One of the best examples can be seen in the writings of the wildlife ecologist Aldo Leopold. Leopold is considered the founder of the science of game management, which regulates the abundance of animal species that are hunted, trapped, or fished, maximizing their instrumental value. Beyond simple game management, however, Leopold was one of the first to see the necessity of ecosystem management, showing that game species could only be maintained if understood in their full ecological context.[34,35]

The richness of Leopold's ecological wisdom promoted a continuous interaction, seen in his work and writings, between the different ways that biodiversity can be valued. The dedication of Leopold's book *Game Management* (1933) was to his father, "Pioneer in Sportsmanship." The young Leopold was an avid hunter; in the preface to *Game Management*, he states, "Control [of wildlife] comes from the co-ordination of science and use."[36] Yet Leopold changed as he grew older; his awareness of less human-centered values grew, as did his application of ecological science to game management and conservation. In his essay "Thinking Like a Mountain," from his book *A Sand County Almanac* (1949), he described the day when he began to rethink his earlier premises and prejudices. He had just shot a female wolf who was playing with her pups by a river. Wolves eat deer; as a deer hunter he felt that wolves were fair game. Then he climbed down to the riverbank to look at his kill. "We reached the old wolf in time to watch a fierce green fire dying in her eyes. I realized then, and have known ever since, that there was something new to me in those eyes—something known only to her and to the mountain. I was young then, and full of trigger- itch; I thought that because fewer wolves meant more deer, that no wolves would mean hunters' paradise. But after seeing the green fire die, I sensed that neither the wolf nor the mountain agreed with such a view.... Perhaps this is behind Thoreau's dictum: In wildness is the salvation of the world. Perhaps this is the hidden meaning in the howl of the wolf, long known among mountains, but seldom perceived among men."[37]

The interplay between intrinsic and instrumental valuing of nature is complex here. Leopold came to realize, as a wildlife manager, that too many deer would overgraze and ruin the mountain ecosystem, ultimately resulting in fewer deer, a decrease in the ecosystem's instrumental value to hunters. A healthy deer population, he believed, needs wolves. The other part of his thinking, however, was also clear: the wolf had an intrinsic value of her own, a value that deserved his respect. And beyond the wolf was the mountain, the functional ecological health of the land community.

In the essays incorporated in the posthumously published *A Sand County Almanac*, Leopold showed that the instrumental and intrinsic values of biodiversity can interact, forming a seamless whole. He wrote: "Most members of the land community have no economic value.... Yet these creatures are members of the biotic community, and if (as I believe) its stability depends on its integrity, they are entitled to continuance." In this statement, he claims that an overall instrumental value of biodiversity, ecosystem resilience, is made up of the collective values of its species, many of which have no individual economic worth. In his essay "A Biotic View of Land,"[38] Leopold had written: "The emergence of ecology has placed the economic biologist in a peculiar dilemma: with one hand he points out the accumulated findings of his search for utility, or lack of utility, in this or that species, with the other he lifts the veil from a biota so complex that no man can say where utility begins or ends.... The old categories of 'useful' and 'harmful' have validity only as conditioned by time, place, and circumstance. The only sure conclusion is that the biota as a whole is useful, [including] not only plants and animals, but soils and waters as well."

Does this mean that ultimately the highest value of biodiversity is its instrumental ecosystem stability? Not necessarily. Leopold goes on to write, using the word "land" as we would use biodiversity, "It is inconceivable to me that an ethical relation to land can exist without love, respect, and admiration for land, and a high regard for its value. By value, I of course mean something far broader than mere economic value, I mean value in the philosophical sense."[37]

The values of biodiversity are a complex mix of the economic and the religious or philosophical. In the long history of thought about the values of nature, these two strands have been both separated in conflict and fused in coexistence, according to circumstances. This pattern seems likely to continue in the foreseeable future.

References

1. *Nasr, S.H. Religion and the Order of Nature: The 1994 Cadbury Lectures at the University of Birmingham; Oxford University Press: USA, 1996; 310 pp.*
2. Lunde, P. Science: the Islamic legacy: the book of animals. Saudi Aramco World **1982**, *33* (3), 1–4.
3. *Mayr, E. The Growth of Biological Thought: Diversity, Evolution, and Inheritance; Harvard University Press: Cambridge, Mass, 1982; 974 pp.*
4. Ray, J. *The Wisdom of God Manifested in the Works of the Creation,* 10th ed.; Innys, W., Manby, R., Eds.; London, 1735; Vol. 18, 1–17.
5. Cruickshank, H.G., ed. *John and William Bartram's America*; Devin-Adair: New York, 1957.
6. Van Doren, M. *Travels of William Bartram*; Dover: New York, 1928.
7. Von Humboldt, A.; Otte, E.C.; Bohn, H.G. *Views of Nature: or Contemplations on the Sublime Phenomena of Nature;* Henry G.B., Eds.; London, 1850; Vol. 210, 459.
8. *Stone, C.D. Should Trees have Standing? Toward Legal Rights for Natural Objects, rev. Ed.; Avon: New York, 1974; 102.*
9. Sierra Club v. Morton -405 U.S. 727 (1972).
10. *Elton, C.S. The Ecology of Invasions by Animals and Plants; Methuen: London, 1958; 659–666.*
11. Ehrenfeld, D. *The Arrogance of Humanism;* Oxford University Press: Oxford, 1978; 286 pp.
12. JPS. *Hebrew-English Tanakh,* 2nd Ed.; Jewish Publication Society: Philadelphia, 1999.
13. Freudenstein, E. *Ecology and the Jewish Tradition;* Atid: New York, 1970.
14. Helfand, J. Ecology and the Jewish tradition: a postscript. Judaism **1971**, *20,* 330–335.
15. White, G. *The Natural History of Selbourne;* Penguin Books: London, 1977; 320 pp.
16. Marsh, G.P. *Man and Nature,* Lowenthal, O., Ed.; Harvard University Press: Cambridge, Massachusetts, 1965.
17. Cardinale, B.J.; Duffy, E.; Gonzalez, A.; Hooper, D.U.; Perrings, C.; Venail, P.; Narwani, A.; Mace, G.M.; Tilman, D.; Wardle, D.A.; Kinzig, A.P.; Daily, G.C.; Loreau, M.; Grace, J.B.; Larigauderie, A.; Srivastava, D.S.; Naeem, S. Biodiversity loss and its impact on humanity. Nature **2012**, *486,* 59–67.
18. Ehrlich, P.R.; Kareiva, P.M.; Daily, G.C. Securing natural capital and expanding equity to rescale civilization. Nature **2012**, *486,* 68–73.
19. Conniff, R. A bitter pill. Conservation **2012**, *13* (1), 18–23.
20. Leonardi-Bee, J.; Pritchard, D.; Britton, J. Asthma and current intestinal parasite infection: systematic review and meta-analysis. Am. J. Respir. Crit. Care Med. **2006**, *174* (5), 514–523.
21. Yazdanbakhsh, M.; Kremsner, P.G.; van Ree, R. Allergy, parasites, and the hygiene hypothesis. Science **2002**, *296,* 490–494.
22. Ege, M.J.; Mayer, M.; Normand, A.-C.; Genuneit, J.; Cook- son, WO.; Braun-Fahrlander, C.; Heederik, D.; Piarroux, R.; von Mutius, E. Exposure to environmental microorganisms and childhood asthma. N. Eng. J. Med. 2011, *364,* 701–709.
23. Mager, D.L. Bacteria and cancer: cause, coincidence or cure? A review. J. Transl. Med. **2006**, *4,* 14.
24. http://www.translational-medicine.com/content/471/14.

25. *Dunn, R. The Wild Life of our Bodies: Predators, Parasites, and Partners that Shape Who We are Today; Harper Collins: New York, 2011; 290.*
26. *Louv, R. Last Child in the Woods: Saving our Children from Nature-Deficit Disorder; Algonquin Books of Chapel Hill: Chapel Hill, NC, 2005; 335.*
27. Wilson, E.O. *Biophilia: The Human Bond with Other Species;* Harvard University Press: Cambridge, MA, 1984; 157.
28. Kellert, S.R.; Wilson, E.O. *The Biophilia Hypothesis;* Island Press: Washington, DC, 1993; 234–240.
29. Iltis, H.H. To the taxonomist and ecologist: whose fight is the preservation of nature? BioScience **1967**, 17, 886–890.
30. Ehrenfeld, D. *Conserving Life on Earth;* Oxford University Press: New York, 1972; 264.
31. Editorial. *Hetch-Hetchy's Past and Future.* The New York Times, Feb. 17, 2012; A26.
32. Schwartz, J.D. Should we put a dollar value on nature? http://www.time.com/time/business/article/0,8559,1970173,00.html (accessed March 2010).
33. McCauley, D.J. Selling out on nature. Nature **2006**, *443*, 27–28.
34. *Ehrenfeld, D. Becoming Good Ancestors: How We Balance Nature, Community, and Technology; Oxford University Press: New York, 2009; 97–127.*
35. *Meine, C. Correction Lines: Essays on Land, Leopold, and Conservation; Island Press: Washington, DC, 2004; 296.*
36. Meine, C. and Knight, R.L. *The Essential Aldo Leopold: Quotations and Commentaries*; University of Wisconsin Press: Madison, WI, 1999; 384 pp.
37. Leopold, A. *Game Management;* Charles Scribner's Sons: New York, 1933; 481 pp.
38. *Leopold, A.A. Sand County Almanac: With Other Essays on Conservation from Round River; Oxford University Press: New York, 1966; 242.*
39. Leopold, A. A biotic view of land. J. For. **1939**, *37*, 727–730.

Bibliography

Rolston, H., III. *A New Environmental Ethics: The Next Millennium for Life on Earth;* Routledge: New York, 2012; 244.

Rolston, H., III. *Environmental Ethics*; Temple University Press: Philadelphia, 1988; 400.

8

Biomes

Paul Reich and
Hari Eswaran
U.S. Department of Agriculture (USDA-NRCS)

Introduction 75
Major Biomes 75
Soils of the Major Biomes • Polar Biome • Boreal Forest Biome • Temperate Grassland and Forest Biome • Desert Biome • Tropic Biome
References 79

Introduction

A biome is defined as "a community of organisms interacting with one another and with the chemical and physical factors making up their environments."[1] For most purposes, the term biome is used to identify the natural habitat conditions around the world. Depending on the purpose, the global ecosystem is divided into units each characterized by a specific combination of climatic factors. Two major determinants of biome type are precipitation (total and its distribution) and air temperature. These two elements of climate have been commonly used to define the major biomes of the world. A third variable that affects the habitat type is the soil. This section presents a general overview of the soils characterizing the major biomes of the world.

Detailed maps showing the biomes of the world are not available due to conceptual differences of definitions and reliable global databases. There are many excellent and detailed studies of specific habitats around the world, and using these and the global soils and climate database of the world, [2] a map showing the distribution of the major biomes was drawn. The terms used to describe the biomes are common in use, but their subdivisions are based on important differentiating factors.

Major Biomes

The five major biomes of the world and their subdivisions are listed in Table 8.1 and their distribution is shown in Figure 8.1 Some geographers have used elevation as a differentiating factor and recognized a "montane biome." This biome's small extent precludes its description in this section. Few geographers recognize the Mediterranean biome as presented here as a subdivision of the Temperate biome. Within each biome, some major distinction is made. In the Polar biome, an attempt is made to differentiate areas with permafrost and warmer areas with intermittent permafrost. The latter is termed "interfrost." The Boreal Forest biome has a humid and a semiarid counterpart similar to the temperate and tropical zones. The semiarid vegetation is generally grass or shrubs. The desert areas are divided into hot, cool, and cold equivalents.

Deserts occupy about 30% of the global ice-free land-mass. Potential evapotranspiration exceeds precipitation during most days of the year, and the biome shows the greatest contrast in temperature, both diurnal and seasonal. Desert vegetation is very sparse with a few isolated shrubs dominated by succulents and annuals. Grasses and forbs become more dominant at the margins. The tropics occupy

TABLE 8.1 Major Global Biomes with Subdivisions and Their Areas

		Global Land Area			
Biome	Subdivision	Remarks	In Thousand km^2	In%	In%
Polar	Permafrost	Mosses and lichens	10,547	8.06	15.46
	Interfrost	Shrubs and stunted trees	9,685	7.40	
Boreal	Semi arid	Shrub	3,552	2.72	9.94
	Humid	Forest	9,446	7.22	
Temperate	Semiarid	Grassland	7,348	5.62	15.14
	Humid	Forest	12,453	9.52	
	Mediterranean warm	Shrubs and forbs	3,624	2.77	3.37
	Mediterranean cold	Forbs and shrubs	791	0.60	
Desert	Hot	Barren	4,418	3.38	29.47
	Cool	Barren	28,521	21.81	
	Cold	Barren	5,599	4.28	
Tropical	Semiarid	Grassland, savanna	20,299	15.52	26.62
	Humid	Forests	14,514	11.10	
Total			130,796	100.00	

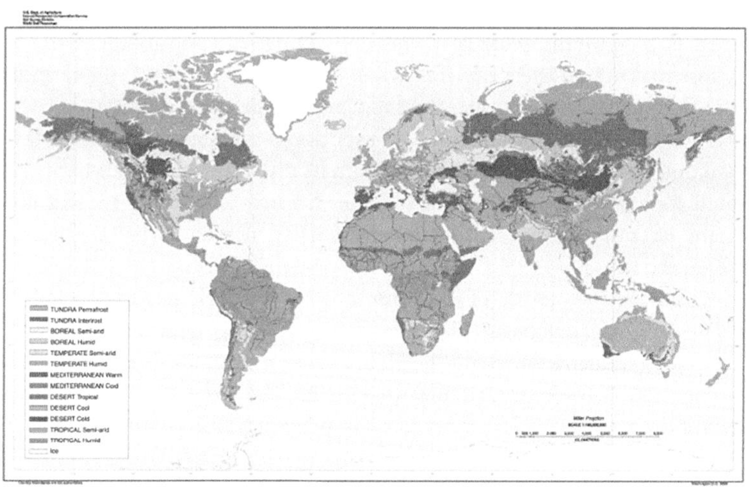

FIGURE 8.1 Major biomes.

about 27% of the landmass and are characterized by only a small variation in temperature during the year. The soil moisture conditions range from semiarid to humid. The availability of soil moisture determines the biota. The temperate grassland and forest biome occupies about 15%. The cold biomes, the boreal and polar, are mainly in the northern latitudes with small areas in South America. The low temperatures control the flora and fauna, and vegetation is scarce to nonexistent in the extreme cold polar regions. Each of these areas has specific kinds of soils that have developed as a response to the specific soil moisture and temperature conditions. The diversity in vegetation has its equivalent in fauna with animal species adapting to the specific bio-climatic conditions. Within each of the biomes, there are specific conditions that promote unique flora and fauna. Examples of such localized systems are volcanic hot springs, wetlands, and oases.

Soils of the Major Biomes

Soil temperature and moisture, with their seasonal and annual variations, are an integral part of the soil classification system called Soil Taxonomy[3] which is used here. As Figure 8.1 and Table 8.1 are based on soil moisture and temperature conditions, there is an implied link between the biomes and their soil resource endowments. The purpose of this introductory section is to present the general geographic distribution of the biomes and the major soils characterizing each. The major soil orders are listed in Table 8.2. The subsequent sections elaborate on the soil resources of each of the six biomes listed in Table 8.1.

Polar Biome

Characterized largely by a mean annual soil temperature of less than 0°C, the Polar biome occupies a large land mass adjacent to the Arctic circle to about 60°N latitude. This is generally the ice-free land of the northern latitudes. The sub-zero soil temperatures that prevail during most of the year are conducive to the formation of permafrost. The dominant soils of the region are Gelisols, occupying 59% of the Polar biome. Turbels freeze and thaw once or more during a year which triggers cryoturbation or physical mixing of the soil material. The Orthels are generally shallower or drier soils and have little or no cryoturbation. The low temperatures and periodic moisture saturation promote the accumulation of organic matter. The Histels are characterized by the high organic matter and occupy about 5% of the Polar biome. These organic soils form the largest contiguous extent of such soils in the world and serve as an important sink of CO_2. Other soils that occur in the Tundra zone are Inceptisols 33%, Spodosols 6%, while Mollisols and Entisols are each 1% of the area.[4]

Boreal Forest Biome

Bordering the southern flank of the Polar biome is the Boreal Forest biome in the northern hemisphere. The high volcanic area between Chile and Argentina has similar climatic regimes but the nature of the soil and physiography may result in a different set of habitat conditions. There are about 243,000 km^2 of Andisols or volcanic ash soils characterizing the Boreal Forest biome in the southern hemisphere. Such soils only occur as a small area in the Kamchatka Peninsula in the Northern Hemisphere. In Northern Europe and Siberia, the Boreal Forest biome has a humid part and is bordered in its southern periphery by a semiarid to arid part. In Canada, the semiarid part is in the middle of the continent. A range of soils are present and the most extensive are the Spodosols, which occupy 2.7 million km^2 (20.6%). The Spodosols are mostly in the humid part of the Boreal Forest biome and form under acid vegetation. Histosols are present in the depressions in this cold region.

Temperate Grassland and Forest Biome

The Temperate Grassland and Forest biome extends from about 25 to 55°N with counterparts in the Southern Hemisphere. Large areas in this belt are also deserts. Due to the favorable climate and soil endowments of this biome, much of the land is used for agriculture and native habitats are local and sporadic. About 50% of the zone (12.1 million km^2) is occupied by Mollisols and Alfisols (grasslands), and Ultisols in the forests. Their general good fertility and tilth have made them in great demand for grain production. Due to the long history of civilization in the Temperate Grassland and Forest biome of Europe and China, pristine ecosystems are rare. In large areas, there have been successive replacements and changes in the floral and faunal composition. Within the Temperate biome, are areas characterized by moist winters and dry summers. These Mediterranean conditions are conducive to unique ecosystems. These areas occur around the Mediterranean Sea and small areas in the western U.S. and southern Australia.

TABLE 8.2 Dominant Soils in the Major Biomes (Area in Thousand km²)

		Polar		Boreal		Temperate				Desert			Tropics	
Soil Order	Ice-Freeland	Permafrost	Interfrost	Semiarid	Humid	Semiarid	Humid	Mediterranean Warm	Mediterranean Cold	Warm	Cool	Cold	Semiarid	Humid
Gelisols	11,869	10,476	1,393	—	—	—	—	—	—	—	—	—	—	—
Histosols	1,526	—	—	299	686	17	82	1	8	—	23	93	37	281
Spodosols	4,596	55	1,058	227	2,566	105	450	39	6	—	—	28	—	61
Andisols	975	6	56	47	196	26	135	32	—	17	10	14	198	235
Oxisols	9,811	—	—	—	—	9	184	—	—	27	4	—	3,387	6,199
Vertisols	3,160	—	—	2	15	595	303	99	0	239	651	0	1,170	87
Aridisols	15,629	1	—	60	76	353	18	203	23	1,824	11,014	2,013	44	1
Ultisols	11,053	—	—	1	14	266	3,122	20	18	—	—	—	4,371	3,240
Mollisols	9,161	9	147	1,059	511	1,280	1,232	874	323	49	1,110	2,255	136	176
Alfisols	12,621	—	—	1,002	1,816	1,867	2,149	860	125	—	—	—	4,165	638
Inceptisols	21,805	—	6,751	167	3,333	1,636	3,078	683	89	—	—	—	3,566	2,502
Entisols	21,467	—	280	172	232	1,173	1,697	807	45	1,708	10,693	342	3,225	1,093
Miscellaneous	7,122	—	—	516	1	20	4	6	153	553	5,016	853	—	—
Total	130,796	10,547	9,685	3,552	9,446	7,348	12,453	3,624	791	4,418	28,521	5,599	20,299	14,514

Desert Biome

Deserts occupy about 38.6 million km^2 and may be distinguished as warm (or tropical), cool (or temperate) and cold (or boreal) kinds of habitats. Though the biome is characterized by lack of moisture for normal vegetative growth of most plants, the ambient temperature conditions further distinguish the habitat conditions. About 38.6% of the Desert biome is occupied by Aridisols, which by definition have some kind of subsurface horizon. Shifting sands or moving dunes occupy about 13.7% and Entisols, which are very recent deposits, occupying about 33% of the Desert biome. The harshness of the environment has resulted in plants and animals with special adaptive features. Many of the Aridisols are increasingly used for agriculture when irrigation facilities are made available. In this fragile ecosystem, both soil and habitat conditions are drastically altered when irrigation is introduced.

Tropics Biome

An absence of winter and summer temperature extremes characterizes the Tropics biome, which occupy about 34.8 million km^2. Availability of moisture separates the semiarid from the humid tropics. As a corollary to the desert, in the humid tropics there are the perhumid areas where the potential evapotranspiration never exceeds the precipitation during any month of the year. This is a biomic condition that deserves much greater detailed studies. The Tropics biome is the home to the Oxisols (9.6 million km^2), which are unique to this biome. These are highly weathered soils where the original vegetation is closed or open forests. Most of the plant nutrients are concentrated in the top 5 cm of the soil and recycled through plant uptake and leaf fall. Ultisols also occur in this biome. Apart from these soils, there are small areas of most of the other soil orders. The wetlands of the tropics present a separate and unique habitat condition and those along the coast have some special soils.

References

1. Tootil, E., Ed.; *Dictionary of Biology*; Intercontinental Kook Productions, Ltd.: Maidenhead, Berkshire, England, 1980.
2. Eswaran, H.; Beinroth, F.H.; Kimble, J.; Cook, T. Soil diversity in the tropics: implications for agricultural development. In *Myths and Science of Soils ofthe Tropics*; Lal, R., Sanchez, P.A., Eds.; Soil Sci. Soc. Am. Spec. Publ. **1992;** *29*, 1–16.
3. *Soil survey staff. In Soil Taxonomy: A Basic System of Soil Classification for Making and Interpreting Soil Surveys*, 2nd Ed.; Natural Resources Conservation Service. U.S. Department of Agriculture, Handbook 436; U.S. Government Printing Office: Washington, DC, 1999.
4. *Keys to Soil Taxonomy*, 9th Ed.; U.S. Government Printing Office: 2003.

9

Decision-Making and Monitoring Strategies in Natural Resource Management and Conservation

Brian D. Gerber
University of Rhode Island

Perry J. Williams
University of Nevada

Introduction 81
Structured Decision-Making
Monitoring Strategies 83
Model-Free Sampling Designs • Model-Based Sampling Designs
Conclusions 85
References 85

Introduction

Natural resource management is an exercise in decision-making: choosing to act, how to act, or even to not act. These decisions are rarely simple, because of uncertainty in (i) the current state of the system, (ii) ecosystem processes, (iii) the effect of management actions on system processes, and (iv) implementing the intended action. For example, let's consider these uncertainties in the context of sustainably managing the harvest of wildlife and fisheries populations. To decide on the annual allowable harvest, we should ideally understand the size and distribution of the population to be harvested (i). Yet, counting wild animals is often challenging, especially at large spatial scales, commonly leading to sampling errors [1] that require explicit consideration [2,3]. Further, we should know about factors that affect population dynamics, such as abiotic and biotic processes (ii) and harvest decisions (iii). The how and why of population change has been a central focus of ecology [4], including the creation of global databases on demography [5]. Yet, there are still substantial knowledge gaps across many taxonomic groups and geographic areas [5], including many well-known species of conservation concern [6]. Last, controlling the number of animals harvested is not always straightforward (iv). We may only be able to control harvest indirectly through local, state, or federal regulation, such as the length of time harvest is allowed, or the method of harvest (e.g., wildlife: archery, rifle, shotgun; fisheries: long-line, seine net, trawl). These regulations partially control the number of harvested animals, but there is a lot of variability in the number of harvested animals that can occur within these regulations [7].

An additional complexity in natural resource management is that many decisions are recurrent, such that current decisions will affect the system and thus all future decisions [8]. For example, consider a manager aiming to conserve grassland breeding birds within a wildlife refuge. They may be managing for open grasslands by using prescribed fire and need to decide on a fire-return interval [9]. The timing of the prescribed fire will affect the grassland, the soil, the breeding birds, and thus all future decisions of when to burn. Further, resource managers almost always have multiple (and potentially competing) objectives motivated by a diverse set of policies and stakeholders [10]. The manger may also be considering the frequency and timing of the burn to minimize the financial costs and disruption to recreational users of the refuge, such that a short burn-interval will be more costly and disruptive than a longer one; however, a shorter interval may be more beneficial for grassland birds. Making a decision that balances the effects on birds, people, and budgets now and in the future is not intuitive.

Managing renewable natural resources for human exploitation requires recurrent decision-making, such as deciding annual fishing and wildlife harvest quotas, the interval between forest cuttings, drawing down of water reservoirs, and agricultural crop rotations. Managing non-renewable natural resources (e.g., oil and natural gas) can also lead to recurrent decisions when managing extraction rates to maximize profit with fluctuating commodity prices. The compounding of recurrent state, process, action, and control uncertainties (i–iv, respectively) along with competing multiple objectives can be overwhelming for decision-makers. This may lead to inaction despite a strong need for immediate action [11]. This may also lead to the complacency for an institutionalized decision process that ignores uncertainties, favoring simplicity over realism. The result of which may lead to decisions that jeopardize meeting long-term management goals [12].

Overcoming the complexities of a decision process requires embracing uncertainty [13] as an inherent reality of perhaps all natural resource management decisions. We suggest management questions be framed in a formal decision process, which is a powerful strategy to deconstruct the decision into essential and manageable elements that can be examined using decision-analytic tools [14,15]. Here, we briefly outline the essential elements for making empirical, logical, and transparent decisions in the face of uncertainty, specifically using a structured decision-making (SDM) framework [16,17]. Second, we focus on sampling strategies to monitor state variables of interest (i.e., measures of the natural resource being managed) as a necessary element of any decision-making process. Monitoring is the key to decrease state, process, and action uncertainty, and thus improve future decisions [18–20]. Furthermore, there is perhaps a lack of appreciation in natural resource management for sampling options and their relevant advantages and disadvantages.

Structured Decision-Making

The six elements of an SDM process are where the decision-makers (i) explicitly state the management objectives, (ii) list all possible alternative actions, (iii) predict the consequence of each action on the management objectives, (iv) choose an action by optimizing the objectives while considering trade-offs, (v) implement the decision, and (vi) monitor the system to evaluate whether the objectives were achieved [15–17,21]. This separation helps identify the distinct roles of stakeholders (e.g., managers, policymakers, resource users) and scientists, such that stakeholders communicate the objectives and their relative importance [9], while scientists make predictions and evaluate trade-offs; both groups identify potential actions.

Perhaps the most obvious and yet underappreciated aspect of a decision process is that a 'good' or 'best' decision is impossible to identify without clear, explicit, and measurable objectives [16,17]. Objectives are typically quantifiable statements (i.e., via an objective/loss function) that contain at least one measure of system performance that can be maximized (or minimized) [9,10,22]; however, alternative approaches are available that seek satisfactory performance, rather than optimal performance [23]. Predicting the consequences of possible actions entails describing a model of the system. Such a model may be qualitative or intuitive expectations, but often are quantitative [20], which lends to being explicit, transparent, objective, and amenable to analyses; quantitative models can synthetically incorporate empirical and expert knowledge [13]. The means of being able to predict the consequences of potential actions, identify the current state of the resource, and evaluate management performance is often through monitoring data [24]. Monitoring is essential to recurrent decisions and the means to which learning can occur through an adaptive management process to improve decision-making over time [19]. However, monitoring natural resources requires careful sampling considerations to ensure data are able to provide accurate and precise estimates of state variables.

Monitoring Strategies

To evaluate the effectiveness of natural resource management, we require quantitative information on how the system responds to management actions. We acquire this information via monitoring. Monitoring has always been a critical component of scientific investigation and natural resource management, and therefore, monitoring methods have, and continue to be, well studied and described. Monitoring programs designed to inform management and conservation of natural resources are inherently spatial, and therefore, we focus on spatial sampling designs. Spatial sampling designs can be classified into two broad types: *model-free approaches* and *model-based approaches* [25]. What exactly to monitor is problem-specific and depends on the objectives of the study and the processes that are considered in models used to predict the consequences of proposed actions; we provide a high-level overview of spatial sampling concepts that is applicable for monitoring many types of natural resources.

Model-Free Sampling Designs

No specific model is assumed for data when using model-free sampling designs. This is advantageous because we are not required to adhere to any assumptions regarding underlining ecological/ecosystem processes, precluding the possibility of selection bias that can occur when relying on a mis-specified model to collect a sample. Model-free approaches include the relatively well-known *probability-based sampling methods* such as simple random sampling, systematic sampling, stratified sampling, stratified sampling with random tessellation, cluster sampling, and Markov sampling [26]. They also include the lesser-known geometrically based *space-filling designs* [27].

Probability-based sampling methods assume the underlining spatial process of interest is fixed, but randomness is introduced in sample selection using a probability sampling design. A probability sampling design consists of both a sampling frame, a set of all elements in the population of interest, and a probability measure defined on the frame; a systematic method for assigning a probability to the selection of any subset (i.e., any sample) from the frame. The statistical properties (e.g., unbiased-ness) of the estimators used to make inference on the population are based entirely on the probability measure that defines the design. For this reason, inference from probability-based sampling methods is often referred to as *design-based inference*. The objective of design-based inference is to quantify a population parameter of interest. Typically, these include summary statistics such as the mean, total, variance, or proportion. Probability-based sampling methods are one of the most common and widely used sampling methods in natural resources. When probability-based sampling designs are combined with the appropriate estimators, investigators obtain objective, unbiased estimates of the population. Both [26] and [28] provide in-depth discussion of probability-based sampling methods and numerous extensions.

Space-filling designs were developed to sample such that large unsampled gaps in the study area were avoided. When there is little prior spatial information about an ecological process of interest (e.g., how to stratify a sample), an intuitive approach to collecting data might be to select a sample that has good spatial coverage of the area interest. This provides an alternative to simple random sampling, which does not guarantee good spatial coverage. A *minimax design* is a type of space-filling design that chooses a sample that minimizes the maximum distance between points. That is, it guarantees that all points in the sample are within a certain distance of at least one point in the design; there are no spatial gaps with a distance larger than a specified distance. Similarly, a *maximin* design maximizes the minimum distance between points so that points are as far away from each other as possible. Justification for a maximin design is that areas close in proximity are more similar than distant sites, and therefore, samples in close proximity may provide redundant information.

Model-Based Sampling Designs

To fully realize the benefits of the probabilistic sampling designs, appropriate estimators that are associated with the probability measure defining the design must be used to make inference on the population. However, data are increasingly being used to fit models to make model-based inference, regardless of how the data were collected. Using data to fit and evaluate statistical models of ecological processes is fundamental to ecology and management. Yet, monitoring designs for collecting data, and the statistical models used to analyze the data, are typically developed independently of each other. When monitoring designs are developed independently of statistical models, monitoring effort can be inefficient and fail to capture essential spatial, temporal, or spatio-temporal variability of an ecological process [20]. Formally linking monitoring designs with statistical modeling using an *optimal monitoring design* permits several advantages over traditional model-free sampling designs (when a model is going to be used to make inference) including increased efficiency, more precise parameter estimation, and reduced prediction uncertainty. These improvements translate to increased reliability of scientists to evaluate the consequences of management actions and thus lead to improved decisions. Optimal monitoring designs permit extraction of the most information from the data that can be affordably collected, helping to improve our understanding of the ecological system, despite limited or shrinking financial resources.

Broadly speaking, a model-based sampling design consists of several steps, with each step being conceptually straightforward. These steps are similar to the steps described above for SDM. In fact, optimal monitoring is a type of SDM, where the decision is the optimal place to collect data to help inform our understanding of some process of interest. In the first step of optimal monitoring, an ecological process such as the occurrence or abundance of a species or resource is modeled using baseline data. Second, using the model from the first step, a statistical forecast of a process of interest is made. The forecast provides a basis for examining potential monitoring designs that could be implemented in the future. Third, managers and stakeholders identify the objectives or criteria they wish to achieve with their monitoring, and formalize this using an objective function [9]. Objectives typically include minimizing prediction uncertainty or minimizing variance of parameter estimates, but could also include minimizing multi-modal uncertainty or cost. Fourth, after a forecast is made, and an objective function is chosen, a monitoring design that optimizes the objective function, relative to the ecological forecast is chosen. Fifth, data are collected using the optimal monitoring design. The model is then updated with the new data, and ecological learning occurs. The process can be iterated through time (a recurrent decision), increasing our understanding of the ecological mechanisms governing the data (a process analogous to adaptive resource management), and lead to improved evaluation of management actions.

While there are several advantages of model-based sampling described in the previous paragraph, there are two potential disadvantages. The first disadvantage is computational and analytic complexity required to obtain an optimal model-based design. It is simple to develop a simple/stratified/systematic random sample. However, identifying an optimal model-based sample requires a large amount of computation, and the ability to forecast sometimes-complex ecological processes. The second potential disadvantage is the optimal survey design is conditional on the selected model used to make inference. If a mis-specified model is used for inference, the resulting errors in mis-specification may be propagated through to the data, resulting in data that are sub-optimal. To address the second point, several researchers have developed 'hybrid designs' that consist of both an optimal, model-based sample, and a model-free sample [29]. Model-based sampling designs have been used disproportionately less in natural resources than model-free designs. However, in recent years, there have been several natural resource applications that have used model-based or hybrid sampling designs [20,29–31].

Conclusions

Managing natural resources can be complex, necessitating careful consideration about the process to which information is gathered and decisions are made. We outline the essential elements of making empirical, logical, and transparent decisions in the face of common uncertainties and when designing sampling strategies for monitoring natural resources. We encourage resource managers tasked with managing public and valued resources (e.g., air, water, soil, and plant and animal species) to strive towards adopting an empirically based logical and transparent decision process informed via a robust monitoring strategy.

References

1. Pollock, K.H.; Nichols, J.D.; Simons, T.R.; Farnsworth, G.L.; Bailey, L.L.; Sauer, J.R. Large scale wildlife monitoring studies: statistical methods for design and analysis. *Environmetrics* 2002, 13, 105–119.
2. Gerber, B.D.; Kendall, W.L. Evaluating and improving count-based population inference: a case study from 31 years of monitoring Sandhill Cranes. *Condor Ornithol. Appl.* 2017, 119, 191–206.
3. Williams, P.J.; Hooten, M.B.; Womble, J.N.; Esslinger, G.G.; Bower, M.R.; Hefley, T.J. An integrated data model to estimate spatiotemporal occupancy, abundance, and colonization dynamics. *Ecology* 2017, 98, 328–336.
4. Caughley, G. *Analysis of Vertebrate Populations*. John Wiley and Sons: New York, 1977.
5. Salguero-Gómez, R.; Jones, O.R.; Archer, C.R.; Bein, C.; de Buhr, H.; Farack, C. et al. COMADRE: a global data base of animal demography. *J. Anim. Ecol.* 2016, 85, 371–384.
6. Tyler, E.M.; Somerfield, P.J.; Berghe, E.V.; Bremner, J.; Jackson, E.; Langmead, O. et al. Extensive gaps and biases in our knowledge of a well-known fauna: implications for integrating biological traits into macroecology. *Glob. Ecol. Biogeogr.* 2012, 21, 922–934.
7. Nichols, J.D.; Runge, M.C.; Johnson, F.A.; Williams, B.K. Adaptive harvest management of North American waterfowl populations: a brief history and future prospects. *J. Ornithol.* 2007, 148, 343–349.
8. Williams, B.K.; Szaro, R.C.; Shapiro, C.D. *Adaptive Management: The US Department of the Interior Technical Guide*. Adaptive Management Working Group, U.S. Department of Interior: Washington, DC, 2009.
9. Williams, P.J.; Hooten, M.B. Combining statistical inference and decisions in ecology. *Ecol. Appl.* 2016, 26, 1930–1942.
10. McGowan, C.P.; Lyons, J.E.; Smith, D.R. Developing objectives with multiple stakeholders: adaptive management of horseshoe crabs and red knots in the Delaware Bay. *Environ. Manage.* 2015, 55, 972–982.
11. Gifford, R. The dragons of inaction: psychological barriers that limit climate change mitigation and adaptation. *Am Psychol.* 2011, 66, 290.
12. Regan, H.M.; Ben-Haim, Y.; Langford, B.; Wilson, W.G.; Lundberg, P.; Andelman, S.J. et al. Robust decision-making under severe uncertainty for conservation management. *Ecol. Appl.* 2005, 15, 1471–1477.
13. Gerber, B.D.; Converse, S.J.; Muths, E.; Crockett, H.J.; Mosher, B.A.; Bailey, L.L. Identifying species conservation strategies to reduce disease-associated declines. *Conserv. Lett.* 2018, 11, e12393.
14. Clemen, R.T.; Reilly, T. *Making Hard Decisions: An Introduction to Decision Analysis*. Duxbury Resource Center: Pacific Grove, CA, 2001.
15. Hammond, J.S.; Keeney, R.L.; Raiffa, H. *Smart Choices: A Practical Guide to Making Better Decisions*. Harvard Business School Press: Brighton, MA, 2015.

16. Conroy, M.J.; Peterson, J.T. *Decision Making in Natural Resource Management: A Structured, Adaptive Approach*. John Wiley & Sons: New York, 2013.
17. Gregory, R.; Failing, L.; Harstone, M.; Long, G.; McDaniels, T.; Ohlson, D. *Structured Decision Making: A Practical Guide to Environmental Management Choices*. John Wiley & Sons: New York, 2012.
18. Williams, B.K. Adaptive management of natural resources-framework and issues. *J. Environ. Manage*. 2011, 92, 1346–1353.
19. Gerber, B.D.; Kendall, W.L. Adaptive management of animal populations with significant unknowns and uncertainties: a case study. *Ecol. Appl*. 2018, 28, 1325–1341.
20. Williams, P.J.; Hooten, M.B.; Womble, J.N.; Esslinger, G.G.; Bower, M.R. Monitoring dynamic spatio-temporal ecological processes optimally. *Ecology* 2018, 99, 524–535.
21. Gregory, R.S.; Keeney, R.L. Making smarter environmental management decisions. *J. Am. Water. Resour. Assoc*. 2002, 38, 1601–1612.
22. Williams, P.J.; Kendall, W.L. A guide to multi-objective optimization for ecological problems with an application to cackling goose management. *Ecol. Model*. 2017, 343, 54–67.
23. Ben-Haim, Y. *Info-Gap Decision Theory: Decisions under Severe Uncertainty*. Academic Press: Oxford, 2006.
24. Lyons, J.E.; Runge, M.C.; Laskowski, H.P.; Kendall, W.L. Monitoring in the context of structured decision-making and adaptive management. *J. Wildl. Manag*. 2008, 72, 1683–1692.
25. Zhu, Z.; Evangelou, E. *Optimal Designs for Spatial Models. Handbook of Design and Analysis of Experiments*. CRC Press: Boca Rotan, FL, 2015.
26. Cochran, W.G. *Sampling Techniques*. John Wiley & Sons: New York, 2007.
27. Johnson, M.E.; Moore, L.M.; Ylvisaker, D. Minimax and maximin distance designs. *J. Stat. Plan. Infere*. 1990, 26, 131–148.
28. Thompson, S.K. *Sampling*. John Wiley & Sons: Hoboken, NJ, 2012.
29. Hooten, M.B.; Wikle, C.K.; Sheriff, S.L.; Rushin, J.W. Optimal spatio-temporal hybrid sampling designs for ecological monitoring. *J. Veg. Sci*. 2009, 20, 639–649.
30. Reich, B.J.; Pacifici, K.; Stallings, J.W. Integrating auxiliary data in optimal spatial design for species distribution modelling. *Methods Ecol. Evol*. 2018, 9, 1626–1637.
31. Wikle, C.K.; Royle, J.A. Dynamic design of ecological monitoring networks for non-Gaussian spatio-temporal data. *Environmetrics* 2005, 16, 507–522.

10

Diversity: Species

Introduction 87
Quantifying Species Diversity 87
Patterns of Species Diversity 88
Maintenance of Species Diversity 89
Threats to Species Diversity 89
Benefits of Species Diversity 90
References 91

Jeffrey D. Corbin and
Brittany L. Oakes
Union College

Introduction

Species diversity is one of the three scales at which diversity can be measured, intermediate between genetic diversity and ecosystem (or habitat) diversity. The concept of biodiversity is often considered to encompass all three of these scales of diversity. In practice the two terms—species diversity and biodiversity—are often used interchangeably, particularly in popular usage or in the context of environmental conservation.

Mora and colleagues used statistical analysis to estimate that there are 8.7 million (±1.3 million SE) eukaryotic species on Earth; thus far, only 1.2 million have been identified and catalogued.[1] Prokaryotic diversity is less well-known but likely to add significantly to this estimate. Invertebrate species are the largest contributors to current estimates, with insect diversity reaching one million species and arachnids around 100,000 species.[2] Plant species rank second to invertebrate richness, owing to the notably high diversity of flowering plants (~280,000 species).[2]

Quantifying Species Diversity

Species diversity is most simply quantified by counting the number of species, also often called species richness. Quantitative diversity indices include species' abundances along with the number of species. Among the most commonly used quantitative diversity indices are the inverse Simpson's index

$$\lambda = 1 / \Sigma p_i^2$$

and the Shannon–Wiener index

$$H' = -\sum p_i \log p_i$$

where p_i = proportion of species i in the total sample of individuals. Each index is differentially sensitive to the presence of rare species, so often more than one index is calculated in order to give a more complete picture.[3]

Patterns of Species Diversity

Tropical habitats, including mesic (or wet) forest and coral reefs, are renowned for their high species diversity. Tropical rainforests may have as many as 10,000 species per hectare; at smaller spatial scales, nutrient poor grasslands, such as those in the Coastal Plains of southeastern North America, are notable for having as many as 50 species per m^2. By contrast, habitats at northern latitudes, including boreal forest and tundra, and salt marshes contain relatively few species at both large (e.g., km^2) and small spatial scales.

At large scales, the number of species of a variety of taxa peaks at the lower (i.e., tropical) latitudes and decreases as latitude increases toward the poles.[4] Several hypotheses have been proposed to explain the latitudinal gradient in species richness. These hypotheses include the evolutionary age of the habitats, habitat heterogeneity, climatic stability, available environmental energy, higher rates of evolution in the tropics, and a sampling effect due to the arrangement of landmasses near the equator (i.e., the "mid-domain" effect).[5]

Endemic species, or those restricted to a specific geographic location and found nowhere else on the Earth, are an important component of species diversity both within a specific habitat and as a contributor to global species diversity. Myers and colleagues overlapped areas with a significant number of endemic species (at least 0.5% of the world's total vascular plants, or 1,500 plant species, as endemics) with areas experiencing high rates of habitat loss (at least 70% of its original vegetation lost) to develop the concept of biodiversity hotspots[6,7] (Figure 10.1). The 34 biodiversity hotspots that are currently recognized cover 2.3% of the Earth's surface; most are located in tropical zones. Benefits of identifying such hotspots include the potential to prioritize conservation funding toward areas that are both threatened by habitat loss and that would result in a significant loss of unique species.[6] This approach has been criticized, however, in that it does not take a variety of other components of diversity into account, including smaller-scale diversity or diversity of taxa besides plants.[8]

At a local scale, Whittaker differentiated between alpha (within-habitat), beta (between-habitat), and gamma (regional or between-ecosystem) diversity.[9] The number of species typically increases as the area sampled increases.[10,11] This species-area relationship has been described as a power function:

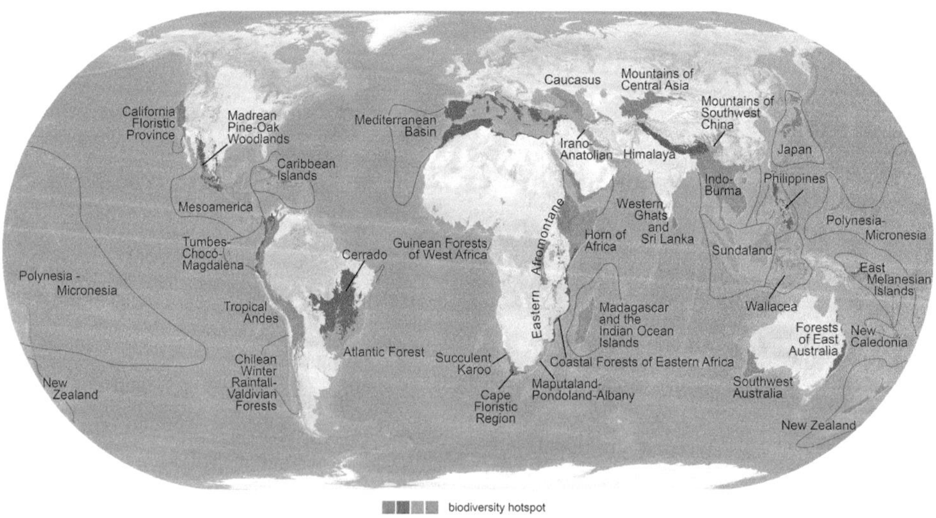

FIGURE 10.1 **(See color insert.)** Map of global biodiversity hotspots.
Source: Courtesy of Conservation International.[7]

FIGURE 10.2 The linear transformation (log S = log c + z log A) of the species-area power function, $\mathbf{S = cA^z}$, where S is species richness, A represents the area sampled, and c and z are constants fit to the data.

$$S = cA^z$$

where S is species richness, A is area sampled, and c and z are constants fit to the data. Graphically, it is often portrayed in logarithmic space, with the logarithm of species number having a linear relationship with the logarithm of are (Figure 10.2).

Maintenance of Species Diversity

Understanding factors that maintain species diversity—that is, how multiple species can coexist in the same habitat rather than one species "winning" and competitively excluding others—has been a central focus of ecological theory since at least G. E. Hutchison's famous paper "Homage to Santa Rosalia, or why are there so many animals?"[12] Models of species coexistence and, therefore, maintenance of species diversity can be roughly divided into equilibrium and non-equilibrium models.[13] Equilibrium models of species coexistence assume that competitive exclusion of one species over another does not take place because direct competition is avoided, for example, because of niche differentiation (e.g., species are utilizing different resources) or because of spatial or temporal heterogeneity (e.g., resources and competitive advantages vary in space or time). Non-equilibrium models assume that periodic events such as predation (e.g., keystone predators)[14] or disturbances[13] prevent equilibrium and, therefore, competitive exclusion.[15]

Threats to Species Diversity

Periods of mass species extinctions have occurred at least five times in the Earth's history, but indications are that human activities have ushered a sixth episode. The impacts of human population on a variety of measures including land use, global climate, and biodiversity have led many to define the current epoch as the "Anthropocene."[16] Human activities may have led to the extinction of 5–20% of the species in some species groups,[17] well above background levels of extinctions that would naturally occur.[18] Change in land use, including habitat loss, conversion of rainforest habitat, and fragmentation,

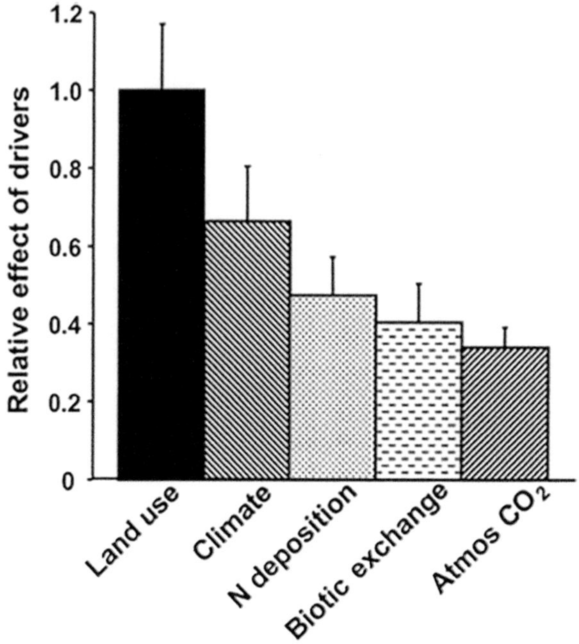

FIGURE 10.3 Relative effect of major drivers of changes on biodiversity: Land use change, climate change, atmospheric nitrogen deposition and acid rain, biotic exchange including invasive species, and elevated atmospheric CO_2. Expected biodiversity change for each biome for the year 2100 was calculated as the product of the expected change in drivers times the impact of each driver on biodiversity for each biome. Values are averages of the estimates for each biome and they are made relative to the maximum change, which resulted from change in land use. Thin bars are standard errors and represent variability among biomes.
Source: Adapted from Sala et al.[19]

is considered the leading driver of modern extinctions. Other causes include overexploitation and invasion of non-native species. Forecasts of future drivers of species extinction suggest that climate change will be a significant factor (Figure 10.3).[19,20]

Benefits of Species Diversity

One of the most often-cited motivations for conserving species diversity is the benefit that biodiversity and nature provide to humans. Such benefits have been quantified as "ecosystem services" and include provisioning of food, fiber, pharmaceutical products, fresh water, etc.; regulation of flood waters, agricultural pests, or atmospheric carbon; cultural uses including recreation, enjoyment of the aesthetics, or spiritual values; and indirect benefits such as maintenance of nutrient cycling, pollination, or primary productivity.[21] In this framework, biodiversity provides the foundation for all other services even if it provides relatively little direct benefit to humans.

Species diversity is also believed to influence an ecosystem's stability. Both theory and empirical evidence suggest that ecosystems with more species have a greater ability to resist and recover from perturbations.[22] It is unclear, however, whether diversity is the driver of this pattern or merely a correlation with other ecological factors. Instead, stability seems to arise from the greater likelihood that a diverse community will include highly productive species or functional groups, or those that can respond differentially to perturbations and buffer ecological changes.[22]

Finally, a principal motivation for conserving nature and species diversity derives from the appreciation that they have value in their own right. To say that a species, or nature in general, has a value of its own regardless of the benefits that humans may derive from them is to say that it has *intrinsic value*.[23] The writings of Aldo Leopold and John Muir gave rise to such an environmental ethic that places a value on species whether or not they provide any benefits to humans.

References

1. Mora, C.; Tittensor, D.P.; Adl, S.; Simpson, A.G.B.; Worm, B. How many species are there on Earth and in the ocean? PLoS Biol. **2011**, *9* (8), 1–8.
2. The World Conservation Union. Summary statistics for globally threatened species. Table 1: Numbers of threatened species by major groups of organisms (1996–2010). In *IUCN Red List of Threatened Species*; IUCN; UK, 2010.
3. Peet, R.K. Relative diversity indices. Ecology **1975**, *56* (2), 496–498.
4. Hillebrand, H. On the generality of the latitudinal diversity gradient. Am. Nat. **2004**, *163* (2), 192–211.
5. Pianka, E.R. Latitudinal gradients in species diversity: A review of concepts. Am. Nat. **1966**, *100* (910), 33–46.
6. Myers, N.; Mittermeier, N.A; Mittermeier, C.G.; da Fonseca, G.A.B.; Kent, J. Biodiversity hotspots for conservation priority. Nature **2000**, *403* (6772), 853–858.
7. Conservation International, 2013. Biodiversity Hotspots Map. http://www.conservation.org/where/priority_areas/hotspots/DocumentsZCI_Biodiversity-Hotspots_2013_Map.pdf (accessed January 2013).
8. Kareiva, P.; Marvier, M. Conserving biodiversity coldspots. Am. Sci. **2003**, *91* (4), 344–351.
9. Whittaker, R.H. Evolution and measurement of species diversity. Taxon **1972**, *21* (2/3), 213–251.
10. Williams, C.B. Patterns in the Balance of Nature and Related Problems in Quantitative Ecology; Academic Press: New York, 1964.
11. Rosenzweig, M.L. *Species Diversity in Space and Time*; Cambridge University Press: Cambridge, 1995.
12. Hutchinson, G.E. Homage to Santa Rosalina or why are there so many kinds of animals? Am. Nat. **1959**, *93* (870), 145–159
13. Connell, J.H. Diversity in tropical rainforests and coral reefs. Science **1978**, *199* (4335), 1302–1310.
14. Paine, R.T. Food web complexity and species diversity. Am. Nat. **1966**, *100* (910), 65–75.
15. Huston, M. A general hypothesis of species diversity. Am. Nat. **1979**, *113* (1), 81–101.
16. Vince, G. An epoch debate. Science **2011**, *334* (6052), 32–37.
17. Pimm, S.L.; Russell, G.J.; Gittleman, J.L.; Brooks, T.M. The future of biodiversity. Science **1995**, *269* (5222), 347–350.
18. Lawton, J.H.; May, R.M. *Extinction Rates*; Oxford University Press: Oxford, 1995.
19. Sala, O.; Chapin, F.S., III; Armesto, J.J.; Berlow, E.; Bloomfield, J.; Dirzo, R.; Huber-Sanwald, E.; Huenneke, L.F.; Jackson, R.B.; Kinzig, A.; Leemans, R.; Lodge, D.M.; Mooney, H.A.; Oesterheld, M.; Poff, N.L.; Sykes, M.T.; Walker, B.H.; Walker, M.; Wall, D.H. Global biodiversity scenarios for the year 2100. Science **2000**, *287* (5459), 1770–1774.
20. Thomas, C.D.; Cameron, A.; Green, R.E.; Bakkenes, M.; Beaumont, L.J.; Collingham, Y.C.; Erasmus, B.F.N.; de Siqueira, M.F.; Grainger, A.; Hannah, L.; Hughes, L.; Huntley, B.; van Jaarsveld, A.S.; Midgley, G.F.; Miles, L.; Ortega-Huerta, M.A.; Peterson, A.T.; Phillips, O.L.; Williams, S.E. Extinction risk from climate change. Nature **2004**, *427* (6970), 145–148.
21. Millennium Ecosystem Assessment. *Our Human Planet: Summary for Decision-Makers*; Island Press: Washington, D.C., 2005.
22. McCann, K.S. The diversity-stability debate. Nature **2000**, *405* (6783), 228–233.
23. Vilkka, L. *The Intrinsic Value of Nature*; Rodopi: Amsterdam, the Netherlands, 1997.

11

Exotic and Invasive Species

Jan Pergl and
Petr Pyšek
Academy of Sciences of the Czech Republic (CAS)

Exotic or Invasive: Who Is Who? 93
What Makes a Species Invasive? 94
Impact of Exotic Species 96
Prevention, Control, Management, and Risk Assessment 98
Exotic Species Lists and Data Sources 100
Conclusions 100
Acknowledgments 101
References 101

Exotic or Invasive: Who Is Who?

What do ecologists mean when they describe a species as native or exotic? Is making this distinction important? The vast majority of invasion ecologists and environmental managers consider the status of a species (native or exotic) as the key attribute that we need to know. Although opinions may differ, most people agree that native (indigenous) species are those that originated in a given area. Conversely, exotic (alien; introduced; non-native; non-indigenous) species are present in a given area due to intentional or unintentional human involvement. The classification as native or exotic depends on the scale at which the assessment is made. For example, a species can be considered alien to the whole continent if it arrived from overseas or to its part if it is native in a different region on the same continent. Distinction between native and exotic is not, however, fully consistent in different parts of the world and changes according to dates of human intervention. In Europe, species are considered native if they were present before the Neolithic period (the onset of widespread crop growing), but in the Americas, the year 1492 (the discovery of America by Europeans) is generally considered the demarcation point. Invasion ecologists in Europe further distinguish exotic plants into archaeophytes, introduced since the beginning of Neolithic agriculture until the end of the middle Ages, and neophytes, the modern invaders introduced after the discovery of America in 1492. This division is supported by the fact that both groups have distinct habitat characteristics, a response to climate and invasion patterns; most serious invasive plants of today recruit from neophytes, whereas archaeophytes are often agricultural weeds. For pragmatic reasons, native species expanding their distribution ranges and spreading into new habitats are not called invasive (this term is reserved for species introduced by humans) but instead are described as expansive. The proper terms referring to their behavior are migration, spread, range expansion, or range extension.

There are two approaches to defining invasive species. The first one, more scientifically rigorous, breaks the process of a biological invasion into a series of consecutive stages (the so-called introduction–naturalization–invasion continuum), defined on the basis of barriers a species needs to negotiate or overcome to reach the next stage. The process begins with transport and introduction, followed by the establishment of self-reproducing populations, and in the final stage, the species spreads through the landscape and reaches a high abundance in invaded communities. Within this framework, an invasive

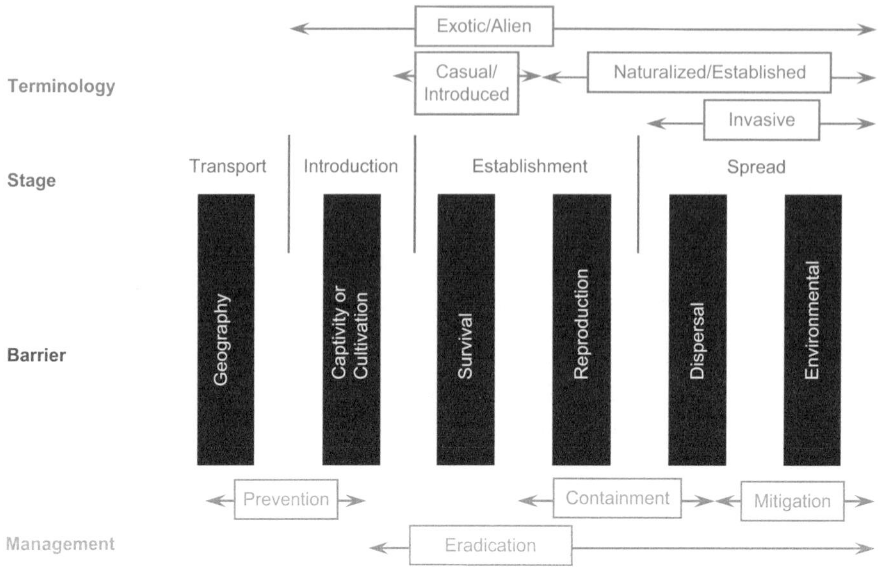

FIGURE 11.1 A scheme showing individual geographic and environmental barriers with the corresponding stages a species passes along the introduction–naturalization–invasion continuum, as its status changes from casual to naturalized to invasive. Management intervention possibilities along the stages of the process are indicated.
Source: Redrawn from Blackburn et al.[1]

species is defined based on ecological criteria and species population dynamics, but the definition does not distinguish the type of habitat invaded (natural vs. human made). The second perspective, more pragmatic, considers biodiversity [e.g., Convention on Biological Diversity, Article 8 (h)] and labels invasive only those exotics that threaten native ecosystems, habitats, species, or humans and their activities. This definition, however, relies on detection of a negative impact by the invading exotic species which is problematic to define and difficult to measure (see section Impact of exotic species). In this entry, we follow the ecological definition.[1–3]

Not all introduced species are successful; only a fraction of each species reaches the next stage of invasion as they overcome barriers or filters (Figure 11.1). This is reflected in a general rule of thumb called the "Tens Rule," which posits that approximately 10% of introduced species are able to escape from cultivation or captivity, 10% of the escapees can establish, and only 10% of these established species exert economic impact (termed "weeds" or "pests").[4] Exotic species that rely on repeated introductions for their persistence, flourish and reproduce occasionally outside cultivation but eventually die out because they do not form self-replacing populations and are called casuals (subspontaneous, occasional escapes, ephemeral taxa). Those that overcome abiotic and biotic barriers to survival and regular reproduction are called naturalized (established). Invasive species are a subset of naturalized species that are able to spread far from their place of introduction and often become abundant in invaded communities.

In addition, the anthropocentric view generates more terms like pests, weeds, harmful species, problem species, or noxious species for those that are not necessarily alien but grow in sites where they are not wanted and have detectable economic and/or environmental impact.[1–3]

What Makes a Species Invasive?

There are three classical questions in invasion biology: Which traits make the species invasive? What are the attributes of a community that make it more or less prone to invasions? And what is the impact of an alien species and how can it be diminished? A number of hypotheses and frameworks have emerged in the last two decades that attempt to identify why some species become successful and others do not.[5,6] The most sophisticated hypotheses reflect the fact that establishment and invasion are complex processes that involve many factors (Figure 11.2). Some of the concepts relate to species invasiveness, whereas others address habitat invasibility. The most popular of the former is the Enemy Release Hypothesis, which suggests that upon introduction, exotic species are freed from their natural enemies with common evolutionary history, such as parasites, pathogens, herbivores, or predators that exert pressure and thus regulate their populations in the native range.[7] This provides them with a competitive advantage in the new range. It has also been suggested and widely tested that populations respond to the lack of enemies in the process of postinvasion evolution, by allocating resources no longer needed for defense to growth and reproduction, thus increasing fitness [formalized in the Evolution of Increased Competitive Ability (EICA) Hypothesis].[8] The phenomenon of enemy release is a basis for classical biocontrol where natural enemies from the native range are deliberately introduced to the new range to control invading populations.[9,10]

The Fluctuating Resources Hypothesis suggests that invasion of a new species is only possible when resources are available. The level of resources in an ecosystem is determined by availability resulting from disturbances and uptake from resident species. Therefore, high levels of disturbance combined with high levels of ecosystem productivity may lead to successful invasion or even expansion by native species, especially in the early stages of succession. By relating the availability of resources to the ability of a species to capitalize on free resources, the hypothesis introduces a link between the invasibility of a community and species traits.[11]

Since Darwin, it has been proposed that species-rich communities do not harbor as many exotic species as species-poor communities. This relationship between species diversity and invasibility has been thoroughly investigated within the framework of the Biotic Resistance Hypothesis.[12] It is based on

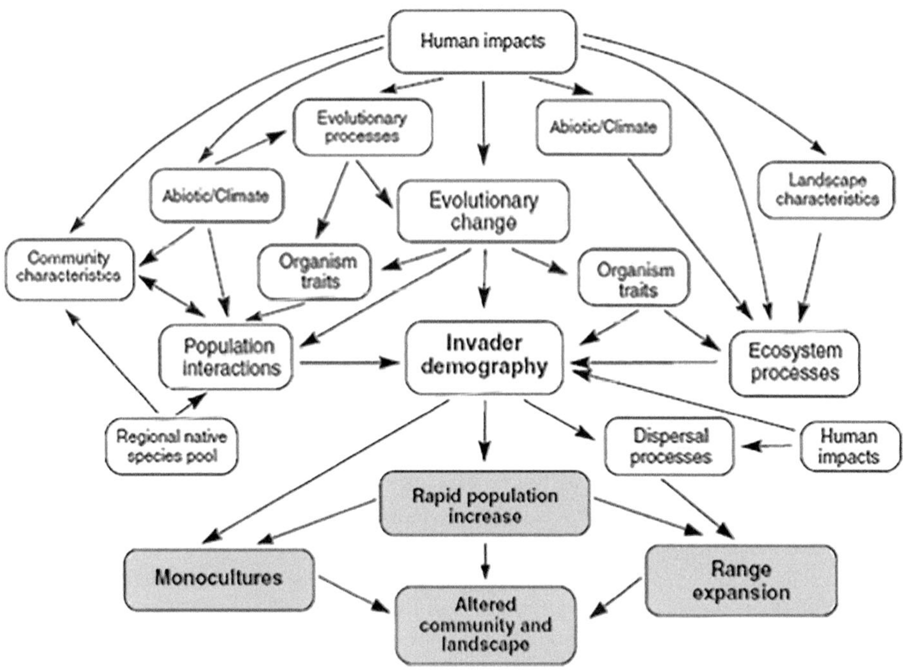

FIGURE 11.2 A conceptual synthetic invasion framework capturing fundamental ecological and evolutionary processes that interact to determine the outcome of invasion. The demography of the invader is central to the concept; in successful invaders, it results in population increase and range expansion and affects community and landscape characteristics.
Source: From Gurevitch et al.[6]

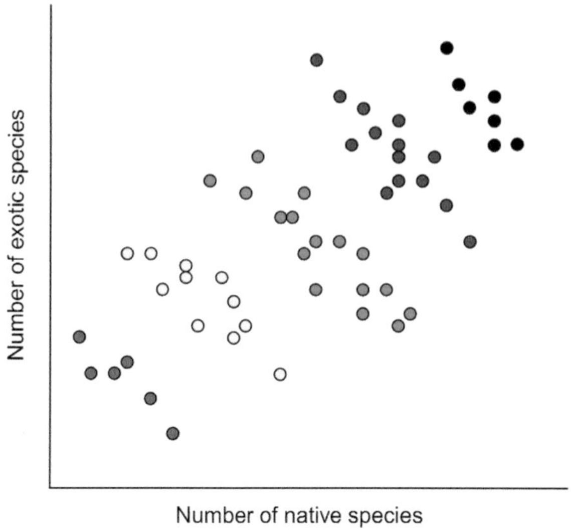

FIGURE 11.3 Alien–native species richness relationship. Within clusters with the same environmental conditions, the correlation is negative but in general, the relationship is positive across the clusters.
Source: From Shea & Chesson.[14]

a notion that high diversity of native species results in lower niche opportunities for invaders, making the invasion less likely. However, it is now understood that the relationship between native and exotic species richness is scale dependent. Competition and niche availability results in negative relationship at the community scale, but on larger scales, there is a positive relationship because both groups of species respond to major environmental factors affecting species richness in the same way (Figure 11.3).[13,14]

The search for biological and ecological traits that distinguish successful exotic species from unsuccessful ones, or exotic from native species, has been an important line of research since the onset of invasion ecology in the 1980s and was therefore transformed into modern and sophisticated pest- or weed-risk assessment schemes (PRA, WRA).[15–17] It is now widely acknowledged that species traits related to reproduction, growth, and dispersal affect the outcome of the invasion. Recent analyses of large data sets point out that invasive and non-invasive plant species differ in traits such as those related to physiology, leaf-area allocation, shoot allocation, growth rate, size, and fitness.[18–20]

However, the role of species traits is context dependent. Different traits are important at different stages of the process, and their effects interact with many other factors related to invasion dynamics such as initial population size, residence time (how long ago the species was introduced), propagule pressure (the intensity with which the species is introduced to the region, e.g., the number of introduction events, extent of planting), climate match between source and target area, availability of habitats determining the spatial distribution, or the vectors and pathways of introduction that may differ in their efficiency. It is a complex interaction between these factors that codetermines whether a species becomes invasive. For example, it is now widely acknowledged that the effect of residence time needs to be controlled for in analyses because the longer the alien species is present in a territory, the greater their chance to pass successfully through the stage of casual occurrence and become naturalized, more widely distributed, and invade over a larger range.[21–24]

Impact of Exotic Species

As predicted by the Tens Rule, the majority of exotic species have little or no measurable impact. In Europe, about 5% of naturalized (established) exotic species are documented to have impact (Table 11.1)[25] and it is estimated that 20–30% of the species introduced into the United States, United

TABLE 11.1 Total Numbers and Percentages of Alien Species Known to Have an Ecological or Economic Impact for Different Taxonomic Groups in Europe

		Species with	
Taxonomic Group	No. of Exotic Species	Ecological Impact	Economic Impact
Terrestrial plants	5789	326 (5.6)	315 (5.4)
Terrestrial invertebrates	2481	342 (13.8)	601 (24.2)
Terrestrial vertebrates	358	109 (30.4)	138 (38.5)
Freshwater flora and fauna	481	145 (30.1)	117 (24.3)
Marine flora and fauna	1071	172 (16.1)	176 (16.4)

Source: From Vilà et al.[25]
Abbreviation: DAISIE, Delivering Alien Invasive Species In Europe.
Numbers in brackets show percentages from recorded species in the DAISIE database (http://www.europe-aliens.org).

Kingdom, Australia, India, South Africa, and Brazil are pests and cause major environmental problems. Similarly, only relatively few exotic species become serious pests with significant damage to natural and managed ecosystems or cause health problems to humans.[26] However, reliable information on how many invasive species exert significant impact is still lacking for most regions of the world due to the difficulties linked to defining what an impact actually is and how to detect it. This is because there is no universal measure of impact that is context dependent and multifaceted.[27] For example, the relatively small impact of the tall and fast-growing Asian plant Impatiens glandulifera invading in Europe can be attributed to the fact that the riparian communities colonized are dominated by native tall nitrophilous species, and therefore the impact of the invader on species diversity is not much different from the competitive effect of these native dominants on resident species. However, by affecting the behavior of insect pollinators, this species decreases the fitness of co-occurring native plants.[28,29]

The impacts of exotic species range from those on the diversity of native communities to the survival of rare species or ecosystem functioning to human health, predation by introduced pests, habitat transformation, or fire regime (Figure 11.4).[2,19,25,27] Some of the impacts are possible to quantify easily irrespective of whether they affect native or exotic species or are directly affecting the economic sphere.

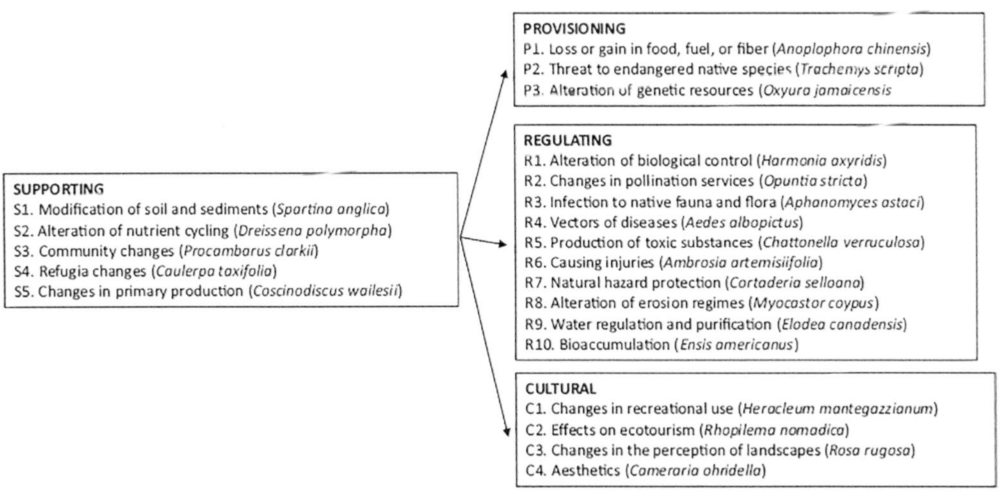

FIGURE 11.4 Impact of exotic species on ecosystem services classified into four categories. Individual types of impact are followed by corresponding example species.
Source: From Vilà et al.[25]

Although the ecological and economic impacts of alien species are usually studied separately, they are closely correlated. Because of the lack of existing data on impacts on other services, current economic valuations focus primarily on provisioning ecosystem services)[30] Overall, the impact of exotic species transformed into economic terms has been estimated to reach 5% of the gross domestic product (GDP),[26] and estimates of costs of invasions by exotic species also exist for world regions. For Europe, they exceed €12.7 billion annually)[31] and for the United States, the figure of $30 billion per year was reported, with most of the costs associated with agriculture)[26] It should be noted that introduced species with the highest documented negative impact and resulting costs are human pathogens, including the bubonic plague or HIV causing AIDS).[30]

Prevention, Control, Management, and Risk Assessment

Pathways, the different ways by which alien species become introduced, not only affect the probability of successful species invasion but also represent a promising tool to prevent invasions by pathway management. Some species are associated with a unique pathway, whereas others are introduced by multiple means)[32,33] The knowledge of invasion pathways is therefore crucial for developing preventive methods such as screening systems, interception programs, early warning strategies, and import regulations.[34–36] Alien species may enter the new region as a direct or indirect result of human activities that can be classified under three broad mechanisms: the importation of a commodity, the arrival of a transport vector, and/or natural spread from a neighboring region where the species is alien. The three mechanisms have been incorporated into a framework consisting of six principal pathways that reflect the degree of human involvement: exotic species are introduced by release, escape, contaminant, stowaway, corridor, and unaided (Figure 11.5). Such a framework is applicable to a wide range of taxa

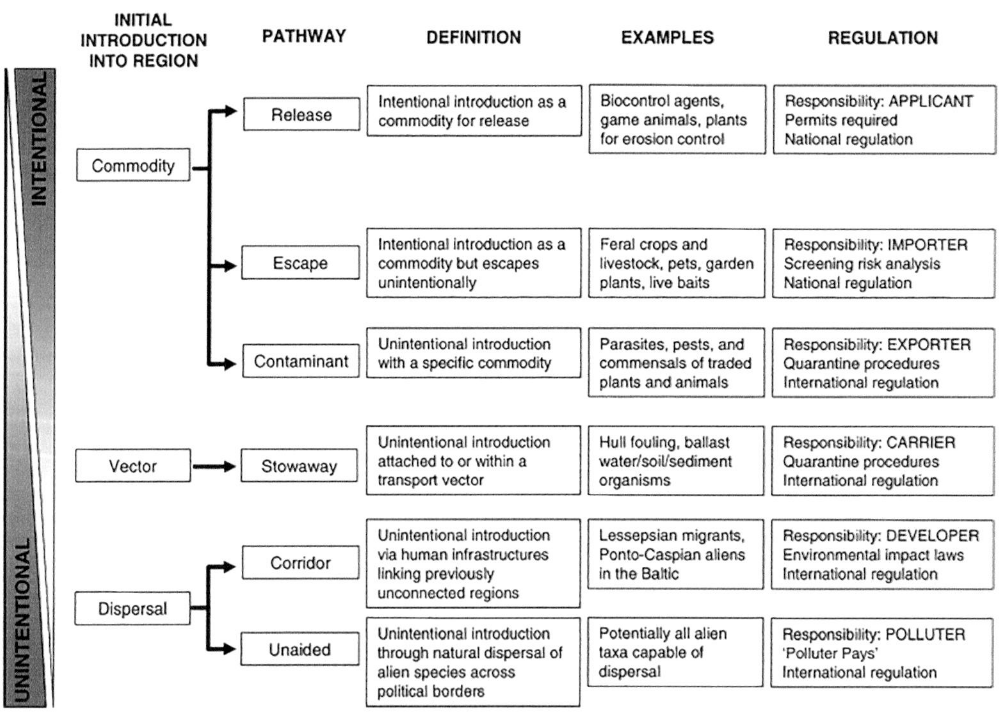

FIGURE 11.5 A framework categorizing existing pathways of introduction of alien species with indication of the mechanisms of introduction, brief description of the pathway and possible ways of regulation.
Source: From Hulme et al.[35]

and invaded environments and can be used as a basis for developing legislation and codes of practice for different sectors responsible for alien introductions.[35]

Prevention is the cheapest and most preferable method of defense against alien species. Although only a few exotic species become environmental or economic pests, it would be plausible to screen all potential invaders at national borders or before approving import, should those imposing a high risk of invasion be identified. However, the huge numbers of species being introduced for ornamental purposes by horticultural trade makes full screening on a regular basis impossible. Similarly, the magnitude of transported material over borders greatly reduces the efficiency of preventive screening controls. Nevertheless, a range of methods are generally used in assessing the invasion risk; the most common are based on traits of exotics that proved to successfully invade elsewhere, taking also the climate and habitat matching between the native and invaded distribution ranges into account. Risk assessments (RAs) aim at determining the likelihood of risks associated with the introduction of a species to a new region as well as the probability of it becoming naturalized and exerting significant impacts. In the last several years, the precision of RA schemes have improved due to accumulation of data in large species databases so that they are used in some regions in preborder screening protocols or become a part of legal instruments and policies.[1,10,17,37,38]

When dealing with a suite of exotic species that have successfully established in a region, it is important to realize that it is impossible and implausible to attempt at managing all exotics within the given area. This points to the necessity of differentiating between exotic species in general and the subset of those that are invasive and exert impact; most invasion ecologists are aware that only the latter need to be targeted by management measures and that there is no reason for xenophobia against all exotics or a fundamentalistic fight against all species of a foreign origin. This does not mean that the precautionary principle should be discarded, especially when assessing new introductions. Here again, RA represents a cost-effective approach toward deciding which species, where and how should be eradicated or mitigated. In the first place, rapid response to emerging invasions is crucial for eradication success; small and isolated populations are easier to manage.[39,40] Full eradication, i.e., the extirpation of an entire population of a species including its dormant stages such as soil seed banks of invasive plants, is usually not feasible; therefore, partial mitigation or eradication from a small defined area (e.g., a nature reserve) is often the target. If eradication is not possible, containment should be considered as another management option to minimize the impact of the target alien species.

Several methods of managing invasive exotic species are in use; mechanical (e.g., in plants: mowing, cutting, pasturing; animals: hunting, physical removing of alien pests) and chemical (plants: usage of herbicides; animals: poisoning or usage of protective coat) control or their combination, and biocontrol. Biocontrol is a deliberate import and release of potential natural enemies of the targeted pest or weed, termed biocontrol agents. Many exotic species lack their natural enemies in a new area, especially introduced crop or ornamental plant species are imported free of insects, mites, pathogens, and diseases. Although biocontrol has been widely used in greenhouses around the world, and there are numerous success stories in the wild, too, it is still regarded as controversial in open environments, especially in some continents. The reason for this skepticism is that an additional exotic species introduced into a system may itself become invasive, attack non-target native species, and exert various side effects. Examples of animals getting out of control and becoming invasive themselves include the cane toad (Rhinella marina), which was introduced to Australia to control sugar cane pests, or the harlequin ladybug (Harmonia axyridis), which was introduced to North America and Europe for the control of aphids and coccids. In the field of plant biocontrol, there is a well-known case of head weevil (Rhinocyllus conicus) that attacked the native Cirsium species in the United States or that of the cactus moth (Cactoblastis cactorum) impacting native Opuntia cacti in Florida and Mexico.[41,42] On the other hand, compared to the classical mechanical or chemical methods, biocontrol provides a sustainable and long-term management tool because once the agent is released, it reproduces itself in the ecosystem and regulates the pest. Even if the biocontrol does not lead to complete eradication of the targeted pest or weed, it may lower its abundance to an acceptable level.[43] It has to be borne in mind, however, that eradication or containment of the invading species creates an empty niche in the invaded community and that restoration of the managed ecosystem needs to follow eradication to prevent reinvasion and secondary invasions of other species. The restoration aims at reintroduction of native species through improving habitat conditions, supporting the survival and establishment of introduced native species, or directing succession toward desired target communities.[44]

Exotic Species Lists and Data Sources

There are two major kinds of lists of alien species: more or less complete inventories for defined regions, and shorter and focused species lists of important invaders, depending on expert opinion and management viewpoint. The former are very useful for early warning systems, prioritization of management, or developing the decision schemes for PRA or WRA. The latter lists of the "worst" invasive species that pose a major threat to native diversity or interfere with human interests include global accounts such as the IUCN (International Union for Conservation of Nature) list of the world's 100 worst invasive alien species[45] or continental datasets such as the DAISIE (Delivering Alien Invasive Species In Europe) list of "one hundred of the most invasive alien species in Europe" for Europe, or the SEBI (Streamlining European Biodiversity Indicators) list of 166 "worst invasive alien species threatening biodiversity in Europe."[46,47] The aim of such lists is to highlight the problems associated with invasive species and illustrate various pathways of introduction, taxon groups, invaded habitats, and impacts. Besides complete lists of alien species, the so-called "black" or "gray lists" are being compiled, including invasive species that are banned to be sold, planted, or used in forestry or agriculture. Such lists of noxious weeds have a long history of use in the United States and Australia at both state and federal levels.[48,49]

Several milestones have led to the establishment of invasion biology itself and have initiated and stimulated the development of the field. Although Elton's 1958 book The Ecology of Invasions by Animals and Plants is generally thought to have sparked the interest in biological invasions, the real stimulus was the SCOPE (Scientific Committee on Problems of the Environment) program addressing invasive species in the 1980s that generated synthetic publication from many regions of the world.[50] Currently, there are many books dealing with biological invasions ranging from popular science to textbooks. Several scientific journals specialize in biological invasions and their consequences (e.g., Biological Invasions, Diversity and Distributions, NeoBiota, International Journal of Plant Management, Weed Management, BioControl). Many other journals also publish articles on biological invasions and the total number of papers published on this topic each year is estimated at several thousands, and still increasing.

Conclusions

Considering the fact that both intentional as well as unintentional introductions are important for all taxonomic groups, we can expect that based on current levels of international trade, human travel, and quarantine procedures, the trend of increasing alien introductions will continue at least at current levels. Even if new introductions cease, we need to be aware that some of the already-present not-yet-established casual exotics will become naturalized and some of the currently naturalized species will became invasive pests and weeds.[51] This reflects the phenomenon of lag phases typically associated with invasion dynamics, and that of the so-called invasion debt, reflecting that exotic species respond to human activities with a delay of several decades.[52,53] Other important factors that are likely to influence the trends of species introductions in the future and their ability to naturalize are climate and land-use changes together with agronomy and forestry, which are responsible for the introduction of new drought-resistant crops, as well as increasing stress on renewable resources of energy (new fast-growing species).[54] Actual risks that have to be considered with respect to newly introduced species and nature protection are the increased use of genetically modified organisms (GMOs), which creates the possibility of hybridization with their wild relatives or transfer of herbicide resistance to weeds. Therefore, with respect to the distribution and impact of already naturalized species, it is crucial to prioritize management, enhance the role of early warning systems and rapid response, and be more careful in the sense of precautionary measures when intentionally introducing species.[10] A key challenge is to improve the cooperation between different bodies responsible for introduction and management of alien species (agriculture, forestry, and fisheries) and those whose role is to protect nature both at the national and international levels.[51]

Acknowledgments

We thank two anonymous reviewers for valuable comments and Laura Meyerson for editing our English. Work on this contribution was funded by the long-term research development project no. RVO 6798593, Praemium Academiae award from the Academy of Sciences of the Czech Republic and project P504/11/1028.

References

1. Blackburn, T.M.; Pyšek, P.; Bacher, S.; Carlton, J.T.; Duncan, R.P.; Jarošik, V.; Wilson, J.R.U.; Richardson, D.M. A proposed unified framework for biological invasions. Trends Ecol. Evol. 2011, 26, 333–339.
2. Lockwood, J.L.; Hoopes, M.F.; Marchetti, M.P. Invasion Ecology; Blackwell Publication: 2007; 466 pp.
3. Richardson, D.M.; Pyšek, P.; Rejmánek, M.; Barbour, M.G.; Panetta F.D.; West, C.J. Naturalization and invasion of alien plants: concepts and definitions. Divers. Distrib. 2000, 6, 93–107.
4. Williamson, M.; Fitter, A. The varying success of invaders. Ecology 1996, 77, 1661–1666.
5. Catford, J.A.; Jansson, R.; Nilsson, C. Reducing redundancy in invasion ecology by integrating hypotheses into a single theoretical framework. Divers. Distrib. 2009, 15, 22–40.
6. Gurevitch, J.; Fox, G.A.; Wardle, G.M.; Inderjit; Taub, D. Emergent insights from the synthesis of conceptual frameworks for biological invasions. Ecol. Let. 2011, 14, 407–418.
7. Blossey, B.; Nötzold, R. Evolution of increased competitive ability in invasive nonindigenous plants: a hypothesis. J. Ecol. 1995, 83, 887–889.
8. Willis, A.J.; Memmott, J.; Forrester, R.I. Is there evidence for the post-invasion evolution of increased size among invasive plant species? Ecol. Let. 2000, 3, 275–283.
9. McFadyen, R.E.C. Biological control of weeds. Ann. Rev. Entomol. 1998, 43, 369–393.
10. Pyšek, P.; Richardson, D.M. Invasive species, environmental change and management, and health. Ann. Rev. Envir. Res. 2010, 35, 25–55.
11. Davis, M.A.; Grime, J.P.; Thompson, K. Fluctuating resources in plant communities: a general theory of invasibility. J. Ecol. 2000, 88, 528–536.
12. Elton, C.S. The Ecology of Invasions by Animals and Plants. (reprinted 2000, University of Chicago Press): London, 1958.
13. Herben, T. Species pool size and invasibility of island communities: a null model of sampling effects. Ecol. Let. 2005, 8, 909–917.
14. Shea, K.; Chesson, P. Community ecology theory as a framework for biological invasions. Tr. Ecol. Evol. 2002, 17, 170–176.
15. Gordon, D.R.; Riddle, B.; Pheloung, P., Ansari, S.; Buddenhagen, C.; Chimera, C.; Daehler, C.C.; Dawson, W.; Denslow, J.S.; Jaqualine, T.N.; LaRosa, A.; Nishida, T.; Onderdonk, D.A.; Panetta, F.D.; Pysek, P.; Randall, R.P.; Richardson, D.M.; Virtue, J.G.; Williams, P.A. Guidance for addressing the Australian Weed Risk Assessment questions. Plant Prot. Q. 2010, 25, 56–74.
16. Kolar, C.S.; Lodge, D.M. Ecological predictions and risk assessment for alien fishes in North America. Science 2002, 298, 1233–1236.
17. Pheloung, P.C.; Williams, P.A.; Halloy, S.R. A weed risk assessment model for use as a biosecurity tool evaluating plant introductions. J. Environ. Manag. 1999, 57, 239–251.

18. Grotkopp, E.; Rejmánek, M.; Rost, T.L. Towards a causal explanation of plant invasiveness: seedling growth and life- history strategies of 29 pine (Pinus) species. Am. Nat. 2002, 159, 396–419.
19. Pyšek, P.; Richardson, D.M. Traits associated with invasiveness in alien plants: where do we stand? In Biological Invasions; Nentwig, W., Ed.; Springer-Verlag: Berlin, Heidelberg, 2007; 97–125.
20. van Kleunen, M.; Weber, E.; Fischer M. A meta-analysis of trait differences between invasive and non-invasive plant species. Ecol. Let. 2010, 13, 235–245.
21. Phillips, M.M.; Murray, B.R.; Leishman, M.R.; Ingram, R. The naturalization to invasion transition: Are there introduction-history correlates of invasiveness in exotic plants of Australia? Austral Ecol. 2010, 35, 695–703.
22. Pyšek, P.; Jarošík, V. Residence time determines the distribution of alien plants. In Invasive Plants: Ecological and Agricultural Aspects; Inderjit, Ed.; Birkhäuser Verlag-AG: Basel, 2005; 77–96.
23. Pyšek, P.; Křivánek, M.; Jarošík, V. Planting intensity, residence time, and species traits determine invasion success of alien woody species. Ecology 2009, 90, 2734–2744.
24. Williamson, M.; Dehnen-Schmutz, K.; Kühn I.; Hill, M.; Klotz, S.; Milbau, A.; Stout, J.; Pyšek, P. The distribution of range sizes of native and alien plants in four European countries and the effects of residence time. Divers. Distrib. 2009, 15, 158–166.
25. Vilà, M.; Basnou, C.; Pyšek, P.; Josefsson, M.; Genovesi, P.; Gollasch, S.; Nentwig, W.; Olenin, S.; Roques, A.; Roy, D.; Hulme, P.E.; DAISIE partners How well do we understand the impacts of alien species on ecological services? A pan-European cross-taxa assessment. Front. Ecol. Envir. 2010, 8, 135–144.
26. Pimentel, D.; McNair, S.; Janecka, J.; Wightman, J.; Simmonds, C.; O'Connell, C.; Wong, E.; Russel, L.; Zern, J.; Aquino, T.; Tsomondo, T. Economic and environmental threats of alien plant, animal, and microbe invasions. Agric. Ecosyst. Envir. 2001, 84, 1–20.
27. Levine, J.M.; Vilà, M.; D'Antonio, C.M.; Dukes, J.S.; Grigulis, K.; Lavorel, S. Mechanisms underlying the impacts of exotic plant invasions. Proc. R. Soc. London, B 2003, 270, 775–781.
28. Chittka, L.; Schürkens, S. Successful invasion of a floral market. Nature (London) 2001, 411, 653.
29. Hejda, M.; Pyšek, P.; Jarošík V. Impact of invasive plants on the species richness, diversity and composition of invaded communities. J. Ecol. 2009, 97, 393–403.
30. Perrings, C.; Mooney, H.; Williamson, M. Bioinvasions & globalization. Ecology, Economics, Management, and Policy; Oxford University Press: 2010; 247–258.
31. Kettunen, M.; Genovesi, P.; Gollasch, S.; Pagad, S.; Starfinger, U.; ten Brink, P.; Shine, C. Technical support to EU strategy on invasive species (IAS)—Assessment of the impacts of IAS in Europe and the EU (final module report for the European Commission). Institute for European Environmental Policy (IEEP), Brussels, Belgium; 2008.
32. Nentwig, W. Pathways in animal invasions. In Biological Invasions, Ecological Studies, Vol. 193; Nentwig, W., Ed.; Springer-Verlag: Berlin Heidelberg, 2007; 11–27.
33. Gollasch, S. Is ballast water a major dispersal mechanism for marine organisms? In Biological Invasions; Nentwig, W., Ed.; Springer-Verlag: Berlin, Heidelberg, 2007; 48–57.
34. Carlton, J.T.; Ruiz, G.M. Vector science and integrated vector management in bioinvasion ecology: conceptual frameworks. In Invasive Alien Species; Mooney, H.A.; Mack, R. N.; McNeely, J.A.; Neville, L.E.; Schei, PJ.; Waage, J.K., Eds.; Island Press: Washington, DC, 2005; 36–58.
35. Hulme, P.E.; Bacher, S.; Kenis, M.; Klotz, S.; Kühn, I.; Minchin, D.; Nentwig, W.; Olenin, S.; Panov, V.; Pergl, J.; Pysek, P.; Roque, A.; Sol, D.; Solarz, W.; Vilà, M. Grasping at the routes of biological invasions: a framework for integrating pathways into policy. J. Appl. Ecol. 2008, 45, 403–414.
36. Lockwood, J.L.; Cassey, P.; Blackburn, T.M. The more you introduce the more you get: the role of colonization pressure and propagule pressure in invasion ecology. Divers. Distrib. 2009, 15, 904–910.
37. Touza, J.; Dehnen-Schmutz, K.; Jones, G. Economic analysis of invasive species policies. In Biological Invasions; Nentwig, W., Ed.; Springer-Verlag: Berlin, Heidelberg, 2007; 353–366.

38. Hallman, G.J. Phytosanitary measures to prevent the introduction of invasive species. In Biological Invasions; Nentwig, W., Ed.; Springer-Verlag: Berlin, Heidelberg, 2007; 367–384.
39. Pluess, T.; Cannon, R.; Jarošík, V.; Pergl, J.; Pyšek, P.; Bacher, S. When are eradication campaigns successful? A test of common assumptions. Biol. Invas. 2012, 14, 1365–1378.
40. Veitch, C.R.; Clout, M. Turning the tide: the eradication of invasive species. IUCN SSC Invasive Species Specialist Group, Gland and Cambridge, 2002.
41. Frank, J.H.; Van Driesche, R.G.; Pitcairn, M.J. Biological control of animals. In Encyclopedia of Biological Invasions; Simberloff, D.; Rejmánek, M., Eds.; University of California Press: 2011; 58–63.
42. Hoddle, M.S.; McCoy, E.D. Biological control of plants. In Encyclopedia of Biological Invasions; Simberloff, D.; Rejmánek, M., Eds.; University of California Press: 2011; 63–70.
43. Knochel, D.G.; Seastedt, T.R. Sustainable control of spotted knapweed. In Management of Invasive Weeds; Inderjit, Ed.; Springer: 2009; 211–225.
44. Hobbs, R.J.; Richardson, D.M. Invasion ecology and restoration ecology: parallel evolution in two fields of endeavour. In Fifty Years of Invasion Ecology: The Legacy of Charles Elton; Richardson, D.M., Ed.; Blackwell: 2011; 61–69.
45. Lowe, S.; Browne. M.; Boudjelas, S.; De Poorter, M. 100 of the World's worst invasive alien species. A selection from the Global Invasive Species Database. ISSG, 2000.
46. EEA. SEBI Assessing biodiversity in Europe: the 2010 report. EEA, Copenhagen, 2010.
47. DAISIE. Handbook of Alien Species; Springer; Europe, 2009.
48. http://plants.usda.gov/java/noxiousDriver (accessed March 2012)
49. http://www.weeds.org.au/noxious.htm (accessed March 2012)
50. Simberloff, D. Charles Elton: neither founder nor siren, but prophet. In Fifty Years of Invasion Ecology: The Legacy of Charles Elton; Richardson, D.M., Ed.; Blackwell, 2011; 11–24.
51. Hulme, P.; Pyšek, P.; Nentwig, W.; Vilà, M. Will threat of biological invasions unite the European Union? Science 2009, 324, 40–41.
52. Aikio, S.; Duncan, R.P.; Hulme, P.E. Lag-phases in alien plant invasions: Separating the facts from the artefacts. Oikos 2010, 119, 370–378.
53. Essl, F.; Dullinger, S.; Rabitsch, W.; Hulme, P.E.; Hülber, K.; Jarošík, V.; Kleinbauer, I.; Krausmann, F.; Kühn, I.; Nentwig, W.; Vilà, M.; Genovesi, P.; Gherardi, F.; Desprez- Lousteau, M.-L.; Roques, A.; Pyšek, P. Socioeconomic legacy yields an invasion debt. Proc. Natl. Acad. Sci. USA 2011, 108, 203–207.

12

Keystone and Indicator Species

Barry R. Noon
Colorado State University

Introduction 105
Indicator Species 106
Keystone Species 107
Expected Values and Benchmarks for Ecological Indicators 108
Conclusions 109
References 110

Introduction

A surrogate-based approach to assessing the effects of land use and land change on biological diversity is a pragmatic necessity. It is impossible to manage or monitor changes directly in the state of populations of all the species that we are legally obligated to conserve or that are important to us for economic, aesthetic, or ethical reasons. Current approaches to sustain plant and animal diversity at broad spatial scales combine coarse-filter and fine-filter strategies.[1] The coarse-filter approach, operative at a landscape scale, proposes to sustain biological diversity indirectly by maintaining the diversity of habitats (i.e., vegetation communities and their successional stages) expected under the historic range of variability for that landscape. In contrast, the fine-filter approach is focused at the species level and is intended to address the needs of species not adequately protected by the coarse-filter strategy.[2] Indicator and keystone species concepts are examples of fine-filter strategies.

The great diversity of plant and animal species defies any simple prioritization. For managers and conservation biologists charged with sustaining biological diversity in human-dominated landscapes, some simplifying characterizations are needed. One uncomplicated classification is to characterize species by 1) their utilitarian value, 2) their information content—what do changes in the status and trends of their population tell us about the state of the ecosystems to which they belong, or 3) their functional role in ecological systems—how, and to what extent, does a given species contribute to key processes. Those with obvious utilitarian value would include species with direct economic benefits such as many commercial marine fish species and species hunted for subsistence consumption or recreation. Species that provide information beyond their own measurement may act as sentinels of ecosystem change and thus provide early warning of impending adverse system effects prior to unacceptable or irreversible state changes. Species with known and important functional roles would include, for example, dominant plant species with high levels of primary productivity, consumer species which provide top-down regulation of food webs, and species that alter the structure and composition of ecosystems and thereby promote high levels of species diversity.

The goal of this entry is to discuss the characteristics of two categories of species mentioned in the preceding text. First, those species that provide significant insights into the state of the ecological systems in which they reside, often indicating impending state changes and generically referred to as indicator species. Second, those species with highly significant functional roles in the structuring and stability of ecological systems, referred to as keystone species.[3]

Indicator Species

I view the indicator species concept broadly—that is, they are species whose status and trends provide insights into environmental conditions that extend beyond their own measurement.[3,4] Indicator species can be used as surrogates for ecosystem processes or conditions that are too difficult or costly to measure directly. The primary role of ecological indicators, in general, and indicator species in particular, is to measure the response of the ecosystem to anthropogenic disturbance.[5] Selection of indicator species is a critical aspect of most environmental monitoring programs. For example, indicator species with known stress–response relationships are often chosen as early warning sentinels of impending stressor effects so as to avoid undesirable or irreversible environmental outcomes.

There is no topology that clearly discriminates among the various types of indicator species identified by environmental scientists—for example, keystone, umbrella, focal, ecological engineers, and flagship species all have attributes that both overlap with one another[6] and can possibly qualify them as ecological indicators. Selection of indicator species is difficult and is often based more on their ease of measurement or historic precedent than their ecological importance or responsiveness to management.[7] Fortunately, there is now an extensive literature on the process of indicator species selection applicable to a wide range of resource management issues.[4,7–10]

The concept of indicator species was originally associated with the assumption that the status and trends of the indicator reflected the dynamics of species with similar ecological requirements.[11] This assumption has been strongly criticized,[6,11–14] and current understanding is that species surrogacy can only be assumed in limited cases where a set of species share a common set of environmental drivers.[14] Though surrogacy for unmeasured species within a study or management area is no longer a tenable assumption for indicator species, they remain useful to the extent that they reflect the biotic or abiotic state of the environment, reveal the impacts of environmental change, or provide indirect insights into the diversity of other species in the area.[5,15] For many natural resource agencies, the judicious selection of indicator species is an essential component of their environmental monitoring programs to assess the effects of management.[1]

Not all candidate indicator species are equally informative—one of the key challenges to environmental assessment and monitoring is to select for measurement species that will provide insights into ecological integrity at the scale of the ecosystem. An ecosystem with integrity is defined as one that has the capacity to support and maintain an integrated, adaptive community of organisms having a full range of components (genes, species, communities) and processes (birth, survival, movement, biotic interactions, energy dynamics) expected from the natural habitat of the region.[16,17] Table 12.1 lists a set of indicator species properties that reflect changes in a system's ecological integrity and make them useful to policy makers.[3]

TABLE 12.1 Desirable Properties of Indicator Species

- Species dynamics parallel those of the ecosystem to which they belong
- Species show a short-term but persistent response to change in ecosystem state
- Species status and trends can be accurately and precisely estimated
- Significant changes in the state of the indicator species can readily be distinguished from change induced by background variation in ecosystem state
- Are measurable in a cost-effective manner
- Are relevant to important societal values
- Are understandable by multiple audiences (scientists, policy-makers, managers, and the public)

In diverse ecosystems, multiple indicator species will often be selected for measurement[18] Collectively, the set of species selected for monitoring should be complementary and comprehensive.[4] Indicator species are complementary if their information content is largely independent of other species in the set. The set is comprehensive if they collectively have ecologies and life history attributes that span the biological diversity of the ecosystem and encompass the system's temporal and spatial domain. These properties are difficult to achieve in a minimal indicator set but are useful goals to strive for. Given the complexity and dimensionality of an ecosystem, a surrogate-based approach to ecosystem assessment based on a small number of indicator species is often a pragmatic requirement.[6,19]

Selection of the indicator species set is the first step. Use of an indicator species in a monitoring program, for example, requires that a specific variable reflecting the state of the indicator species be identified and measured. That is, will the state of the indicator species be assessed by its distribution, abundance, density, birth rate, or some other demographic attribute? Interpreting the state of an indicator species so as to make inferences about the state of the environment in which it resides also requires a statistically defensible survey design.[3,20–22]

Keystone Species

It is obvious that abundant species play major ecological roles in controlling the rates of ecosystem processes simply as a result of their numerical or material dominance. Keystone species differ from dominant species in that their effects are much greater than would be predicted from their abundance.[23] Power et al.[23] measure a species' impact by its community importance (CI)—that is, the observed change in a community or ecosystem trait (e.g., primary productivity, species richness) per unit change in the abundance (or biomass) of the putative keystone species. CI can be expressed mathematically as where p is the proportional abundance (or biomass) of the keystone species experiencing a change.

$$CI = \left[\frac{d(trait)}{dp}\right]\left[\frac{1}{trait}\right]$$

A set of experiments conducted by Robert Paine,[24] who first proposed the keystone species concept, illustrates this mathematical model. Working in the intertidal zone of Washington state, he found a sea star to have a keystone effect on macroinvertebrate species diversity mediated by its predation on a dominant mussel consumer species. When he experimentally deleted the sea star (lowering p), mussel abundance greatly increased and macroinvertebrate diversity (the ecosystem trait) substantially decreased.

Operationally, it is difficult to estimate whether a species is a keystone. However, management "experiments" often inadvertently remove or greatly reduce the abundance of an assumed keystone species. A measure of a species' "keystoneness" can be computed as the change in the ecosystem trait before and after its elimination,[18]

$$CI = \left[\frac{(t_n - t_p)}{t_n}\right]\left[\frac{1}{p_i}\right]$$

where tn and tp are quantitative measures of an ecosystem trait (e.g., species diversity) before and after deletion of the keystone species, respectively, and pi is the relative abundance of the keystone before deletion. Hurlbert,[25] also proposed a heuristic model for estimating the relative importance of species i (Ii) in a community of S species as,

$$I_i = \sum_{j=1}^{s} \left| A_{j,t=1} - A_{j,t=0} \right|$$

where, Aj is the abundance of species j (or some other trait) before ($t = 0$) and after ($t = 1$) removal of the ith species.

The keystone concept has been heavily criticized, initially by Hurlbert and more recently by Caro.[6] Hurlbert's criticism focuses on the fact that both the I and CI metrics are not measurable for even a single species, let alone an entire community of species. Thus, these metrics have theoretical but not practical value. Hurlbert concludes that the keystone species notion lacks a clear definition and has never been convincingly demonstrated. Caro's criticisms focuses on the context dependence of the keystone concept—under one set of conditions, a species may play a keystone role but under another, it has no substantial effect on any ecosystem trait.

Recently, the keystone concept has been recast as a search for strongly or weakly interacting species measured by the strength of their trophic links within food webs.[26–28] In deterministic analyses of food web dynamics, importance indices[25] indicate that either top predators or producers are of primary importance.[27] However, when the variance of food web responses to species deletions are evaluated in stochastic simulations, strongly interacting species can appear at multiple trophic levels including the middle of food webs.[28]

The loss of strongly interacting species from food webs has been used to explain the emergence of trophic cascades, broadly defined as the propagation of impacts by apex consumers on their prey down through food webs.[29,30] Even in the context of the practical imitations of the keystone concept, recent reports on the adverse effects of the loss of top- down consumers from freshwater, marine, and terrestrial ecosystems, and subsequent trophic cascades, supports the idea that species are often unequal in their functional importance to ecosystem processes and resilience.[29] Recent research on the unequal role of species in food web dynamics may have revitalized the keystone species concept.

Expected Values and Benchmarks for Ecological Indicators

An essential component of an environmental monitoring program is the expected values (e.g., baselines) or temporal trends of the ecological indicators. Only by comparing observed with expected values or trends can a determination be made about the current state of the ecosystem or the effectiveness of management practices. The close approach to, or the passing of, an expected value is the threshold point that triggers a change in management practices. In the ecological literature, these expected values have been referred to as reference conditions—"the condition in the absence of human disturbance which is used to describe the standard, or benchmark, against which the current condition is compared."[30] It is often difficult, however, to standardize reference conditions across indicator species. Stoddard et al.[31] have categorized ecosystem reference conditions usefully along a continuum from minimally disturbed to historical condition to least disturbed to best attainable. In terms of indicator or keystone species, these concepts translate into expected demographic attributes including birth and survival rates, population sizes, and geographic distributions.

Estimating reference conditions for indicator species is difficult and imprecise for several reasons: (1) our limited understanding of the demographic patterns expected in pristine, undisturbed ecosystems to provide insights into benchmark conditions; (2) an incomplete understanding of the relationship between the state of an indicator species and the desired ecosystem state(s); (3) inadequate knowledge of the expected variability, over time and space, of the indicator species' populations; (4) the nonlinear relationships between indicator values and ecosystem processes, including the existence of sharp threshold regions;[32] and (5) the fact that indicator benchmarks may be best represented by probability distributions rather than single target values.[3]

Expected demographic values and thresholds for indicator species implicitly assume that an ecosystem will evolve to (or was historically at) a dynamic steady state with regard to its species composition, disturbance regimes, and climatic processes. This assumption is often false. The dynamic nature of ecosystems argues for specifying a distribution of indicator values (e.g., population sizes, birth rates, survival rates, etc.) rather than an expected value at a single point in time or space. That is, the current state of an ecosystem is best described by a probability distribution rather than a specific target value. In practice, this makes the management decision process less certain, but the probability concept is a more realistic characterization of nature.

Determining threshold values and reference distributions requires the selection of both a spatial and temporal scale to assess the status and trends of the indicator species. If the spatial scale is a point in space, an indicator threshold may be specified as a single value (e.g., a population density >75 individuals/ha). In practice, it will be uncommon for the state of an indicator species, and thus the integrity of the ecosystem to which it belongs, to be characterized and interpreted in terms of a single target value. More commonly, the spatial scale of interest would be large landscapes, and indicator values would be estimated at multiple sample locations. Replicate estimates of population size, for example, would provide a distribution of indicator values. This distribution is critically dependent on the spatial scale and temporal window over which the distribution is estimated.

Once measured, the observed distribution of indicator values would be compared to the expected (benchmark) distribution to detect both the magnitude and pattern of deviation from desired conditions. The concept of a spatial distribution of indicator values as the appropriate evaluative statistic is critical to the monitoring and assessment of ecological systems.[3,31]

Despite the importance of establishing benchmark distributions, specifying reference conditions is subject to some degree of arbitrariness. For example, there is uncertainty about how benchmark conditions are to be estimated.[31] There is no clear guidance on how far back in time, or to what spatial extent, one should go to find an appropriate point of reference or what constitutes pristine conditions. There may be no current analogs for past conditions. In addition, it is clear that the concept of benchmarks for ecological indicators can be reconciled with the dynamic nature of ecosystems only in terms of frequency distributions, especially when viewed over long temporal or broad spatial scales.[3] Thus, all decisions about the proximity of the status or trend of an indicator or keystone species to its expected value (e.g., abundance, density) must be made in probabilistic terms.

Conclusions

Assessing the status and trends of natural resources requires monitoring the components of ecosystems that give rise to these resources. The key reason to assess ecological systems at the species level is that species are the fundamental agents of production and transfer of matter and energy in ecosystems. However, rarely can all the components of biological diversity be monitored—some prioritization of species must occur.[1,33] The challenge is to strategically select a small subset of species to monitor, whose change in state provides insights into a broader set of species or ecological processes.[1]

Because the dynamics of ecosystems are often driven by a small number of species that have uneven effects on ecosystem processes, a surrogate-based approach is justified.[26–29]

Knowledge of changes in the status and trends of biological diversity is essential to effective ecosystem management. Many empirical and theoretical studies have demonstrated that more diverse plant and animal communities support less variable ecosystem outputs and provide more ecosystem services.[34] Ecosystem resilience to contemporary environmental stressors and future climate change is strongly related to native species diversity.[34–36] Finally, species are useful early-warning indicators of impending environmental change, because they integrate the effects of multiple stressors across broad temporal and spatial scales. Multiple categories of species provide useful insights into ecosystem integrity and the ability of the ecosystem to sustainably provide natural resources that people depend on.[6] Indicator and keystone are two logical species categories whose change in state can be directly linked to management impacts, environmental stressors, and ecological consequences.

References

1. Noon, B.R.; McKelvey, K.S.; Dickson, B.G. Multispecies conservation planning on U.S. federal lands. In *Models for Planning Wildlife Conservation in Large Landscapes;* Millspaugh, J.J., Thompson, F.R. III, Eds.; Academic Press: New York, 2009; 51–84.
2. Haufler, J.B.; Meh, C.A.; Roloff, G.J. Using a coarse-filter approach with species assessment for ecosystem management. Wildl. Soc. Bull. **1996,** *24,* 200–208.
3. Noon, B.R. Conceptual issues in the monitoring of ecological resources. In *Monitoring Ecosystems: Interdisciplinary Approaches for Evaluating Ecoregional Initiatives*; Busch, D.E., Trexler, J.C., Eds.; Island Press: Washington, DC, 2003; 27–71.
4. Noon, B.R.; McKelvey, K.S. *The Process of Indicator Selection. USDA Forest Service Proceedings RMRS-P-42CD;* U.S. Department of Agriculture Forest Service: Washington, DC, 2006; 944–951.
5. Niemi, G.J.; McDonald, M.E. Application of ecological indicators. Annu. Rev. Ecol. Syst. **2004,** *35,* 89–111.
6. Caro, T.M. *Conservation by Proxy: Indicator, Umbrella, Keystone, Flagship, and Other Surrogate Species*; Island Press: Washington, DC, 2010.
7. Tulloch, T.; Possingham, H.P.; Wilson. K. Wise selection of an indicator for monitoring the success of management actions. Biol. Conserv. **2011,** *144,* 141–154.
8. Carignan, V.; Villard, M.A. Selecting indicator species to monitor ecological integrity: A review. Environ. Manag. Assess. **2002,** *78,* 45–61.
9. Niemeijer, D.; DeGroot, R.S. A conceptual framework for selecting environmental indicator sets. Ecol. Indicators **2008,** *8,* 14–25.
10. Rochet, M-J.; Rice, J.C. Do explicit criteria help in selecting indicators for ecosystem-based fisheries management? ICES J. Mar. Sci. **2005,** *62,* 528–539.
11. Landres, P.B.; Verner, J.; Thomas, J.W. Ecological use of vertebrate indicators: A critique. Conserv. Biol. **1988,** *2,* 316–328.
12. Landres, P.B. Ecological indicators: Panacea or liability? In *Ecological Indicators;* Hyatt, D.E., McDonald, V.J., Eds.; Elsevier Applied Science: New York, 1992; Vol. 2, 1295–1318.
13. Rolstad, J.; Gjerde, I.; Gunderson, V.S.; Sætersdal, M. Use of indicator species to assess forest continuity: A critique. Conserv. Biol. **2002,** *18,* 76–85.
14. Cushman, S.A.; McKelvey, K.S.; Noon, B.R.; McGarigal, K. Use of the abundance of one species as a surrogate for the abundance of others. Conserv. Biol. **2010,** *24,* 830–840.
15. Lawton, J.H.; Gaston, K.J. Indicator species; In *Encyclopedia of Biodiversity;* Levin, S., Ed.; Academic Press: California, 2001; 437–450.
16. Karr, J.R. Assessment of biological integrity using fish communities. Fisheries **1981,** *6,* 21–27.
17. Karr, J.R. Ecological integrity and ecological health are not the same. In *Engineering Within Ecological Constraints;* Schultz, P.C., Ed.; National Academy Press: Washington, DC, 1996; 97–109.
18. Hierl, L.A.; Franklin, J.; Deutschman, D.H.; Regan, H.M.; Johnson, B.S. Assessing and prioritizing ecological communities for monitoring regional habitat conservation plans. Environ. Manag. **2008,** *42,* 165–179.
19. Wiens, J.A.; Hayward, G.D.; Holthausen, R.S.; Wisdom, M.J. Using surrogate species and groups for conservation and management. Bioscience **2008,** *58,* 241–252.
20. Osenberg, C.W.; Schmitt, R.J.; Holbrook, S.J.; Abu-Saba, K.E.; Flegal, A.R. Detection of environmental impacts: Natural variability, effect size, and power analysis. Ecol. Appl. **1994,** *4,* 16–30.
21. Thompson, W.L.; White, G.W.; Gowan, C. *Monitoring Vertebrate Populations*; Academic Press: New York, 1998.
22. Noon, B.R., Bailey, L.L.; Sisk, T.D.; McKelvey, K.S. Efficient species-level monitoring at the landscape scale. Conserv. Biol. **2012,** *26,* 432–441.
23. Power, M.E.; Tilman, D.; Estes, J.A; Menge, B.A.; Bond, W.J.; Mills, L.S.; Daily, G.; Castilla, J.C.; Lubchenco, J.; Paine, R.T. Challenges in the quest for keystones. Bioscience **1996**, *46,* 609–620.

24. Paine, R.T. Food web complexity and species diversity. Am. Nat. **1966,** *100,* 65–75.
25. Hurlbert, S.H. Functional importance vs keystoneness: Reformulating some questions in theoretical biocenology. Aust. J. Ecol. **1997,** *22,* 369–382.
26. Soule, M.E.; Estes, J.A.; Miller, B.; Honnald, D.L. Strongly interacting species: Conservation policy, management, and ethics. Bioscience **2005,** *55,* 168–176.
27. Libralato, S.; Christense, V.; Pauly, D. A method for inden- tifying keystone species in food web models. Ecol. Model. **2006,** *195,* 153–171.
28. Livi, C.M.; Jordan, F.; Lecca, P.; Okay, T.A. Identifying key species in ecosystems with stochastic sensitivity analysis. Ecol. Model. **2011,** *222,* 2542–2551.
29. Estes, J.A.; Terborgh, J.; Brushares, J.S.; Power, M.E.; Berger, J.; Bond, W.J.; Carpenter, S.R.; Essington, T.E.; Holt, R.D.; Jackson, J.B.C.; Marquis, R.J.; Oksanen, L.; Oksanen, T.; Paine, R.T.; Pikitch, E.K.; Ripple, W.J.; Sandin, S.A.; Scheffer, M.; Schoener, T.W.; Shurin, J.B.; Sinclair, A.R.E.; Soulé, M.E.; Virtanen, R.; Wardle, D.A. Trophic down-grading of planet Earth. Science **2011,** *333,* 301–306.
30. Sanchez-Montoya, M.M.; Vidal-Abarca, M.R.; Punti, T.; Poquet, J.M.; Prat, N.; Rieradevall, M.; Alba-Tercedor, J.; Zamora-Muñoz, C.; Toro, M.; Robles, S.; Álvarez, M.; Suárez, M.L. Defining criteria to select reference sites in Mediterranean streams. Hydrobiologica **2009,** *619,* 39–54.
31. Stoddard, J.; Larsen, J; Hawkins, D.; Johnson, R.K.; Norris, R.H. Setting expectations for the ecological condition of streams: the concept of reference condition. Ecol. Appl. **2006,** *16,* 267–1276.
32. Groffman, P.M.; Baron, J.S.; Blett, T.; Gold, A.J.; Goodman, I.; Gunderson, L.H.; Levinson, B.M.; Palmer, M.A.; Paerl, H.W.; Peterson, G.D.; Poff, N.L.; Rejeski, D.W.; Reynolds, J.F.; Turner, M.G.; Weathers, K.C.; Weins, J. Ecological thresholds: the key to successful environmental management, or an important concept with no practical application? Ecosystems **2006,** *9,* 1–13.
33. Possingham, H.P.; Andelman, S.J.; Noon, B.R.; Trombulak, S.; Pulliam, H.R. Making smart conservation decisions. In: *Conservation Biology: Research Priorities for the Next Decade;* M.E. Soule, G.H. Orians, Eds.; Island Press: Washington, D.C., 2001; 225–244.
34. Naeem, S., Bunker, D.E.; Hector, A.; Loreau, M.; Perrings, C. *Biodiversity, Ecosystem Functioning, and Human Wellbeing: An Ecological and Economic Perspective*; Oxford University Press: New York, 2009.
35. Cottingham, K.L.; Brown, B.L.; Lennon, J.T. Biodiversity may regulate the temporal variability of ecological systems. Ecol. Lett. **2001,** *4,* 72–85.
36. Hooper, D.U.; Chapin, F.S.; Ewel, J.J.; Hector, A.; Inchausti, P.; Lavorc, S.; Lawton, J.H.; Lodge, D.M.; Loreau, M.; Naeem, S.; Schmid, B.; Setälä, H.; Symstad, A.J.; Vander- meer, J.; Wardle, D.A. Effects of biodiversity on ecosystem functioning. A consensus of current knowledge. Ecol. Monogr. **2005,** *75,* 3–35.

13

Population Genetics

Introduction 113
Evolutionary Forces That Drive DNA Sequence Change 113
Mutation: The Source of Genetic Novelty 114
Genetic Random Drift: The Role of Chance in Evolutionary Change ... 114
The Neutrality Hypothesis: An Interaction between Mutation and Drift 115
Natural Selection: The Engine of Adaptive Change 115
Recombination as Both Randomizing and Creative Force 116
The Coalescent: Looking Backward in Time 116
Migration: The Spatial Dimension of Genetic Change 117
Conclusion 117
References 117

Michael T. Clegg
University of California

Introduction

Genes are the fundamental hereditary units transmitted from parent to offspring in every generation. Population genetics is concerned with the statistical rules that govern the transmission of genes within collections of interbreeding individuals over time. Changes in the frequencies of genes within populations are the basis for evolutionary change, as well as the basis for the genetic improvement of plants and animals that provide human sustenance.

Population genetics employs mathematical models to describe trajectories of genetic change under various scenarios. These models are also used to design experiments to dissect the processes of evolution. The basic mathematical models of population genetics were developed long before it was established that DNA is the chemical basis of heredity. The challenge today is to analyze and learn from the rapidly accumulating body of genome sequence data. The basic models of population genetics are highly relevant to this enterprise because they are general, in the sense that change in gene frequency can be substituted with change in individual nucleotide frequency without any alteration in the underlying mathematics. This entry provides a brief outline of the theory of population genetics as it applies to changes in the composition of DNA molecules over time. Several excellent books provide a comprehensive introduction to population genetics theory, including works by Hartl and Clark and Hedrick.

Evolutionary Forces That Drive DNA Sequence Change

Five forces determine genetic change.[1,2] These begin with mutation, defined as a permanent heritable change in a gene. A second force is recombination, which is a randomizing force that can also lead to permanent heritable changes in a gene. Genetic random drift, defined as the small fluctuations in gene frequencies caused by the sampling of individuals in reproduction, is third. The fourth force is migration, or the movement of genes among different populations. Fifth is natural selection, in which some individuals are more successful in transmitting their genes owing to superior adaptation to their environment and hence improved survival and reproduction. Natural selection is the force that leads to adaptive genetic change; it essentially sorts among mutations to eliminate those that are deleterious and increase the frequency of those that are advantageous.

Mutation: The Source of Genetic Novelty

Gene or nucleotide sequences are subject to constant change. To begin, mutation takes many forms, each of which has some likelihood of occurring at each DNA replication cycle. From this it is apparent that any DNA molecule is subject to mutational erosion over time; this erosion process is a form of evolutionary change. DNA sequences that code for functionally important molecules are expected to be conserved because natural selection rejects most deleterious mutational changes in functionally important genes. Current genome sequencing efforts suggest that much of the DNA of organisms does not code for proteins or regulatory signals—these regions may be evolving relatively rapidly.

Genetic Random Drift: The Role of Chance in Evolutionary Change

Once a new mutation has arisen in a single DNA molecule (to create a new allelic form of a gene or a new allele), it may be transmitted to the next generation or it may be lost. Assuming the new mutation has no effect on fitness, the probability of transmission to the next generation is $1 - [1 - 1/2N]^{2N}$ in diploid organisms, where N is the population size. This result follows from the fact that a new mutation will initially occur in a single individual (frequency = $1/2N$). The pool of gametes is assumed to be much larger than the N reproductive individuals in the population, so sampling with replacement is appropriate. (This expression is the probability of not being lost in the first generation, derived assuming binominal sampling.) In the vast majority of cases a new mutation will be lost through the vagaries of sampling within a generation or two. A very small proportion will drift to high frequency and be fixed (i.e., become the sole allele in the population or species).

An important result of mathematical genetics indicates that the probability of ultimate fixation of a new neutral mutation is $1/2N$ (the concept of neutrality is discussed later). Because there is a constant flux of new mutations, there will also be a constant rate of fixation of new allelic variants over the long term. In fact, the rate of fixation of new alleles is simply μ, the mutation rate. This result provides a formal justification for the molecular clock hypothesis, because it indicates that the rate of mutational change should be constant over time if the mutation rate is constant.

A second important theoretical result from population genetics demonstrates that the average time until the fixation of a new mutation is $T \sim 4N_e$, so for large N_e, many mutant alleles are expected to be in transition to fixation at any point in time. (N_e is a population size adjusted to take account of variation in reproductive output among different members of a breeding population.) For example, many plant species must have population sizes in the order of 10^6 to 10^7. The rate of mutation per nucleotide site per generation is approximately $\mu = 5 \times 10^{-9}$. If a typical gene is 10^3 nucleotide sites in length, the total probability of a new mutation in the gene in any single generation is 5×10^{-6}. For a species size ($2N$) of 10^6, five new mutations in this particular gene are expected in the total species per generation. Moreover, the average new mutation that is destined to be fixed will drift in the species of size 10^6 for 4,000,000 generations. The probability of drawing two different alleles of the gene in a population or species can be shown to be approximately $1–4N_e\mu/(1+ 4N_e\mu) = 19/21 = 0.90$ for this numerical example, so almost every pair of genes sampled is expected to be different. Thus, high levels of molecular polymorphism are expected in species. Current efforts to define all human single nucleotide polymorphisms (SNPs) verify these theoretical predictions.

The Neutrality Hypothesis: An Interaction between Mutation and Drift

The theoretical results cited in the preceding discussion are based on the assumption that selection is absent. In the 1970s, Motoo Kimura[3] and associates advanced a theory called the neutrality hypothesis that claims that a large proportion of molecular evolutionary change is neutral to natural selection. This hypothesis was in part motivated by an important calculation by J. B. S. Haldane,[4] who showed that there is an upper bound to the rate of evolutionary change under natural selection. Because selective change requires differential survival and reproduction, the reproductive excess of the species defines this upper bound. Stated differently, some or all of the reproductive excess of a species is consumed by the process of selection. Empirical evidence on rates of molecular change appear to be too high to be consistent with this upper bound, especially for long-lived low-fecundity organisms like the human. As a consequence, it was postulated that a large proportion of molecular change is not driven by selection, but by random drift and mutation.

The test of the neutral theory is to ask how well observed data fit the theoretical calculations outlined earlier. Because the neutral theory provides precise mathematical predictions, the theory is testable; moreover, these tests provide a very useful context for identifying particular genes that are selected. This is because it is known precisely how neutral genes should behave, and departures from neutrality are likely to arise from selection. To date, available data do suggest that many polymorphisms are neutral or only very weakly selected. However, the statistical power to detect departures from neutrality increases with database size, which are growing very rapidly at present.

Natural Selection: The Engine of Adaptive Change

Natural selection operates at the level of phenotypic differences, so an allelic change must alter some aspect of phenotype to be perceived by selection. The phenotypic dimensions of organisms, and the ways these may be perceived by various environmental factors, are very complex. These range from subtle changes in rates of flow through a metabolic pathway, to resistance to disease, to the ability to withstand extreme stress environments, to aspects of behavior, and so on. The list is virtually endless. Moreover, each aspect of phenotype may be determined by a number of different genes and by the way these genes interact in development and metabolism. It is similarly difficult to identify the precise facets of an environment that may act as a selective agent. As a consequence, selection is often identified retrospectively. That is, the analytical tools of population genetics may support the inference that selection has affected a genetic locus based on the pattern of change through time. Such an analysis rarely tells the reasons why a particular change is adaptive.

A number of mathematical results that provide useful predictions about the statics and dynamics of selective genetic change are available from population genetic theory. We will mention just one result from this large body of work. The probability of fixation of a new mutant that is favored by selection (assuming the heterozygous genotype is intermediate in its fitness between the two homozygous genotypes) is $\sim 2s$, where s is the selective intensity favoring the new mutant allele. So if $s = 0.02$ (implying a 2% selective advantage), the favored allele will still be lost 96% of the time. This result demonstrates that most new favored mutations will still be lost owing to drift.

The average time to fixation of a new favored mutant is approximately $T \sim (2/s)\mathrm{Ln}(2N_e)$ generations (where Ln denotes the natural logarithmic function). Numerical iterations of this equation, compared to the neutral case, reveal that the time to fixation of selectively favored alleles is much shorter than for a neutral allele. Indeed, fixation of a selected allele can take place in orders of magnitude less time than a neutral allele. This calculation suggests that in contrast to the high levels of neutral polymorphism expected from the neutral theory outlined in the fore going, it may be relatively rare to find a selected gene in transition when one allele is favored over the other allele. The exception to this statement is the case where two or more alleles are favored, either because they lead to a higher fitness among heterozygotes or because their advantage is frequency-dependent (frequency dependent means that *s* is not constant, but instead depends on allele frequencies). Both of these situations are referred to as balanced polymorphism. Classic examples of a balanced polymorphism are often associated with disease resistance, where the resistance allele causes some other disadvantage or where there is a constant arms race between the disease agent and the target organism favoring new mutations at a resistance locus (a case of frequency dependence). An example of such a balanced polymorphism arises from the multiple major histocompatibility polymorphism (MHC) alleles in human populations.[5] In this case, disease resistance is strongly implicated because the MHC locus is involved in the recognition of potential disease agents.

Recombination as Both Randomizing and Creative Force

Recombination refers to the rearrangement of genes with respect to one another on chromosomes. Recombination is a randomizing force, but it is also much more. It is a creative force that operates at many levels within plant genomes.[6] Thus, molecular evidence has revealed that recombination is responsible for the generation of many gene novelties. For example, studies of different allelic sequences of a gene often show traces of past intragenic recombination events where new allelic forms have arisen, owing to the exchange of stretches of sequence between two different alleles. This reciprocal exchange can yield a new allele that is different from both parental forms.

The Coalescent: Looking Backward in Time

An important theoretical construct from population genetics is known as the coalescent. Consider a set of DNA sequences that are drawn from a single locus within a species. The sample of sequences must have all descended from a common ancestor. The point where all alleles trace back to the most recent common ancestor is known as the coalescent. It is straightforward to estimate the genealogical history of the sample by using phylogeny estimation tools. If the nucleotide sequence diversity in the sample is the result of a drift/mutation process at equilibrium, then the arrival times of new mutation events are given by an exponential process.[7] This result provides a useful framework for testing sequence samples to see if they conform to a neutral process.

A particularly important statistic calculated from a sample of sequences is θ, a measure of the nucleotide sequence diversity. A different statistic that also measures nucleotide sequence diversity is π. (Both of these quantities are related to the probability of drawing two different nucleotides at a single site averaged over all sites examined.) If we calculate π and θ as an average over many independent replications of the evolutionary process that led to the observed sample, and if the underlying evolutionary process is the result of a drift/mutation process at equilibrium, then both θ and π should be approximately equal to $4N_e\mu$. Thus, the difference between π and θ should be zero under the drift/mutation assumption. A test statistic introduced by Tajima[8] based on a function of the difference between π and θ allows one to ask whether a data set conforms to the drift/mutation assumption by testing whether this difference is zero. If the difference is nonzero, then other forces such as selection or demographic change must be invoked. Many standard computer programs for sequence analysis implement these calculations (i.e., DnaSP).

Another statistic that features importantly in coalescence theory is T_{MARCA} (time to the most recent common ancestor of the sample). Mathematical calculations show that T_{MARCA} is also approximately equal to $4N_e$ when the drift/mutation assumption is satisfied. Finally, if μ is known it is possible to estimate N_e for the species or population from which the data came. Estimates of N_e from a few important crop plants vary over nearly an order of magnitude from 2×10^5 (wild barley) to 3×10^5 (pearl millet) to 18×10^5 (maize).[9] These data are derived from broad samples of the entire species so the estimates can be viewed as the best estimates of long-term effective population size for each species.

Migration: The Spatial Dimension of Genetic Change

Every new mutation must have originated in a single individual at a specific location in space. This means that the spread of a mutation through a species is a spatial process, and it is possible to gain some insight into historical migration patterns through the analysis of geographic samples of gene sequences. Moreover, the coalescent framework has been refined to include spatial considerations, so migration rates can be quantified. Efforts to measure migration are still in very early stages, but contemporary data suggest that plants and their genes have been quite mobile over long periods of time.

There is great contemporary concern over the spread of so-called transgenes from agricultural plants into wild or weedy relatives through migration. This question has stimulated much work on the direct measurement of migration in plants. Current results suggest that gene migration rates may be higher than previously believed.[10]

Conclusion

Although population genetics is a mature science with a history of nearly 100 years, it is more relevant today than ever. The theoretical models of population genetics provide the conceptual and statistical tools to address important questions ranging from concerns about transgene escape to detection of the influence of selection on specific genes. As the body of genome sequence data expands, these tools will become ever more important in interpreting and mining this new resource.

References

1. Hartl, D.L.; *Clark, A.G. Principles of Population Genetics,* 3rd Ed.; Sinauer Associates Inc.: Sunderland, MA, 1997; 542.
2. Hedrick, P.A. *Genetics of Populations,* 2nd Ed.; Jones and Barlett: Boston, MA, 2000; 553.
3. Kimura, M. *The Neutral Theory of Molecular Evolution*; Cambridge University Press: Cambridge, England, 1983; 367.
4. Haldane, J.B.S. The cost of natural selection. J. Genet. **1957**, *55*, 511–524.
5. Nei, M. DNA Polymorphism and Adaptive Evolution. In *Plant Population Genetics, Breeding and Genetic Resources*; Brown, A.H.D., Clegg, M.T., Kahler, A., Weir, B.W., Eds.; Sinauer Assoc.: Sunderland, MA, 1990; 128–142.
6. Clegg, M.T. The Role of Recombination in Plant Genome Evolution. In *Evolutionary Processes and Theory*; Waser, S.P., Ed.; Kluwer Academic Publishers: the Netherlands, 1999; 33–45.
7. Hudson, R.R. Gene genealogies and the coalescent process Oxf. Surv. Evol. Biol. **1990**, *7*, 1–44.
8. Tajima, F. Statistical method for testing the neutral mutation hypothesis by DNA polymorphism. Genetics **1989**, *123*, 585–595.
9. Cummings, M.P.; Clegg, M.T. Nucleotide sequence diversity at the alcohol dehydrogenase I locus in wild barley (*Hordeum vulgare* ssp. *spontaneum*): An evaluation of the background selection hypothesis. Proc. Natl. Acad. Sci. U. S. A. **1998**, *95*, 5637–5642.
10. Ellstrand, N. When transgenes wander, should we worry? Plant Physiol. **2001**, *125*, 1543–1545.

14

Prioritizing Conservation Sites for Biodiversity Protection in Uganda

Samuel Ayebare, Grace Nangendo, and Simon Nampindo
Wildlife Conservation Society

Introduction 119
Biodiversity Prioritization: Key Biodiversity Areas (KBAs) 120
KBA National Coordination Group 122
Conservation and Development 123
Game Conservation in Uganda 124
Discussion 125
References 127

Introduction

Uganda is an important country for biodiversity conservation supporting a diversity of flora and fauna species of global and national conservation concern [1,2]. It is ranked among the top ten most biodiverse countries in Africa [3]. A total of 1,742 vertebrates have been recorded, representing more than 50% of Africa's bird species, 39% of Africa's mammals, 19% Africa's mammal species, and 14% of Africa's reptile species [1,2,4]. The high species diversity in Uganda is due to the presence of five major biomes, namely, Sudan-Guinea Savanna, Guinea-Congo Forests, Lake Victoria Basin biome, Afrotropical Highlands biome, and Somali-Masai biome, and a high variation in altitudinal range from 600 m above sea level to 5110 m at the peak of the Rwenzori mountains. As such, the fauna and flora of Uganda is representative of moist Congo basin forest species and the dry savannah species of northern Kenya and Somalia [5]. Figure 14.1 shows Uganda's protected area system. Protected areas in Uganda are managed by the Uganda Wildlife Authority—UWA (i.e. 10 national parks, 12 wildlife reserves, 5 community wildlife areas, 10 wildlife sanctuaries), the National Forestry Authority—NFA (506 central forest reserves), and District Forest Services (200 local forest reserves).

In this chapter, we describe the identification of priority areas for biodiversity conservation in Uganda based on the Key Biodiversity Areas (KBAs) identification and delineation criteria. We also argue in favor of discouraging industrial development projects and other competing forms of land use within protected areas, KBAs, irreplaceable conservation sites, and wetlands of high conservation value.

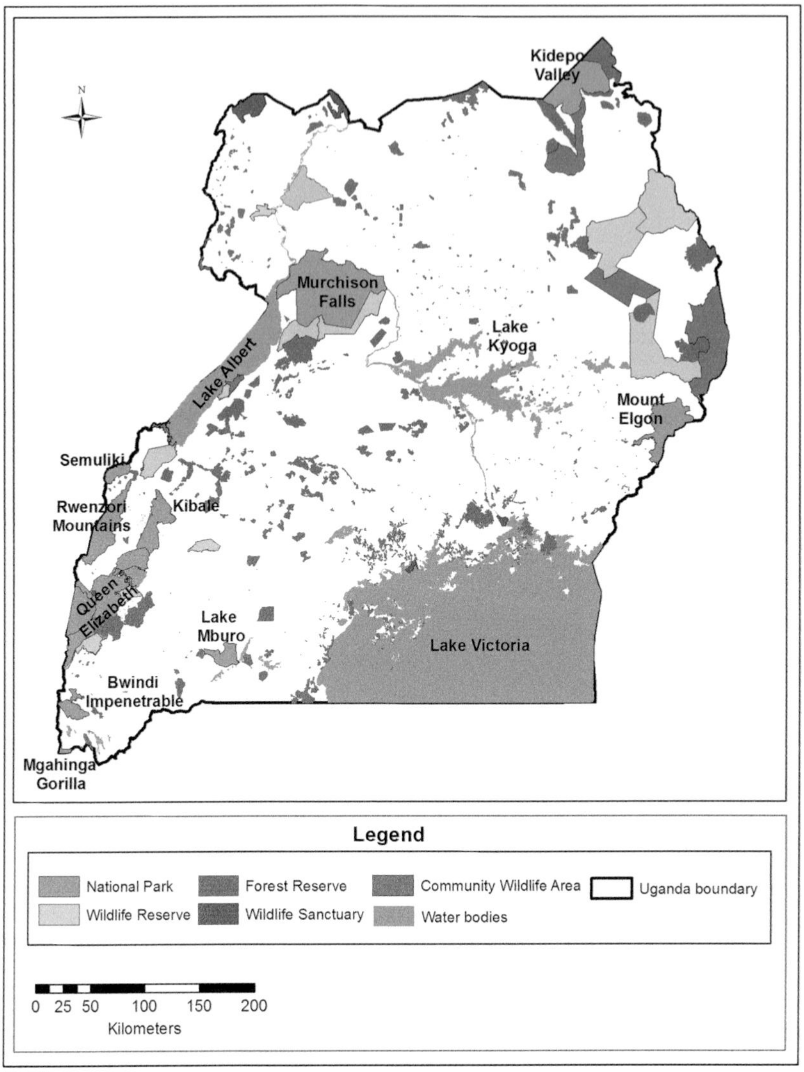

FIGURE 14.1 Protected areas in Uganda.

Biodiversity Prioritization: Key Biodiversity Areas (KBAs)

Priority areas for conservation of species of national and global importance in Uganda were identified based on the KBA criteria [4]. KBAs are 'sites contributing significantly to the global persistence of biodiversity in terrestrial, inland water and marine environments' [6]. The KBA delineation standard builds upon various initiatives that have been developed and used in the past to identify and priority sites for biodiversity conservation across various taxa at different scales. These include ecoregions, hotspots, Important Bird and Biodiversity Areas (IBAs), Alliance for Zero Extinction (AZE) sites, B-ranked sites, Important Fungus Areas, and Important Plant Areas [6].

Through a worldwide consultative process, the International Union for the Conservation of Nature (IUCN) in partnership with other conservation agencies adopted a global standard (A Global Standard for Identification Key Biodiversity Areas) for the identification of KBAs that includes definitions, criteria, thresholds, and delineation procedures [6]. The KBA criteria are provided in Table 14.1.

TABLE 14.1 KBA Criteria

KBA Criteria Higher Level	Criteria Code	Sub-criteria
Threatened biodiversity	A	A1: Threatened species
		A2: Threatened ecosystem types
Geographically restricted biodiversity	B	B1: Individual geographically restricted species
		B2: Co-occurring geographically restricted species
		B3: Geographically restricted assemblages
		B4: Geographically restricted ecosystem types
Ecological integrity	C	C: Intact ecological communities
Biological processes	D	D1: Demographic aggregations
		D2: Ecological refugia
		D3: Recruitment sources
Irreplaceable through quantitative analysis	E	E: High irreplaceability for global persistence of species

TABLE 14.2 List of Terrestrial Key Biodiversity Areas in Uganda

ID	Site Name and Area (km^2)	Number of KBA Trigger Species and Criteria	ID	Site Name and Area (km^2)	Number of KBA Trigger Species and Criteria
1	Budongo Forest Reserve (825)	5, (3A1, 2B1)	19	Nabugabo wetland (777)	2, (B1)
2	Bugoma Forest Reserve (400)	4, (2A1, 2B1)	20	Timu extension (47.7)	
3	Bwindi Impenetrable National Park (331)	8, (4 A1, 5B1)	21	Queen Elizabeth National Park (including Kigezi Wildlife Reserve) (2243)	2, (A1,B1)
4	Echuya Forest Reserve (34)	3, (A1)	22	Rwenzori Mountains National Park (998)	14 (5A1, 10B1)
5	Kasyoha-Kitomi Forest Reserve (433)	5, (A1, 4B1)	23	Semuliki National Park (220)	1 (B1)
6	Kibale National Park (766)	3, (A1, 2B1)	24	Mardiopei—South Moyo (26)	1, (A1)
7	Kidepo Valley National Park (1442)	1, (A1	25	Mpanga Falls (2.4)	1, (A1)
8	Kyambura Wildlife Reserve (154)	2, (D1, B1)	26	East of Thurston Bay (2.5)	1, (A1)
9	Lake Bisina (542)	1, (B1)	27	Tororo Rock (9.4)	1, (A1)
10	Lake Mburo National Park	1, (B1)	28	Kyenjojo-Mubende inselberg (0.36)	1, (B1)
11	Lake Opeta	1, (B1)	29	Sango bay Area (540)	1, (D1)
12	Lutembe Bay (0.98)	1 (D1)	30	Itwara Forest Reserve (87)	2 (B1)
13	Mabira Forest Reserve (300)	5 (A1, 4B1)	31	Kalinzu Forest Reserve (147)	1 (A1)
14	Mgahinga Gorilla National Park (33.9)	3 (2A1, B1)	32	Morungole Forest Reserve	1, (B1)
15	Mount Elgon National Park (1145)	5 (A1, 4B1)	33	Mubende—Kasanda (2.4)	1, (B1)
16	Mount Moroto Forest Reserve (483)	363624	34	Ogili Forest Reserve (53)	1, (B1)
17	Mount Otzi Forest Reserve (188)	1 (B1)	35	Kome—Sesse Islands (77.5)	1, (B1)
18	Murchison Falls National Park (3893)	2 (A1,B1)	36	Bugala Island—Sesse Islands (41)	2 (B1)

An assessment of KBAs in Uganda was carried out by the Wildlife Conservation Society (WCS) and its partners using criteria A, B, and D [4]. The initial assessment list for KBAs was generated mainly based on the IBAs and reinforced by considering the conservation status of other KBA trigger species [4]. A total of 36 terrestrial KBAs were identified in Uganda, of which, 24 are protected areas [4]. Table 14.2 shows a list of the identified terrestrial KBAs in Uganda, and Figure 14.2 shows the location of the proposed terrestrial KBAs.

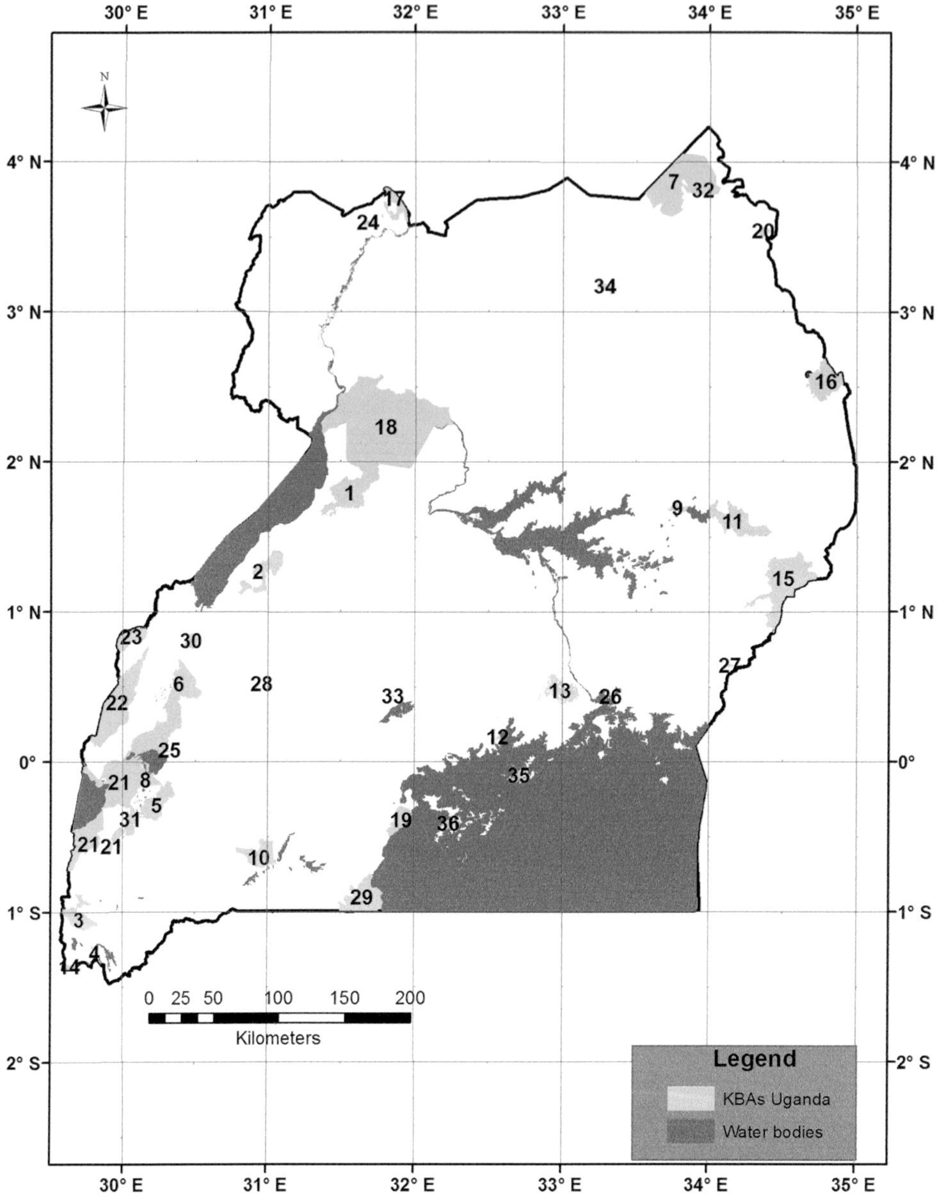

FIGURE 14.2 Proposed terrestrial Key Biodiversity Areas in Uganda.

KBA National Coordination Group

A KBA National Coordination Group (NCG) was set up to coordinate the activities that relate to the identification, documentation and delineation of KBAs, and institutionalization of this conservation planning framework in Uganda. The Terms of Reference for the NCG were adopted from the KBA secretariat (http://www.keybiodiversityareas.org/home). Members of the KBA NCG include non-governmental organizations (e.g., WCS, World Wildlife Fund (WWF), Fauna & Flora International (FFI), IUCN, Nature Uganda), government institutions (e.g., National Environment Management Agency, Ministry of Tourism Wildlife and Antiquities, UWA, NFA), Academia, and the private sector.

The Wildlife Conservation Society Uganda program is the current chair of the KBA NCG. Thematic group sub-committees within the NCG were set up with the following roles and responsibilities:

1. Data management and sharing
   i. Data mapping, sourcing, and acquisition
   ii. Data vetting across different taxa and ecosystems
   iii. Develop protocols for data sharing and use
   iv. Constantly identify and fill data gaps
   v. Prioritize species to be researched on under KBA trigger species.
2. Fundraising
   i. Identify possible options for financing KBA identification, mapping, and conservation management.
   ii. Engage in proposal writing
   iii. Initiate themes upon which proposals will be based
   iv. Identify and populate a list of funders and funding streams.
3. Stakeholder engagements
   i. Identify and map different stakeholders
   ii. Engage funding bodies, relevant government institutions, businesses, and civil society organizations
   iii. Input and advise the KBA Committee on the dissemination and use of KBA data and information
   iv. Inform and influence public policy and private sector decision-making that impacts, and/or benefits KBAs management and conservation
   v. Promote KBA identification and raise their profile in support for increased investment in their management and conservation.
4. KBA site-level management
   i. Help to establish site-level KBA management committees
   ii. Monitor the health and conservation status of KBAs
   iii. Site-level data vetting across taxa and ecosystems
   iv. Communicate needs and challenges in the use and application of KBA data
   v. Collaborate to conserve, protect, and safeguard these important places
   vi Support the process to develop and implement KBA site management plans.

Conservation and Development

The establishment of protected areas and the identification of priority sites for conservation are fundamental in mitigating biodiversity loss. The existence and functioning of protected areas, however, is hampered by competing land uses and resource use pressures [7]. An assessment of the state of Uganda's biodiversity shows a negative trend, mainly outside protected areas [8]. It's estimated that Uganda lost about 50% of its overall biodiversity between 1975 and 1995 [8]. The main drivers of biodiversity loss in Uganda are habitat conversion for agriculture and human settlement, and overexploitation of resources such as fuelwood, charcoal, bushmeat, and timber. In addition, the increase in infrastructure developments, particularly roads, railways, and extractive industries, e.g., mining, gas and oil, and hydropower generation coupled with climate change and invasive species, is expected to make it worse [1]. The majority of Ugandans (70%) are dependent on agriculture for their livelihoods [9]. The dependency on agriculture coupled with the increasing human population, and soil fertility decline due to poor farming practices have intensified the conversion of natural vegetation cover to agricultural farmland. According to the National Development Plan II (NDP-II; 2015/2016–2019/2020), the government of Uganda prioritized agriculture, tourism, minerals, oil and gas, infrastructure, and human capital development. Figure 14.3 shows the overlap between protected areas and existing, ongoing, and potential development areas.

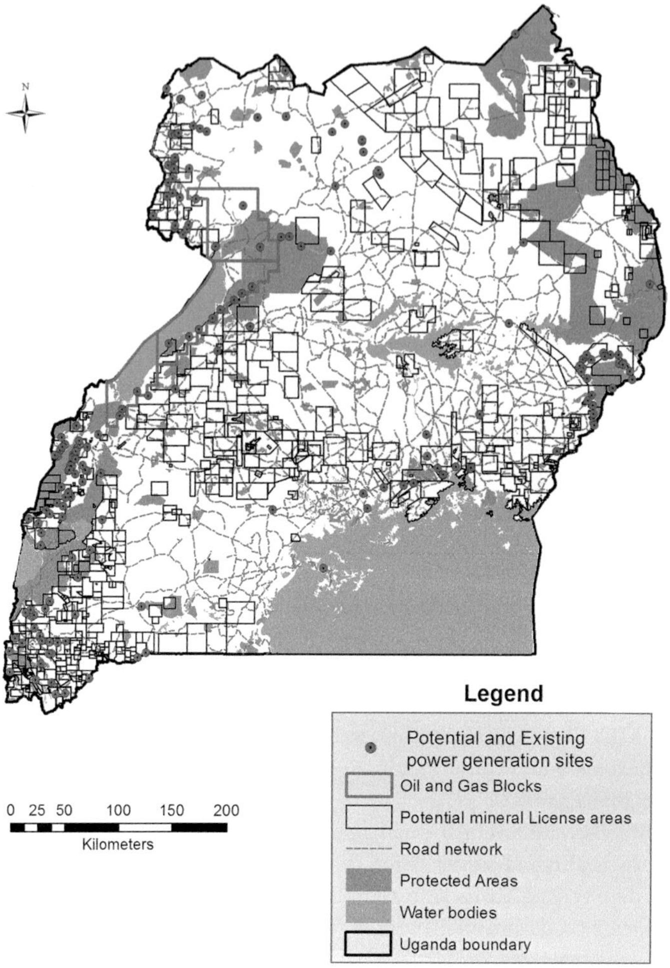

FIGURE 14.3 Overlap between protected areas and existing, ongoing, and potential development areas.

Game Conservation in Uganda

The first game reserve (area north of Lake George) in Uganda was gazetted in 1906 by the Protectorate administration to regulate hunting and the expansion of wheat and cotton fields [10]. The outbreak of sleeping sickness in 1911 up to 1930 leads to the evacuation of people around the Lake George and Ishasha areas [10]. As such, resettling people who were affected by sleeping sickness heavily contributed to the expansion of the protected area system in Uganda [11,12].

It is estimated that 70% of Uganda was occupied by game animals in the 1920s [5,11]. However, as the human population increased, there was an increase in human–wildlife conflicts. As such, the Protectorate administration established the Elephant Control Department in the 1920s, which was tasked with confining wildlife to game reserves [11]. Although a number of species were subjected to control operations, elephants were the most prominent, and on average, 999 elephants were shot per year during the period of 1927–1958 totaling to 31,966 individuals mainly to prevent crop raiding and to maintain the carrying capacity for the country [5,12]. Most of the elephants (67%) were killed in the districts of Bunyoro, Mubende-Mengo, and Toro under the control program [12]. In addition, 8,170 elephants were killed by licensed hunters during the same period [12]. Currently (2019), there are 5,564 elephants in Uganda mainly in Queen Elizabeth National Park, Murchison Falls National Park, Kibale National Park, and Kidepo Valley National Park [13,14], and a few individuals left in Bwindi Impenetrable National Park, Katonga Wildlife Reserve, and Semliki Wildlife Reserve. There was widespread poaching in the 1970s and 1980s due to civil strife that led to a reduction of wildlife inside and outside protected areas [11]. However, with the return of law and order, most of the wildlife numbers in Uganda's protected areas increased steadily and are currently stable or showing a positive trend [11,14].

Most species had lost about 75% of their original range by 1962 due to

I. Poaching by use of wire snares, traps, pitfalls, fire arms
II. Increase in human population and associated settlements resulting in habitat loss and degradation
III. Agricultural extension resulting habitat loss and degradation.

The reduction in ranging patterns for most species resulted into the disruption of seasonal migration routes (Atlas of Uganda, 1962, Brooks and Buss, 1962) and predator–prey disruptions. Elephants used to migrant along well-established routes between Bunyoro, Buganda, Acholi, Toro, Bugisu, and Ankole regions in search for new pastures and breeding grounds. Due to pressure from other competing land uses, wildlife has retreated into protected areas with little or no connectivity to maintain gene flow. Figure 14.4 shows the distribution of wildlife in Uganda in the 1950s. The ranges were approximated from location records showing the distribution of game animals in the Uganda Atlas of 1962.

Discussion

The long-term existence and functioning of protected areas in Uganda are likely to be negatively impacted by the proposed and ongoing, industrial, and infrastructure development projects, and commercial agriculture to feed the 42 million Ugandans. Uganda has already lost about 75% of its biodiversity to human influence compared to natural causes in reference to the 1920s. Studies quantifying the responses of large mammals towards industrial activities in protected areas have shown that there is direct and indirect habitat and species loss [15–17].

As such, we propose that any industrial developments within and outside protected areas should follow the mitigation hierarchy and also apply the International Finance Corporation performance Standard 6 (PS6) on Biodiversity Conservation and Sustainable Management of Living Natural Resources [18,19]. The mitigation hierarchy seeks to limit the negative direct and indirect impacts of developmental projects on biodiversity and ecosystem services guided by the principle of achieving no net loss, and if possible a net positive impact to biodiversity and society.

The mitigation hierarchy is based on four key steps that are implemented sequentially:

1. Avoidance
2. Minimization of impacts
3. Restoration of impacts
4. Offsetting residual impacts.

Avoidance is the most important and effective step of the mitigation hierarchy. Measures taken during the early stages of implementing the mitigation hierarchy (avoidance and minimization) are likely to be more effective in reducing the anticipated project impacts compared to the later stages of restoration and offsetting. When the development projects are allowed to take place in protected areas and KBAs, extreme caution should be taken to avoid the most sensitive habitats at project design stage by considering project alternative sites. More so, irreplaceable sites hosting endemic species such as Bwindi Impenetrable National Park with 53% of the remaining mountain gorilla population in the world and Lake Victoria with endemic fisheries should never be developed at all.

If development projects of high public interest and avoidance are not possible, then measures should be taken to minimize the duration and extent of impacts through locating camps outside protected areas, traffic control, using advanced technologies, and domesticating international good practices

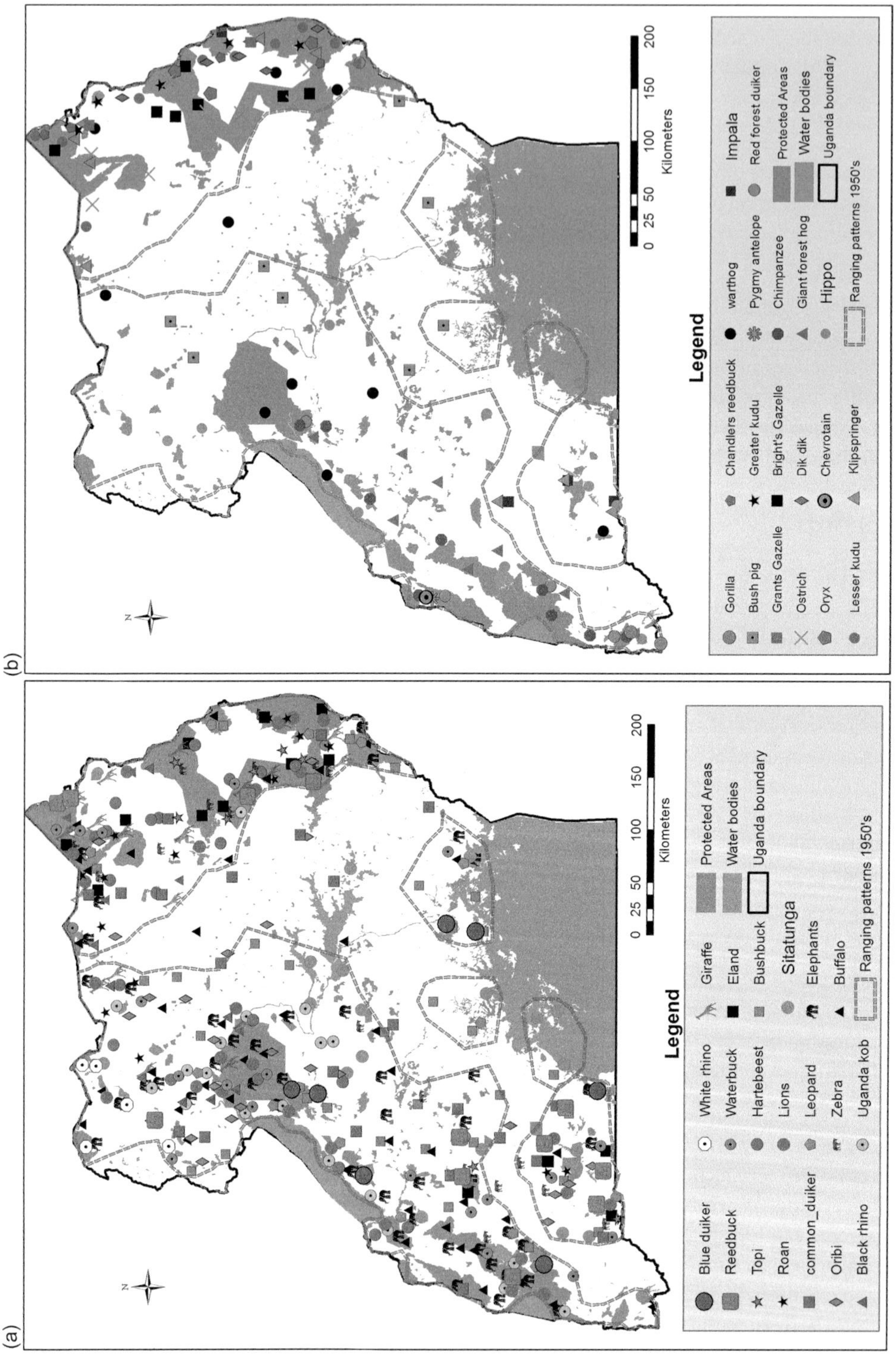

FIGURE 14.4 **(See color insert.)** Distribution of wildlife in Uganda in the 1950s (a) and the current protected area system (b).

for a particular industrial sector. Following the implementation of the avoidance and minimization measures on project impacts, the third step in the mitigation hierarchy is restoration. The target for restoration measures is to repair the damage on biodiversity and ecosystem services that could have remained after the application of avoidance and minimization measures. Some of restoration measures include revegetating the project footprint with native vegetation types.

The conservation outcome for the application of the mitigation hierarchy is to achieve no net loss or preferably net gain. However, given that the Uganda has lost about 75% of its biodiversity compared to the 1920 level and industrial project developments will continue to contribute to the decline in biodiversity, it may not be possible to achieve the conservation outcome of no net loss or net gain in biodiversity nationally. As such, industrial companies should avoid operating within protected areas, KBAs, wetlands of high conservation value, and other irreplaceable conservation sites in Uganda.

References

1. NEMA (2016). National Biodiversity Strategy and Action Plan II (2015–2025). https://nema.go.ug/sites/all/themes/nema/docs/NBSAP%20Uganda%202015%20-%20Re-designed.pdf
2. Winterbottom, B. and Eilu, G., (2006). Uganda Biodiversity and Tropical Forest Assessment, EPIQ II Task Order No. 351, International Resources Group, Washington, DC.
3. Mongabay (2016). The top 10 most biodiverse countries. In *What Are the Most Biodiverse Countries?* Retrieved from https://news.mongabay.com/2016/05/top-10-biodiverse-countries/.
4. Plumptre, A.J., Ayebare, S., Behangana, M., et al. (2019) Conservation of vertebrates and plants in Uganda: Identifying Key Biodiversity Areas and other sites of national importance. *Conservation Science and Practice* 1:e7. doi:10.1111/csp2.7.
5. Department of Lands and Survey, Uganda (1962). Atlas of Uganda; Game Conservation. Government of Uganda, Entebbe, Uganda.
6. IUCN (2016). *A Global Standard for the Identification of Key Biodiversity Areas*. Version 1.0 (p. 37). IUCN: Gland, Switzerland.
7. Mena, C.F., Barbieri, A.F., Walsh, S.J., Erlien, C.M., Holt, F.L., and Bilsborrow, R.E., (2006). Pressure on the Cuyabeno Wildlife Reserve: development and Land use/cover change in the northern Ecuadorian amazon. *World Development* 34(10): 1831–1849
8. Pomeroy, D. Tushabe, H. and Loh, J. (2017). *The State of Uganda's Biodiversity 2017.* National Biodiversity Data Bank. Makerere University: Kampala.
9. World Bank (2018). *Closing the Potential-Performance Divide in Ugandan Agriculture*. World Bank. © World Bank: Washington, DC.
10. Blomley, T. (2000). Woodlots, woodfuel, and wildlife. Lessons from QENP. Unpublished report: CARE, Uganda.
11. Olupot, W., Parry, L., Gunnes, M., and Plumptre, A., (2010). *Conservation Research in Uganda's Savannas: A Review of Park History, Applied Research, and Application of Research to Park Management*. Nova Science Publishers, Inc: New York.
12. Brooks, A.C., and Buss, I.O., (1962). Past and present state of the elephant in Uganda. *Journal of Wildlife Management* 26: 30–50.
13. UWA (2016). Elephant conservation action plan for Uganda 2016–2026. Digiprint Systems (U) Ltd.
14. UWA (2018). State of Wildlife Resources in Uganda 2018. Uganda Wildlife Authority (UWA), Kampala.
15. Ayebare, S. (2011). Influence of Industrial Activities on the Spatial Distribution of Wildlife in Murchison Falls National Park, Uganda. Unpublished MSc thesis, University of Rhode Island, South Kingstown, RI, USA.
16. Mudumba, T. and Jingo, S., (2014). Murchison Falls National Park Lions: population structure, ranging and key threats to their survival. Report to USAID through the WILD Program of WCS.

17. Plumptre, A.J., Ayebare, S. and Mudumba, T. (2015). Impacts of Oil Exploration on Elephants: an assessment in Murchison Falls National Park. Unpublished Report for Total E&P, Uganda.
18. Ekstrom, J., Bennun, L. and Mitchell, R. A. (2015). Cross-sector Guide for Implementing the Mitigation Hierarchy.Cross Sector Biodiversity Initiative. The Biodiversity Consultancy. Available at: http://www.ipieca.org/resources/good-practice/a-cross-sector-guide-for-implementing-the-mitigation-hierarchy/.
19. IFC (2012). Guidance Note 6: Biodiversity Conservation and Sustainable Management of Living Natural Resources. International Finance Corporation, World Bank Group.

15

Weeds: Seed Banks and Seed Dormancy

Introduction 129
Characteristics of the Soil Seed Bank 129
Periodicity and Intermittent Germination 130
Weed Seeds and Depth of Burial 130
Persistence of Weed Seeds in Soil 130
Environmental Factors That Affect Seed Dormancy 131
Temperature • Light
Conclusion 131
References 132

Lynn Fandrich
Oregon State University

Introduction

The relationship between seed dormancy and success of a plant as an agricultural weed is significant. Holm et al.[1] estimated that there are about 250 significant weeds in world agriculture. If the top 40 weeds classified as "serious" are selected by ranking them according to the number of countries in which they are considered "serious" weeds, it can be shown, without exception, that each one of the 40 species has dormancy. If weed seed were forever dormant, or simply nondormant, the complexities of weed management would be greatly reduced. However, weed seeds vary widely with respect to degree, duration, and source of dormancy. The existence of large populations of weed seed with varying degrees and states of dormancy is the basis for the annual weed problem. Dormancy allows a weed seed to avoid germination when conditions are not appropriate to complete its life cycle and prolongs population survival in the form of a soil seed bank.

Characteristics of the Soil Seed Bank

All of the viable seed present on the surface and in the soil is usually described as the soil seed bank. The number of seeds in the seed bank is determined by the rate of input in the seed rain, minus the losses resulting from disease, predation, and germination. The seed bank consists of new seeds recently shed by a parent plant, as well as older seeds that have persisted in the soil for several years. All of these seeds will vary in depth of burial and state of dormancy. Only a small proportion of seeds, varying between and among species, germinate when conditions are favorable. It is generally assumed that annual species contribute up to 95% of the seeds in the agricultural seed bank and only 2 to 10% of weed seeds emerge as seedlings each year.

Because most weeds are capable of producing a high number of seeds, especially compared to cultivated species, a few uncontrolled plants can rapidly increase the number of seeds in the soil seed bank. Data collected from plants grown in monoculture showed that wild oat (*Avena fatua*) produced 372–623 seeds,[2] common lambsquarters (*Chenopodium album*) produced 75,600–150,400 seeds,[3] and Palmer amaranth (*Amaranthus palmeri*) produced 200,000 to 600,000 seeds per plant.[4] It should be noted that biotic (inter- and intraspecific competition) and abiotic (climate) factors affect seed production, and most plants in agricultural situations do not often approach these high seed production values.

Large numbers of seeds are dispersed both horizontally and vertically in the soil. Plowing, chiseling, disking, and other forms of tillage are major ways that seeds are moved and covered with soil. After a number of years of soil disturbance, a relatively stable vertical distribution of seeds is reached in the tillage zone. Early estimates of seed numbers in agricultural soils exceeded 100,000 seeds per hectare in the tillage layer. Samples of soil on well managed farms in Minnesota averaged 16 million seeds per hectare, and as many as 45 million per hectare have been counted.

Periodicity and Intermittent Germination

Most weed species exhibit annual periodicity in germination and emergence restricted to certain times of the year. This behavior was part of the original evidence for annual cycles in the relief and imposition of seed dormancy. Observations on seed germination and seedling emergence from soil samples placed in the greenhouse revealed that there was a typical periodicity, as well as a period of maximum germination for more than 100 weed species.[5] Exhumed seeds of corn spurry *(Spergula arvensis)* showed clear seasonal changes in dormancy during three successive years of germination tests.[6] Often, seeds are conditionally dormant, i.e., the appropriate conditions for germination are not present; seeds of winter annual species will not germinate in late spring, and spring annual species will not germinate in late fall.

Weed Seeds and Depth of Burial

Seed germination and emergence of seedlings from various depths below the soil surface depend on the degree of dormancy present and conditions of the immediate environment. However, dormancy-breaking stimuli are most effectively encountered and germination is rapid when seeds remain at or near the soil surface. There is an overall agreement that the number of seedlings and rate of seedling emergence decrease proportionately with increases in the depths of seed burial. Examination of seeds recovered from 12 cm deep showed that approximately 85% were completely dormant.[7] The results of these investigations also revealed a strong association between seed mass and maximum depth of emergence, with the heaviest seeds emerging from the greatest depths.

Persistence of Weed Seeds in Soil

It is quite common for seeds to be dormant when they are fully mature on the parent plant and for dormancy to be lost after they are shed. The term "after-ripening" has frequently been used to denote the interaction between environment and seed over time that leads to a loss of dormancy.[8] The length of the after-ripening period varies between and among species and by seed location on the plant. Most weed species show one or two periods of high percent germination in seed produced during any given year, and the percent germination in subsequent years can be quite low. The mean length of the dormancy period in populations drives turnover in the seed bank. Grime[9] makes a distinction between "transient" (complete turnover in less than one year) and "persistent" seed banks (some seed remains in the bank longer than one year). The term "aging" is sometimes used incorrectly to describe after-ripening. Aging reduces seed viability and seedling vigor.

Environmental Factors That Affect Seed Dormancy

Temperature

Temperature is the major environmental factor to cause a change in seed dormancy. Temperatures may break dormancy but also send seeds into dormancy when other environmental conditions are limited. Seeds of summer and winter annual weeds may be conditionally dormant due to temperature influences. Experimentally, the effect of chilling is tested through exposure of imbibed seeds to low temperatures, usually around 0 to 1°C, for a substantial period. Seeds of bur buttercup (*Ranunculus testiculatis*) germinated only at very cold incubation temperatures, and maximum germination occurred at 5°C.[10]

Seeds of winter annual species must be exposed to high summer temperatures for several months to germinate at fall temperatures. If lack of moisture prevents seeds of obligate winter annuals from germinating in the fall, low winter temperatures may induce dormancy. Consequently, viable seeds that fail to germinate in the fall cannot germinate in spring because they are dormant. Seeds of the winter annual Japanese brome (*Bromus japonicus*) were forced into dormancy with winter temperatures and would not germinate during the spring and summer, despite favorable germination conditions.[11]

Under natural conditions in the soil, seeds are subjected to fluctuating temperatures. It has been observed that random fluctuations in temperature, in and above a low temperature, are as effective in promoting germination as exposure to constant low temperatures. It has also been found that seeds which remain dormant at constant temperatures may be induced to germinate by a diurnal fluctuation of temperatures. Thompson and Grime[12] investigated the effect of fluctuating temperatures on the germination of 112 species of herbaceous plants. Forty-six of the species examined were found to have their germination stimulated by temperature fluctuations in the light.

Light

Exposure to light breaks dormancy in many weed species, but there may be situations in which light has no effect or even inhibits germination. Baskin and Baskin[13] observed that among 142 species, germination of 107 was promoted by light, 32 were unaffected by the difference in light and dark conditions, and 3 were inhibited by light. Wesson and Wareing[14] concluded that "under natural conditions in the field, the germination of buried seeds following a disturbance of the soil is completely dependent upon exposure of seed to light." Even a short exposure or "light break" was sufficient to significantly increase the total number of seedlings emerging from soil.

Conclusion

Understanding the processes of dormancy loss and germination is essential to the development of good weed management programs. Currently, most management programs aim to control emerged weed populations and affect the quantity of seed returned to the soil, but it should be recognized that almost all agronomic practices also affect weed seed dormancy and germination through manipulation of the soil environment. Because many agricultural weeds exhibit periodicity of emergence, tillage treatments, planting dates, and other cultural practices can be altered to follow peak germination times and maximize depletion of the seed bank. A natural population of viable weed seeds in the field declined at a rate of approximately 50% per year with frequent cultivation.[15] A review of the utilization of specific agronomic practices to manipulate weed seed dormancy and germination requirements for weed management is available from Dyer[16] and others.[17,18]

References

1. Holm, L.; Pancho, J.V.; Herberger, J.P.; Plucknett, D.L. *A Geographical Atlas of World Weeds*; John Wiley and Sons: New York, 1979.
2. O'Donnell, C.C.; Adkins, S.W. Wild oat and climate change: The effect of CO_2 concentration, temperature, and water deficit on the growth and development of wild oat in monoculture. Weed Sci. **2001**, *49*, 694–702.
3. Colquhoun, J.; Stoltenberg, D.E.; Binning, L.K.; Boerboom, C.M. Phenology of common lambsquarters growth parameters. Weed Sci. **2001**, *49*, 177–183.
4. Keeley, P.E.; Carter, C.H.; Thullen, R.J. Influence of planting date on growth of Palmer amaranth *(Amaranthus palmeri)*. Weed Sci. **1987**, *35*, 199–204.
5. Brenchley, W.E.; Warrington, K. The weed seed population of arable soil. I. Numerical estimation of viable seeds and observations on their natural dormancy. J. Ecol. **1930**, *18*, 235–272.
6. Bouwmeester, H.J.; Karssen, C.M. The effect of environmental conditions on the annual dormancy pattern of seeds of *Spergula arvensis*. Can. J. Bot. **1993**, *71*, 64–73.
7. Roberts, H.A. Studies on the weeds of vegetable crops. II. Effect of six years cropping on the weed seeds in the soil. J. Ecol. **1962**, *50*, 803–813.
8. Simpson, G.M. *Seed Dormancy in Grasses;* Cambridge University Press: Cambridge, England, 1990.
9. Grime, J.P. The role of seed dormancy in vegetation dynamics. Ann. Appl. Biol. **1981**, *98*, 555–558.
10. Young, J.A.; Martens, E.; West, N.E. Germination of bur buttercup seeds. J. Range Manag. **1992**, *45*, 358–362.
11. Baskin, J.M.; Baskin, C.C. Ecology of germination and flowering in the weedy winter annual grass *Bromus japonicus*. J. Range Manag. **1981**, *34*, 369–372.
12. Thompson, K.; Grime, J.P. A comparative study of germination responses to diurnally-fluctuating temperatures. J. Appl. Ecol. **1983**, *20*, 141–156.
13. Baskin, C.C.; Baskin, J.M. Germination ecophysiology of herbaceous plant species in a temperate region. Am. J. Bot. **1988**, *75*, 286–305.
14. Wesson, G.; Wareing, P.F. The role of light in the germination of naturally occurring populations of buried weed seeds. J. Exp. Bot. **1969**, *20*, 402–413.
15. Burnside, O.C.; Wilson, R.G.; Wicks, G.A.; Roeth, F.W.; Moomaw, R.S. Weed seed decline and buildup in soils under various corn management systems across Nebraska. Agron. J. **1986**, 78, 451–454.
16. Dyer, W.E. Exploiting weed seed dormancy and germination requirements through agronomic practices. Weed Sci. **1995**, *43*, 498–503.
17. Forcella, F. Real-time assessment of seed dormancy and seedling growth for weed management. Seed Sci. Res. **1998**, 8, 201–209.
18. Mulugeta, D.; Stoltenberg, D.E. Weed and seedbank management with integrated methods as influenced by tillage. Weed Sci. **1997**, *45*, 706–715.

II

Ecosystem Type, Function, and Service

16

Community Forestry: Sustainability and Equity Issues

Bill Buffum
University of Rhode Island

Introduction 135
Operation of Community Forests 135
Sustainability of Silviculture Practices in CFs 136
Equity in Community Forests 137
Attributes of CFs and National Forestry Policies That Result in Successful Management of CFs 138
Conclusions 139
References 139

Introduction

In recent years, many governments around the world have given local communities the authority to manage blocks of national forest [1–3]. By 2016, governments had given communities legal rights to manage 732 million hectares of national forest in 62 countries, representing 28% of the forests in these countries [4]. This approach has been particularly widespread in South Asia, where Nepal and India have the oldest and largest programs. In Nepal, 22,266 community forest (CF) user groups were managing 22 million hectares of national forest in 2016 [5], while in India, 84,632 forest protection committees were managing 17 million hectares of national forests in 2011 [6]. The experience of Nepal and India in the 1980s and 1990s influenced the development of forestry programs in many other countries throughout the world [7]. Overall, community forestry programs have been highly successful at providing a range of economic and social benefits to the participating user groups [1–3]. However, concerns have been raised about the ability of user groups to manage their CFs on a sustainable basis and share the CF products in an equitable manner. This entry reviews the literature related to the sustainability and equity of forest management in CFs.

Operation of Community Forests

Community forestry has recently been defined as "forestry practices which directly involve direct forest users in common decision making processes and implementation of forestry activities" [8, p. 3]. There are many variations on how CFs operate in different countries. The key feature is that a group of traditional users of a national forest is given legal rights to manage a specific block of national forest under the guidance of the national forestry department (FD). Typically, the users must prepare a community forestry management plan for approval by the FD. Then, a CF management committee elected by the members assumes responsibility for management of the CF. The CF management committee organizes regular meetings of the users to implement the CF management plan. An important activity in all CFs is patrolling to ensure that illegal harvesting of non-timber forest products or forest grazing is not taking place. In some CFs, one or more forest guards are hired, while in other CFs, the members take turns patrolling the forest on a voluntary basis. Besides patrolling, other management activities include nursery management, tree-planting in degraded areas, thinning and tending operations, and harvesting, as well as grazing and fire protection. When members require wood products, they request the CF management committee to issue a permit. The management committee verifies whether the member has already received his/her allocation of products according to the guidelines in the management plan. The CFs generate income from royalty payments from the harvesting of wood products and contributions from visitors. The funds are used for a variety of purposes, including providing loans and emergency support to members, constructing CF offices, providing meals during work days and meetings, and contributing to local development activities. In some countries, the CF user group is obligated to share some of the revenue with the FD.

Sustainability of Silviculture Practices in CFs

Community forestry programs have had a positive impact on forest cover and production of wood products. CFs have been found to exhibit greater increases in forest cover than government forests, according to studies based on analyses of satellite imagery and aerial photography [9–11] and forest inventories [12,13]. Community forestry has been linked to reduced rates of deforestation in countries such as Mexico [14]. Community forestry may support carbon sequestration as a function of increased timber stocks: a study of CFs in Nepal estimated that annual forest growth resulted in the sequestration of 4,700 tCO_2e per year in 630 hectares [15]. Community forestry also supports the efforts of governments in forest protection by increasing community participation in forest management: illegal logging on public lands has been reported to cost governments US$10–15 billion per year [16].

However, there are concerns that some CF user groups may lack the technical skills and discipline to harvest their CFs in a sustainable manner [3,16–18]. Not all CFs are successfully managed: "members of some groups fail to perceive the growing scarcity of their local forests, fail to create effective rules to counteract the incentives to overharvest, and fail to enforce their own rules" [19, p. 228]. In Nepal, it was reported that some CF user groups did not understand the forest inventory process, and the lack of simple approaches to regulating the harvests led to overharvesting, especially in CFs containing high-value timber [20]. Community ownership has been reported to have resulted in overharvesting and increased rates of deforestation in Papua New Guinea [16]. On the other hand, studies in Bhutan found that user groups were managing their CFs in accordance with their management plans without any indications of overharvesting [21,22] or negative impacts on species composition, regeneration, and productivity [23].

There is concern that user group harvesting approaches, which are usually based on single-tree selection, may not promote natural regeneration of preferred species and maintain forest diversity. This issue has rarely been studied in CFs, but in other forests, single-tree selection was reported to reduce diversity and favor shade tolerant species [24,25]. Another study reported that regeneration and stem density of important species had not recovered two decades after selective logging in India [26]. However, other studies have reported that single-tree selection caused less damage to forests than harvesting systems that create large gaps in the forest canopy [27], maintained tree diversity [28–31], and was operationally more efficient than group selection [32]. Decisions about whether single-tree selection is appropriate for a particular CF will have to be based on the silvics of the concerned tree species.

Additionally, the concern is that user groups may high-grade the CFs through excessive removal of the straightest stems of economically valuable species, resulting in residual stands containing crooked and otherwise deformed overstory trees and genetic selection of inferior genotypes [33–35]. This process of "high-grading," which occurs when forest managers maximize current profits rather than long-term productivity has been well documented in North America [36,37], but is widespread in many other regions.

Surprisingly, the sustainability of harvesting approaches in CFs has rarely been addressed by researchers: a recent review of participatory forest management (PFM) concluded that "the most conspicuous absence from the PFM literature is a thorough analysis of forestry itself, the technology of intervention in a forest ecosystem to provide a reliable supply of the products and benefits required by the users" [1, p. 2]. A number of authors have noted that professional foresters have not yet been able to provide silvicultural guidance to meet the diverse needs of community forestry user groups [1,3,38–40]. Victor [39, p. vi] summed up this issue as follows: "Unfortunately, the development and spread of alternative forest management practices (barefoot silviculture) for community forest management has yet to really begin. Most foresters' knowledge is limited to mensuration and silviculture developed for timber production at an industrial scale... As a result, it is difficult for government foresters to readily provide appropriate advice to users and to recognize good and bad community forestry practice or to suggest feasible improvements."

With the rapid expansion of community forestry in developing countries, it is important to assess the sustainability of CF harvesting practices, which are generally based on indigenous knowledge and traditional practices of selection cutting. If these harvesting practices are not sustainable, there is an urgent need for technical assistance and training.

Equity in Community Forests

Despite the many positive achievements of CFs, there is evidence that community forestry programs do not always provide benefits equally to all social groups, and in some cases, CFs may actually increase inequality [7,41]. The emerging equity issues are part of what the donor and practitioner literature refers to as the "second-generation" issues of community forestry [1]. Two aspects of equity apply to community forestry: political equity, which involves participation in decision-making and the ability of stakeholders to express their ideas and concerns; and economic equity, which involves the distribution of benefits [42]. In the context of community forestry, equity is usually understood to refer to "fairness" rather than "equality": the management of a CF is considered to be equitable if the members view their share of use and decision-making rights as being fair [43].

Several studies have investigated the lack of political equity (participation in decision-making) in CFs. Menzies [2, p. 17] noted that "traditional societies are rarely equitable and participatory... there may be aspects of traditional social capital that do not favor 'just and wise' management of forest resources." Management committees in many countries were dominated by political, social, or economic elites and thus resulted in "elite capture" of the CF management process by men and high caste groups [17,44,45]. Cultural norms have discouraged women from participating in CF meetings [46], and women were excluded from decision-making even in seemingly participatory user groups [47]. However, there are also examples of positive political equity in CF: in contrast to the above studies, some authors have reported the positive impacts of increased understanding and collaboration in CFs between dominant and disadvantaged ethnic groups, and between men and women [48–51], and other studies have noted that involvement of elites in community development activities does not necessarily lead to elite capture [52,53].

Other studies have documented a lack of economic equity (distribution of benefits) in some CFs. For example, some management committees in Nepal practiced favoritism in distributing products [46,54], CF rules were biased towards meeting the needs of richer households [55], and influential members were allowed to graze their animals in CFs even when the bylaws of the CF prohibited forest grazing [56]. In other CFs, the poorest members of the user groups had less access to CF products than the richer and higher caste members [57–61], and benefited less from loans from CF user group funds [62].

Equity has environmental consequences in addition to social consequences: a study of CFs in the Indian Himalayas found that improved forest condition was associated with reduced levels of conflict among users and greater involvement of women in decision-making—the CFs that did not have internal conflicts were better able to make effective decisions about forest protection, and women were able to make sound management decisions due to their strong involvement in the collection of forest products [63]. Thus, national governments have many reasons for adopting community forestry programs that ensure equity in CFs. The linkage between equity and conservation is reflected in the objectives of the UN Convention on Biological Diversity, which include conservation of biological diversity, sustainable use of its components, and fair and equitable benefit sharing [64].

Attributes of CFs and National Forestry Policies That Result in Successful Management of CFs

Several authors have conducted multi-country studies to identify the attributes of successful user groups in an attempt to understand why some forestry user groups are better able than others to manage their forests sustainably and equitably [43,65–67]. These studies are complicated by different views of what constitutes success in a CF. The most common indicator of successful forest management is improved forest condition, which was used in recent analyses of forest governance in the Indian Himalaya [63] and Nepal [68]. However, other indicators of success, such as equity of benefit sharing or fulfillment of local needs, may be equally important [66]. For example, some CFs in Nepal exhibited significant increase in forest cover but produced few benefits for the user groups, and the CF products were unfairly distributed among the members, especially to poor users who lacked access to forest products on private land [69].

The set of positive attributes prepared by Ostrom [65] (Table 16.1) is more detailed than other recent analyses of forestry user groups [63,66,67,70]. Ostrom's list includes attributes of user groups that are conducive to a community adopting or changing rules limiting access to their forests [65, p. 4]. These attributes are relevant because CF user groups in many countries must initiate the process of forming a CF, even though the government subsequently plays a significant role in supporting the user group. Ostrom's list also includes several attributes of CFs—including that the forest should not be so degraded that it is useless to organize, or so underutilized that there is little advantage from organizing. Governments in many countries have been unwilling to hand over well-stocked forests for community management [2,3]. For example, forest policies in Nepal since 2000 have restricted the handover of well-stocked forests in the Terai/Churia regions while continuing to promote the handover of relatively low value forests in the mid-hills [48,71]. The Forest Department in India reportedly retained control of the most productive forest land and allocated fragmented and degraded patches for community management [71]. In Bhutan, however, after initially only handing over degraded

TABLE 16.1 Summary of Ostrom's Attributes of Successful User Groups [65]

The forest should not be so degraded that it is useless to organize, or so underutilized that there is little advantage from organizing.
Reliable and valid information about the general condition of the resource should be available at reasonable costs.
The availability of resource units should be relatively predictable.
The resource should be sufficiently small, given the transportation and communication technology in use, so that users can develop accurate knowledge of external boundaries and internal microenvironments.
Users should be dependent on the resource for a major portion of their livelihood or other variables of importance to them.
Users should have a shared image of the resource and how their actions affect each other and the resource.
Users should have a sufficiently low discount rate in relation to future benefits to be achieved from the resource.
Users with higher economic and political assets should be similarly affected by a current pattern of use.
Users should trust each other to keep promises and relate to one another with reciprocity.
Users should be able to determine access and harvesting rules without external authorities countermanding them.
Users should have learned at least minimal skills of organization through participation in other local associations or learning from neighboring groups.

forests to communities, the government decided to give villagers a stronger incentive to participate by stipulating that CFs should be approximately 50% well stocked [72,73]. This provision was probably influenced by the abundance of forest resources in Bhutan, but appears to reflect a genuine willingness to give up control of some national forests in order to promote economic development at the village level [74].

Conclusions

Community forestry programs are expanding throughout the world. The experience to date has largely been positive: CFs in many countries have been able to provide a range of social and economic benefits, and many CF user groups have demonstrated that they are capable of effectively managing the forest resources in their CFs. However, there are still questions about the sustainability of harvesting and management systems in CFs, and the lack of political and economic equity in some CFs is troubling. It is crucial that additional research be conducted on these issues to provide guidance for the development of national community forestry policies and programs so that appropriate technical support is provided to CF user groups to maximize their ability to manage CFs in a sustainable and equitable manner.

References

1. Lawrence, A., Beyond the second generation: Towards adaptiveness in participatory forest management. *Veterinary Science, Nutrition and Natural Resources*, 2007. **2**(28): p. 15.
2. Menzies, N.K., 'Global Gleanings' Lessons from six studies of community based forest management, *Report for Ford Foundation's Environment and Development Affinity Group (EDAG)*. 2002, Ford Foundation: Berkley.
3. Carter, J., Recent Experience in Collaborative Management - CIFOR Occasional Paper No. 43. 2005, Centre for International Forestry Research: Bogor.
4. Gilmour, D., *Forty Years of Community-Based Forestry: A Review of its Extent and Effectiveness*. 2016, FAO: Rome.
5. Bhandari, P.K., et al., Importance of community forestry funds for rural development in Nepal. *Resources*, 2019. **8**(2): p. 85.
6. Bhat, A W., Forest conservation and livelihood generation through joint forest management in India. *Forest*, 2018. **3**(2).
7. Hobley, M., Building state-people relationships in forestry, In *Forest Policy and Environment Programme: Grey Literature*, G. Shepard et al., Editors. 2005, Forest Policy and Environment Programme, Overseas Development Institute: London. p. 7.
8. Maryudi, A., et al., Back to basics: Considerations in evaluating the outcomes of community forestry. *Forest Policy and Economics*, 2012. **14**(1): pp. 1–5.
9. Gautam, A.P., G.P. Shivakoti, and E.L. Webb, Forest cover change, physiography, local economy, and institutions in a mountain watershed in Nepal. *Environmental Management*, 2004. **33**(1): pp. 48–61.
10. Gautam, A.P., E.L. Webb, and A. Eiumnoh, GIS assessment of land use/land cover changes associated with community forestry implementation in the Middle Hills of Nepal. *Mountain Research and Development*, 2002. **22**(1): pp. 63–69.
11. Sakurai, T., et al., Efficiency of timber production in community and private forestry in Nepal. *Environment and Development Economics*, 2004. **9**: pp. 539–561.
12. Branney, P. and K.P. Yadav, *Change in CF Condition and Management*. 1998, Nepal-UK Community Forestry Project: Kathmandu.
13. Yadav, N.P., et al., Forest management and utilization under community forestry. *Journal of Forest and Livelihood*, 2003. **3**(1): pp. 37–50.

14. Barsimantov, J. and J. Kendall, Community forestry, common property, and deforestation in eight Mexican states. *The Journal of Environment & Development*, 2012. **21**(4): pp. 414–437.
15. Bhattarai, T., et al., The carbon sequestration potential of community based forest management in Nepal. *International Journal of Climate Change*, 2012. **3**(2): pp. 233–254.
16. White, A. and A. Martin, *Who Owns the World's Forests? Forest Tenure and Public Forests in Transition*. 2002, Center for International Environmental Law: Washington, DC. p. 30.
17. Kellert, S.R., et al., Community natural resource management: Promise, rhetoric, and reality. *Society & Natural Resources*, 2000. **13**(8): pp. 705–715.
18. Roberts, E.H. and M.K. Gautam, *Community Forestry Lessons for Australia: A Review of International Case Studies*. 2003, The Australian National University: Canberra. p. 16.
19. Gibson, C.C., M.A. McKean, and E. Ostrom, Some initial theoretical lessons, In *People and Forests: Communities, Institutions, and Governance*, C.C. Gibson, M.A. McKean, and E. Ostrom, Editors. 2000, MIT Press: Cambridge, MA. pp. 227–242.
20. Maharjan, M.R., Yield regulation techniques for community forest management in Nepal: Opportunities and constraints, In *Cultivating Forests - Alternative Forest Management Practices and Techniques for Community Forestry*, M. Victor and A. Barash, Editors. 2001, RECOFTC: Bangkok. pp. 189–197.
21. Buffum, B., G. Gratzer, and Y. Tenzin, The sustainability of selection cutting in a late successional broadleaved community forest in Bhutan. *Forest Ecology and Management*, 2008. **256**: pp. 2084–2091.
22. Buffum, B., et al., *Equity and Sustainability of Community Forestry in Bhutan: Analysis of Three Village Case Studies*. 2005, Royal Government of Bhutan: Thimphu.
23. Moktan, M.R., L. Norbu, and K. Choden, Can community forestry contribute to household income and sustainable forestry practices in rural area? A case study from Tshapey and Zariphensum in Bhutan. *Forest Policy and Economics*, 2016. **62**: pp. 149–157.
24. Jenkins, M.A. and G.R. Parker, Composition and diversity of woody vegetation in silvicultural openings of southern Indiana forests. *Forest Ecology and Management*, 1998. **109**: pp. 57–74.
25. Schuler, T.M., Fifty years of partial harvesting in a mixed mesophytic forest: Composition and productivity. *Canadian Journal of Forest Research*, 2004. **34**: pp. 985–997.
26. Ganesan, R. and P. Davidar, Effect of logging on the structure and regeneration of important fruit bearing trees in a wet evergreen forest, southern Western Ghats, India. *Journal of Tropical Forest Science*, 2003. **15**(1): pp. 12–25.
27. Rice, R.E., R.E. Gullison, and J.W. Reid, Can sustainable management save tropical forests? *Scientific American*, 1997. **276**(4): pp. 44–49.
28. Cannon, C.H., D.R. Peart, and M. Leighton, Tree species diversity in commercially logged Bornean rainforest. *Science*, 1998. **281**(5381): pp. 1366–1368.
29. Dekker, M. and N.R. de Graaf, Pioneer and climax tree regeneration following selective logging with silviculture in Suriname. *Forest Ecology and Management*, 2003. **172**(2–3): pp. 183–190.
30. Montagnini, F., et al., Tree regeneration and species diversity following conventional and uniform spacing methods of selective cutting in a subtropical humid forest reserve. *Biotropica*, 1998. **30**(3): pp. 349–361.
31. Webb, E.L. and R. Peralta, Tree community diversity of lowland swamp forest in Northeast Costa Rica, and changes associated with controlled selective logging. *Biodiversity and Conservation*, 1998. **7**(5): pp. 565–583.
32. Suadicani, K. and D. Fjeld, Single-tree and group selection in montane Norway spruce stands: Factors influencing operational efficiency. *Scandinavian Journal of Forest Research*, 2001. **16**: p. 79–87.
33. Clatterbuck, W.K., *Treatments for Improving Degraded Hardwood Stands*. Publication FOR-104. 2006, University of Kentucky: Knoxville, TN.

34. Nyland, R.D. Rehabilitating cutover stands: Some ideas to ponder. In *Proceedings of the Conference on Diameter-Limit Cutting in Northeastern Forests*, 2006, Northeastern Research Station, U.S. Forest Service: Amherst, MA.
35. Oliver, C.D. and B.C. Larson, *Forest Stand Dynamics*. Update ed. 1996, Wiley: New York. p. 520.
36. Kenefic, L.S. and R.D. Nyland, *Diameter-Limit Cutting and Silviculture in Northeastern Forests*. 2005, Forest Service, United States Department of Agriculture: Newtown Square.
37. Kenefic, L.S. and R.D. Nyland. Facilitating a dialogue about diameter-limit cutting. In *Proceedings of the conference on diameter-limit cutting in northeastern forests,* 2006, Northeastern Research Station, U.S. Forest Service: Amherst, MA.
38. Donovan, D.G. Where's the forestry in community forestry? In *Cultivating Forests: Alternative Forest Management Practices and Techniques for Community Forestry,* M. Victor and A. Barash, Editors. 2001. RECOFTC: Bangkok.
39. Victor, M. The search for a barefoot silviculture: Too little, too early? In *Cultivating Forests: Alternative Forest Management Practices and Techniques for Community Forestry,* M. Victor and A. Barash, Editors. 2001. RECOFTC: Bangkok.
40. Pokharel, R.K., et al., Assessing the sustainability in community based forestry: A case from Nepal. *Forest Policy and Economics*, 2015. **58**: pp. 75–84.
41. Schreckenberg, K., C. Luttrell, and C. Moss, Participatory forest management: An overview, In *Forest Policy and Environment Programme: Grey Literature,* G. Shepard et al., Editor. 2006, Forest Policy and Environment Programme, Overseas Development Institute: London.
42. Mahanty, S., et al., Equity in community-based resource management, In *Hanging in the Balance: Equity in Community-Based Natural Resource Management in Asia,* S. Mahanty, et al., Editors. 2006, RECOFTC and East-West Center: Bangkok. pp. 1–13.
43. McKean, M.A., Common property: What is it, what is it good for, and what makes it work? In *People and Forests : Communities, Institutions, and Governance,* C.C. Gibson, M.A. McKean, and E. Ostrom, Editors. 2000, MIT Press: Cambridge, MA. pp. 27–55.
44. Uprety, D.R., Conflicts in natural resource management: Examples from community forestry. *Jahrbuch der Österreichischen Gesellschaft für Agrarökonomie*, 2006. **15**: pp. 143–155.
45. Race, D., R. Thwaites, and B. Devkota, Community forestry and community development in Nepal, In *Community Forestry in Nepal,* R. Thwaites, R. Fisher, and M Poudel, Editors. 2017, Routledge: London. pp. 82–107.
46. Springate-Baginski, O., et al., Community forest management in the middle hills of Nepal: The changing context. *Journal of Forest and Livelihood*, 2003. **3**(1): pp. 5–10.
47. Agarwal, B., Participatory exclusions, community forestry, and gender: An analysis for South Asia and a conceptual framework. *World Development*, 2001. **29**(10): pp. 1623–1648.
48. Bhattarai, B., Widening the gap between Terai and hill farmers in Nepal: The implications of the new forest policy 2000, In *Hanging in the Balance: Equity in Community-Based Natural Resource Management in Asia*, S. Mahanty, et al., Editors. 2006, RECOFTC and East-West Center: Bangkok. pp. 143–161.
49. Buffum, B., A. Lawrence, and K.J. Temphel, Equity in community forests in Bhutan. *International Forestry Review*, 2010. **12**(3): pp. 187–199.
50. Luintel, H., Do civil society organizations promote equity in community forestry? A reflection from Nepal's experiences, In *Hanging in the Balance: Equity in Community-Based Natural Resource Management in Asia*, S. Mahanty, et al., Editors. 2006, RECOFTC and East-West Center: Bangkok. pp. 122–142.
51. Luintel, H., et al., The effect of the Nepal community forestry program on equity in benefit sharing. *The Journal of Environment & Development*, 2017. **26**(3): pp. 297–321.
52. Dasgupta, A. and V.A. Beard, Community driven development, collective action and elite capture in Indonesia. *Development and Change*, 2007. **38**(2): pp. 229–249.

53. Mansuri, G. and V. Rao, Community-based and -driven development: A critical review. *The World Bank Research Observer*, 2004. **19**(1): pp. 1–39.
54. Varalakshmi, V., Joint forest management and conflict in Haryana, India, In *Community-Based Forest Resource Conflict Management*, K. Means and C. Josayma, Editors. 2002, Food and Agriculture Organization, United Nations: Rome. pp. 321.
55. Adhikari, B., Poverty, property rights and collective action: Understanding the distributive aspects of common property resource management. *Environment and Development Economics*, 2005. **10**: pp. 7–31.
56. Pandit, B.H. and G.B. Thapa, Poverty and resource degradation under different common forest resource management systems in the mountains of Nepal. *Society and Natural Resources*, 2004. **17**: pp. 1–16.
57. Adhikari, B., S. Di Falco, and J.C. Lovett, Household characteristics and forest dependency: Evidence from common property forest management in Nepal. *Ecological Economics*, 2004. **48**(2): pp. 245–257.
58. Chhetry, B., et al., A framework for the analysis of community forestry performance in the Terai. *Journal of Forest and Livelihood*, 2005. **4**(2): pp. 1–16.
59. Malla, Y., H.R. Neupane, and P.J. Branney, Why aren't poor people benefiting more from community forestry? *Journal of Forest and Livelihood*, 2003. **3**(1): pp. 78–90.
60. Nightingale, A., Nature-society and development: Social, cultural and ecological change in Nepal. *Geoforum*, 2003. **34**(4): pp. 525–540.
61. Sarin, M., *Who is Gaining? Who is Losing? Gender and Equity Concerns in Joint Forest Management.* 1998, Society for the Promotion of Wastelands Development: New Delhi.
62. Pokharel, B. and M. Nurse, Forest and people's livelihood: Benefiting the poor from community forestry. *Journal of Forest and Livelihood*, 2004. **4**(1): pp. 19–30.
63. Agrawal, A. and A. Chhatre, Explaining success on the commons: Community forest governance in the Indian Himalaya. *World Development*, 2006. **34**(1): pp. 149–166.
64. Knapp, S., Dynamic diversity. *Nature*, 2003. **422**: p. 475.
65. Ostrom, E., Self-governance and forest resources, *CIFOR Occasional Paper No.* 20. 1999, CIFOR: Jakarta. p. 15.
66. Pagdee, A., Y.-s. Kim, and P.J. Daugherty, What makes community forest management successful: A meta-study from community forests throughout the world. *Society and Natural Resources*, 2006. **19**(1): pp. 33–52.
67. Persha, L., A. Agrawal, and A. Chhatre, Social and ecological synergy: Local rulemaking, forest livelihoods, and biodiversity conservation. *Science*, 2011. **331**(6024): pp. 1606–1608.
68. Nagendra, H., Tenure and forest conditions: Community forestry in the Nepal Terai. *Environmental Conservation*, 2002. **29**(4): pp. 530–539.
69. Shrestha, K.K. and P. McManus, The politics of community participation in natural resource management: Lessons from community forestry in Nepal. *Australian Forestry*, 2008. **71**(2): pp. 135–146.
70. Nagendra, H., Drivers of reforestation in human-dominated forests. *Proceedings of the National Academy of Sciences*, 2007. **104**(39): pp. 15218–15223.
71. Agrawal, A. and E. Ostrom, Collective action, property rights, and decentralization in resource use in India and Nepal. *Politics & Society*, 2001. **29**(4): pp. 485–514.
72. Chhetri, B.B. Review of experiences of social forestry and forestry extension in Bhutan. In *Social Forestry and Forestry Extension - Report of the National Workshop on Social Forestry and Forestry Extension*. 1992. Royal Government of Bhutan: Lingmethang.
73. Upadhyay, K. Review of community forest/community protected forest rules, In *Social Forestry and Forestry Extension - Report of the National Workshop on Social Forestry and Forestry Extension*. 1992. Royal Government of Bhutan: Lingmethang.
74. Buffum, B., Why is there no tragedy in these commons? An analysis of forest user groups and forest policy in Bhutan. *Sustainability*, 2012. **4**(7): pp. 1448–1465.

17

Ecosystem Services: Evaluation

James Boyd
Resources for the Future

Introduction 143
Biophysical Evaluation 144
The Importance of Spatial Analysis 145
Economic Evaluation 145
What Affects the Value of Ecosystem Services? 147
Coordinated Biophysical and Social Evaluation 147
Conclusion 148
References 148

Introduction

The term "ecosystem services" conveys the principle that natural systems provide socially and economically valuable goods and services deserving of protection, restoration, and enhancement. Ecosystem goods and services include the ecological features, qualities, or commodities we value, such as food, timber, clean drinking water, available water for irrigation, transportation, and industry, clean air, scenic beauty, and species important to us for recreational, ethical, or cultural reasons. Ecosystem goods and services also relate to nature's ability to protect us from harm. For example, wetlands and natural coastlines provide a service by protecting against flood damages. And trees and other plants provide a service by sequestering carbon and thereby reducing the costs of greenhouse gas emissions.

Despite the fact that we do not typically trade them in markets or see their value via prices, ecosystem goods and services are valuable inputs to most aspects of our economy and well-being. In fact, they are the foundation on which all economic activity and social welfare depend. Wise decisions about ecosystem goods and services thus require both biophysical and economic evaluation. Not only that, the biophysical and economic evaluation must be coordinated. The relationship between changes in natural systems and corresponding changes in human welfare is today a focus of both ecology and economics.[1–8]

Many government, nongovernmental organization (NGO), and private sector decisions affect the delivery of ecosystem goods and services. Carbon sequestration practices, water allocations, land uses, air and water quality regulations, and resource management choices all affect nature's delivery of ecosystem goods and services. When we protect or conserve natural resources, we are protecting and conserving ecosystem goods and services.

Evaluation of ecosystem goods and services is necessary if we are to protect the social wealth they provide. To counter the economic invisibility of ecosystem goods and services, natural and social scientists are working together to depict nature's role in our social and economic well-being.

Evaluation requires extensive knowledge of biophysical systems, the ways in which human activity alters those systems, and our ability to influence them positively via restoration, management, and protection. It also requires knowledge of nature's role in our economic and social lives and the value of those goods and services.

Biophysical Evaluation

The evaluation of ecosystem services involves two linked missions, one biophysical and the other, economic and social. Biophysical evaluation is associated with the natural sciences including ecology, hydrology, biology, oceanography, and soil science. If we want clean air and water, healthy and abundant species populations, pollination, irrigation, and protection from floods and fires, how can we take action to preserve these things? The natural sciences describe how ecosystems function, the factors that threaten ecological functions, and our ability to restore, enhance, or protect functions in decline. This knowledge is necessary if we are to manage, protect, or enhance the goods and services ecosystems provide.

Any natural process, by definition, transforms a set of inputs into a different set of outputs, much like an industrial process transforms inputs like labor and capital into outputs like cars and loaves of bread. Hydrological processes transform rainfall into ground and surface water. Biological and chemical processes transform water of one quality into water of a different quality. Reproductive, forage, and migratory processes relate biotic and physical conditions to the abundance of species. Food webs convert one form of biomass into another. Wetland processes transform the scale, location, and speed of flood pulses. Sequestration processes affect the release and in some cases, transformation of chemical inputs into water or the atmosphere. Even a process as simple as "shading" relates tree canopies to the regulation of water temperature.

When ecologists or other natural scientists speak of ecological processes or functions, they are referring to the transformation of one set of biophysical conditions into another. Ecologists and economists refer to these processes as *biophysical production functions*.[9–11] Biophysical production functions provide the causal link between ecological stresses—from land development, climate change, and invasive species—and losses in ecosystem goods and services. They also provide the link between ecological protection or enhancement—from conservation, regulation, and restoration—and improvements in ecosystem services.

For example, consider a wetland ecosystem. A wetland ecosystem has a structure, size, and boundaries that contain a bundle of biophysical inputs including soils, water volumes, water quality, and plant and animal populations. The socially valuable outputs of this ecosystem include more abundant species, cleaner water, and lower flood risks. Biophysical production functions describe how changes in the wetland system's inputs (size, boundaries, soil type, water volume, and vegetation) affect its ability to support species, absorb floodwater, and remove contaminants (the system's goods and services). Knowledge of these production functions is necessary to answer questions like the following: How many wetland acres do we need to protect downstream communities from flood damage? Or, what kinds of wetlands do we need to prevent the extinction of some bird or amphibian species? Production functions provide the link between actions we can take as resource managers and conservationists and the socially desirable biophysical benefits those actions create.

Measurement of these production functions requires investment in evidence-based ecology that explicitly focuses on the relationship between observable stresses to ecological systems and outputs (goods and services) that are socially important. Like clinical medicine, ecosystem services scientists are developing ways to consistently relate stresses (e.g., impervious land cover) and interventions (e.g., stream bank restoration) to ecosystem service outcomes (e.g., water quality, fish abundance). An important feature of these activities is that they are inherently interdisciplinary, requiring collaboration among disciplines such as hydrology, water chemistry, population ecology, and ecotoxicology.

The Importance of Spatial Analysis

Biophysical evaluation of ecosystem services has another important characteristic: it almost always involves spatial analysis because both the production of biophysical functions and the social determinants of the service's benefits depend upon the landscape context in which those functions and services arise.[12,13] Numerous efforts are underway to map where ecosystem goods and services are provided and are most threatened.[14,15] From the biophysical perspective, geographic context matters for several broad reasons. First, ecological production can exhibit non-linearities in scale and configuration—for example, where a whole protected landscape produces more than the sum of the unconnected parts. Second, natural systems are often characterized by movement: air circulates, water runs downhill, species migrate, and seeds and pollen disperse. Moreover, the movement of one biophysical feature—say water—tends to trigger the movement of other things— like birds and fish.[16] In general, the ecosystem goods and services we care about often depend on biophysical conditions and processes that occur over large distances.

Put another way, if we want to enjoy ecosystem services in one location, we must usually protect, enhance, or manage ecosystems in other locations. Evaluation thus involves understanding the spatial phenomena and interdependencies. Examples of these spatial production relationships describe the dependence of the following: species on the configuration of lands needed for their reproduction, forage, and migration; surface and aquifer water volume and quality on land cover configurations and land uses; flood and fire protection services on land cover configurations; soil quality on climate variables and land uses; and air quality on pollutant emissions, atmospheric processes, and natural sequestration.

In particular, the pattern of conservation of lands on the landscape can significantly impact the value of both the ecological and economic attributes.[17] It should also be noted that temporal analysis of production functions is also often important. First, many ecosystem processes are time dependent (e.g., runoff from melting snowpacks in spring, the need for forage food when migratory species are present). Second, the value of ecosystem goods and services is often a function of when those services are provided (e.g., irrigation water during the growing season).

Economic Evaluation

Ecosystem services evaluation also involves some form of social assessment to measure or communicate the importance of ecosystem services to households, communities, or other decision-makers. Often, social assessment of ecosystem services takes the form of economic, monetary evaluation of the services' value. It should be noted that there are also non-economic, non-monetary approaches to ecosystem services evaluation, a review of which is beyond the scope of this entry (but for examples, see Sherrouse et al.,[18] Blamey et al.,[19] Hagerhall,[20] Pereira et al.,[21] and Bender et al.[22]).

The economic value of natural resources and ecological systems goes well beyond the value of consuming natural resources such as water, lumber, and food. It also includes the value associated with non-consumptive natural resource uses (e.g., for aesthetic or recreational enjoyment), the intrinsic value of protecting nature for its own sake, and the bequest value of passing a healthy environment along to our descendants. In principle, *any* contribution of nature to our objective or subjective well-being can be expressed in economic terms.

Economic valuation is a way to depict something's importance or desirability. Economic values can be thought of as rankings, weights, or priorities. Values are detected or measured by examining people's choices. Whenever we choose one thing over another, as individuals do everyday, we are engaged in valuation. Choices are a particularly reliable form of evidence when it comes to detecting preferences and values. When we make choices, we reveal our preferences for one thing over another. Paying for something is a choice. When we pay for something, we are deliberately choosing that something over the amount of money paid. Assuming people are rational, they will only pay the price if the thing they are buying is worth at least that much to them. The higher the price paid, the higher the valuation we can infer.

Prices are desirable, not because they are the ideal measure of value, but because they are readily available. We can use market prices to value some ecosystem goods and services, but only those that are bought and sold as private goods. Examples of ecosystem goods that are bought and sold in private markets include timber, commercial fish harvests, and carbon sequestration credits (if a credit market exists). Often, however, ecosystem goods and services are public, non-market commodities for which there is no market price. Without market prices, economists must resort to so-called non-market valuation methods, including the methods described in the following text.

Hedonic valuation methods examine the prices people pay for things that have an environmental component. For example, when people purchase a home near an aesthetically pleasing ecosystem, home prices reflect that environmental amenity.[23] The price premium of living near the ocean, having a mountain view, or being in close proximity to urban parks can be measured via statistical analysis. Similarly, farm values are related to the availability of groundwater, precipitation, and soil quality. The premium due to those features can be estimated by controlling for other factors that affect farm value.

Evidence of conservation value can also be inferred from political choices such as prohibitions on drilling, development, and other land use changes associated with public lands or conservation referenda that approve local or state financing of land acquisitions.[24]

Travel cost methods examine the costs people are willing to bear to enjoy natural resources. When we spend time and money to enjoy nature, we are revealing something about its value. Again, if we are willing to pay the price (the cost), we must value the experience, enjoyment, or use of the resource more than the cost.[25] The travel cost method requires data and analysis linking the number of trips to a site to its quality, size, or location. Changes in these attributes can be valued if there is a perceptible change in the number, length, or cost of trips taken to the site.

Another technique is to examine costs avoided by the presence of an ecological feature or service. For example, if we lose wetlands and lose their water purification or flood damage reduction benefits, we may have to invest in water treatment facilities or levees and dams. If instead we protect the wetlands, we avoid the costs associated with built infrastructure alternatives. Similarly, private firms can conduct engineering and economic analyses that calculate the costs associated with, for example, the loss of surface waters for cooling, where the cost might be associated with new refrigeration technologies.

Another approach, called the stated preference method, is to present people with hypothetical scenarios that ask them to choose—in a survey format—between ecosystem goods or services and something with a clear dollar value, such as an increase in property tax. To pass academic muster, these studies are much more structured and carefully designed than simple opinion polling.[26] Stated preference methods are controversial because people's choices are undisciplined by the need to spend their own, real money, which in principle may lead them to overstate their willingness to pay. Care must also be given to clearly defining and isolating the goods or services in question and framing the choice problem in a way that does not bias people's responses. Nevertheless, stated methods are an improvement relative to evaluation techniques that ignore social preferences.[27]

Finally, mention should be made of benefit transfer methods. Benefit transfer methods take existing valuations derived from any of the aforementioned methods and transfer them to new landscape and resource contexts. Benefit transfer studies are desirable because they avoid the costs of conducting original valuation research. However, the transfer of valuations from one ecological and social context to another is dangerous because ecosystem values are highly dependent on location. Benefit transfer involves statistical methods designed to control for similarities and differences in spatial contexts and adjust the transferred valuation accordingly.

A wide range of ecosystem goods and services, in specific spatial and social contexts, has been given economic values by researchers. A review of existing valuation studies is beyond the scope of this entry (see Boyd and Krupnik,[28] the Environmental Valuation Reference Inventory,[29] and Kumar[30] for a review and organized collection of studies, respectively). In general, the non-market valuation methods described earlier have a long history and are considered within economics to be valid, if imperfect, approaches to the missing prices problem associated with public environmental goods.[31]

What Affects the Value of Ecosystem Services?

From an economic perspective, we can make several broad statements about the value of ecosystem goods and services: the scarcer an ecological feature, the greater its value; the scarcer are substitutes for an ecological feature, the greater its value; the more abundant are complements to an ecological feature, the greater its value; the larger the population benefiting from an ecological feature, the greater its value; and the larger the economic value protected or enhanced by the feature, the greater its value. All of these factors relate to the geographic context in which the services are delivered and enjoyed.

As economic commodities, ecosystem goods and services resemble real estate, rather than cars or bottles of dish soap. The value of real estate is highly dependent on its location—the features of the surrounding neighborhood. This is because a given house or building cannot be easily transported to another neighborhood. In contrast, cars or soap can be easily moved around (shipped from one location to another); so, their value tends to be independent of their geographic location.

For example, the value of irrigation and drinking water quality depends on how many people depend on the water—a function of proximity to the water. Flood damage-avoidance services are more valuable, the larger the value of lives, homes, and businesses protected from flooding in a given watershed. Species important to recreation (for anglers, hunters, and birders) are more valuable when more people have proximity and access to recreational lands.

The scarcity of an ecosystem service is also usually a function of geography. Scarcity—and the availability of substitutes—matters to economic assessment because, all else equal, value increases with scarcity. For example, the value of irrigation water depends on the availability (and hence location) of alternative water sources. If wetlands are plentiful in an area, then a given wetland may be less valuable as a source of flood pulse attenuation than it might be in a region in which it is the only such resource. Finally, many ecosystem goods and services are valuable only if they are bundled with certain man-made assets. These assets are called "complements" because they complement the value of the ecosystem service. Recreational fishing and kayaking require docks or other forms of access. For example, a beautiful vista yields social value when people have access to it. Access may require infrastructure— roads, trails, parks, and housing, all of which are spatially configured.

Coordinated Biophysical and Social Evaluation

This entry has emphasized that ecosystem services evaluation involves both biophysical and economic analyses. An important corollary is that those two types of analyses need to be linked and coordinated, so that the delivery of ecosystem goods and services can be described in economic terms. Biophysical analysis describes nature's condition and how that condition changes in response to interventions or stresses. Social analysis describes the benefits or costs of changes in those biophysical conditions.

One way to facilitate this coordination is by having ecological and economic researchers codefine *ecological end points*.[32] Ecological end points are a distinct subset of the larger universe of biophysical outcome measures. The test of such measures is that they are meaningful and understandable to communities, businesses, households, planners, and other stakeholders. In other words, they are biophysical outcome measures that require little further biophysical translation to make clear their relevance to human welfare.

Many outcomes of interest to natural science do not have this property. Outcomes like biotic integrity indices, chemical water quality concentrations, hydrogeomorphic classifications, and biological productivity measures are of scientific interest, are related to ecosystem services measurement, and establish the scientific basis for accurately modeling ecosystem functions and services. But their effect on people's well-being is unclear.

Consider a commonly used measure such as an "aquatic macroinvertebrate community index."[33] This index is a useful ecological measure because it acts as a proxy for food availability for fish and birds, indicates level of chronic stress on aquatic organisms, and may be correlated with abundance of fish species or fish diversity. But this measure does not satisfy the conditions of an end point: it is technical and not directly meaningful to non-ecologists. Consequently, it is by itself almost impossible to evaluate socially or economically. To be clear, that does not mean the information conveyed by the measure is not useful or important. On the contrary, such information may be extremely relevant to our understanding of ecosystem functions. But alone, this kind of measure is difficult to interpret in social or economic terms.

Note, though, that these indicators are used because they are thought to be proxies for the goods and services we can evaluate socially (such as fish abundance or reduced risks of waterborne illness). What are needed are biophysical production functions that translate technical outcome measures into ecological end points amenable to social evaluation. Other examples of measures that require further biophysical translation to foster social evaluation include, but are not limited to, criteria such as carbon sequestration potential, dissolved oxygen, nutrient, and other chemical concentrations, bacterial and other pathogen concentrations, biotic tissue burdens and pathological measures, and biodiversity indices.

Conclusion

Ecosystem services evaluation links biophysical and social analyses to measure the economic benefits and costs of decisions that affect natural resources. Biophysical evaluations describe the ways in which environmental stresses (e.g., from land use or climate change) or protections (e.g., conservation or restoration) change ecosystems' ability to deliver beneficial outcomes such as cleaner air, water for irrigation, drinking, and commerce, more productive soils, and more abundant species. Economic evaluations measure the value of these ecological outcomes (or costs of their loss) in monetary or other social terms. Together, the two forms of analyses allow decision-makers to describe the economic consequences of ecological change. Because ecological losses often lead to economic costs, ecosystem service evaluation can help society perceive, and thus hopefully manage and avoid those costs.

References

1. *Daily, G. Nature's Services: Societal Dependence on Natural Ecosystems; Island Press: Washington, DC, 1997.*
2. Wainger, L.; Dennis, K.; Salzman, J.; Boyd, J. Wetland value indicators for scoring mitigation trades. Stanford Environ. Law J. **2001**, *20* (2), 413–478.
3. Ricketts, T.H.; Daily, G.C.; Ehrlich, P.R.; Michener, C.D. Economic value of tropical forest to coffee production. Proc. Natl. Acad. Sci. USA **2004**, *101* (34), 12579–12582.
4. Polasky, S.; Nelson, E.; Lonsdorf, E.; Fackler, P.; Starfield, A. Conserving species in a working landscape: land use with biological and economic objectives. Ecol. Appl. **2005**, *15* (4), 1387–1401.
5. Boyd, J. The nonmarket benefits of nature: what should be counted in green GDP? Ecol. Econ. **2006**, *61* (4), 716–723.
6. Carpenter, S.R.; DeFries, R.; Dietz, T.; Mooney, H.A.; Polasky, S.; Reid, W.V.; Scholes, R.J. Millennium ecosystem assessment: research needs. Science **2006**, *314* (5797), 257–258.
7. Boyd, J.; Banzhaf, S. What are ecosystem services? Ecol. Econ. **2007**, *63* (2–3), 616–626.
8. Barbier, E.B.; Koch, E.W.; Silliman, B.R.; Hackery, S.D.; Wolanski, E.; Primavera, J.; Granek, E.F.; Polasky, S.; Aswani, S.; Cramer, L.A.; Stoms, D.M.; Kennedy, C.J.; Bael, D.; Kappel, C.V.; Perillo, G.M.; Reed, D.J.; Coastal ecosystem-based management with nonlinear ecological functions and values. Science **2008**, *319* (5861), 321–323.
9. U.S. EPA (Environmental Protection Agency) Science Advisory Board. Valuing the Protection of Ecological Systems and Services: A Report of the EPA Science Advisory Board. EPA-SAB-09–012. U.S. EPA: Washington, DC, 2009.
10. Daily, G., Matson, P. Ecosystem services: from theory to implementation. Proc. Natl. Acad. Sci. USA **2008**, *105* (28), 9455–9456.

11. Boyd, J. Counting Non Market, Ecological Public Goods: The Elements of a Welfare Significant Ecological Quantity Index. Discussion paper 07 42. Resources for the Future: Washington, DC, 2007.
12. Bockstael, N. Modeling economics and ecology: the importance of a spatial perspective. Am. J. Agric. Econ. **1996**, *78*, 1168.
13. Potschin, M.; Haines-Young, R. Ecosystem services: exploring a geographical perspective. Prog. Phys. Geography. **2011**, *35* (5), 579–598.
14. Raymond, C.M.; Bryan, B.A.; MacDonald, D.H.; Cast, A.; Strathearn, S.; Grandgirard, A.; Kalivas, T. Mapping community values for natural capital and ecosystem services. Ecol. Econ. **2009**, *68*, 1301–1315.
15. Egoh, B., Reyers, B.; Rouget, M.; Richardson, D.M.; LeMaitre, D.C.; van Jaarsveld, A.S.; Mapping ecosystem services for planning and management. Agriculture Ecosystems Environ. **2008**, *127*, 135–140.
16. Semmens, D.; Diffendorfer, J.E.; Lopez-Hoffman, L.; Shapiro, C.; Accounting for the ecosystem services of migratory species: quantifying migration support and spatial subsidies. Ecol. Econ. **2011**, *70*, 2236–2242.
17. Polasky, S.; Nelson, E.; Camm, J.; Csuti, B.; Fackler, P.; Lonsdorff, E.; Montgomery, C.; White, D.; Arthur, J.; Garber-Yonts, B.; Haight, R.; Kagan, J.; Starfield, A.; Tobalske, C.; Where to put things? spatial land management to sustain biodiversity and economic returns. Biol. Conserv. **2008**, *141*, 1505–1524.
18. Sherrouse, B.C.; Clement, J.M.; Semmens, D.J. A GIS application for assessing, mapping, and quantifying the social values of ecosystem services. Appl. Geography **2011**, *31* (2), 748–760.
19. Blamey, R.K.; James, R.F.; Smith, R.; Niemeyer, S. Citizens' Juries and Environmental Value Assessment. Research School of Social Sciences, Australian National University: Canberra, 2000.
20. Hagerhall, C.M. Consensus in landscape preference judgements. J. Environ. Psychol. **2001**, *21* (1), 83–92.
21. Pereira, E.; Queiroz, C.; Pereira, H.; Vicente, L. Ecosystem services and human well-being: a participatory study in a mountain community in Portugal. Ecol. Soc. **2005**, *10* (2), Art. 14.
22. Bender, O.; Boehmer, H.J.; Jens, D.; Schumacher, K.P. Using GIS to analyse long-term cultural landscape change in Southern Germany. Landscape Urban Plan. **2005**, *70* (1–2), 111–125.
23. Brent, M.; Polasky, S.; Adams, R. Valuing Urban Wetlands: A property price approach. Land Econ. **2000**, *76* (1), 100–113.
24. Banzhaf, H.; Wallace Oates, S.; Sanchirico, J. Success and design of local referenda for land conservation. J. Policy Anal. Manag. **2010**, *29*, 769–798.
25. McConnell, K. On-site time in the demand for recreation. Am. J. Agric. Econ. **1992**, *74*, 918.
26. Kopp, R.J.; Pommerehne, W.; Schwarz, N., Eds.; *Determining the Value of Non-Marketed Goods*. Kluwer, Boston, MA, 1997.
27. Carson, R.; Flores, N.; Meade, N. Contingent valuation: controversies and evidence. Environ. Resource Econ. **2001**, *19*, 173–210.
28. Boyd, J.; Alan, K. *The Definition and Choice of Environmental Commodities for Nonmarket Valuation*. Resources for the Future, Washington, D.C., USA, DP-09–35, 2009.
29. The Environmental Valuation Reference Inventory. http:// www.evri.ca/Global/Splash.aspx.
30. *Kumar, P. The Economics of Ecosystems and Biodiverstiy: Ecological and Economic Foundations. Earthscan, London: Washington, DC, 2010.*
31. *Freeman, A.M. The Measurement of Environmental and Resource Values: Theory and Methods. Resources for the Future: Washington, D.C., 1993.*
32. Boyd, J.; Krupnick, A. *The Definition and Choice of Environmental Commodities for Nonmarket Valuation*. Resources for the Future Washington, D.C., USA, DP-09–35, 2009.
33. Plafkin, J.L.; Barbour, M.T.; Porter, K.D.; Gross, S.K.; Hughes, R.M. *Rapid Bioassessment Protocols for Use in Streams and Rivers: Benthic Macroinvertebrates and Fish*. U.S. Environmental Protection Agency, Office of Water Regulations and Standards, EPA 440–4–89-001: Washington, D.C., 1989.

18

Ecosystem Services: Land Systems Approach

Richard Aspinall
James Hutton Institute

Introduction 151
Land System as a Coupled Human–Environment System 151
Ecosystem Goods and Services 152
Land Systems Change and Ecosystem Services 152
Conclusion 153
References 153

Introduction

Land systems are a category of coupled human–environment system in which the functions and purposes of land are understood and viewed as the results of human activities and environmental processes working together. Ecosystem services are the goods and services that societies recover from both natural and managed ecosystems. Many ecosystem goods and services are, therefore, a direct product of management and use of land systems. Most analyses of ecosystem services focus on natural processes and functions that are based in biodiversity, although a wide variety of biogeochemical, hydrological, and other processes, and the use of technologies, financial capital, and labor and energy, influence the functioning of ecosystems and the quantity and quality of ecosystem services recovered. Ecosystem services can be usefully considered as the product of land systems, since land systems reflect not only ecosystem processes and functions, but also human management and financial, social, and technological inputs that influence processes and functions, especially in terrestrial ecosystems.

Land System as a Coupled Human–Environment System

Land systems are a category of coupled human and environment system,[1,2] also known as a socio-ecological system, that incorporates both human and environment systems and their interaction. As such, they represent not only the structures and processes that link human and environment systems, but also the driving factors and processes producing change and dynamics[3] and that influence the multiple goods and services societies derive from natural resources. A land system approach necessarily embeds the different spatial, temporal, and organizational scales that the environment[4,5] and human[6] system processes to provide an understanding of local, regional, and global changes.[7,8] The environment system includes the Earth system, which is made up of biological, biogeochemical, hydrological, and other physical environment sub-systems. The human system includes economic, institutional, demographic, technological, and political factors and subsystems. Processes in the environment system are largely driven by biodiversity, energy, and material flows, while processes in the human system are driven by flows of finance, ideas, innovation, and human activity, including choices and decisions. Human and environment systems link most directly through the interaction of land cover and land use,[8,9] and through the influence and impacts of land management.[10] Land systems are studied with methods of interdisciplinary research including Earth observation and monitoring, survey, and modeling.[11]

A land systems approach emphasizes the system-level interactions, not only of factors that affect land but also of the different activities and processes that have an impact on food, water, and other ecosystem goods and services provided by land systems. Land use and land-produced ecosystem services are thus not solely the product of environmental systems; active land management is central to the delivery of ecosystem services.

Ecosystem Goods and Services

Ecosystem goods and services are defined as the benefits society receives from environmental systems.[12] Global-scale environmental changes have raised the profile and importance of ecosystem goods and services,[13] leading to reviews of the state of ecosystems and the goods and services at global[12] and national[14] scales. Terrestrial ecosystems are recognized as the primary source of goods and services that support human well-being and society.[12,15]

There are numerous taxonomies of ecosystem goods and services. The Millennium Ecosystem Assessment (MEA), for instance, has popularized their classification into four main types: supporting, regulating, provisioning, and cultural services (Table 18.1). This taxonomy has been continued in other classifications, including the Common International Classification of Ecosystem Services, which also explicitly recognizes a variety of services from the abiotic elements of the environment.[16]

Ecosystem assessment, a social process of evaluating ecosystem services and their value,[14] is generally based on a conceptual model that emphasizes a foundation in biodiversity and natural capital.[12,17–20] In this approach, human and technological capitals are considered exogenous factors. Because ecosystem goods and services are not, however, produced solely by biodiversity and natural capital, there is increasing recognition of the importance of evaluating ecosystem goods and services using a land systems approach that explicitly includes land management.[2,20]

Land Systems Change and Ecosystem Services

Land use change and the consequences of land management activities are global in their scope and scale of impact.[7,21,22] About 140 million km^2 of the Earth, some 28% of its total surface area, is land.[12] Cropland and pastures/rangelands together occupy about 43 million km^2,[23] or about 40% of the ice-free land area.[24]

TABLE 18.1 Definitions and Examples of the Four Main Categories of Ecosystem Goods and Services

Provisioning Services: The goods and other products obtained from ecosystems.
Examples include food, fiber, fuel, fresh water, genetic and ornamental resources, biochemicals, natural medicines, and pharmaceuticals. Since the environment also provides sources of energy, renewable and fossil sources of energy can also be considered among provisioning services.
Regulating Services: The goods, services, and benefits obtained from environmental regulation functions provided by ecosystem processes.
Examples include air quality, climate regulation, flood and hazard regulation, erosion regulation, regulation of diseases and pests, regulation of water quality (water purification), and waste treatment.
Supporting Services: The underpinning system functions necessary for the production of all other ecosystem goods and services. Supporting services refer to the products and results of biogeochemical, physical, chemical, and biological functions that operate in environmental systems.
Examples include soil formation, photosynthesis, primary production, nutrient cycling, and water cycling.
Cultural Services: The non-material benefits people obtain from ecosystems through education, exercise and recreation, spiritual enrichment, cognitive development, reflection, and aesthetic experience.
The UK National Ecosystem Assessment provides a rich suite of definitions of cultural services that expand the range of cultural services defined by the MEA.[34]

Source: Adapted from Millennium Ecosystem Assessment,[12] UK National Ecosystem Assessment,[14] EEA,[16] and Church et al.[33]

Human land use and management practices and activities have increased the intensity of resource use and impact,[25] and dominate biogeochemical[26,27] and water[28,29] fluxes. Human use or approximately 24% (15.6 Pg C yr^{-1}) of the total potential net primary productivity in terrestrial systems.[30]

These demands on land systems, combined with global population growth and changing climate, lead to international and national concerns over food, water, and energy security. Although land use and land use change are phenomena with global scope and impact, land use and land management are subject to the influences of a wide variety of human and environmental processes operating at a range of scales.[31] These influences include global-scale political and economic forces. Internationally, the land systems science community and land managers recognize that global forces become the main determinants of land use change primarily through their ability to amplify or attenuate local factors[3,8] and via global teleconnections.[32,33] Local factors, including land management practices, are, therefore, of primary importance not only in driving land changes that are local, regional, and global in scope and impact, but also in resolving land use issues and the delivery of ecosystem services.

Conclusion

Land systems offer a basis for a more holistic understanding of the generation and recovery of ecosystem services by individuals and society. Land systems include both human and environment system influences on the dynamics that produce ecosystem services, linking natural, technological, financial, and human capital and flows, as well as land management and decision-making to ecosystem services.

References

1. Foresight Land Use Futures Project. *Land Use Futures: Making the Most of Land in the 21st Century*, Final Project Report; London, 2010.
2. *Global Land Project: Science Plan and Implementation Strategy*, IGBP Report No. 53/IHDP Report No. 192005; IGBP Secretariat: Stockholm, 2005; p 64.
3. Lambin, E.F.; Turner, B.L.; Geist, H.J.; Agbola, S.B.; Angelsen, A.; Bruce, J.W.; Coomes, O.T.; Dirzo, R.; Fischer, G.; Folke, C.; George, P.S.; Homewood, K.; Imbernon, J.; Leemans, R.; Li, X.B.; Moran, E.F.; Mortimore, M.; Ramakrishnan, P.S.; Richards, J.F.; Skanes, H.; Steffen, W.; Stone, G.D.; Svedin, U.; Veldkamp, T.A.; Vogel, C.; Xu, J.C. The causes of land-use and land-cover change: Moving beyond the myths. Global Environ. Change-Hum. Policy Dimens. **2001**, *11* (4); 261–269.
4. Chave, J.; Levin, S. Scale and scaling in ecological and economic systems. Environ. Resour. Econ. **2003**, *26* (4), 527–557.
5. Levin, S.A. The problem of pattern and scale in ecology. Ecology **1992**, 73 (6), 1943–1967.
6. Gibson, C.C.; Ostrom, E.; Ahn, T.K. The concept of scale and the human dimensions of global change: A survey. Ecol. Econ. **2000**, *32* (2), 217–239.
7. Foley, J.A.; DeFries, R.; Asner, G.P.; Barford, C.; Bonan, G.; Carpenter, S.R.; Chapin, F.S.; Coe, M.T.; Daily, G.C.; Gibbs, H.K.; Helkowski, J.H.; Holloway, T.; Howard, E.A.; Kucharik, C.J.; Monfreda, C.; Patz, J.A.; Prentice, I.C.; Ramankutty, N.; Snyder, P.K. Global consequences of land use. Science **2005**, *309* (5734), 570–574.
8. Lambin, E.F.; Geist, H., Eds. Local processes and global impacts. In *Land-Use and Land-Cover Change*; Springer-Verlag: Berlin, 2006; 222.
9. Gutman, G.; Janetos, A.C.; Justice, C.O.; Moran, E.F.; Mustard, J.F.; Rindfuss, R.R.; Skole, D.; Turner, B.L.; Cochrane, M.A., Eds. *Land Change Science. Observing, Monitoring and Undertanding Trajectories of Change on the Earth's Surface*; Kluwer Academic Publishers: Dordrecht/Boston/London, 2004; 457.
10. Lambin, E.F.; Rounsevell, M.D.A.; Geist, H.J. Are agricultural land-use models able to predict changes in land-use intensity? Agric. Ecosyst. Environ. **2000**, *82* (1–3), 321–331.

11. Turner, B.L.; Lambin, E.F.; Reenberg, A. The emergence of land change science for global environmental change and sustainability. Proc. Natl. Acad. Sci. U.S.A. **2007**, *104* (52), 20666–20671.
12. Millennium Ecosystem Assessment. *Ecosystems and Human Well-being: Synthesis*; Island Press: Washington, DC, 2005; 137.
13. Daily, G.C.; Matson, P.A. Ecosystem services: From theory to implementation. Proc. Natl. Acad. Sci. U.S.A. **2008**, *105* (28), 9455–9456.
14. UK National Ecosystem Assessment. *The UK National Ecosystem Assessment Technical Report 2011*; UNEP-WCMC: Cambridge.
15. Costanza, R.; dArge, R.; de Groot, R.; Farber, S.; Grasso, M.; Hannon, B.; Limburg, K.; Naeem, S.; ONeill, R.V.; Paruelo, J.; Raskin, R.G.; Sutton, P.; van den Belt, M. The value of the world's ecosystem services and natural capital. Nature **1997**, *387* (6630), 253–260.
16. EEA. Proposal for a Common International Classification of Ecosystem Goods and Services (CICES) for Integrated Environmental and Economic Accounting. In Paper prepared for the EEA by Centre for Environmental Management; Background document, ESA/STAT/AC.217, UNCEEA/5/7/Bk, Fifth Meeting of the UN Committee of Experts on Environmental-Economic Accounting, NewYork, June 23–25, 2010; University of Nottingham: United Kingdom, 2010.
17. Mace, G.M.; Bateman, I.; Albon, S.; Balmford, A.; Brown, C.; Church, A.; Haines-Young, R.; Pretty, J.; Turner, K.; Vira, B.; Winn, J. Conceptual framework and methodology. In *The UK National Ecosystem Assessment Technical Report*; UK National Ecosystem Assessment, Ed.; UNEP-WCMC: Cambridge, 2011.
18. Pe'er, G.; McNeely, J.A.; Dieterich, M.; Jonsson, B.-G.; Selva, N.; Fitzgerald, J.M.; Nesshover, C. IPBES: Opportunities and challenges for SCB and other learned societies. Conserv. Biol. **2013**, 27 (1), 1–3.
19. Turnhout, E.; Bloomfield, R.; Hulme, M.; Vogel, J.; Wynne, B. Conservation policy: Listen to the voices of experience. Nature **2012**, *488* (7412), 454–455.
20. Carpenter, S.R.; Mooney, H.A.; Agard, J.; Capistrano, D.; DeFries, R.S.; Diaz, S.; Dietz, T.; Duraiappah, A.K.; Oteng-Yeboah, A.; Pereira, H.M.; Perrings, C.; Reid, W.V.; Sarukhan, J.; Scholes, R.J.; Whyte, A. Science for managing ecosystem services: Beyond the Millennium Ecosystem Assessment. Proc. Natl. Acad. Sci. U.S.A. **2009**, *106* (5), 1305–1312.
21. Rockstrom, J.; Steffen, W.; Noone, K.; Persson, A.; Chapin, F.S., III; Lambin, E.F.; Lenton, T.M.; Scheffer, M.; Folke,
22. Schellnhuber, H.J.; Nykvist, B.; de Wit, C.A.; Hughes, T.; van der Leeuw, T.; Rodhe, H.; Sorlin, S.; Snyder, P.K.; Costanza, R.; Svedin, U.; Falkenmark, M.; Karlberg, L.; Corell, R.W.; Fabry, V.J.; Hansen, J.; Walker, B.; Liverman,
23. Richardson, K.; Crutzen, P.; Foley, J.A. A safe operating space for humanity. Nature **2009**, *461* (7263), 472–475.
24. Vitousek, P.M.; Mooney, H.A.; Lubchenco, J.; Melillo, J.M. Human domination of Earth's ecosystems. Science **1997**, 277 (5325), 494–499.
25. Ramankutty, N.; Evan, A.T.; Monfreda, C.; Foley, J.A. Farming the planet: 1. Geographic distribution of global agricultural lands in the year 2000. Global Biogeochem. Cycles **2008**, 22, GB1003, doi:10.1029/2007GB002952.
26. Ramankutty, N.; Foley, J.A. Estimating historical changes in global land cover: Croplands from 1700 to 1992. Global Biogeochem. Cycles **1999**, *13* (4), 997–1027.
27. Monfreda, C.; Ramankutty, N.; Foley, J.A. Farming the planet: 2. Geographic distribution of crop areas, yields, physiological types, and net primary production in the year 2000. Global Biogeochem. Cycles **2008**, *22*, GB1022, doi:10.1029/2007GB002947.
28. Matson, P.A.; Parton, W.J.; Power, A.G.; Swift, M.J. Agricultural intensification and ecosystem properties. Science **1997**, 277 (5325), 504–509.
29. Vitousek, P.M.; Aber, J.D.; Howarth, R.W.; Likens, G.E.; Matson, P.A.; Schindler, D.W.; Schlesinger, W.H.; Tilman, D. Human alteration of the global nitrogen cycle: Sources and consequences. Ecol. Appl. **1997**, *7* (3), 737–750.

30. Gleick, P.H. Water use. Annu. Rev. Environ. Resour. **2003**, *28*, 275–314.
31. Hoekstra, A.Y.; Mekonnen, M.M. The water footprint of humanity. Proc. Natl. Acad. Sci. U.S.A. **2012**, *109* (9), 3232–3237.
32. Haberl, H.; Erb, K.H.; Krausmann, F.; Gaube, V.; Bondeau, A.; Plutzar, C.; Gingrich, S.; Lucht, W.; Fischer-Kowalski, M. Quantifying and mapping the human appropriation of net primary production in earth's terrestrial ecosystems. Proc. Natl. Acad. Sci. U.S.A. **2007**, *104* (31), 12942–12945.
33. Seto, K.C.; Reenberg, A.; Boone, C.G.; Fragkias, M.; Haase, D.; Langanke, T.; Marcotullio, P.; Munroe, D.K.; Olah, B.; Simon, D. Urban land teleconnections and sustainability. Proc. Natl. Acad. Sci. U.S.A. **2012**, *109* (20), 7687–7692.
34. Lambin, E.F.; Geist, H.J.; Lepers, E. Dynamics of land-use and land-cover change in tropical regions. Annu. Rev. Environ. Resour. **2003**, *28*, 205–241.

19

Ecosystem Services: Pollinators and Pollination

Introduction 157
Ecosystem Services in Natural Resource Systems 157
Pollination 158
Defining Pollination • Defining Pollination Services • The Pollination of Ecosystem Services • Supportive Services • Provisional Services • Regulating Services • Cultural Services
Maintaining and Supporting Pollination Services 162
Conclusion 162
References 162
Bibliography 163

Victoria A. Wojcik
Pollinator Partnership

Introduction

Pollination is unique as an individual ecosystem service in and of itself, as well as a supportive factor to other ecosystem services. The sustainability of terrestrial ecosystems relies on the maintenance of plant reproductive cycles, evolutionary patterns, and ecosystem functions.[1] The vast majority of the botanical diversity that constitutes ecosystems, defines biomes, and structures faunal communities is dependent on animal vectors for successful reproduction.[2–4] This direct role in plant reproduction makes pollinators keystone species.

In addition to their direct role in plant reproduction and food production in both wild and managed landscapes, pollinators support the full scope of global ecosystem services. The supportive roles of pollinators in larger global systems include climate regulation, nutrient cycling, water quality management, and air filtration.

Ecosystem Services in Natural Resource Systems

Ecosystem services are those benefits provided by the biosphere that maintain, support, and enhance life on Earth. Ecosystem services are defined as the set of direct and indirect benefits that are derived from healthy, sustainable, and functioning ecosystems.[5,6] Efforts to create a tangible, manageable system for classification have sorted ecosystem services into groups of functions that interact with all aspects of life. *Supportive services* connect ecosystems and facilitate the movement of energy and matter. *Provisional services* are responsible for the bounty of food and resources provided to humans and animals. *Regulating services* provide resilience to the environment, regulating and buffering against extreme shifts and supporting the status quo. *Cultural services* support creativity and esthetic landscapes, provide the inspiration for arts, literature, belief systems, and deliver opportunities for recreation.[7]

The spatial extent of ecosystem services offered by the biosphere, and pollinators, can be further subdivided into their geographic scale of interaction with humans and other species.[8] Services can be *proximal* (local), derived and applied at a source, or *distal* (global), derived at a source and distributed broadly. Global services include the regulation of climate and the atmosphere, the transition of energy and materials from one site to another, and the general esthetic value of diversity to human culture. Local services provide environmental regulation at the local scale, mediating disturbance and local climate, filtering water, buffering pollution, regulating pests and disease, and providing habitat.

Pollination

Defining Pollination

What is pollination? By definition, and in the most robust sense, it is the transfer of a pollen grain, the mobile gamete, to the ovary that contains the immobile gamete. Plant species that reproduce by sexual means (either exclusively or in addition to vegetative growth) must have this movement of pollen facilitated.[1] Pollen is a male gamete that can be compared to sperm, but unlike sperm cannot move on its own. The pollen vector can be abiotic (water or wind) or biotic (an animal vector). Animal pollen vectors include bees, birds (hummingbirds, honey creepers), bats, butterflies, moths, beetles, flies, wasps, small scansorial mammals (honey possums), larger fruit-eating mammals (lemurs, perhaps even chimpanzees), and even geckos.[9] To be a pollinator an animal must somehow be able to transport pollen; this usually occurs through contamination when an animal visits a flower for a nectar meal.

Defining Pollination Services

An examination of the origins of products the interconnections of natural systems quickly reveals the support that pollinators provide to our natural landscapes and resources. Many of the direct and indirect services that humans and all other living beings benefit from are a result of intimate long-standing pollinator–plant relationships.

The Pollination of Ecosystem Services

Pollinators and pollination have primary or direct roles in ecosystem service provisioning and secondary or indirect roles in ecosystem service delivery. Tertiary, or associated, roles can also be defined in some systems. For example, pollinators can have a distant role in water purification in ecosystems where riparian vegetation or other ground cover (that depends on pollinators for reproduction) provides the main role in sediment retention. Table 19.1 lists various terrestrial ecosystem services, their reliance on pollinators, and their spatial extent. This list is comprehensive, but by no means exhaustive, as some services can be segregated into further subcategories and future study will likely reveal the connection of pollinators to other services. The aim of this table is to highlight this complex network of energy transfer and support.

Supportive Services

Supportive ecosystem services provided or enhanced by pollinators primarily include the structuring of ecosystems that define local ecotypes, wildlife patterns, and resource bases. Between 67% and 96% of all terrestrial species of angiosperms require animal pollinators for reproduction.[10] Considering the act of pollination itself, even when vectored by wind or water, the entirety of sexually reproducing plants (angiosperms and gymnosperms) that are not capable of self fertilization are dependent on pollination. It follows that all consumers of natural resources are reliant on pollination.

TABLE 19.1 Terrestrial Ecosystem Services and Their Connections to Pollinators and Pollination Services

Service Type	Service	Service Scale	Pollinator Role
Supportive services	Nutrient cycling	Local	Primary/direct
	Seed dispersal	Local/global	Tertiary/associated
	Genetic diversity	Local	Primary/direct
	Primary production	Local/global	Tertiary/associated
Provisioning services	Pollination[a] – plant reproduction	Local	Primary/direct
	Crop production[a] (food for agrarian human population)	Local	Primary/direct
	Food production[a] (food-gathering human populations, wildlife)	Local	Primary/direct
	Biomass[a]	Local	Primary/direct
	Pharmaceuticals	Local	Primary/direct
	Industrial products (hardwoods and plant fibers)	Local	Primary/direct
	Maintenance of genetic diversity	Local	Primary/direct
Regulating services	Carbon sequestration	Local/global	Secondary/indirect
	Climate regulation[b]	Local/global	Primary/direct and secondary/indirect
	Pest/disease control	Local	Tertiary/associated
	Air purification[b]	Local/global	Primary/direct and secondary/indirect
	Water purification	Local	Secondary/indirect
Cultural services	Recreational experiences	Local	Primary/direct and secondary/indirect
	Cultural inspiration	Local	Primary/direct and secondary/indirect

Source: From Ollerton et al. 2011.[10]

[a] Between 67% and 96% of all terrestrial plants are believed to require pollinators for reproduction, or see significant increases in yields/seed development, when pollinators are present.

[b] The spatial extent of pollinators' roles in some services can shift depending on the time of year and nearness to equatorial regions.

Diversity and local endemism are natural resources that define biomes, local communities, and cultures. The plant communities that define our natural systems and structure our landscape are reliant on pollination. For example, the Sonoran Desert is filled with vast plains spotted with saguaro cactus (*Carnegiea gigantea*). The giant saguaro is a keystone species that supports life in this arid land. But the saguaro is incapable of self-fertilization, dependent on bats, such as the endangered lesser long-nose bat, *Leptonycteris curasoae yerbabuenae* (Figure 19.1), bees, and birds for pollination. Like the pollinator visitors of succulents in arid environments, pollinators across the globe help perpetuate local diversity by ensuring the reproduction of unique and keystone plants.

The cycling of nutrients is another primary supportive service. Some pollinator guilds (functional groups), such as ground nesting bees and wasps, assist in nutrient cycling and soil aeration by their digging actions. About 60% of bee species nest in the ground.[11,12] For example, large ground-nesting bees in the genus *Habropoda* (pictured emerging from a nest in Figure 19.2) tunnel into hillsides and slopes, loosening soil. Pollinators also have a role in maintaining reproduction in plants that fix and cycle nitrogen as many legumes are bee-pollinated.[13]

Provisional Services

The pollination of food crops is the primary and best known example of pollinator services. Two of every three bites of food consumed by humans are the direct result of pollinator activities.[9] Corn, wheat, and rice—the carbohydrate stables of most cultural diets—are wind pollinated, but the food products

FIGURE 19.1 (See color insert.) The Lesser long-nose bat, *Leptonycteris curasoae yerbabuenae*, feeding at flowers of *Agave palmeri palmeri* blossom during the night in the Sonoran Desert. *L. curasoae yerbabuenae* is also an essential pollinator of saguaro cactus *(Carnegiea gigantea)* that defines the Sonoran ecosystem.
Source: Photo credit Steven Buchmann.

FIGURE 19.2 (See color insert.) A male *Habropoda pallida* emerging from a ground nest, which is a network of tunnels beneath the surface that house brood chambers. Ground nesting aids in the aeration of soil and nutrient cycling.
Source: Photo credit Steven Buchmann.

that provide the necessary and essential nutrients for the human diet require pollinator visitation for fruiting and reproduction.[14] The variety of agricultural species that are dependent on pollinators is high, nearly 1000 of the 1200 common crop varieties planted across the globe.[13,15] The roles of pollinators and pollination services in food production systems are often more directly quantifiable than those in resource-based systems. The global value of pollination to agriculture has been estimated at €153 billion ($210 billion).[16] The individual contribution of insect pollinators is $57 billion (€41 billion) to the agricultural economy of the United States.[17]

The influence of pollinators on the daily lives of humans extends to the products that we interact with. Many of the resources and substrates that are used to develop and advance societies are derived from the living components of ecosystems. Hardwood products, fibers, textiles, dyes, scents, plant-derived chemical, and pharmaceutical products are all dependent on reproductive lineages of flowering plants,[5,6] the majority of which are dependent on animal pollinators. Softwood timber species are not dependent on animal vectors for pollination, but require pollination for continued reproduction. Without pollination we would see a decline in woody species that provide lumber, fibers, and other resources that have helped humans industrialize.

Regulating Services

Regulating services, like the maintenance of climatic patterns, carbon sequestrations, air and water purification, and resistance to disturbances, are all characteristics of biomes that are maintained by the actions of pollinators. The sequestration of carbon by terrestrial plants depends on the continued growth and reproduction of these species. Tropical forest systems play a key role in air filtration and carbon regulation, especially in winter periods when deciduous species in temperate regions are in a resting stage. These tropical forest communities are dominated by pollinator-dependent species, with an estimate of nearly 95% dependence.[10]

Regulating services that prevent or buffer against environmental stochasticity are supported by healthy pollinator communities. The California chaparral ecosystem is one example. Plant species responsible for soil stabilization along hillsides require pollinators for reproduction.[18] A loss of pollinators, and the eventual senescence of these plant species can cause vulnerability to landslides during annual wet periods. Similarly, coastal mangrove communities that provide protection during storm events need pollinators for population maintenance.[19]

Cultural Services

Scientist and philosopher E.O. Wilson has written extensively on the topic of biophilia, meaning the love of the living world.[20] Humans are enriched and simply happier when surrounded by an abundance of living things. Species richness and landscape diversity inspire creativity. Cultural, recreational, and spiritual enhancement is provided by ecosystems that depend on pollinators. Recreational landscapes frequented by hikers, wildlife viewers, and hunters have primary or secondary dependencies on pollinators. Every few years, massive wildflower blooms in Death Valley National Park (Californian, USA) draw tourists from around the world to see the phenomenon of life in an unforgiving landscape. These desert species bloom each year because pollinators maintain their population. Another example is the indirect role that pollinators play in supporting seed dispersal by supporting populations of songbirds. These seed-dispersing birds feed directly off of many insect pollinators (adults and larvae), especially during juvenile development; thus pollinators support the growth and development of bird populations that eventually play a role in seed dispersal. Similarly, common game species such as deer, wild turkey, pheasant, and quail rely on angiosperm-dominated landscapes for habitat and food; threatened and rare species have the same dependence.

While pollinators and pollination services are generally restricted to terrestrial ecosystems there are some examples of interactions with aquatic systems. For example, a salmon fishery is dependent on the quality of upland river systems for spawning; as mentioned above, pollinators have indirect roles in maintaining water quality by maintaining natural filtration systems. Additionally, many aquatic fly larvae (the larvae of some pollinators) are eaten by salmon fry—a case can be made for the role that pollinators have in supporting salmon fisheries. These examples, while relevant, are difficult to quantify, but certainly should be qualified when the total value of pollination to natural resource systems is considered.

Maintaining and Supporting Pollination Services

When productivity and service provision are concerned, increased biodiversity corresponds to increased stability, resistance, longevity, and product derivation.[5,21–23] The same is true of pollinator-dependent ecosystem services. Ecosystems with compromised or reduced diversity are susceptible to regime shifts that can change their composition and evolutionary trajectory.[24] For example, disturbances of ecosystems by anthropogenic factors (habitat loss, pollution, climate change, invasive species introduction, etc.) that consequently remove large numbers of species can eliminate functional groups, even entire trophic levels. A loss of biodiversity and the resultant loss in ecosystem function can have substantial effects on the output of ecosystem services.[25,26] In resource-based economic systems, a loss of function translates into a loss of productivity and a further loss of livelihood.[27] Losing pollinators will first result in decreased food production;[28] subsequently, each ecosystem service provided or supported by pollinators will suffer.

Conclusion

Our natural resources are those environmental products or elements on which we rely for survival. They support and enhance our lives. As consumers of natural resource we are reliant on pollination and pollinators. Understanding the local and global, as well as the primary, secondary, and tertiary connections that natural systems have with pollinators enhances our ability to manage and maintain them.

Healthy and diverse pollinator communities are an important factor for the functioning and sustainability of ecosystems and for the future of our societies.

References

1. Kevan, P. Pollination: a plinth, pedestal, and pillar for terrestrial productivity: the why, how, and where of pollination protection, conservation, and promotion. Bees and Crop Pollination-Crisis, Crossroads, and Conservation, 2001; 7–68.
2. Bierzychudek, P. Pollinator limitation of plant reproductive effort. Am. Nat. **1981**, *117*, 838–840.
3. Kevan, P.G.; Baker, H.G. Insects as flower visitors and pollinators. Ann. Rev. Entomol. **1983**, *28*, 407–453.
4. Knight, T.M.; Steets, J.A.; Varmosi, J.C.; Mazer, S.J.; Burd, M.; Campbell, D.R.; Dudash, M.O.; Johnston, R.J.; Mitchell, R.J., Ashman, T.L. Pollen limitation of plant reproduction: Pattern and process. Ann. Rev. Ecol. Evol. Syst. **2005**, *36*, 467–497.
5. *Daily, G.C., Ed.; Nature's Services: Societal Dependence On Natural Ecosystems. Island Press: Washington DC, 1997.*
6. Costanza, R.; d'Arge, R.; de Groot, R.; Faber, S.; Grasso, M.; Hannon, B.; Limburg, K.; Naeem, S.; O'Neill, R.V.; Paruelo, J.; Raskin, R.G.; Sutton, P.; van den Belt, M. The value of the world's ecosystem services and natural capital. Nature **1997**, *387*, 253–260.
7. Millennium Ecosystem Assessment (MEA). *Ecosystems and Human Well-Being: Synthesis;* Island Press: Washington, 2005; 155 pp.
8. Costanza, R. Ecosystem services: Multiple classification systems are needed. Biol. Conserv. **2008**, *141*, 350–352.
9. Buchmann, S.; Nabhan, GP. *The Forgotten Pollinators*; Island Press: New York, 1996.
10. Ollerton, J.; Winfree, R.; Tarrant, S. How many flowering plants are pollinated by animals? Oikos **2011**, *120*, 321–326.
11. Linsley, EG. The ecology of bees. Hilgardia **1958**, *27*, 543–599.
12. Michener, CD. *The Bees of the World;* Johns Hopkins University Press: Baltimore, 2000.
13. Free, JB. *Insect Pollination of Crops;* Academic Press: London, UK, 1993.

14. Kevan, P.; Wojcik, VA. Pollinator services. In: *Managing Biodiversity in Agricultural Ecosystems;* Jarvis, D.I., Padoch, C., Cooper, HD., Eds.; Columbia University Press: New York, 2007; 200–249 pp.
15. Klein, AM.; Vaissière, B.E.; Cane, J.H.; Steffan-Dewenter. I.; Cunnigham, S.A.; Kremen, C.; Tscharntke, T. Importance of pollinators in changing landscapes for world crops. Proc. R. Soc. **2007**, *274*, 303–313.
16. Gallai, N.; Salles, J.M.; Settele. J.; Vaissière, B.E. Economic valuation of the vulnerability of world agriculture confronted with pollinator decline. Ecol. Econ. **2009**, *68*, 810–821.
17. Losey, J.E.; Vaughan, M. The economic value of ecosystem services provided by insects. BioScience **2006**, *56* (4), 311–323.
18. Kremen, C.; Bugg, R.L.; Nicola, N.; Smith, S.A.; Thorp, R.W.;Williams, N.M. Native bees, native plants, and crop pollination in California. Fremontia **2002**, *3-4* (30), 41–49.
19. Tomlinson, T.B. *The Botany of Mangroves.* Cambridge University Press: New York, 1986.
20. Wilson, E.O. *Biophilia.* Harvard University Press: Cambridge 1986; 1–176 pp.
21. Ehrlich, R.; Ehrlich, A.H.; Holdren, J.P. *Ecoscience: Population, Resources, Environment;* Freeman, W.H.: San Francisco, 1977.
22. Aylward, B.A.; Barbier, E.B. Valuing environmental functions in developing countries. Biodivers. Conserv. **1992**, *1*, 34.
23. Costanza, R.; Daly, H.E. Natural capital and sustainable development. Conserv. Biol. **1992**, *6*, 37–46.
24. Folke, C.; Carpenter, S.; Walker, B.; Scheffer, M.; Elmqvist, T.; Gunderson, L.; Holling, C.S. Regime shifts, resilience, and biodiversity in ecosystem management. Annu. Rev. Ecol. Evol. Syst. **2004**, *35*, 557–581.
25. Perrings C. The economic value of biodiversity. In *The Economic Value of Biodiversity;* Heywood, V.H., Gardner, K., Eds.; The United Nations Environment Programme by the Press Syndicate of the University of Cambridge: Cambridge, UK, **1995**; 822–913 pp.
26. Perrings, C.; Walker, B. Biodiversity, resilience and the control of ecological-economic systems: the case of fire-driven rangelands. Ecol. Econ. **1997**, *22*, 73–83.
27. Kevan, PG.; Phillips, TP. The economic impacts of pollinator declines. Conserv. Ecol. **2001**, *5*, 8. http://www.consecol.org/vol5/issl/art8.
28. Aizen, M.A.; Garibaldi, L.A.; Cunningham. S.A.; Klein, A.M. How much does agriculture depend on pollinators? Lessons from long-term trends in crop production. Ann. Bot. **2009**, *103*, 1579–1588.

Bibliography

Eardley, C.; Roth, D.; Clarke, J.; Buchmann, S.; Gemmill, B.; Eds.; Pollinators and Pollination: A Resource Book for Policy and Practice; African Pollinator Initiative (API): Johannesburg. 2006; viii-77 pp.

National Research Council (NRC): Committee on the Status of Pollinators in North America NRC. Status of Pollinators in North America. The National Academies Press, Washington, D.C., 2007.

20

Ecosystems: Diversity

Introduction 165
Biodiversity at the Ecosystem Level 166
Ecosystem Types 166
World Ecosystems 167
Forest Ecosystems 168
Tropical Rainforests 168
Tropical Seasonal Forests 168
Deciduous Forests 169
Grasslands 169
Deserts 169
Wetlands 171
Polar Ecosystems 171
Agroecosystems 172
Urban Ecosystems 172
Ecosystem Management 173
References 173

L. M. Chu
Chinese University of Hong Kong

Introduction

As the basic functional unit of nature, an ecosystem is a community of living organisms (microorganisms, plants, and animals) that interact among themselves and with the abiotic components of their environment.[1,2] Cultural components may also be included as part of this community, as ecosystems are shaped by humans. Ecosystems are defined by the network of such interactions[3] and form the basic level at which the properties required to support life exist. Ecosystems have system boundaries but may differ in their spatial and temporal scale and components to form systems with unique compositions and properties. They are adaptive units as they are evolving together with the species within them. They have the attributes of structure, function, heterogeneity, and process dynamics. They range in size, productivity, and complexity, but usually encompass specific, limited spaces.[4] However, ecosystems have also been viewed as functional, not spatial, units, as defined from a process-function perspective.[5] Due to biological hierarchy and functional continuity, the boundaries of ecosystems can be chosen arbitrarily such that ecosystems exist at all scales of organization and complexity, and should include communities and landscapes.[6,7]

Ecosystems can be broadly divided into two major types—terrestrial and aquatic. Examples include forests, grasslands, ponds, and estuaries. Occasionally, they can be grossly categorized into lowland and upland ecosystems based on elevations. Ecosystems usually exist as spatially heterogenous land areas of interconnected systems where natural processes and human activities interact to produce ever-changing mosaics,[8] such as a pond in a forest with adjacent grassland. Although borders between ecosystems may be clear and well defined, such as a freshwater lake, the transition from one ecosystem to another can also be gradual. Compared with aquatic ecosystems, in particular marine ecosystems, terrestrial ecosystems are characterized by greater light availability, higher gaseous concentration, lower water content, and greater seasonal and diurnal temperature fluctuations.

Biodiversity at the Ecosystem Level

Ecological diversity refers to the species richness in a biological community—it is the diversity of a site at the level of the ecosystem. Together with genetic and species diversity, it forms the third level of biodiversity. In general, the diversity and functional complexity of an ecosystem are governed by the population and diversity of the species that are present, the interactions between these species and their interactions with the environment, and the complexity of the physical environment.[9]

The diversity of interacting types at the species or population level maintains ecological complexity and stability. It is known that species diversity has pronounced effects on ecosystem structures and functions. However, ecosystems are dynamic in the sense that they experience periodic perturbations from external forces such as environmental stochasticity and catastrophic events, and internal processes such as nutrient cycling. They have the capacity of resistance and/or resilience to disturbances, which is a function of their ecological attributes. Seasonality is a driving force behind annual changes in community composition and is important in shaping the life history of individual species in terms of breeding behavior and habits. Plant and animal communities constantly and gradually alter and develop over time through a directional, non-seasonal process called succession, as a result of endogenic processes and exogenic influences, thus changing the diversity and structural complexity of the ecosystem. Succession drives natural ecosystem development and assists in ecosystem restoration, which can be manipulated with human interference. Primary succession occurs naturally on glacier moraines or volcanic ashes, which are changed by weathering, plant colonization, organic matter accumulation, and soil formation, which subsequently affect the community and ecosystem levels.

Ecosystem properties are influenced by the diversity of species and their functional characteristics (functional traits, functional types, and functional diversity). Resource partitioning and niche differentiation allow the coexistence of species competing for similar resources. This coexistence increases the combinations of species that are complementary, and hence increases the biodiversity, productivity, and nutrient cycling of the ecosystem. Nevertheless, biological richness may differ in ecosystems that have similar basic processes of matter and energy fluxes.

Ecosystem Types

Ecological diversity should be distinguished from ecosystem diversity, which relates to the assortment of biological communities and their habitats, and the physical environment and ecological processes within an ecosystem and between different ecosystems. Ecosystem diversity can be defined as the variety and relative extent of ecosystem types, including their composition, structure, and processes. It refers to the spectrum of ecosystems—i.e., the diversity of ecosystems, sometimes called ecodiversity—present in the biosphere. Ecosystem diversity is high in Southeast Asia, Africa, and South America, which have a massive land area with variable climate, complex geomorphic types, a long shoreline, and an extensive river system.

Our biosphere is viewed as a giant system that contains a diversity of ecosystems. Ecosystems can be divided into two distinct groups—natural and man-made—that vary in different parts of the world. Terrestrial ecosystems can be classified on the basis of the physiognomy and physical environment of their vegetation, defined by the external appearance and the structure of the dominant vegetation, such as deciduous forests, coniferous forests, savannas, and deserts. More often, ecosystems are classified using multi-factor systems that include a combination of biotic and abiotic factors such as vegetation,

TABLE 20.1 Ecosystem Categories according to External Forcing Functions

Category	Forcing Functions	Unique Properties	Examples
1	Natural	Buffer capacity, biodiversity, ecosystem health	Polar terrestrial ecosystems, tundra, deserts, dunes, alpine forests, caves
2	Pollution	Buffer capacity, biodiversity, adaptation	Boreal forests, savannas, steppes, prairies, rainforests
3	Pollution and climate changes	Buffer capacity, adaptation, resilience	Estuaries, lakes, marshes, floodplains, swamps, mangroves, rivers
4	Anthropogenic	Management-oriented	Agroecosystems, plantations, urban ecosystems, landfills, quarries

Source: Adapted from Jorgensen.[10]

climate, soil, geology, physiography, and hydrology. Ecosystems can also be categorized in terms of their forcing functions (external variables) that determine the ecosystem conditions (Table 20.1).[10]

There are many different ecosystems—ponds and coral reefs, deserts and grasslands, tundra and rainforests. The types of ecosystems found on Earth are largely determined by climatic factors, which are influenced by latitude and altitude and affect the formation of the dominant vegetation. Ecosystems can be classified by latitude, e.g., temperate forests and tropical forests, which causes variations in temperature and precipitation. These variations give rise to ecosystems with different magnitudes of productivity. There are spatial gradients in biodiversity and energy flux among ecosystems with different climates and geographical locations. Nonetheless, the Earth's ecosystems are linked by global atmospheric and water circulation to form the biosphere. The occurrence of an organism in a particular habitat is a reflection of its adaptation to environmental stresses or disturbances, regardless of whether or not its population distribution is restricted, though dispersal ability is as well important, e.g., high salinity in salt marshes or mangroves, heat and drought in hot deserts, physiological drought in montane ecosystems, shade in forest understory, and cold in polar ecosystems. Because stresses impose different effects on plants and animals in different ecosystems, living organisms that survive will have different adaptation strategies to cope with/exploit these stresses. These specialist organisms may pay an ecological price—e.g., extreme biomass turnover rate and shifted reproductive mode—for their adaptation to those high-stress ecosystems. An important outcome of this adaptation is that stressed ecosystems provide a refuge for these specialists, resulting in greater overall biodiversity than if all organisms were generalists.

World Ecosystems

The biosphere is occupied by biomes, which represent distinctive plant and animal communities at the continental level. The geographical pattern of biomes on a global scale is determined by primary productivity, which is governed mainly by the temperature and rainfall in the terrestrial environment. Biomes are regional-scale assemblages of ecosystems (continental-level ecosystems). The biome approach to ecosystem diversity is linked directly to the spatial pattern of ecosystems throughout the biosphere. The distribution of biomes is heavily influenced by climate—precipitation produces humid, semihumid, semiarid, and arid ecosystems and thus has an overriding effect. Where rainfall is plentiful, vegetation is abundant and tall trees dominate. With decreasing rainfall, the biomass of vegetation is reduced and becomes scattered and is dominated by shrubs and grasses. Temperature is another limiting factor, and its ecological effects are manifested at higher latitudes and altitudes. Nival, alpine, subalpine, and montane ecosystems vary by altitude, while polar, boreal, temperate, subtropical, and tropical ecosystems vary by latitude. Interestingly, there are two types of rainforests—tropical and temperate—and grasslands are found on all continents except Antarctica.

Forest Ecosystems

Forest ecosystems are characterized by their vertical structure, which usually consists of several layers: overstoreys of trees and understoreys of shrubs and herbs. The overstoreys can differentiate to form a non-uniform crown canopy. Forest ecosystems can be pure or mixed stands. Young ecosystems are usually open stands, while mature ones generally have closed crown canopies. Tree species assemblages vary between regions, with the crown canopy structure greatly affected by age, the prevailing climatic conditions, and ecosystem management. Changes in forest vegetation are strongly associated with climatic factors due to latitudinal and elevation gradients as well as edaphic factors such as geology and soil. Forest ecosystems provide various ecological services, although the focus is often on their value for timber production. Many forests disappear as a result of mismanagement, causing carbon dioxide release and exacerbating global warming. Despite global and local regulatory measures to halt the negative effects of forest degradation and deforestation, their efficacy is yet to be proven.

Tropical Rainforests

Tropical rainforests are equatorial evergreen rainforests located in humid tropical areas in Central and South American, Southeast Asia, and West Africa. The annual rainfall in such areas exceeds 3000–4000 mm and is evenly distributed throughout the year. Humidity and temperature are relatively and constantly high. They have the greatest biodiversity, highest primary productivity, and most complex structure of all terrestrial ecosystems. They contribute 30% of the total net primary productivity on land.[11] The Amazonian tropical rainforests are megadiverse hotspots, with over 5000 vascular plant species per 10,000 km^2.[12] Tropical rainforests are the most complex terrestrial ecosystems, as they are species-rich and stratified vertically. Their canopies are thick and continuous, and competition for light is intense. There are typically three to four vertical layers—forest floor, understory, main canopy, and emergent. Most trees are about 30 m tall, with some reaching 60–80 m, and they can support massive epiphytic growths on their trunks; this stratification creates a variety of microhabitats in terms of light intensity, moisture content, and wind velocity. Plant species are adapted to deal with the overwhelming competition from tall trees for light, water, and nutrients. Soils are surprisingly thin and nutrient poor. Nutrient loss due to leaching by heavy rains is severe and decomposition is rapid, keeping the organic matter content of the soil low. However, these soils can sustain high biomass and biodiversity as rainforest plants are adapted to low levels of nutrients and can recycle nutrient efficiently due to the presence of mycorrhizae in the roots, which enhance nutrient absorption. Faunal diversity is enormous and includes insects, amphibians, reptiles, bats, and primates, most of which are arboreal and canopy-dwellers and are supported by the diverse food sources from trees and epiphytes.

Tropical Seasonal Forests

These are commonly called tropical dry forests and are located in regions further away from the equator, between 10° and 25° latitude north and south of the Amazon rainforests, northern Mexico, north and south of the central African rainforests, India, Indochina, northern Australia, and some tropical islands. They are tropical forests with an annual rainfall cycle of a 5–7-month wet season with pulses of torrential rains, and a 5–7-month dry season. Soil is nutrient-rich but highly susceptible to erosion. Rainfall seasonality is a dominant ecological force that drives biological activities, with tall deciduous trees forming a dense canopy during the wet season, and their height highly correlated with average rainfall.[13] Trees are shorter and shed their leaves during dry months. Seasonal migration is a phenomenon in animal populations, which move according to water availability in dry seasons. Many tropical dry forests are increasingly threatened by human activities: habitat fragmentation and fire are important in African forests, agricultural conversion is most notable in Eurasia, and climate change is more influential in the Americas. Globally, they are considered the most endangered tropical ecosystem and should be accorded higher conservation priority.[14]

Deciduous Forests

Deciduous forest is a common ecosystem that used to cover most of the more humid areas at temperate latitudes. The primary productivity of temperate deciduous forests is about 40% of that of tropical rainforests, due to lower growth and activity during the cold winter months. They exhibit strong seasonality in terms of foliage cover and flowering, with distinct vegetative and dormant periods. Summer-green forests are found in regions with pronounced seasonal temperature changes, and contain tall broad-leafed trees that shed their leaves in winter. Their biodiversity is lower and their vegetation structure less complex. Conifers may comprise a considerable portion of deciduous forests as we move northward in latitude or upward in elevation. The original vegetation in this type of ecosystem has been severely altered by humans as a result of logging, livestock grazing, agriculture, and plantations, independent of successional changes.

Grasslands

Grasslands are ubiquitous ecosystems, occupied exclusively by low-growth vegetation that contributes to its open nature. Grasses and sedges are the dominant species, together with flowering herbs, including many kinds of composites and legumes. Grasslands are usually devoid of woody species. However, while temperate grasslands are usually treeless, tropical grasslands have scattered small trees. Temperate species are C3 plants while tropical species are C4 plants based on their photosynthetic pathways. Grasslands can be classified by the unidirectional approach of successional linear replacement into types such as pioneer and climax, or less commonly into increasers, decreasers, neutrals, invaders, and retreaters,[15] depending upon their responses to factors such as grazing, drought, and fire. Annuals or short-lived perennials are pioneer grassland species that readily invade and colonize exposed soils and thrive in full sunlight. Grasslands are found in areas where the moisture content of soil is too low for a closed forest canopy, and grass-dominated communities are maintained through vegetative growth cycles in a seasonal pattern. Grassland areas include prairie in North America, pampas in South America, steppe in central Asia, and savanna in tropical Africa, Australia, South America, and Asia. They are the grazing grounds for both wild and domestic herbivores. Their primary production supports diverse fauna. Plants respond morphologically, structurally, and chemically to herbivory, and herbivores evolve by developing different feeding strategies regarding food source, body size, and feeding behavior. Plant-herbivore interactions and the effects of grazing by large generalist herbivores are generally accounted for by predation, disturbances arising from the evolutionary history and the selective pressures of grazing,[16] and the feedback mechanism between plants and grazers that is manifested in the switching capabilities of plant species and competition modes, especially in grasslands with a long grazing history.[17] Grasses are highly adapted to repetitive grazing and fire, which stimulate shoot regeneration from leaf meristems near the soil surface or by tillering after defoliation. Grassland can be maintained by moderate grazing intensity or fire, which prevents fire-intolerant woody species from successional recovery, and by mowing and other management practices. The long-term sustainability of pastures is threatened by grassland degradation attributable to climate change, desiccation, and over-grazing.

Deserts

Dryland ecosystems constitute the largest biome on Earth, occupying about one-third of its surface. Major deserts are found on all continents except Europe and are geographically located at 25–40° North and South on continental inlands, though there are also coastal deserts in Central America

TABLE 20.2 Arid Lands Categorized by Bioclimatic Aridity Resulting from Precipitation (P) and Evapotranspiration (ET)

	Bioclimatic Aridity	Rainfall	Vegetation
Hyper-arid	P/ET < 0.03	Low, irregular, aseasonal	Perennial; shrubs in riverbeds
Arid	0.03 < P/ET < 0.2	80–350 mm (annual)	Perennial; woody succulents; thorny or leafless
Semi-arid	0.2 < P/ET < 0.5	300–800 mm (summer) 200–500 mm (winter)	Shrubs; steppes, savannas, tropical scrub
Sub-humid	0.5 < P/ET < 0.75		Tropical savanna, chaparral, steppes

Source: From UNESCO.[18]

and Southern Africa. They range from semi-arid rangelands to semi-deserts, deserts, and hyper-arid deserts, with decreasing rainfall intensity which restricts primary productivity (Table 20.2).[18] Deserts are characterized by bioclimatic aridity, defined by the ratio of precipitation to evapotranspiration, rainfall seasonality, erosion, and deposition. Climatologically, deserts are influenced by descending dry air currents that reduce rainfall formation. They are found in areas with annual precipitation of less than 250 mm. Rainfall is usually highly irregular, both spatially and temporally.

Deserts can be broadly divided into two groups based on their geological structure and elevational setting: a) shield and platform deserts found mainly in Africa, Arabia, Australia, and India; and b) basin and mountain deserts located in Asia and America. Geographically, they are categorized into four main types: subtropical, cool coastal, rain shadow, and interior continental.[19] Deserts may be devoid of vegetation when rainfall is totally absent. High temperature is usually associated with aridity, which further limits productivity. The composition and structure of the vegetation are largely determined by the landscape's ecology, lithology, and geomorphology,[20,21] which are a function of the topography, gradient, and aspect.

Water supply in deserts is erratic and sporadic, and is aggravated by the high evaporation rate. The resulting water shortage severely restricts life and drives adaptive evolution. Desert organisms are usually tolerant to drought and heat. Desert plants have a scattered distribution and are adapted to arid conditions by means of escape, evasion, resistance, and endurance to drought. Plants that become established usually develop water-conserving features to reduce moisture loss and deep roots to access water at greater soil depths. Only drought-resistant plants such as aloe and sagebrush and succulents such as cacti can survive. Desert shrubs and trees are morphologically adapted to enhance stem flow and root channelization.[22] Plants may also develop ecophysiological strategies to overcome water scarcity, which to a certain extent may generate more efficient photosynthesis. Desert animals survive by escape, burrowing, physiological resistance, and aestivation. They have evolved a range of structural, physiological, and behavioral adaptations in body temperature regulation to cope with the heat of the desert.

Water is the major limiting factor, as illustrated by the "pulse-reserve" life strategy,[23] which responds to abiotic conditions modified by biotic interactions. Many desert plants are ephemerals, which complete their life cycles by taking advantage of short wet periods. Some species may germinate, grow, flower, and set seeds after a single rainfall event. Competition and predation have major effects on water–nutrient availability, seed dispersal, and herbivory, which alter carbon allocation patterns and plant morphologies. Most organisms live at or near the threshold for surviving the desert's climatic extremes and resource availability. Competitive interactions determine the patch-mosaic pattern and dynamics, resulting in a patchy distribution of plants and animals.[24] The association between plants, though may be weak, provides subcanopy microsites for seed germination and plant establishment. Patches are attributed to differential germination and establishment on sites modified by the subcanopy microclimate and animal activities (animal dispersal). Nurse plant–commensal relationships facilitate a number of cacti and stem succulents, and some grasses. Processes such as herbivory affect vegetation patchiness as a result of the differential spatial pattern of plant material consumption and fecal matter accumulation. The spatial distribution of plant life forms is also governed by the elevation, soil texture and soil depth, and the availability of nutrients and water. Despite low primary productivity, biodiversity can be higher than expected. Arid land is expanding as a result of desertification caused by climate change and poor land management.

Wetlands

Wetland broadly refers to areas of marshland, fens, peatlands, or water, whether natural or artificial, permanent or temporary. Water may be static or flowing, fresh, brackish, or salt, including marine water areas in which the depth at low tide does not exceed 6 m.[25] Wetland ecosystems are a transition between terrestrial and aquatic systems, with hydric soil that results in prolonged anaerobic conditions. Wetland ecosystems can be classified into coastal and inland wetlands. Coastal wetlands include tidal salt marshes, tidal freshwater marshes, and mangrove wetlands, while inland wetlands consist of freshwater marshes, freshwater swamps, riparian forests, and peatlands. Freshwater wetlands can be further divided into herbaceous wetlands that include lacustrine (ponds and lakes), riverine, and palustrine (marshes and peatlands) wetlands, as well as woody wetlands, i.e., forested wetlands such as swamps. Vegetation is dominated by hydrophytes, which include submerged, floating, and emergent plants. Floating hydrophytes have their roots in water. Emergents include reed, cattail, and cypress; they have their rhizomes and roots in substrates, while their shoots protrude into the air.

Wetlands are highly productive and wetland plants are efficient in absorbing nutrients and retaining them. They have been described as the kidneys of the Earth because of their capacity to purify the water that flows through them.[26] Wetlands play significant roles in modifying the hydrology and reducing flood flows. They act like a sponge during periods of high water inflow, which has important implications for downstream ecosystems. Wetland vegetation retains sediments and reduces erosion. Wetlands are of international conservation value and are significant in carbon sequestration. They have high biodiversity and attract a great variety of fauna, including migratory birds that use them as fueling stations.

Mangrove ecosystems are forested wetlands found in estuaries and sheltered inlets exclusively in the intertidal zones of the tropical and subtropical coastline regions. True mangroves are taxonomically diverse and comprise about 60 species. They are woody species that, together with the mangrove associates, make up the mangals in tidal sediment habitats. Their mechanisms for salinity tolerance are exclusion, secretion, and/or accumulation of salts. To adapt to waterlogging and fluctuating salinity, mangroves have developed complex structural and physiological specializations such as aerial roots, pneumatophores, salt glands in their leaves, and viviparous floating propagules, allowing them to survive in brackish water. Their growth and productivity as well as litter dynamics are governed by hydroperiod and nutrient biogeochemistry, which produce different stress combinations associated with coastal wetlands.[27] Mangroves support a variety of food webs that involve a wide range of halophytes and create a habitat for a rich heterogeneous faunal community, and are used by a vast array of marine and terrestrial animals as feeding and breeding grounds. They have great economic and educational value and protect the shoreline by acting as a natural barrier against water erosion. They are under high deforestation pressure as a result of human impacts; one-fifth of the world's mangrove ecosystems have been lost since 1980.[28]

Polar Ecosystems

Polar ecosystems are characterized by extreme environments because of their high latitudes and low temperatures (below 5°C). Most are glaciated and permanently covered with a thick layer of ice—permafrost that reaches a thickness >600 m.[29] Geographically, the Arctic region in the Northern Hemisphere is primarily an ice-covered ocean surrounded by land, whereas the Antarctic in the Southern

Hemisphere is a landmass surrounded by ocean. The Arctic and Antarctic regions differ in various ways regarding their geography, physical features, vegetation composition, and wildlife diversity. However, both regions have cold, dry, and windy climate, with short-growing seasons and prolonged darkness in the winter. They are sometimes referred to as polar deserts because of the lack of precipitation and low temperatures. Polar life survives through unique adaptations to withstand the cold. They contain 3–5% of global biodiversity and net primary productivity ranges from 1 to 1000 $g/m^2/yr$.[30] Polar ecosystems are typically treeless, and in warmer and wetter places with organic soil, tundra vegetation is found and is dominated by simple plant communities of algae, lichens, mosses, and liverworts. They contain large mammals such as bears, caribou, wolves, foxes, walruses, and seals, and a variety of birds including penguins and migratory species. Most of them are homeotherms that have adaptations to allow them to thrive in cold environments. Polar ecosystems are vulnerable to human disturbances such as global climate change, pollution, and resource extraction. Global warming has facilitated the invasion of exotic species to polar regions, which used to be physiological filters.[31,32]

Agroecosystems

Agroecosystems are managed ecosystems designed for the production of crops and livestock. Conventional agroecosystems have lower species diversity, a more homogenous ecological structure, and modified growth dynamics and agricultural input. To overcome the climatic and soil constraints, the productivity of agroecosystems can be enhanced through the use of machinery and human practices such as irrigation and the application of fertilizers, pesticides, herbicides, and fungicides. Arable fields are composed of crop species that are cultivated for biomass harvest and usually have a short life-cycle and small size; weeds, which are unwanted species that compete with the crops for sunlight, space, water, and nutrients; and pests such as insects that cause harm to crops in terms of health and productivity. The organic matter turnover and nutrient cycling of agricultural fields is manipulated. Biodiversity can be remarkably low in monocultures, in contrast with the general relationship between biodiversity, productivity, and ecological stability. Agricultural systems can be classified according to farming intensity, the degree of mechanization and the use of agrochemicals. Industrial farming is a conventional system, while integrated and organic farming use less or no pesticides and chemical fertilizers. The majority of agricultural land is formed by the conversion of forests. In addition to food production, agricultural land provides ecosystem services such as organic recycling, carbon sequestration, and landscape attributes.

Urban Ecosystems

Urban areas are essentially built environment covered by concrete structures and asphalt surfaces, which drastically change their metabolisms, material flows, and energy cycling.[33] With the concentration of populations, cities usually have high ecological loads and footprint with respect to food and water consumption, waste and wastewater disposal, as well as pollutant discharges. Urban ecosystems have a human dimension and are created as a result of urbanization, which is considered a major driving factor in the extinction of many fauna and flora. They have a unique community structure and ecosystem function due to their altered soil and water status, climate change, and local air pollution. Pollution problems are widespread particularly for those with heavy industrial and traffic emissions to result in ecosystem responses typically along a rural–urban gradient.[34,35] Urban growth and sprawl are converting natural habitats to isolated fragments with great reduction in biodiversity. Biogeochemically, urban ecosystems have modified soil carbon pools and fluxes;[36] physically, they have elevated temperatures and are urban heat islands that have ecological, energy, and climate implications. Urban landscapes and green areas are becoming important, and the more common use of alien species that fill up their specific urban niches have significantly modified urban floristic diversity and the ecological processes in such artificial ecosystems.[37,38]

Urban ecosystems range from roof gardens, parks, and greenbelts to urban forests and entire cities; they can be very diverse and include cemeteries, golf courses, and slopes. They sometimes encompass man-made ecosystems that are formed by the rehabilitation of degraded and damaged habitats such as mine sites, quarries, landfills, and coal ash disposal sites. However, these habitat patches represent a barrier to animal and plant movement unless they are connected by ecological corridors to maintain biological interactions. Cities will have distinctive species assemblages and support special fauna and flora that are highly adapted to the urbanized environmental conditions.[39,40] Wildlife– human conflicts do exist and wildlife management is an integral part of urban ecosystem management.[41] There are indeed variations in their management, which affect their role in city greening and biodiversity conservation. Overall, urban ecosystems benefit human populations through the provision of ecosystem services, among which the regulation of air quality, city climate, and water cycling are the most important.

Ecosystem Management

Ecosystem diversity is the focal point of biodiversity, and ecosystem function is regarded as a core element in biodiversity assessment.[42] Likewise, the ecosystem approach to conservation and sustainable development is a strategy for the integrated management of water, land, atmosphere, and biological resources at the appropriate spatial and temporal scales in a fair and equitable manner. This requires the adaptive management of ecosystems. However, ecosystems of all kinds are being threatened worldwide. Many types of forests, grasslands, coastal and lowland areas, and wetlands have been under severe pressure from human destruction or degradation. These pressures affect biodiversity in the ecosystems and ecosystem properties which in turn depend on biodiversity and ecosystem functioning.

A coarse filter conservation approach has been proposed to maintain ecosystem diversity, which is the key to ecological sustainability. It is argued that preserving a good representation of all ecosystems helps to meet the habitat requirements of all native species in a locality, hence protecting biodiversity on both a local and regional scale.[43] Ecosystems include both physical and biological dimensions, and may be expanded to include economic, social, cultural, and ethical factors. With the challenges from a multitude of development threats, ecosystem management is vital for setting priorities and accommodating stressors, whether natural or anthropogenic, to sustain our planet as a provider of resources and other essential ecological services. Our knowledge of the relationship between biodiversity and ecosystem functioning will certainly benefit nature conservation and management as a whole.

References

1. Tansley, A.G. The use and abuse of vegetational concepts and terms. Ecology **1935**, *16* (3), 284–307.
2. Likens, G.E. *The Ecosystem Approach: Its Use and Abuse*; Ecology Institute: Oldendorf/Luhe, Germany, 1992.
3. Schulze, E.D.; Beck, E.; Müller-Hohenstein, K. *Plant Ecology*; Springer: Berlin, 2005.
4. Chapin, F.S., III; Matson, P.A.; Mooney, H.A. *Principles of Terrestrial Ecosystem Ecology*; Springer: New York, 2002.
5. O'Neill, R.V.; DeAngelis, D.L.; Waide, J.B.; Allen, T.F.H. *A Hierarchical Concept of Ecosystem*; Princeton University Press: Princeton, 1986.
6. Haufler, J.B.; Crew, T.; Wilcove, D. Scale considerations for ecosystem management. In *Ecological Stewardship. A Common Reference for Ecosystem Management*; Szaro, R.C., Johnson, N.C., Sexton, W.T., Malk, A.J., Eds.; Elsevier: Oxford, 1999; Vol. 2, 331–342.
7. Lugo, A.E.; Baron, J.S.; Frost, T.P.; Cundy, T.W.; Dittberner, P. Ecosystem processes and functioning. *In Ecological Stewardship. A Common Reference for Ecosystem Management*; Szaro, R.C., Johnson, N.C., Sexton, W.T., Malk, A.J., Eds.; Elsevier, Oxford, 1999; Vol. 2, 219–254.
8. Forman, R.T.T. *Land Mosaics: The Ecology of Landscapes and Regions*; Cambridge University Press: Cambridge, 1995.

9. Smith, R.L. *Ecology and Field Biology*, 4th Ed.; Harper Collins: New York, 1990.
10. Jorgensen, S.E. Overview of ecosystem types, their forcing functions and most important properties. *In Ecosystem Ecology*; Jorgensen, S.E., Ed.; Elsevier: Amsterdam, 2009; 140–141.
11. Whittaker, R.H. *Communities and Ecosystems*, 2nd Ed.; The Macmillan Co.: Toronto, 1975.
12. Barthlott, W.; Hostert, A.; Kier, G.; Küper, W.; Kreft, H.; Mutke, J.; Rafiqpoor, D.; Sommer, J.H. Geographic patterns of vascular plant diversity at continental to global scales. Erdkunde **2007**, *61* (4), 305–315.
13. Murphy, P.G.; Lugo, A.E. Ecology of tropical dry forest. Annu. Rev. Ecol. Syst. **1986**, *17*, 67–88.
14. Miles, L.; Newton, A.C.; DeFries, R.S.; Ravilious, C.; May, I.; Blyth, S.; Kapos, V.; Gordon, J.E. A global overview of the conservation status of tropical dry forests. J. Biogeogr. **2006**, *33* (3), 491–505.
15. Vogl, R.J. The effects of fire on the vegetational composition of bracken-grasslands. Trans. Wis. Acad. Sci. Arts Lett. **1964**, *53*, 67–82.
16. Fox, J.F. Intermediate-disturbance hypothesis. Science **1979**, *204* (4399), 1344–1345.
17. Milchunas, D.G.; Sala, O.E.; Lauenroth, W.K. A generalized model of the effects of grazing by large herbivores on grassland community structure. Am. Nat. **1988**, *132* (1), 87–106.
18. United Nations Educational, Scientific and Cultural Organization (UNESCO). *World Map of Arid Regions*; UNESCO: Paris, 1977.
19. Cloudsley-Thompson, J.L. *Man and the Biology of Arid Zones*; Edward Arnold: London, 1977.
20. Schulz, E.; Whitney, J.W. Vegetation in north-central Saudi Arabia. J. Arid Environ. **1986**, *10* (3), 175–186.
21. Wondzell, S.M.; Cunningham, G.L.; Bachelet, D. Relationship between landforms, geomorphic processes and plant communities on a watershed in the northern Chihuahuan Desert. Landscape Ecol. **1996**, *11* (6), 351–362.
22. DeSoyza, A.G.; Whitford, W.G.; Martinez-Meza, E.; Van Zee, J.W. Variation in creosotebush *(Larrea tridentata)* canopy morphology in relation to habitat, soil fertility and associated annual plant communities. Am. Midl. Nat. **1997**, *137* (1), 13–26.
23. Noy Meir, I. Desert ecosystems: Environment and producers. Annu. Rev. Ecol. Syst. **1973**, *5*, 195–214.
24. Whitford, W. *Ecology of Desert Systems*; Academic Press: San Diego, 2002.
25. Ramsar Convention. Convention on Wetlands of International Importance especially as Waterfowl Habitat, http://www.ramsar.org/cda/en/ramsar-documents-texts-convention-on/main/ramsar/1-31-38%5E20671_4000_0, 1971 (accessed November 2012).
26. Mitsch, W.J.; Gosselink, J.G. *Wetlands*, 2nd Ed.; John Wiley & Sons: Hoboken, New Jersey, 2007.
27. Twilley, R.R.; Rivera-Monroy, V.H. Developing performance measures of mangrove wetlands using simulation models of hydrology, nutrient biogeochemistry, and community dynamics. J. Coast. Res. **2005**, SI40, 79–93.
28. Spalding, M.; Kainuma, M.; Collins, L. *World Atlas of Mangroves*; Routledge: Oxford, 2010.
29. Stonehouse, B. *Polar Biology*; Chapman and Hall: New York, 1998.
30. Callaghan, T.V. Polar terrestrial ecology. In *Ecosystem Ecology*; Jorgensen, S.E. Ed.; Elsevier: Amsterdam, 2009; 339–342.
31. Walther, G.-R.; Post, E.; Convey, P.; Menzel, A.; Parmesan, C.; Beebee, T.J.C.; Fromentin, J.-M.; Hoegh-Guldberg, O.; Bairlein, F. Ecological responses to recent climate change. Nature **2002**, *416* (6879), 389–395.
32. Robinson, S.A.; Wasley, J.; Tobin, A.K. Living on the edge – Plants and global change in continental and maritime Antarctica. Glob. Chang. Biol. **2003**, *9* (12), 1681–1717.
33. Kennedy, C.; Cuddihy, J.; Engel-Yan, J. The changing metabolism of cities. J. Ind. Ecol. **2008**, *11* (2), 43–59.
34. Gilbert, O.L. *The Ecology Of Urban Habitats*; Chapman and Hall: London, 1989.
35. McDonnell, M.J.; Pickett, S.T.A.; Groffman, P.; Bohlen, P.; Pouyat, R.V.; Zipperer, W.C.; Parmelee, R.W.; Carreiro, M.M.; Medley, K.Y. Ecosystem processes along an urban-torural gradient. Urban Ecosyst. **1997**, *1* (1), 21–36.

36. Pouyat, R.; Groffman, P.; Yesilonis, I.; Hernandex, L. Soil carbon pools and fluxes in urban ecosystems. Environ. Pollut. **2002**, *116* (Suppl. 1), S107–S118.
37. Kowarik, I. On the role of alien species in urban flora and vegetation. In *Plant Invasions: General Aspects and Special Problems*; Pysek, P., Prach, K., Rejmanek, M., Wade, M., Eds.; SPB Academic Publishing: Amsterdam, 1995; 85–103.
38. Savard, J.-P.L.; Clergeau, P.; Mennechez, G. Biodiversity concepts and urban ecosystems. Landscape Urban Plann. **2000**, *48* (3–4), 131–142.
39. Blair, R.B. Birds and butterflies along an urban gradient: Surrogate taxa for assessing biodiversity? Ecol. Appl. **1999**, *9* (1), 164–170.
40. McKinney, M.L. Effects of urbanization on species richness: A review of plants and animals. Urban Ecosyst. **2008**, *11* (2), 161–176.
41. Adams, C.E.; Lindsey, K.J.; Ash, S.J. *Urban Wildlife Management*; CRC Press: Boca Raton, 2006.
42. United Nations Environment Program (UNEP). *Global Biodiversity Assessment*; Cambridge University Press: Cambridge, 1996.
43. Hunter, M.L.; Jacobson, G.L.; Webb, T. Paleoecology and the coarse-filter approach to maintaining biological diversity. Conserv. Biol. **1988**, *2* (4), 375–385.

21

Ecosystems: Forest Nutrient Cycling

Neil W. Foster and
Jagtar S. Bhatti
Canadian Forest Service

Introduction 177
Influence of Climate 178
Influence of Biotic Factors 179
Influence of Abiotic Factors 179
Role of Disturbance 180
References 180

Introduction

Nutrients are elements or compounds that are essential for the growth and survival of plants. Plants require large amounts of nutrients such as nitrogen (N), phosphorus (P), carbon (C), hydrogen (H), oxygen (O), potassium (K), calcium (Ca), and magnesium (Mg), but only small amounts of others such as boron (B), manganese (Mn), iron (Fe), copper (Cu), zinc (Zn) and chlorine (Cl) (micronutrients). Forest nutrient cycling is defined as the exchange of elements between the living and nonliving components of an ecosystem.[1] The processes of the forest nutrient cycle include: nutrient uptake and storage in vegetation perennial tissues, litter production, litter decomposition, nutrient transformations by soil fauna and flora, nutrient inputs from the atmosphere and the weathering of primary minerals, and nutrient export from the soil by leaching and gaseous transfers.

Each nutrient element is characterized by a unique biogeochemical cycle. Some of the key features of the major nutrients are shown in Table 21.1. Forest trees make less demand on the soil for nutrients

TABLE 21.1 Features of the Major Nutrient Cycles

Element	Uptake by the Trees	Major Sources for Tree Uptake	Limiting Situations
Carbon	Atmosphere	Atmosphere	Atmospheric concentration may limit growth
Oxygen	Atmosphere	Atmosphere	Waterlogged soils
Hydrogen	Atmosphere	Atmosphere	Extremely acidic and alkaline conditions
Nitrogen	Soluble NO_3 and NH_4; N_2 for nitrogen fixing species	Soil organic matter; atmospheric N_2 for nitrogen fixing species	Most temperate forests, many boreal forests and some tropical forests
Phosphorus	Soluble phosphorus	Soil organic matter; adsorbed phosphate and mineral phosphorus	Old soils high in iron and aluminum, common in subtropical and tropical environment
Potassium, calcium, magnesium	Soluble K^+/soluble Ca^{2+}/ soluble Mg^{2+}	Soil organic matter; exchange complex and minerals	Miscellaneous situations and some old soils

than annual crops because a large proportion of absorbed nutrients are returned annually to the soil in leaf and fine root litter and are reabsorbed after biological breakdown of litter materials. Also, a large portion of nutrient requirement of trees are met through internal cycling as compared with agricultural crops.

Nutrient cycling in forest ecosystems is controlled primarily by three key factors: climate, site, abiotic properties (topography, parent material), and biotic communities. The role of each factor in ecosystem nutrient dynamics is discussed and illustrated with selected examples from boreal, temperate, and tropical zones. The importance of ecosystem disturbance to nutrient cycling is examined briefly, since some nutrients are added or lost from forest ecosystems through natural (e.g., fire, erosion, leaching) or human activity (harvesting, fertilization).

Influence of Climate

Large-scale patterns in terrestrial primary productivity have been explained by climatic variables. In above-ground vegetation, nutrient storage generally increases in the order: boreal < temperate < tropical forests (Table 21.2). In contrast, forest floor nutrient content and residence time increase from tropical to boreal forests, as a result of slower decomposition in the cold conditions of higher latitudes.

In subarctic woodland soils and Alaskan taiga forests, nutrient cycling rates are low because of extreme environmental conditions.[2] Arctic and subarctic forest ecosystems have lower rates of nutrient turnover and primary production because of low soil temperature, a short growing season, low net AET and the occurrence of permafrost. Low temperature reduces microbial activity, litter decomposition rates, and nutrient availability and increases C accumulation in soil.

In contrast with high latitudes, conditions in a tropical forest favor microbial activity throughout the year, which generally results in faster decomposition except in situations with periodic flooding, soil dessication, and low litter quality.[3] Rates of plant material decay are an order of magnitude higher in tropical soils than in subarctic woodland soils. The low storage of C and high amount of litter production in highly productive tropical forests contrasts with the high C storage and low litter production in boreal forests (Table 21.2).

TABLE 21.2 Nutrient Distribution in Different Forest Ecosystems

	Vegetation (Mg ha $^{-1}$)	Forest Floor (Mg ha^{-1})	Soil (Mg ha^{-1})	Residence Time (year)
		Carbon		
Boreal coniferous	78–93	37–113	41–207	800
Temperate deciduous	103–367	42–105	185–223	200
Tropical rain forest	332–359	7–72	2–188	120
		Nitrogen		
Boreal coniferous	0.3–0.5	0.6–1.1	0.7–2.87	200
Temperate deciduous	0.1–1.2	0.2–1.0	2.0–9.45	6
Tropical rain forest	1.0–4.0	0.03–0.05	5.0–19.2	0.6
		Phosphorus		
Boreal coniferous	0.033–0.060	0.075–0.15	0.04–1.06	300
Temperate deciduous	0.06–0.08	0.20–0.10	0.91–1.68	6
Tropical rain forest	0.2–0.3	0.001–0.005	0.06–7.2	0.6
		Potassium		
Boreal coniferous	0.15–0.35	0.3–0.75	0.07–0.8	100
Temperate deciduous	0.3–0.6	0.050–0.15	0.01–38	1
Tropical rain forest	2.0–3.5	0.020–0.040	0.05–7.1	0.2

Influence of Biotic Factors

Nutrient cycles are modified substantially by tree species-specific controls over resource use efficiency (nutrient use per unit net primary production). Species vary widely in their inherent nutrient requirements and use.[4] These effects can be split into two categories: accumulation into living phytomass and production of various types of nutrient-containing dead phytomass. Rapid accumulation of phytomass is associated with a net movement of nutrients from soil into vegetation. More than half of the annual nutrient uptake by a forest is typically returned to forest floor (litterfall) and soil (fine-root turnover). The subsequent recycling of these nutrients is a major source of available nutrients for forest vegetation. The mean annual litterfall from above-ground vegetation increases from boreal regions to the tropics following the gradient of productivity (Table 21.2).

Nutrient availability is strongly influenced by the quantity and quality of litter produced in a forest. A high proportion of the variation in foliar N concentrations at the continental scale has been explained by differences between forest types, which in turn has large impact on litter quality and the nutrient content of forest floors. In many temperate and boreal forest ecosystems, microbial requirement for N increases or decreases with labile supplies of soil C. Increased microbial demand for N may temporarily decrease the N availability to trees during the initial decomposition of forest residues with a wide C/N ratio. Microbes immobilize N from the surrounding soil, relatively rapid for readily decomposable organic matter (needle litter), and more slowly for recalcitrant material (branches, boles).

Rates of net N mineralization are higher and retention of foliar N is lower in temperate and tropical than in boreal forest soils. Nitrogen limitation of productivity, therefore, is weak in tropical forests and increases from temperate to boreal and tundra forest systems. Trees may obtain organic N and P from the soil via mycorrhizae or by relocation from older foliage prior to abscission, and thereby, partly reduce their dependence on soil as a source of inorganic nutrients. Increased understanding of the fundamental relationships between soil properties and plant nutrients requirements will most likely come from examination of plant-fauna-microbe interactions at root surfaces (rhizo-sphere), rather than in the bulk soil.

Influence of Abiotic Factors

Forests have distinctive physiographic, floristic, and edaphic characteristics that vary predictably across the landscape within a climatically homogeneous region. Differences in the elemental content of parent material influence the tree species composition between and within a landscape unit. For example, wind deposited soils, which support hardwood or mixed wood forest, are likely to be fine textured with high nutrient supplying capacity. In contrast, outwash sands that often support pine forests are coarse textured and infertile.

Heterogeneity within the landscape results in sites differing in microclimatic conditions, and physical and chemical properties, which produces different geochemical reaction rates and pools of available nutrients in soil. Soil type and topographic-microclimate interactions are important feedbacks that influence biological processes, such as the rate of N mineralization in soil. Low P availability is a characteristic of geomorphically old, highly weathered tropical, subtropical, and warm temperate soils.[3] The type and age of parent material from which the soil is derived can influence the base status and nutrient levels in soil. Soils in glaciated regions are relatively young and rich in weatherable minerals. Mineral weathering is an important source of most nutrients for plant uptake, with the exception of N. Nutrient availability is regulated by the balance between weathering of soil minerals and precipitation, adsorbtion, and fixation reactions in soil.

Edaphic conditions can exert a strong influence on forest productivity and produce considerable variation in nutrient cycling processes. Soils with low N, P, or pH support trees with low litter quality (high in lignin and tannins that bind N) that decomposes slowly. Edaphic limitations on growth may be compensated for by an increase in rooting density and depth. Some late-succession or tolerant species have a shallower root distribution relative to intolerant pioneer species and are adapted to sites where nutrients and moisture are concentrated at the soil surface. In contrast, nutrient uptake from sub-soil horizons is more important in highly weathered warm temperate soils where nutrient depletion takes place deeper in the soil.

Role of Disturbance

Disturbances such as fire, harvesting, hurricanes, or pests affect nutrient cycling long after the event. In fire-dominated ecosystems, intensive wildfire results in a horizontal and vertical redistribution of ecosystem nutrients. Redistribution results from the combined effects of the following processes: 1) oxidation and volatilization of live and decomposing plant material; 2) convection of ash particles in fire generated winds; 3) water erosion of surface soils; and 4) percolation of solutes through and out of the soil. The relative importance of these processes varies with each nutrient and is modified by differences in fire intensity, soil characteristics, topography, and climatic patterns. Expressed as a percentage of the amount present in vegetation and litter before fire, the changes often follow the order:

$$N > K > Mg > Ca > P$$

Harvesting removes nutrients from the site and interrupts nutrient cycling temporarily. The recovery of the nutrient cycle from harvest disturbance is dependent partly on the rate of re-establishment of trees and competing vegetation. Re-vegetation may occur within months in the tropics, 2–5 years in temperate regions, and longer in boreal and tundra regions.[5] Recovery assumes that the soil's ability to supply nutrients to plant roots has not been altered by disturbance. If nutrients cannot be supplied by the soil at rates sufficient to at least maintain the rate of growth of the previous forest then fertilization may be necessary to maintain site productivity.

Nutrient cycling and the impacts of disturbance on nutrient cycling have been examined thoroughly in many representative world forests. The impact of natural disturbances and management practices on nutrient cycling processes are generally characterized of the stand or occasionally on a watershed basis. There is a growing demand from policy makers and forest managers for spatial estimates on nutrient cycling at local, regional, and national scales.

The availability of N, P, and K in soil largely determines the leaf area, photosynthetic rate, and net primary production of forest ecosystems. Forest management practices that produce physical and chemical changes in the soil that accentuate the cycle of nutrients between soil and trees, may increase forest productivity. Clear-cut harvesting and site preparation practices (mechanical disturbance, slash burning) remove nutrients from soil in tree components and by increased surface runoff, soil erosion, and off-site movement of nutrients in dissolved form or in sediment transport. In the tropics, potential negative impacts associated with complete forest removal and slash burning are greatest because a larger proportion of site nutrients are contained in the living biomass. Environmental impacts associated with clear-cutting and forest management in general, are confounded by climatic, topographic, soil, and vegetation diversity associated with the world's forests. Best forest management practices can be utilized to control negative impacts on nutrient cycling.

References

1. Soil Science Society of America. *Glossary of Soil Science Terms*; Soil Science Society of America: Madison, WI, 1996; 134.
2. Van Cleve, K.; Chapin, F.S. III; Dyrness. C.T. III; Viereck, L.A. Element cycling in taiga forests: State factor control. Bioscience **1991**, *41* (2), 78–88.

3. Vitousek, P.M.; Stanford, R.L. Jr. Nutrient cycling in moist tropical forest. Ann. Rev. Ecol. Syst. **1986**, *17* (1), 137–167.
4. Cole, D.W.; Rapp, M. Elemental cycling in forest ecosystems. In *Dynamic Properties of Forest Ecosystems;* Reichle, E.D. Ed.; Cambridge University Press: London, 1981, 341–409.
5. Keenan, R.J.; Kimmins, J.P. The ecological effects of clear-cutting. Environ. Rev. **1993**, *1* (2), 121–144.

22

Ecosystems: Functions and Services

Ecosystem Services 183
Valuation 184
Case Studies 185
Early Valuation Syntheses
Major World Reports on Ecosystem Services 186
The Ecosystem Services' Partnership and Ongoing Work 186
Integrated Measurement, Modeling, Valuation, and Decision Science in Support of Ecosystem Services • Trade-offs • Accounting and Assessment • Modeling • Bundling • Scaling • Adaptive Management and New Institutions for Ecosystem Services • Property Rights • Scale Matching • Distribution Issues • Information Dissemination • Participation • Science–Policy Interface
Conclusions 188
References 188
Bibliography 190

Robert Costanza
Portland State University

Ecosystem Services

"Ecosystem services" (ESs) are the ecological characteristics, functions, or processes that directly or indirectly contribute to human well-being, the benefits people derive from functioning ecosystems.[1,2] Ecosystem processes and functions may contribute to ESs but they are not synonymous. Ecosystem processes and functions describe biophysical relationships and exist regardless of whether humans benefit from or not.[3,4] ESs, on the other hand, only exist if they contribute to human well-being and cannot be defined independently.

The ecosystems that provide the services are sometimes referred to as "natural capital," using the general definition of capital as a stock that yields a flow of services over time.[5] In order for these benefits to be realized, natural capital (which does not require human activity to build or maintain) must be combined with other forms of capital that *do* require human agency to build and maintain. These include: (1) built or manufactured capital; (2) human capital; and (3) social or cultural capital.[6]

These four general types of capital are all required in complex combinations to produce any and all human benefits. ESs thus refer to the relative contribution of natural capital to the production of various human benefits, in combination with the three other forms of capital. These benefits can involve the use, non-use, option to use, or mere appreciation of the existence of natural capital.

The following categorization of ESs has been used by the Millennium Ecosystem Assessment:[2]

Provisioning services: ESs that combine with built, human, and social capital to produce food, timber, fiber, or other "provisioning" benefits. For example, fish delivered to people as food require fishing boats (built capital), fisherfolk (human capital), and fishing communities (social capital) to produce.

Regulating services: Services that regulate different aspects of the integrated system. These are services that combine with the other three capitals to produce flood control, storm protection, water regulation, human disease regulation, water purification, air quality maintenance, pollination, pest control, and climate control. For example, storm protection by coastal wetlands requires built infrastructure, people, and communities to be protected. These services are generally not marketed but have clear value to society.

Cultural services: ESs that combine with built, human, and social capital to produce recreation, esthetic, scientific, cultural identity, sense of place, or other "cultural" benefits. For example, to produce a recreational benefit requires a beautiful natural asset (a lake), in combination with built infrastructure (a road, trail, dock, etc.), human capital (people able to appreciate the lake experience), and social capital (family, friends, and institutions that make the lake accessible and safe). Even "existence" and other "non-use values" require people (human capital) and their cultures (social and built capital) to appreciate.

Supporting services: Services that maintain basic ecosystem processes and functions such as soil formation, primary productivity, biogeochemistry, and provisioning of habitat. These services affect human well-being *indirectly* by maintaining processes necessary for provisioning, regulating, and cultural services. They also refer to the ESs that have not yet, or may never be intentionally combined with built, human, and social capital to produce human benefits that support or underlie these benefits and may sometimes be used as proxies for benefits when the benefits cannot be easily measured directly. For example, net primary production (NPP) is an ecosystem function that supports carbon sequestration and removal from the atmosphere, which combines with built, human, and social capital to provide the benefit of climate regulation. Some would argue that these "supporting" services should rightly be defined as ecosystem "functions," since they may not yet have interacted with the other three forms of capital to create benefits. We agree with this in principle, but recognize that supporting services/functions may sometimes be used as proxies for services in the other categories.

This categorization suggests a very broad definition of services, limited only by the requirement of a contribution to human well-being. Even without any subsequent valuation, explicitly listing the services derived from an ecosystem can help ensure appropriate recognition of the full range of potential impacts of a given policy option. This can help make the analysis of ecological systems more transparent and can help inform decision makers of the relative merits of different options before them.

Valuation

Many ESs are public goods. This means they are nonexcludable and multiple users can simultaneously benefit from using them. This creates circumstances where individual choices are not the most appropriate approach to valuation. Instead, some form of community or group choice process is needed. Furthermore, ESs (being public goods) are generally not traded in markets. We therefore need to develop other methods to assess their value.

There are a number of methods that can be used to estimate or measure benefits from ecosystems. Valuation can be expressed in multiple ways, including monetary units, physical units, or indices. Economists have developed a number of valuation methods that typically use metrics expressed in monetary units[7] while ecologists and others have developed measures or indices expressed in a variety of nonmonetary units such as biophysical trade-offs.[8]

There are two main methods for estimating monetary values: revealed and stated preferences. Both these methods typically involve the use of sophisticated statistical methods to tease out the values.[9] Revealed preference methods involve analyzing individuals' choices in real-world settings and inferring value from those observed choices. Examples of such methods include production-oriented valuation that focuses on changes in direct use values from products actually extracted from the environment (e.g., fish). This method may also be applicable to indirect use values, such as the erosion control benefits forests provide to agricultural production. Other revealed preference methods include hedonic pricing, which infers ecosystem service values from closely linked housing markets. For example, urban forest ecosystems and wetlands may improve water quality and that may be (partially) captured in property values.[10] The travel cost valuation method is used to value recreation ESs and estimate values based on the resources, money, and time visitors spend to visit recreation sites.

Stated preference methods rely on individuals' responses to hypothetical scenarios involving ESs and include contingent valuation and structured choice experiments. Contingent valuation utilizes a highly structured survey methodology that acquaints survey respondents with ecosystem improvements (e.g., better stream quality) and the ESs they will generate (e.g., increased salmon stocks). Respondents are then asked to value ecosystem improvements usually using a referendum method.[11]

Choice experiments are sometimes called conjoint analysis. This method presents respondents with combinations of ESs and monetary costs and asks for the most preferred combinations. Based on these choices, ecosystem service values are inferred.

A key challenge in any valuation is imperfect information. Individuals might, for example, place no value on an ecosystem service if they do not know the role that the service is playing in their well-being.[12] Here is an analogy. If a tree falls in the forest and there is no one around to hear it, does it still make a sound? Assume in this case that the "sound" is the ecosystem service. The answer to this old question obviously depends on how one defines "sound". If "sound" is defined as the perception of sound waves by people, then the answer is no. If "sound" is defined as the pattern of physical energy in the air, then the answer is yes. In this second case, choices in both revealed and stated preference models would not reflect the true benefit of the ecosystem service. Another key challenge is accurately measuring the functioning of the system to correctly quantify the amount of a given service derived from that system.[13,14]

But recognizing the importance of information does not obviate the limitations of human perception-centered valuation. As the tree analogy demonstrates, perceived value can be a quite limiting valuation criterion, because natural capital can provide positive contributions to human well-being that are either never (or only vaguely) perceived or may only manifest themselves at a future time. A broader notion of value allows a more comprehensive view of value and benefits, including, for example, valuation relative to alternative goals/ends, like fairness and sustainability, within the broader goal of human well-being.[15] Whether these values are perceived or not and how well or accurately they can be measured are separate (and important) questions.

Case Studies

Early Valuation Syntheses

Scientists and economists have discussed the general concepts behind natural capital, ESs, and their value for decades, with some early work as far back as the 1920s. However, the first explicit mention of the term "ESs" was in Ehrlich and Mooney.[16] More than 6000 papers have been published on the topic of ESs since then. The first mention of the term "natural capital" was by Costanza and Daly.[5]

One of the first studies to estimate the value of ESs globally was published in *Nature* entitled "The value of the world's ESs and natural capital."[1] This paper estimated the value of 17 ESs for 16 biomes to be in the range of US$16–54 trillion per year, with an average of US$33 trillion per year, a figure larger than annual GDP at the time. Some have argued that global society would not be able to pay more than their annual income for these services, so a value larger than global GDP does not make sense. However, not all benefits are picked up in GDP and many ESs are not marketed, so GDP does not represent a limit on real benefits.[17]

In this study, estimates of global ESs were derived from a synthesis of previous studies that utilized a wide variety of techniques like those mentioned above to value-specific ESs in specific biomes. This technique, called "benefit transfer," uses studies that have been done at other locations or in different contexts, but can be applied with some modification. See Costanza[18] for a collection of commentaries and critiques of the methodology. Such a methodology, although useful as an initial estimate, is just a first cut and much progress has been made since then.[19,20]

Major World Reports on Ecosystem Services

More recently the concept of ESs gained attention with a broader academic audience and the public when the Millennium Ecosystem Assessment (MEA) was published.[2] The MEA was a 4-year, 1300 scientist study commissioned by the United Nations in 2005. The report analyzed the state of the world's ecosystems and provided recommendations for policymakers. It determined that human actions have depleted the world's natural capital to the point that the ability of a majority of the globe's ecosystems to sustain future generations can no longer be taken for granted.

In 2008, a second international study was published on the Economics of Ecosystems and Biodiversity (TEEB), hosted by United Nations Environment Programme. TEEB's primary purpose was to draw attention to the global economic benefits of biodiversity, to highlight the growing costs of biodiversity loss and ecosystem degradation, and to draw together expertise from the fields of science, economics, and policy to enable practical actions moving forward. The TEEB report was picked up extensively by the mass media, bringing ESs to a broad audience.

The Ecosystem Services' Partnership and Ongoing Work

With such high profile reports being published, ESs have entered not only the public media[21] but also into business. Dow Chemical established a $10 million collaboration with The Nature Conservatory to tally up the ecosystem costs and benefits of every business decision.[22] Such collaborations will provide a significant addition to ES valuation knowledge and techniques. However, there is significant research that is still required. Our scientific institutions can help lead this process through transdisciplinary graduate education, such as the ESs for the Urbanizing Regions program funded by the National Science Foundation's Integrative Graduate Research and Education Traineeship program.[23]

Hundreds of projects and groups are currently working toward better understanding, modeling, valuation, and management of ESs and natural capital. It would be impossible to list all of them here, but the new Ecosystem Services Partnership (ESP; http://www.es-partnership.org) is a global network that does just that and helps to coordinate the activities and build consensus.

The following lays out the research agenda as agreed to by a group of 30 participants at a meeting in Salzau, Germany, in June 2010, at the launch of the ESP.

Integrated Measurement, Modeling, Valuation, and Decision Science in Support of Ecosystem Services

The scientific community needs to continue to develop better methods to measure, monitor, map, model, and value ESs at multiple scales. Ideally, these efforts should take place using interdisciplinary teams and strategies and in close collaboration with ecosystem stakeholders. Moreover, this information must be provided to decision makers in an appropriate, transparent, and viable way, to clearly identify differences in outcomes among policy choices. At the same time, we cannot wait for high levels of certainty and precision to act when confronting significant irreversible and catastrophic consequences. We must synergistically continue to improve the measurements with evolving institutions and approaches that can effectively utilize these measurements.

Trade-offs

Ecological conflicts arise from two sources: (1) scarcity and restrictions in the amount of ESs that can be provided and (2) the distribution of the costs and benefits of the provisioning of the ES. ES science makes trade-offs explicit and, thus, facilitates management and planning discourse. It enables stakeholders to make sound value judgments. ES science thus generates relevant social-ecological knowledge for stakeholders and policy decision makers and sets of planning options that can help resolve sociopolitical conflicts.

Accounting and Assessment

Accounting attempts to look at the flow of materials with relative objectivity, while assessment evaluates a system or process with a goal in mind and is more normative. Both are integrating frameworks with distinctive roles. Both ecosystem service accounting and assessment need to be developed and pursued using a broader socioecological lens. Within the broader lens we also need to balance expert and local knowledge across scales.

Modeling

We need modeling to synthesize and quantify our understanding of ESs and to understand dynamic, nonlinear, spatially explicit trade-offs as part of the larger socioecological systems. Stakeholders should be active collaborators in this model development process to assure relevancy. These models can incorporate and aid accounting and assessment exercises and link directly with the policy process at multiple time and space scales. In particular, modeling can quantify potential shifts in ESs under different environmental and socioeconomic scenarios.

Bundling

Most ESs are produced as joint products (or bundles) from intact ecosystems. The relative rates of production of each service vary from system to system, site to site, and time to time. We must consider the full range of services and the characteristics of their bundling in order to prevent creating dysfunctional incentives and to maximize the net benefits to society[24,25] For example, focusing only on the carbon sequestration service of ecosystems may, in some instances, reduce the overall value of the full range of ESs.

Scaling

ESs are relevant over a broad range of scales in space, time, governance, and complexity, including the legacy of past behavior. We need measurement, models, accounts, assessments, and policy discussions that address these multiple scales, as well as interactions, feedbacks, and hierarchies among them.

Adaptive Management and New Institutions for Ecosystem Services

Given that pervasive uncertainty always exists in ecosystem service measurement, monitoring, modeling, valuation, and management, we should continuously gather and integrate appropriate information regarding ESs, with the goal of learning and adaptive improvement. To do this we should constantly evaluate the impacts of existing systems and design new systems with stakeholder participation as experiments from which we can more effectively quantify performance and learn ways to manage such complex systems.

Property Rights

Given the public goods nature of most ESs, we need institutions that can effectively deal with this characteristic using a sophisticated suite of property rights regimes. We need institutions that employ an appropriate combination of private, state, and common property rights systems to establish clear property rights over ecosystems without privatizing them. Systems of payment for ecosystem services and common asset trusts can be effective elements in these institutions.

Scale Matching

The spatial and temporal scale of the institutions to manage ESs must be matched with the scales of the services themselves. Mutually reinforcing institutions at local, regional, and global scales over short-, medium- and long-time scales will be required. Institutions should be designed to ensure the flow of information across scales, to take ownership regimes, cultures, and actors into account, and to fully internalize costs and benefits.

Distribution Issues

Systems should be designed to ensure inclusion of the poor, since they are generally more dependent on common property assets like ESs. Free riding, especially by wealthier segments of society, should be deterred and beneficiaries should pay for the services they receive from biodiverse and productive ecosystems.

Information Dissemination

One key limiting factor in sustaining natural capital is lack of knowledge of how ecosystems function and how they support human well-being. This can be overcome with targeted educational campaigns that are tailored to disseminate success and failures to both the general public and elected officials and through true collaboration among public, private, and government entities.

Participation

Relevant stakeholders (local, regional, national, and global) should be engaged in the formulation and implementation of management decisions. Full stakeholder awareness and participation not only improves ES analyses but contributes to credible accepted rules that identify and assign the corresponding responsibilities appropriately, and that can be effectively enforced.

Science–Policy Interface

ES concepts can be an effective link between science and policy by making the trade-offs more transparent.[4] An ES framework can, therefore, be a beneficial *addition* to policy-making institutions and frameworks and to integrating science and policy.

Conclusions

Natural capital and ESs are key concepts that are changing the way we view, value, and manage the natural environment. They are changing the framing of the issue away from "jobs vs. the environment" to a more balanced assessment of all the assets that contribute to human well-being. Significant transdisciplinary research has been done in recent years on ESs, but there is still much more to do and this will be an active and vibrant research area for the coming years, because better understanding of ESs is critical for creating a sustainable and desirable future. Placing credible values on the full suite of ESs is key to improving their sustainable management.

References

1. Costanza, R.; d'Arge, R.; de Groot, R.; Farber, S.; Grasso, M.; Hannon, B.; Naeem, S.; Limburg, K.; Paruelo, J.; O'Neill, R.V.; Raskin, R.; Sutton, P.; van den Belt, M. The value of the world's ecosystem services and natural capital. Nature **1997a**, *387*, 253–260.
2. *Millennium Ecosystem Assessment; United Nations, 2005.*
3. Boyd, J.; Banzhaf, S. What are Ecosystem Services? Ecol. Econ. **2007**, *63*, 616–626.
4. Granek, E.F.; Polasky, S.; Kappel, C.V.; Stoms, D.M.; Reed, D.J.; Primavera, J.; Koch, E.W.; Kennedy, C.; Cramer, L.A.; Hacker, S.D.; Perillo, G.M.E.; Aswani, S.; Silliman, B.; Bael, D.; Muthiga, N.; Barbier, E.B.; Wolanski, E. Ecosystem services as a common language for coastal ecosystem-based management. Conserv. Biol. **2010**, *24*, 207–216.

5. Costanza, R.; Daly, H.E. Natural capital and sustainable development. Conserv. Biol. **1992**, *6*, 37–46.
6. Costanza, R.; Cumberland, J.C.C.; Daly, H.E.; Goodland, R.; Norgaard., R. *An Introduction to Ecological Economics;* St. Lucie Press: Boca Raton,1997b; 275.
7. Freeman, A.M. *The Measurement of Environmental and Resource Values: Theories and Methods* 2nd edition; RFF Press: Washington, DC, 2003.
8. Costanza, R. Value Theory and Energy. In *Encyclopedia of Energy.* C. Cleveland Ed.; Elsevier: Amsterdam, 2004, Vol. no. 6: 337–346.
9. Haab, T.; McConnell, K. *Valuing Environmental and Natural Resources: The Econometrics of Non-Market Valuation;* Edward Elgar Publishing Ltd.: Cheltenham, UK, 2002.
10. Phaneuf, D.J.; Smith, V.K.; Palmquist, R.B.; Pope, J.C. Integrating property value and local recreation models to value ecosystem services in urban watersheds. Land Econ. **2008**, *84*, 361–381.
11. Boardman, A.D.; Greenberg, A.; Vining; Weimer, D. *Cost-Benefit Analysis: Concepts and Practice,* 4th Edition; Prentice Hall, Inc: Upper Saddle River, NJ, 2006.
12. Norton, B.; Costanza, R.; Bishop, R. The evolution of preferences: why "sovereign" preferences may not lead to sustainable policies and what to do about it. Ecol. Eco. **1998**, 24, 193–211.
13. Barbier, E.B.; Koch, E.W.; Silliman, B.R.; Hacker, S.D.; Wolanski, E.; Primavera, J.; Granek, E.F.; Polasky, S.; Aswani, S.; Cramer, L.A.; Stoms, D.M.; Kennedy, C.J.; Bael, D.; Kappel, C.V.; Perillo, G.M.E.; Reed, D.J. Coastal ecosystem-based management with non-linear ecological functions and values. Science **2008**, *319*, 321–323.
14. Koch, E.W.; Barbier, E.D.; Silliman, B.R.; Reed, D.J.; Perillo, G.M.E.; Hacker, S.D.; Granek, E.F.; Primavera, J.H.; Muthiga, N.; Polasky, S.; Halpern, B.S.; Kennedy, C.J.; Wolanski, E.; Kappel, C.V. Non-linearity in ecosystem services: temporal and spatial variability in coastal protection. Front. Ecol. Environ. **2009**, *7*, 29–37.
15. Costanza, R. Social goals and the valuation of ecosystem services. Ecosystems **2000**, *3*, 4–10.
16. Ehrlich, P.R.; Mooney, H. Extinction, substitution, and ecosystem services. BioScience **1983**, *33*, 248–254.
17. Costanza, R., d'Arge, R.; de Groot, R.; Farber, S.; Grasso, M.; Hannon, B.; Naeem, S.; Limburg, K.; Paruelo, J.; O'Neill, R.V.; Raskin, R.; Sutton, P.; van den Belt. M. The value of the world's ecosystem services: putting the issues in perspective. Ecol. Eco. **1998**, *25*, 67–72.
18. Costanza, R. The Value of Ecosystem Services. Special Issue Eco. Eco. **1998**, *25* (1), 139.
19. Boumans, R.; Costanza, R.; Farley, J.; Wilson, M.A.; Portela, R.; Rotmans, J.; Villa, F. Grasso, M. Modeling the Dynamics of the Integrated Earth System and the Value of Global Ecosystem Services Using the GUMBO Model. Ecol. Eco. **2002**, *41*, 529–560.
20. U.S. Environmental Protection Agency Science Advisory Board. Valuing the Protection of Ecological Systems and Services: A Report of the EPA Science Advisory Board. EPA-SAB 09–012. Washington, DC: EPA, 2009 http://yosemite.epa.gov/sab/sabproduct.nsf/WebBOARD/ValProtEcolSys&Serv?OpenDocument.
21. Schwartz, J. D. Should We Put A Dollar Value On Nature?; Time Magazine. Time Inc.: http://www.time.com/time/business/article/0,8599,1970173,00.html, 2010.
22. Walsh, B. Paying for Nature. Time Magazine. Time Inc. http://www.time.com/time/magazine/article/0,9171,2048324,00. html, 2011.
23. Ecosystem Services for Urbanizing Regions (ESUR). Portland State University, http://www.pdx.edu/esur-igert, 2011.
24. Nelson, E.; Mendoza, G.; Regetz, J.; Polasky, S.; Tallis, H.; Cameron, D.R.; Chan, K.M.A.; Daily, G.; Goldstein, J.; Kareiva, P.; Lonsdorf, E.; Naidoo, R.; Ricketts, T.H.; Shaw, M.R. Modeling multiple ecosystem services, biodiversity conservation, commodity production, and tradeoffs at landscape scales. Front. Ecol. Environ. **2009**, *7*, 4–11.
25. Polasky, S.; Nelson, E.; Pennington, D.; Johnson, K. The impact of land-use change on ecosystem services, biodiversity and returns to landowners: a case study in the State of Minnesota. Env. Resour. Eco. **2011**, *48*, 219–242.

Bibliography

Daily, G.; Polasky, S.; Goldstein, J.; Kareiva, P.M.; Mooney, H.A.; Pejchar, L.; Ricketts, T.H.; Salzman, J.; Shallenberger, R. Ecosystem services in decision-making: time to deliver. Front. Ecol. Environ. **2009**, *7*, 21–28.

de Groot, R.S.; Wilson, M.A.; Boumans, R. A typology for the classification, description and valuation of ecosystem functions, goods and services. Ecol. Eco. **2002**, *41,* 393–408.

Polasky, S.; Segerson, K. Integrating ecology and economics in the study of ecosystem services: some lessons learned. Annu. Rev. Resour. Econ. **2009**, *1*, 409–434.

Sukhdev, P.; Kumar, P. The economics of ecosystems & biodiversity (TEEB). http://www.teebweb.org/, 2008.

23

Ecosystems: Soil Animal Functioning

Introduction 191
Spatial Distribution of Soil Animals 193
Influence of Soil Animals on Soil Functions 193
Decomposition and Nutrient Cycling
Water Cycling 194
Pest Control 194
Impact of Key Soil Animals 194
Earthworms • Termites • Protozoa
Soil Biodiversity 195
References 195

Alan J. Franzluebbers
U.S. Department of Agriculture (USDA-ARS)

Introduction

Soil animals (i.e., fauna) are represented by a diverse array of creatures living in or on soil for at least a part of their life cycle. Many animals have influences on soil properties, but should not be considered soil dwellers since only a minor portion of their life cycle is spent in the soil (Figure 23.1).

Based on body size, soil animals can be divided into three categories:

1. microfauna (<200 μm length, <100 μm width) including protozoa, rotifers, and nematodes
2. mesofauna (0.2–10 mm length, 0.1–2 mm width) including tardigrades, collembola, and mites
3. macrofauna (>10 mm length, >2 mm width) including millipedes, spiders, ants, beetles, and earthworms

Soil animals can also be classified according to where they inhabit the soil. The aquatic fauna (e.g., protozoa, rotifers, tardigrades, and some nematodes) live primarily in the water-filled pore spaces and surface water films covering soil particles. Earthworms are divided into species that occupy the surface litter of soil (epigeic), that are found in the upper soil layers (endogeic), or that burrow deep into the soil profile (anecic).

A further classification of five groups of soil animals is based on feeding activity, which can be useful in distinguishing how different groups affect soil ecosystem functions:

1. Carnivores feed on other animals. This group can be subdivided into: i) predators (e.g., centipedes, spiders, ground beetles, scorpions, ants, and some nematodes), who normally engulf and digest their smaller prey and ii) parasites (e.g., some flies, wasps, and nematodes), who feed on or within their typically larger host organism.
2. Phytophages feed on living plant materials, including those that feed on above-ground vegetation (e.g., snails and butterfly larvae), roots (e.g., some nematodes, fly larvae, beetle larvae, rootworms, and cicadas), and woody materials (e.g., some termites and beetle larvae).

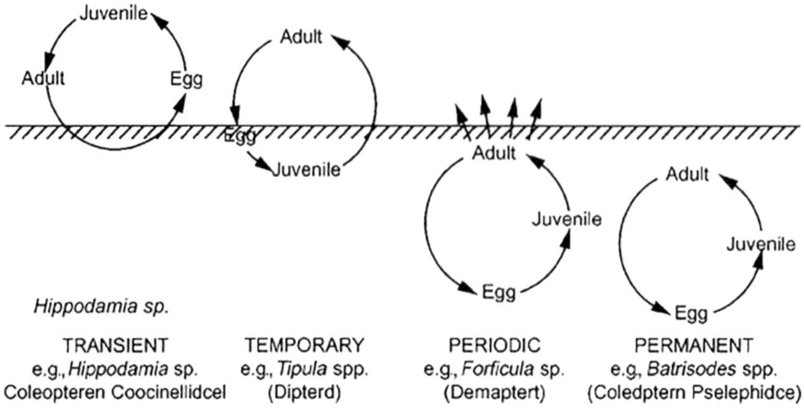

FIGURE 23.1 Categories of soil animals defined according to degree of presence in soil, as illustrated by some insect groups.
Source: From Wallwork.[1]

3. Saprophages feed on dead and decaying organic material and include many of the earthworms, enchytraeids, millipedes, dung beetles, and collembola (or spring- tails). Saprophages are often referred to as scavengers, debris-feeders, or detritivores.
4. Microphytic feeders consume bacteria, fungi, algae, and lichens. Typical microphytic feeders include mites, collembola, ants, termites, nematodes, and protozoa.
5. Miscellaneous feeders are not restrictive in their diet and consume a range of the previously mentioned sources of food. This group includes certain species of nematodes, mites, collembola, and fly larvae.

The arrangement of these feeding groups can be visualized as a soil food web with multiple trophic levels, beginning with the autotrophic flora (Figure 23.2). Trophic levels describe the order in the food chain. The first trophic level is composed of photosynthetic organisms, including plants, algae, and cyanobacteria, which fix CO_2 from the atmosphere into organic compounds. Organisms that consume

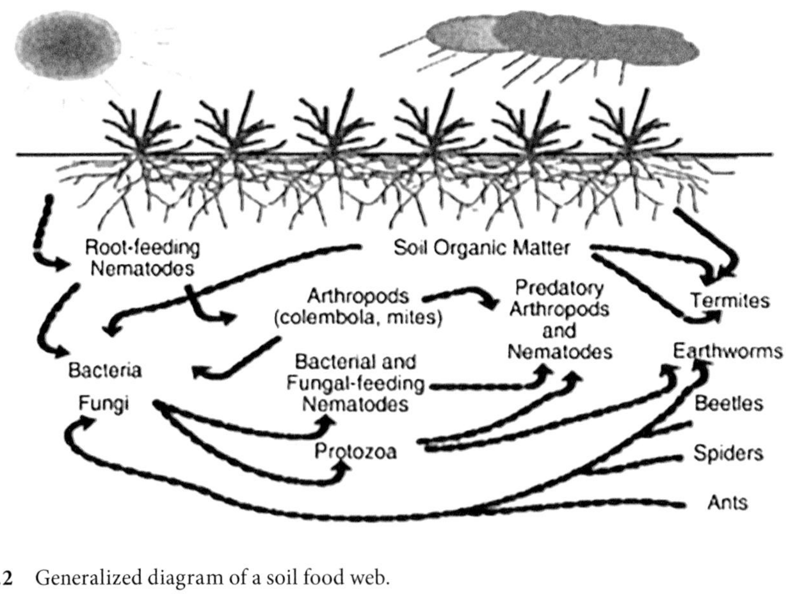

FIGURE 23.2 Generalized diagram of a soil food web.

the photosynthesizers are in the second trophic level, which includes bacteria, actinomycetes, fungi, root-feeding nematodes and insects, and plant pathogens and parasites. The third trophic level feeds on the second trophic level, including many of the dominant soil animals, including bacterial- and fungal-feeding arthropods, nematodes, and protozoa. The soil food web can be continued to include various vertebrates, including amphibians, reptiles, and mammals.

Spatial Distribution of Soil Animals

Soil animals are not uniformly distributed in soil. Unlike the soil microflora, which could be considered ubiquitous, the proliferation of soil animal communities is more sensitive to environmental disturbances and ecological interactions. Gross climatic differences afford opportunities for unique assemblages of organisms. Even within a specific climatic region, large differences occur in the community of organisms present based upon type of vegetation, soil, availability of water, land use, and presence of xenobiotics. Within the confines of a seemingly uniform pedon, "hot spots" of soil organism activity can be isolated based on localized availability of resources and environmental conditions (Figure 23.3).

Influence of Soil Animals on Soil Functions

Decomposition and Nutrient Cycling

Soil animals work directly and indirectly with the soil microflora (i.e., bacteria, actinomycetes, fungi, and algae) to decompose organic matter and mineralize nutrients.[3] The primary consumers of organic materials are the soil microflora. Soil animals, like many of the microflora, are heterotrophs and therefore consume organic materials to gain energy for growth and activity. Soil animals make important contributions to decomposition by

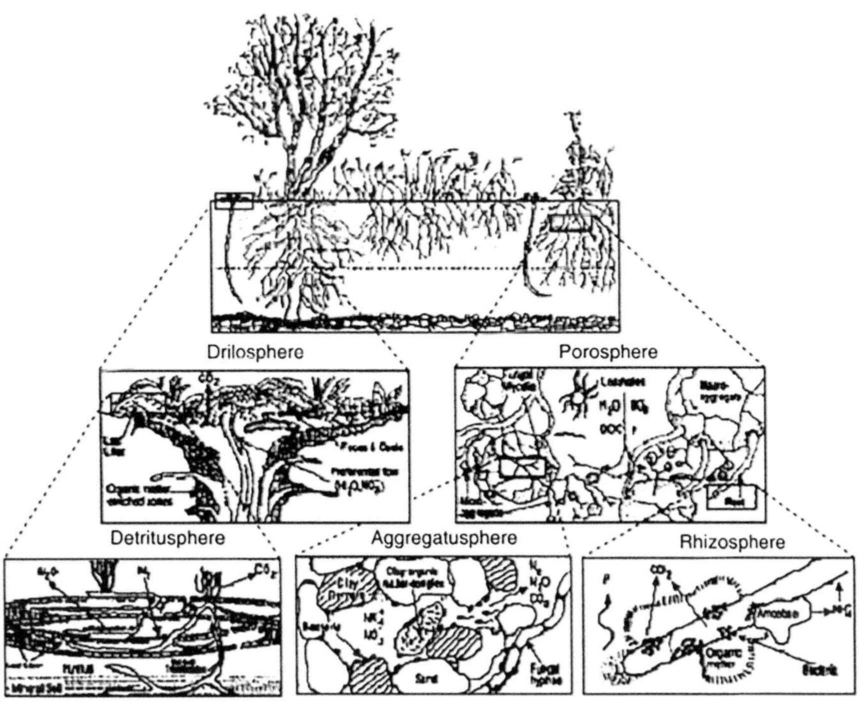

FIGURE 23.3 Key locations of soil organism activity.
Source: From Beare et al.[2]

1. shredding organic materials, thereby exposing a greater surface area for enhancing the activities of other organisms, especially microorganisms;
2. consuming resistant plant materials that would decompose slowly otherwise, such as wood, roots, and dung, and transforming these materials into more decomposable constituents;
3. dispersing soil microorganisms (i.e., inoculation) within the soil profile by transporting them on their bodies and through their intestinal tracts;
4. creating burrows in soil to increase aeration, which stimulates microbial activity;
5. transporting organic materials from the soil surface to deeper in the soil profile, thereby improving environmental conditions for decomposition and increasing biological interactions deeper in the soil profile;
6. consuming bacteria and fungi, thereby releasing nutrients and stimulating the regeneration of microbial populations; and
7. providing unique food sources themselves for consumption by other soil fauna and microflora.

Water Cycling

Soil animals are active participants in the formation of soil structure, which is an important characteristic that influences water infiltration, soil water retention, and percolation.[4] The biochemical activity of soil organisms transforms organic materials into soil-stabilizing cementing agents, which bind the primary soil particles (i.e., sand, silt, and clay) into aggregates. In addition, the burrowing activity of soil animals creates larger pores alongside water-stable aggregates to increase total porosity of soil, which aids water flow without decreasing overall water retention capacity and improves the plant rooting environment.

Both aggregates and porosity are important components of soil structure. Poor soil structure due to disruption of aggregates, which fills pores with disaggregated primary particles and causes crusting of the soil surface, results in more rainfall that runs off land (i.e., less infiltration), potentially carrying with it sediment, nutrients, and pesticides that can contaminate surface waters. Reduced infiltration with poor aggregation reduces available water for plant growth (i.e., reduces net primary productivity and the potential to fix atmospheric CO_2) and reduces percolation of water through the soil profile, essential for purification and recharge of groundwater.

Those animals that create burrows in soil also create conduits for water movement through the soil profile. These biopores can be important for improving water percolation and improving rooting below claypans and other restrictive soil layers.

Many different soil animals deposit fecal pellets, which become stable soil aggregates when the organic material is mixed with soil mineral particles. These aggregates are able to retain more water because of the high water-holding capacity of soil organic matter.

Pest Control

Intense competition among soil organisms keeps an ecosystem healthy by preventing one organism from becoming dominant. Potential plant pathogens, such as root-feeding nematodes, are often held below damaging levels because of consumption by predatory nematodes and arthropods. With a healthy food web rich in species diversity, the predatory activity of many arthropods can keep crop pests below economic thresholds.

Impact of Key Soil Animals

Earthworms

Earthworms are well-known soil animals inhabiting many environments, most prominently found in moist-temperate ecosystems. As earthworms ingest organic materials and mineral particles, they excrete waste as casts, which are a particular type of soil aggregate that is rich with organic matter and mineralizable nutrients. It is estimated that a healthy population of earthworms can consume and aerate a 15 cm surface of soil within one or two decades. Anecic or deep-burrowing species of earthworms can create relatively permanent vertical channels for improving root growth and water transport. Important attributes of earthworm activities are increased surface soil porosity, enhanced water infiltration and nutrient cycling, and distribution of organic matter within the soil profile to increase soil microbial activity.

Termites

Termites are important soil animals in grasslands and forests of tropical and subtropical regions. They often build mounds by excavating subsoil and depositing it above ground to build a city of activity with a complex social system. Termites are able to decompose cellulose in wood because they harbor various microorganisms (protozoa, bacteria, or fungi) to aid in decomposition. Better drainage and aeration of termite mounds may be beneficial to nearby plant growth in soils with a high water table. Stable macrochannels created by termites can improve water infiltration into soils that otherwise would form impermeable surface crusts.

Protozoa

Protozoa are single-celled animals that generally consume bacteria and soluble organic matter. Protozoa are more numerous in marine and freshwater environments, but do occur widely in water films of many soils.[5] Their principal soil function is predation on soil bacteria, which releases nutrients for potential plant uptake; increases decomposition and soil aggregation by stimulating their bacterial prey; and prevents some bacterial pathogens from developing on plant roots.

Soil Biodiversity

There has been a great deal discovered about soils and the organisms that live in them, yet it is estimated that <10% of the species of soil organisms have been identified globally. A rich diversity of genetic information resides in soil. An awaiting challenge is to discover the ecological consequences of soil biodiversity. A critical understanding of how organisms interact has begun, yet there will be much more to learn about how species diversity interacts with functional diversity. There may be soil functions provided by organisms that we are unaware of even today.

How biodiversity relates to ecosystem functioning is an intensive area of current research in both above- and below-ground ecology.[6,7] Recent experimental evidence suggests that loss of species richness due to perturbations may not always lead to loss of ecosystem functioning, especially in initially species-diverse functional groups.[8] Ecosystem functions can be performed by a number of different species within a trophic level, suggesting that functional redundancy is a mechanism to insure stability. However, loss of functional groups in trophic levels closest to the base of the detrital food web would be most detrimental to ecosystem stability. Future work on soil biodiversity should unravel the relative importance of species richness on the resistance and resilience of ecological processes under various short- and long-term stresses and environmental conditions.

References

1. Wallwork, J.A. *Ecology of Soil Animals*; McGraw-Hill: London, 1970; 283 pp.
2. Beare, M.H.; Coleman, D.C.; Crossley, D.A., Jr.; Hendrix, P.F.; Odum, E.P. A hierarchical approach to evaluating the significance of soil biodiversity to biogeochemical cycling. Plant Soil **1995**, *170* (1), 5–22.

3. Coleman, D.C.; Crossley, D.A., Jr. *Fundamentals of Soil Ecology*; Academic Press: San Diego, CA, 1996; 205 pp.
4. Brady, N.C.; Weil, R.R. *The Nature and Properties of Soils,* 12th Ed.; Prentice Hall: Upper Saddle River, NJ, 1999; 881 pp.
5. Dabyshire, J.F., Ed.; *Soil Protozoa*; CAB International: Wallingford, Oxon, UK, 1994; 209 pp.
6. Tilmann, D.; Knops, J.; Wedin, D.; Reich, P.; Ritchie, M.; Siemann, E. The influence of functional diversity and composition on ecosystem processes. Science **1997**, *277,* 1300–1302.
7. Wardle, D.A. How soil food webs make plants grow. Trends Ecol. Evol. **1999**, *14,* 418–420.
8. Laakso, J.; Setälä, H. Sensitivity of primary production to changes in the architecture of belowground food webs. Oikos **1999**, *87,* 57–64.

24

Ecotone

Introduction 197
Ecotone Hierarchies 198
Plant-Scale Ecotones 198
Patch-Scale Ecotones 198
Landscape-Scale Ecotones 200
Spatial and Temporal Dynamics of Ecotones 200
Directional Transitions 200
Stationary Transitions 200
Shifting Transitions 200
Emergent Properties of Ecotones 201
Conclusions 202
References 202

Finn C. Pillsbury and Debra P. C. Peters
U.S. Department of Agriculture (USDA-ARS)

Introduction

Any two adjacent ecological states, be they tundra and taiga, field and forest, or bunchgrass and sand dune, have an area of transition between them. This transition can be gradual and indistinct, like that between shortgrass and tallgrass prairie, or it can be abrupt, like the transition between a hayfield and a neighboring woodlot. These transitions are collectively known as ecotones. Coined by Clements,[1] the term's original conception was of a short transition between stable climax communities, although Weaver and Clements[2] concede that ecotones can be "broad and indefinite." The concept remained something of a mere curiosity until the middle of the 20th century, with the advent of quantitative methods to measure and evaluate transitions between adjacent vegetation associations.[3,4] The essential idea to emerge from this period was that natural systems that are heterogeneous in space can be characterized by a set of ecological gradients underlying those various spatial differences.[5] Adjacent ecological states and the ecotones between them can thus be characterized by a series of gradients in soil characteristics, rainfall, vegetation structure, or other environmental conditions that illustrate the magnitude and abruptness of the ecotone.

Understanding the nature of these gradients is important because ecotones are expected to be sensitive to climate change.[6–8] With changing temperature and precipitation patterns, species' ranges are shifting worldwide, and these shifts are, by definition, showing the greatest flux at range boundaries, which coincide with broad-scale ecotones among biomes.[8] Being able to understand ecotone dynamics will be a key component of conservation planning in the coming decades, as scientists and managers look for effective ways of incorporating climate change into natural resource conservation efforts.[6]

The concept of the ecotone falls into a broader class of "biotic transitions"[9] or "ecological boundaries"[10,11] and has been historically associated with the plant ecology literature because it is conceptually consistent with observed patterns of plant distributions across a landscape. Others have focused on "habitat edges," which are also definable transitions between ecological states.[12]

The difference between the ecotone and edge concepts is in the conceptual frameworks for measuring and evaluating them. Although ecotones are generally defined by their structural properties, edges are usually defined in relation to how they modify the flow of materials and energy from one adjacent state to another. In many ways, it is a purely semantic distinction, but it is illustrative of the fact that ecotones and other types of transitions are, ultimately, defined by the questions and scales relevant to the organisms and processes of interest.

Until the late 20th century, ecotones were assumed to be an "average" of two neighboring vegetation patches, showing a gradient of conditions bounded by those two patch types. However, recent conceptual advances have demonstrated that rather than being simple transitions between adjacent communities, ecotones occur at multiple scales of ecological organization, are dynamic in space and time, and can exhibit emergent properties.[9–11]

Ecotone Hierarchies

The original concept of the ecotone was related to a spatial scale that encompassed two distinct vegetation communities, but recent conceptual development has emphasized the hierarchy of scale within ecotones, from transitions between individual plants (or smaller) to transitions from one biome to another.[9,13] Though there is a continuum of spatial scales at which ecotones occur, we can condense them into a few basic types: plant scale, patch scale, and landscape scale.[9,13] Transitions at these different scales are influenced by distinct ecological patterns and processes and exhibit interactions with transitions at higher and lower levels of the ecotone hierarchy. Although landscape-scale ecotones are generally more gradual than those at plant and patch scales, the relative abruptness of the transition is directly related to the ecological processes and climatological and edaphic context in which that ecotone occurs. To illustrate the hierarchical nature of ecotones, we draw upon transitions within a landscape-scale ecotone between a black grama (*Bouteloua eriopoda*) grassland and a mesquite (*Prosopis glandulosa*) shrubland at the Jornada Experimental Range in southern New Mexico, United States of America.

Plant-Scale Ecotones

The basic unit of the ecotone hierarchy is the individual plant, because it is the nonrandom distribution of individuals, populations, and species that make up extant ecological patterns and from which landscape structure is characterized. In a given area, the location of an individual plant is constrained by an interaction between physiology, microsite characteristics, and competition with other plants. At this spatial scale, an ecotone can be characterized by a transition in microclimate, soil characteristics, or other fine-scale factors. For example, the plant-scale ecotone of a mesquite shrub at the Jornada can be characterized by a gradient in light transmission at a spatial scale of centimeters to a few meters (Figure 24.1A). Light transmission is an abiotic factor that influences soil temperature, soil moisture, and photosynthesis by understory species, therefore playing an important role in soil community differences across the ecotone at the edge of an individual mesquite shrub.

Patch-Scale Ecotones

A landscape can be characterized as a mosaic of different patches of vegetation, where a patch is defined as an area that is compositionally distinct from adjacent areas. Patchscale ecotones are the transitions between adjacent patches and can be characterized by a gradient of population density of a dominant species. An example of a patch-scale ecotone is that between individual patches of black grama and adjacent bare patches at the Jornada. The patch-scale ecotone can be characterized by a gradient in the density of black grama plants across intermediate scales of tens of meters (Figure 24.1B). Plant-scale ecotones are nested within this scale.

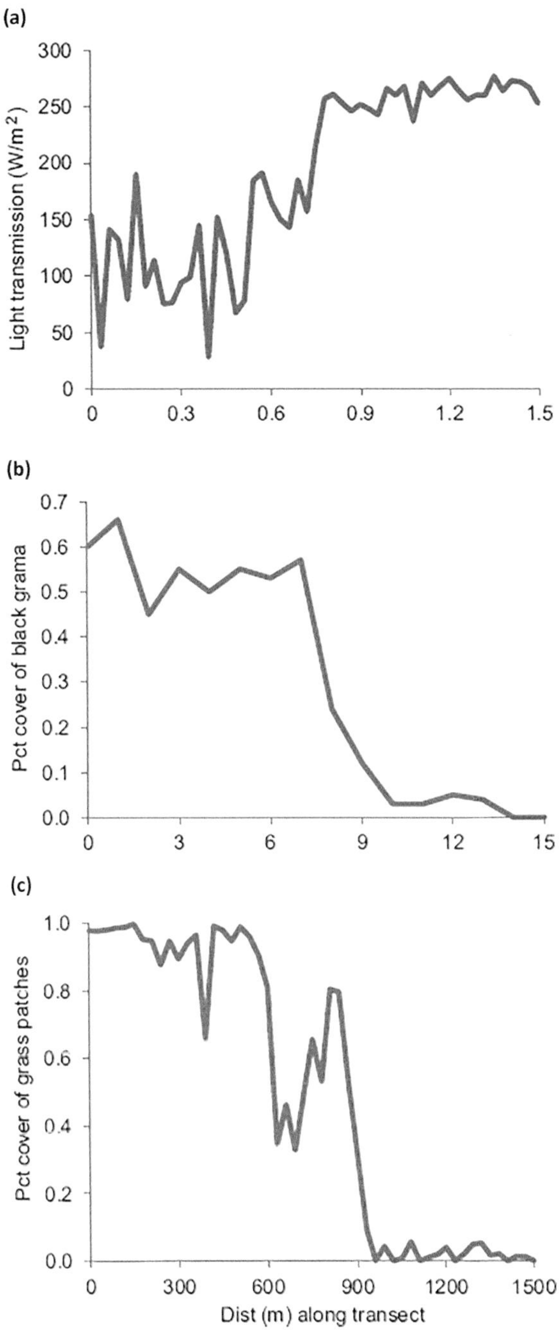

FIGURE 24.1 Examples of different ecotone scales at the Jornada Experimental Range in southern New Mexico. The plant-scale ecotone (**a**) is a measure of light transmission along a 1.5 m transect from the middle of a mesquite (*Prosopis glandulosa*) dune to an adjacent bare gap, demonstrating a fine-scale gradient in an abiotic ecological parameter. The patch-scale ecotone (**b**) is a measure of proportional cover of black grama (*Bouteloua eriopoda*) in 1 m^2 quadrats along a 15 m transect between a patch of black grama and an adjacent bare gap, demonstrating a gradient of cover in a foundational species. The landscape-scale ecotone (**c**) is a measure of proportional cover of grass patches within 900 m^2 grid cells along a 1500 m transect, measured using a digitized satellite image, showing a gradient in the distribution of grass patches across a broad area.

Landscape-Scale Ecotones

Vegetation patches are not distributed randomly in space; they aggregate based on climatic conditions, geomorphology, dispersal, competition, and other factors. A landscape scale ecotone is one that can be characterized by a change in the distribution of patches of a given type across space. The dominant landscape-scale ecotone at the Jornada is between black grama grasslands and mesquite shrublands, which is characterized by a gradient in grass patch cover at scales of hundreds to thousands of meters (Figure 24.1C). Plant- and patch-scale ecotones are nested within this scale.

Spatial and Temporal Dynamics of Ecotones

An ecotone has been described as a "zone of tension" between two adjacent ecological states.[14,15] This tension occurs when there are strong internal dynamics within those two states, produced by positive feedbacks that maintain one state over another. This causes some ecotones to be inherently unstable, because they operate outside the biotic feedbacks that maintain a stable ecological state. Other ecotones, however, do not display such instability; these spatial and temporal dynamics are a key component to understanding ecotone function and future behavior.

Peters et al.[9] identify three basic types of ecotone dynamics in space and time: directional, stationary, and shifting, depending on the relative strength of different abiotic and biotic factors in the system. Abiotic factors could be drivers like precipitation or wind erosion, inherent constraints such as geomorphology, or feedbacks such as soil accumulation or microclimate. Biotic factors in this context are feedbacks such as competition or recruitment.

Directional Transitions

Landscapes undergoing active invasion of one ecological state into another are examples of a directional transition, and this has several consequences for ecotone dynamics in space and time. In this scenario, abiotic constraints are generally weak (i.e., underlying environmental conditions are suitable to both patch types), and the ecotone shifts based on abiotic and biotic feedbacks within the invading patch type (Figure 24.2A). For example, in the northern Chihuahuan Desert, mesquite (*Prosopis glandulosa*) dunes are actively invading historic black grama (*Bouteloua eriopoda*) grasslands. The ecotone between these two states is characterized by a mixture of small grass patches surrounded by bare ground and small shrubs, which facilitates wind erosion and further grass loss (an abiotic feedback). The asymmetry in the strength of these processes maintains a directional shift in landscape structure, exemplified by a directional shift in ecotone location through time.

Stationary Transitions

Where two stable ecological states meet, there is often a stationary ecotone between them. In this situation, each state is maintained by strong abiotic and biotic feedbacks, as well as the presence of strong abiotic constraints limiting the distribution of one patch type or the other (Figure 24.2B). These ecotones occur where there is an abrupt geomorphic transition, like between a valley floor and steep mountainside, or where human activities dominate. For example, the ecotone between a field and forest is stationary when the field is actively grazed by livestock. Dispersal of woody species from the forest edge is kept low by grazing or other human management activities.

Shifting Transitions

Some landscapes exhibit a "shifting mosaic" of patch types through time, when abiotic drivers such as precipitation and fire fluctuate through time (Figure 24.2C). In this scenario, internal feedbacks maintaining one patch type or another are secondary to abiotic drivers. Ecotones between patches can be

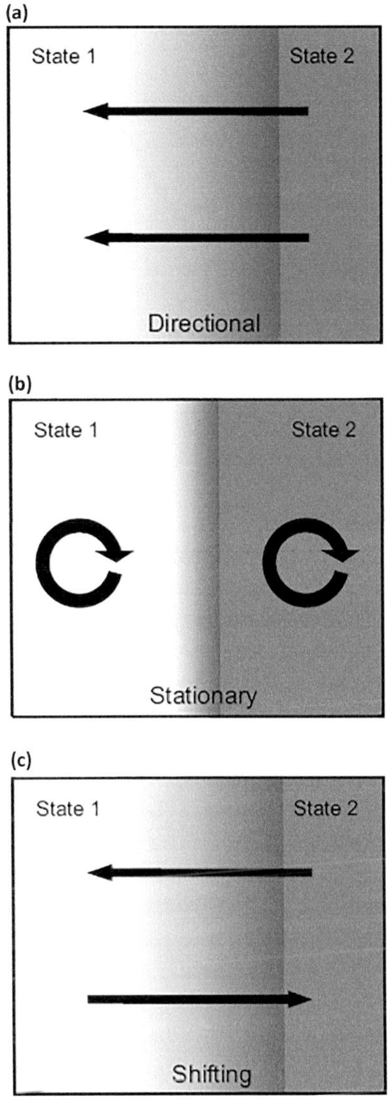

FIGURE 24.2 Ecotone dynamics in time and space. Directional transitions exhibit asymmetry in the strength of drivers and feedbacks between two states. Stationary transitions occur when the internal dynamics of each state are dominated by feedbacks that reinforce the boundary between them. Shifting transitions occur when outside abiotic drivers overwhelm internal patch dynamics.

very abrupt, if spatially variable fire is the dominant driver, or much more indistinct, if multiyear fluctuations in precipitation promote one or another patch type. For example, prescribed fire can be used to maintain a shifting mosaic of different grassland states, wherein an abiotic driver (fire) overwhelms the internal dynamics of each grass patch and maintains very short ecotones between patches.[16]

Emergent Properties of Ecotones

Ecotones modulate flows of material, energy, and organisms across heterogeneous landscapes, which has ramifications for a host of ecological processes such as dispersal, predation, and reproduction.[10,12,17] As a result, ecotones can exhibit behaviors, or emergent properties, that are not simply a combination or

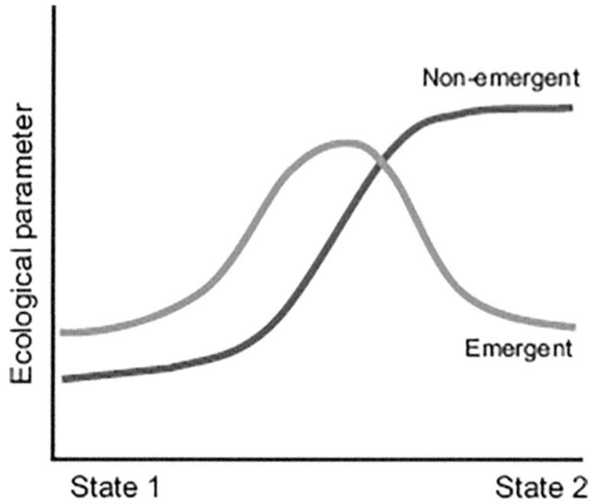

FIGURE 24.3 Defining characteristic of emergent and non-emergent properties of ecotones.

average of the ecological states between which they are situated.[11,18] One common emergent property is an increase in population density of certain species of plants and animals, termed "edge species" because of their association with ecotones between ecological states.[12] Such species might require multiple complementary habitats, and traveling along ecotones offers an efficient means of acquiring resources.[19] Other species might exhibit a preference for ecotones if their forage or prey species are associated with ecotones, a phenomenon known as "resource mapping."[12] Increased abundance of edge species in ecotones may thus have a cascading effect on other ecological properties, making ecotones distinct in numerous ways from adjacent ecological states (Figure 24.3).

Conclusions

Ecotones, the boundaries between adjacent ecological states, are a key component of all landscapes. Much more than being just gradual transitions between states, they are dynamic and exist at multiple scales of ecological organization. They can also exhibit emergent properties distinct from adjacent states, thus making them important landscape components in their own right. Understanding the abiotic and biotic processes associated with ecotones are especially important to natural resource conservation; under ongoing climate change, it is these biotic transitions that will exhibit the leading edge of broader landscape change. Understanding and predicting landscape change at these boundaries is a key component to understanding and predicting broader effects of climate change in an unpredictable future.

References

1. Clements, F.E. *Research Methods in Ecology*; University Publishing Company: Lincoln, NE, 1905; 161–183 pp.
2. Weaver, J.E.; Clements, F.E. *Plant Ecology*; Second Edition. McGraw-Hill Book Company: New York, 1938; 601 pp.
3. Whittaker, R.H. Vegetation of the great smoky mountains. Ecol. Monogr. **1956**, *26*, 1–80.
4. Whittaker, R.H. Gradient analysis of vegetation. Biol. Rev. **1967**, *4*, 207–264.
5. Risser, P.G. The status of the science examining ecotones. BioScience. **1995**, *45*, 318–325.
6. McCarty, J.P. Ecological consequences of recent climate change. Conserv. Biol. **2001**, *15*, 320–331.
7. Parmesan, C; Yohe, G. A globally coherent fingerprint of climate change impacts across natural systems. Nature **2003**, *421*, 37–42.
8. Parmesan, C. Ecological and evolutionary responses to recent climate change. Annu. Rev. Ecol. Evol. Syst. **2006**, *37*, 637–669.

9. Peters, D.P.C.; Gosz, J.R.; Pockman, W.T.; Small, E.E.; Par- menter, R.R.; Collins, S.S.; Muldavin, E. Integrating patch and boundary dynamics to understand and predict biotic transitions at multiple scales. Landscape Ecol. **2006**, *21*, 19–33.
10. Cadenasso, M.L.; Pickett, S.T.A.; Weathers, K.C.; Jones, C.G. A framework for a theory of ecological boundaries. BioScience **2003**, *53*, 750–758.
11. Yarrow, M.M.; Marin, V.H. Toward conceptual cohesiveness: A historical analysis of the theory and utility of ecological boundaries and transition zones. Ecosystem. **2007**, *10*, 462–476.
12. Ries, L.; Fletcher, R.J.; Battin, J; Sisk, T.D. Ecological responses to habitat edges: mechanisms, models, and variability explained. Annu. Rev. Ecol. Evolu. Syst. **2004**, *35*, 491–522.
13. Gosz, J.R. Ecotone hierarchies. Ecol. Appl. **1993**, 3, 369–376.
14. Clements, F.E. Peculiar zonal formations of the Great Plains. Am. Nat. **1897**, *31*, 968–970.
15. Lidicker, W.Z. Responses of mammals to habitat edges: An overview. Landscape Ecol. **1999**, *14*, 333–343.
16. Fuhlendorf, S.D.; Engle, D.M. Restoring heterogeneity on rangelands: ecosystem management based on evolutionary grazing patterns. BioScience **2001**, *51*, 625–632.
17. Wiens, J.A.; Crawford, C.S.; Gosz, J.R. Boundary dynamics: a conceptual framework for studying landscape ecosystems. Oikos **1985**, *45*, 421–427.
18. Strayer, D.L.; Power, M.E.; Fagan, W.F; Pickett, S.T.A; Bel- nap, J. A classification of ecological boundaries. BioScience **2003**, *53*, 723–729.
19. Dunning, J.B.; Danielson, B.J.; Pulliam, H.R. Ecological processes that affect populations in complex landscapes. Oikos **1992**, *65*, 169–175.

25

Environmental Goods and Services: Economic and Non-Economic Methods for Valuing

Daniel E. Campbell
U.S. Environmental Protection Agency (EPA)

Introduction 205
Economic Valuation 206
Non-Economic Valuation 207
Modern Origin of the Two Perspectives on Valuing the Work of the Environment 208
The Relationship between Money and Emergy Flow 208
Market Value and Emergy Value of Environmental Goods and Services 208
Nonmarket Value and Emergy Value of Environmental Goods and Services 209
Summary 209
References 211

Introduction

One of the most important problems facing global society in the 21st century is to accurately determine the value of the contributions that the environment makes to support society. These contributions can be valued by economic methods, both market and nonmarket, as well as by accounting methods that determine the inputs required for the production of a good or service. In the past, the determination of value has been seen as a uniquely economic problem, which has been quantified in neoclassical economics by willingness to pay or accept payment for an item (receiver value). However, at least from the development of Marx's labor theory of value during the 19th century, the idea of determining value from the required inputs to a production process (donor value) has been understood, if not widely accepted among economists. Of course, labor is not the only important input to economic production, and thus Marx's labor theory of value can be easily refuted based on practical arguments. However, in the latter half of the 20th century, H.T. Odum with his students and colleagues used the principles of Energy Systems Theory[1] to develop theory and methods to account for all the various inputs to any production process in terms of the past use of available energy (i.e., energy with the potential to do work) required for the production of any product or service. This donor-based method tracks the transformations of available energy that are required for the production of a product or service. Thus, any ecological, economic, or social production process, for which all the required inputs are known, can be quantified in this manner. Odum's complete donor-based accounting was performed using a new quantity, emergy,[2] expressed with a new unit, the solar emjoule (semj), where "em" is a prefix meaning "within" or "inside" and is sometimes interpreted as a mnemonic for "energy memory." The emergy measure accounts for the solar equivalent joules used in the past, both directly and indirectly, to create a product or service. In this way, the emergy of the inputs to a production process in the economy or in the environment can be summed to determine the total work required to have a product or service as part of the system. This total work requirement is then taken as an estimate of the value of the product or service.

Economic Valuation

Figure 25.1 shows a diagram of economic (i.e., monetary) valuation methods expanded to include emergy evaluation as an example of a quantitative non-economic (i.e., non-monetary) method of valuation. The economic model for value is shown to the left of the dotted line. Note that total economic value (TEV) includes both real money flows, measured directly by markets and real or hypothetical money flows determined by various nonmarket methods. Some nonmarket values are assigned from money flows that are indirectly revealed by markets, that is, hedonic values or travel costs. Other nonmarket methods depend on stated preferences, which are usually measured as hypothetical dollar flows determined by surveys, that is, willingness to pay or accept payment as inferred from the answers given to survey questions. The contingent valuation and contingent choice methods are the most commonly used stated preference methods of determining value, but the revealed preference methods are also commonly used nonmarket methods of valuing ecological goods and services. The economic value model is designed to elicit the total value, including both use and nonuse values, that people assign to a given product or service. The various aspects of TEV are additive, as long as double counting is avoided, and because economic value can be concerned with value received by people, who benefit from the goods or services,

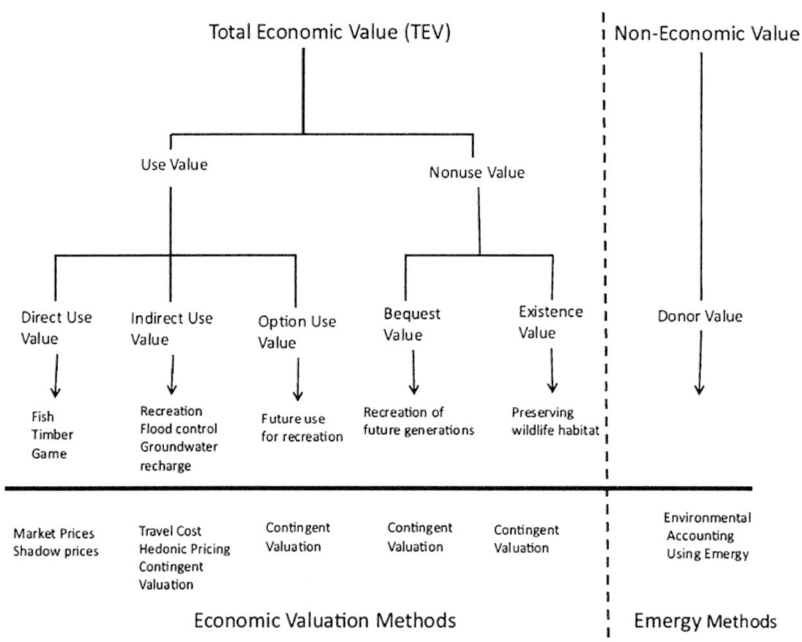

FIGURE 25.1 The elements of total economic value are shown in relationship to one another and in comparison with the primary non-economic method of determining the total value of an item, Environmental Accounting Using Emergy.
Source: Adapted from Bateman and Langford.[3]

regardless of their location, it is not necessarily constrained by the geographic boundaries or area of influence of the system providing the products and services, whose value is being assessed. In practice, double counting within TEV is a difficult problem that is the subject of current research.

The methods of economic valuation have been applied to attempt to quantify the benefits that humans derive from ecosystems (i.e., ecosystem services) and published papers related to this topic have grown exponentially since the early 1980s.[4] Several classification schemes for "ecosystem services" have been proposed;[4,5–8] however, there is still some debate concerning the correctness and usefulness of these ideas.[9–11] While most valuations of "ecosystem services" have been approached using one or more of the economic methods mentioned above, their value can also be determined using donor-based methods.[12]

Non-Economic Valuation

Donor-based quantitative methods are shown to the right of the dotted line in Figure 25.1 Environmental accounting using emergy is the only example given here, although other biophysical accounting methods might be used, for example, energy, exergy, material, or information flow. However, emergy evaluation includes information on all other kinds of flows and thus it is the most complete method of assessing donor value; therefore, emergy will be used to represent all non-economic biophysical methods of determining value. Note that determining the value of an item using emergy relies on the creation and use of the item in a real system. Emergy like economics has nothing to say about the intrinsic value of a thing in itself, that is, an entity divorced from the system of which it is a part and within which it was created.

Odum and Odum[13] demonstrated that emergy evaluation was a more complete method of determining the value of products and services in an environmental system than was market valuation (Figure 25.2). Note that emergy evaluations include the flows of energy, matter, and information from the entire system as well as the flow of products going to market. Markets only capture the money that flows as a countercurrent to the useful products of the environmental system, for example, the logs mentioned in the example below. Campbell and Brown[12] found that economic estimates of the market value of goods and services produced by the U.S. Forest Service (USFS) lands captured 46% of the value

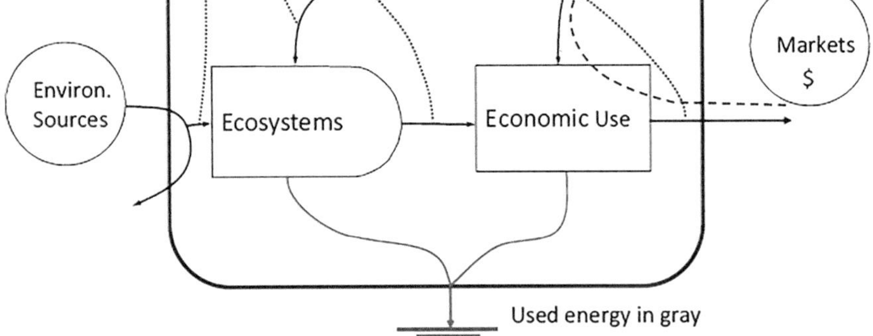

FIGURE 25.2 Environmental system flows evaluated by emergy and economic market methods. The dashed line is money flow; the dotted lines indicate that information is being read.
Source: Modified from Odum and Odum.[13])

obtained by emergy evaluation of the same systems. When the emergy value of nonmarket services was included, only one-third of the value assessed using emergy methods was captured by the economic analyses; however, economic assessments of nonmarket values (e.g., clean air, clean water, gross primary production, etc.) were not available for the USFS system at that time.

Modern Origin of the Two Perspectives on Valuing the Work of the Environment

The origins of environmental economics can be traced to the 1950s when Resources for the Future was established in Washington DC.[14] It is interesting that the origin of the modern field of ecological economics can be traced to about this same time (the late 1950s) in the study that H.T. Odum was doing and sponsoring as Director, Port Aransas Laboratory, University of Texas.[15] Although these two different perspectives on valuing the environment, one rooted in economics and the other in ecology, began to develop at about the same time, there was little or no interaction between them until the ecological economic approach was organized formally around 1989[16] by an Odum student, Robert Costanza,[17] working with Herman Daly, the proponent of Steady-State Economics[18] and others. Costanza transferred many of Odum's synthetic ideas on general systems' theory and ecology[1] to the new field, but one of the key ideas of Energy Systems Theory (i.e., the maximum empower principle) and the emergy evaluation methods were left behind, perhaps because they were too radical for general acceptance at that time. To see some of what was left behind compare Farber et al.[19] with Campbell.[20]

The Relationship between Money and Emergy Flow

Both receiver and donor methods can be used to determine the value of the work of the environment to society, albeit from different perspectives. Since both approaches can be used to inform decision-making and each one uses a different unit for the evaluation (i.e., money flow vs. emergy flow), it is important to understand the relationship between the two methods and the values that they assign to the work of the environment. One method of comparison is to relate money flows in an economy to the emergy flows supporting that economy by converting emergy (semj) to emdollars (Em$). This is accomplished by dividing an emergy flow (semj/yr) by the Emergy to Money Ratio (EMR) of the economy for the year in which the values are being determined. The EMR for an economy is the ratio of the system's emergy flow to the system's money flow (e.g., semj/$ of U.S. GDP) for a given year. If all the emergy flows supporting an economy in a given year are converted to Em$, the dollar value of the GDP is redistributed or prorated according to the fraction of the total emergy flow represented by each individual flow in the system, regardless of whether that flow is from the environment or from the economy. So, the rainfall in Texas in a given year can be assigned an Em$ value, which represents the fraction of the total buying power in the state accounted for by the rain, even though there is no market where rain is bought and sold. Recently, Pulselli et al. suggested comparing the value of ecosystem services determined in monetary units per hectare per year to the renewable emergy flow (semj/ha/yr) needed to support that flow of services as a method for showing the relative efficiency with which ecosystem services are being produced in different systems and as a way to track trajectories of efficiency and evaluate changes in the state of a system, when these values are plotted over time.[21]

Market Value and Emergy Value of Environmental Goods and Services

Emergy evaluations determine value based on a specific system's spatial boundaries and the values of products and services are constrained by the emergy of total resource flows that enter the system from outside and by the internal emergy flows derived from the natural capital storages within the system.

Odum et al.[22] provided the first comparison of economic market value and emergy value of environmental resources by analyzing the water supply system of Texas (Em$ values were obtained by dividing measured emergy flows in rain, groundwater, etc. by the EMR of the United States in 1985, 2.2E+12 semj/$). They found that environmental products like the rain (Em$ 0.035 m^{-3}), rivers (Em$ 0.09 m^{-3}), and ground water (Em$ 0.25 m^{-3}) on the Texas plains represented quantifiable values using emergy, but water flows were not valued by markets until they were extracted for human use. The market value of irrigation water for Texas agriculture in the 1980s ($0.04 m^{-3}) was about one-eleventh of its value as indicated by emergy valuation (Em$ 0.44 m^{-3}). Campbell and Cai [23] recognized that the Em$ value of any item would always be greater than the market value, given that the market was not perturbed by psychological factors. Using data on U.S. forest resources from Tilley,[24] they showed a pattern in which the monetary value of a forest resource was less than the Em$ value of the same resource by the greatest amount at the point of extraction from nature. Money captured progressively more of the value of an item as measured by Em$ as the resource moved through the economic production chain. This relationship was observed because the work required for all products is either from the environment or the economy and only the work of people is tracked by money. When a product is first extracted from nature most of the work required for that product still comes from the environment. For example, the market value (monetary value) of logs harvested from a forest only reflects the work that people supplied to support the harvest; which, in general, is smaller than the environmental work of the rain, soil nutrients, etc. that was required for the growth of the tree over the number of years needed to reach a harvestable size. As the available energy in a raw product (e.g., pulp logs) is transformed as it passes through a chain of industrial and commercial operations (e.g., from wood pulp, to paper, and then to the pages of a book) more and more of the work required for the end product is supplied by people, who are paid for their work. Thus, there is a transition along a manufacturing chain, in which the emergy directly associated with the money paid to workers, investors, and suppliers becomes greater than the emergy of the raw materials (e.g., the logs). Emergy evaluation tracks the work processes associated with environmental and economic sources on an equal basis.

Nonmarket Value and Emergy Value of Environmental Goods and Services

When nonmarket values are assessed so that economic value is measured as TEV rather than market value alone, it is possible for economic methods to assign values to the pathways formerly identified as outside of the market (Figure 25.3). On one hand, since the components of TEV are additive and the boundaries of an economic evaluation are not restricted necessarily to those of the system producing the goods and services, in theory, the sum of market and nonmarket value could exceed the value of the system as determined by emergy evaluation. On the other hand, if information is only available on one or two aspects of TEV (e.g., the hedonic value of housing near a stream or the recreational benefit of fishing) that are then used to assess the value of ecosystem services delivered by a stream, the value ascribed to the ecosystem services could be considerably less than the value determined by emergy methods. Figure 25.3 shows that double counting is a real problem for such comprehensive economic evaluations, as the products sold on the market generate a money flow that may well depend on all of the internal flows of the system. Emergy evaluations avoid double counting by applying the rules of emergy algebra to trace the inflows of emergy through the network[2] to the outputs.

Summary

All commonly used methods of economic valuation are subjective, because they are designed to capture the value that people assign to environmental products and services based on their preferences. In contrast, emergy methods are objective, because they take an accounting perspective, which determines value based on the sum of the normalized energy, material and information inputs required for

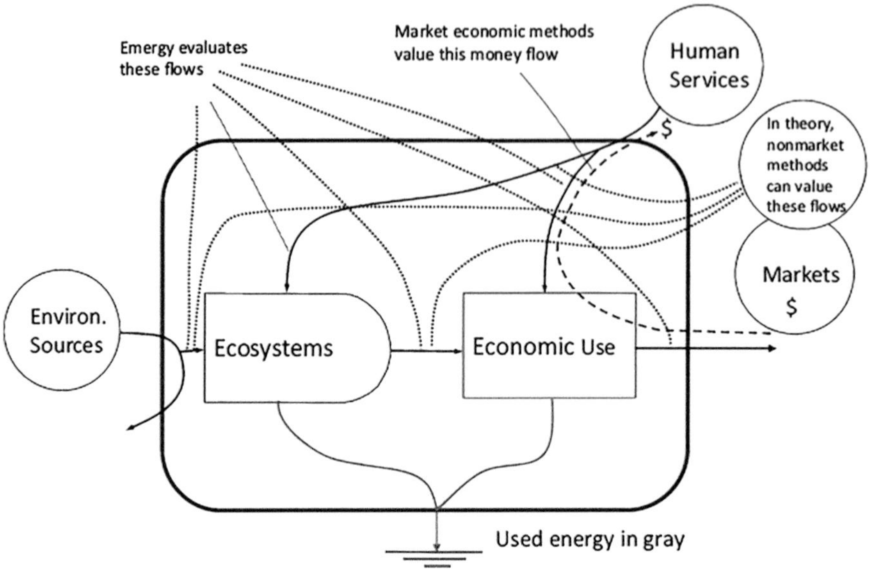

FIGURE 25.3 The application of market and nonmarket economic methods to value ecosystem flows using the analogy of ecosystem services. The dashed line is money flow; the dotted lines indicate that information is being read. **Source:** Modified from Odum and Odum.[13]

the production of a good or service. Furthermore, the principles of Energy Systems Theory indicate that maximizing the emergy flow in a system is nature's decision criterion.[20] In this regard, developing an understanding of the relationship between emergy value and economic value may not be simply a matter of academic interest, but may be critical to improve decision-making and to set reasonable public policy values for environmental products and services.

For example, in comparing the results of analyses using the two different methods, we might say that where emergy and economic methods agree on the value of an environmental good or service, decisions can be made with greater certainty. On the contrary, if they disagree, a cautionary flag should be raised and more research recommended before implementing policies. In this case, disagreement indicates that people's values are not consistent with the thermodynamic and biophysical conditions governing the system, and thus there will be both constraints on how long such an activity can continue and possible negative consequences for the choices made.

In summary, non-economic value as measured by emergy exceeds the market value of the same product or service in a predictable manner with the difference between the two narrowing for more highly processed products and services. When nonmarket value estimates are added to market values to get TEV of an ecosystem's contributions to society, the values obtained can be either more or less than those found from an emergy evaluation of the same system. Furthermore, both market and nonmarket economic methods are useful in identifying the different values that people assign to environmental goods and services. In addition, emergy evaluation provides an independent biophysically based assessment of the value of the same goods and services. Values of environmental goods and services determined through emergy evaluations can promote better decision-making by providing an objective check on the subjective values assigned by economic methods using human perceptions of value, which are invariably based on incomplete information. More research needs to be done to better understand the relationship between these two very different perspectives on determining the value of the environment to society and on their joint application in developing environmental policies that are conducive to both human and environmental well-being in the long run.

References

1. Odum H.T. *Systems Ecology: An Introduction*; John Wiley & Sons: New York, NY 1983; 644 p.
2. Odum, H.T. *Environmental Accounting, Emergy and Environmental Decision Making*; John Wiley & Sons: New York, NY. 1996; 370 p.
3. Bateman, I.J.; Langford, I.H. Non-users' willingness to pay for a national park: An application and critique of the contingent valuation method. Reg. Stud. **1997**, *31* (6), 571–582.
4. Fisher, B.; Turner, R.K.; Moring, P. Defining and classifying ecosystem services for decision making. Ecol. Econ. **2009,** *68,* 643–653.
5. Costanza, R.; d'Arge, R.; de Groot, R.; Farber, S.; Grasso, M.; Limburg, K.; Naeem, S.; O'Neill, R.V.; Paruelo, J.; Raskin, R.G.; Sutton, P.; van den Belt, M. The value of the world's ecosystems services and natural capital. Nature **1997,** *387,* 253–260.
6. De Groot, R.S.; Wilson, M.A.; Boumans, R.M.J. A typology for the classification, description and valuation of ecosystem functions, goods and services. Ecol. Econ. **2002,** *41,* 393–408.
7. MA, Millennium Assessment, *Millennium Ecosystem Assessment;* Island Press: Washington DC, USA, 2005.
8. Wallace, K.J. Classification of ecosystem services: Problems and solutions. Biol. Conserv. **2007,** *139,* 235–246.
9. Costanza, R. Ecosystem services: Multiple classification systems are needed. Biol. Conserv. **2008,** *141,* 350–352.
10. Wallace, K.J. Ecosystem services: Multiple classifications or confusion? Biol. Conserv. **2008,** *141,* 353–354.
11. Peterson, M.J.; Hall, D.M.; Feldpausch-Parker, A.M.; Peterson, T.R. Obscuring ecosystem function with application of the ecosystem services concept. Conserv. Biol. **2009**, *24* (1), 113–119.
12. Campbell, E.T.; Brown, M.T. Environmental accounting of natural capital and ecosystem services for the U.S. National Forest System. Environ. Dev. Sustain. **2012,** DOI 10.1007/s10668-012-9348-6.
13. Odum, H.T.; Odum, E.P. The energetic basis for valuation of ecosystem services. Ecosystems **2000,** *3,* 21–23.
14. Pearce, D. An intellectual history of environmental economics. In Ann. Rev. Energy Environ. **2002,** 57–81 pp, DOI: 10.1146/annurev.energy.27.122001.083429.
15. Kangas, P. Ecological economics began on the Texas bays in the 1950s. Ecol. Model. **2004,** *178* (1,2), 179–181.
16. Ropke, I. An early history of modern ecological economics. Ecol. Econ. **2004,** *50,* 293–314.
17. Costanza, R. What is ecological economics? Ecol. Econ. **1989,** *1,* 1–7.
18. Daly, H.E. *Steady State Economics;* W.H. Freeman and Co.: San Francisco 1977.
19. Farber, S.C.; Costanza, R.; Wilson, M.A. Economic and ecological concepts for valuing ecosystem services. Ecol. Econ. **2002,** *41,* 375–392.
20. Campbell, D.E. Proposal for including what is valuable to ecosystems in environmental assessments. Environ. Sci. Technol. **2001,** *35,* 2867–2873.
21. Pulselli, F.M.; Coscieme, L.; Bastianoni, S. Ecosystem services as a counterpart of emergy flows to ecosystems. Ecol. Model. **2012,** *222,* 2924–2928.
22. Odum, H.T.; Odum, E.C.; Blissett, M. *Ecology and Economy: Emergy Analysis and Public Policy in Texas;* Lyndon B. Johnson School of Public Affairs, University of Texas at Austin: Policy Research Project Report 1987; 78, 178 p.
23. Campbell, D.E., Cai, T.T., Emergy and economic value, 2007, pp 24-1–24-16. In: Brown, M.T., Bardi, E., Campbell, D.E., Huang, S.L., Rydberg, T., Tilley, D.R., Ulgiati, S. Eds.; *Emergy Synthesis 4, Theory and Applications of the Emergy Methodology.* Proceedings of the 4th Biennial Emergy Research Conference, the Center for Environmental Policy, University of Florida, Gainesville, FL.
24. Tilley, D.R. *Emergy Basis of Forest Systems;* Doctoral Dissertation, University of Florida, UMI, Ann Arbor: MI, 1999; 296 pp.

26

Forests: Temperate Evergreen and Deciduous

Lindsay M. Dreiss
Defenders of Wildlife

John C. Volin
University of Connecticut

Introduction 213
Climate 214
Land Formation and History 215
Structure and Composition 216
Temperate Deciduous Forests • Temperate Evergreen Forest
Threats 220
Conclusions 221
References 222

Introduction

Temperate forests represent one of the major biomes on Earth, covering ~14% of Earth's terrestrial land surface [1] as shown in Figure 26.1. Along with boreal and subtropical/tropical wet and dry forests, temperate forests are one of the Earth's dominant forest types. In comparison to other types, temperate forests are intermediate in latitude, temperature, and precipitation. By contrast, they generally exhibit much stronger seasonality, which favors those plant and animal species that were able to adapt to short-term climatic variation. Seasonal changes include extreme fluctuations in temperatures and variable access to moisture (rain or snow), light, and nutrients. As a result, growing conditions throughout the year are ever changing, creating environmental stresses that ultimately result in greater species adaptive opportunities and in unique ecosystem structure and composition [2–4].

Temperate forests are located in some of the most heavily populated and developed regions on Earth, including much of eastern North America, western and central Europe, northeastern Asia, and, to a lesser extent, Australia, New Zealand, South America, and areas of South Africa [5–7] in Figure 26.1. Temperate forests are an established part of the history and culture of these regions, and continue to play an important role in providing ecosystem services to human populations, including timber products, carbon storage, clean drinking water, erosion prevention, recreation, tourism, spiritual inspiration, and others. On the other hand, the unsustainable use of such services has led to negative consequences for temperate forest ecosystems worldwide.

In this entry, we provide an overview of temperate forests from both an ecological and societal perspective. We describe the unique characteristics of temperate deciduous and evergreen forests, the major anthropogenic agents that have impacted them in the past, and some of the emerging threats they face today.

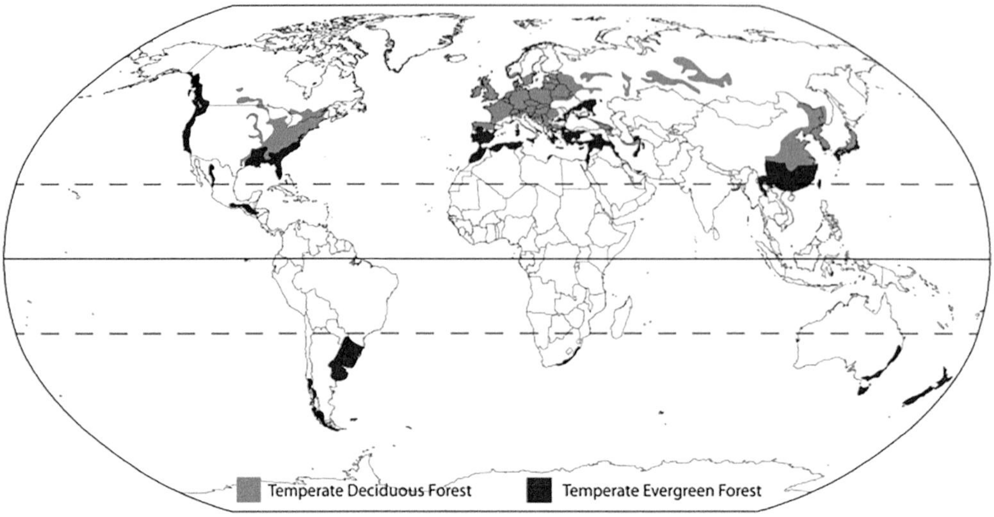

FIGURE 26.1 Global temperate deciduous and evergreen forest biomes.

Climate

Climate is a fundamental factor in defining the Earth's biomes, including the temperate forest biome. Generally, a forest is considered temperate if it is located in a region with hot summers and cold winters. The annual range in temperature can be ±30°C, and the mean annual temperature is between 3°C and 18°C, with mean midwinter temperature below 8°C and mean midsummer temperature about 18°C [8]. In the Northern Hemisphere, the length of the frost-free period ranges from 120 to over 250 days, with growing season and temperature increasing closer to the equator. The climate is milder in the Southern Hemisphere relative to similar latitudes in the Northern Hemisphere, which is why relatively few days of the year are subject to frost below the equator [6].

Temperate forests receive even precipitation year-round averaging 750–1,500 mm/year. Snowfall can range from nonexistent in southern regions to extremely heavy in northern regions. In the Northern Hemisphere, temperate forests are geographically located between the boreal forest/tundra and the tropical/subtropical forests, while in the Southern Hemisphere, they are found south of the tropical/subtropical forests and up to treeline in some areas as highlighted in Figure 26.1. Correspondingly, the climate, productivity, and species diversity of temperate forests are intermediate in relation to other biomes [9].

The seasonal climatic regime in the temperate forest regions has changed very little in the past 1,000–2,000 years. This is reflected in the structure and ecosystem dynamics of the forests [10]. Winter deciduous trees or coniferous evergreen species that are physiologically dormant during winter months dominate the majority of the temperate forests. Genera differ between Northern and Southern Hemispheres. Examples include *Acer*, *Fagus*, *Tilia*, *Quercus*, *Carya*, *Populus*, *Ulnus*, *Betula*, *Fraxinus*, *Magnolia*, *Cornus*, *Robinia*, and *Juglans* in the Northern Hemisphere and *Eucalyptus*, *Acacia*, *Quercus*, and *Nothofagus* in the Southern Hemisphere.

At higher latitudes and altitudes, where temperatures range from −30°C to 20°C annually, coniferous evergreen tree species tend to dominate the temperate forests. These regions are generally drier, with annual precipitation from 300 to 900 mm. Seasons are defined by cold, long snowy winters and warm, humid summers with at least four to six frost-free months [11]. Evergreen temperate forests may also occur in regions with a stronger maritime climatic influence, where precipitation can be very high (typically 700–1,000 mm) and snow is uncommon [12,13]. Summer fog is often an important contributor to moisture uptake, reducing stresses from evapotranspiration. In addition, the growing season is much longer, and sometimes continuous, in forests closer to the equator. In the midst of global climate change, however, a shift is occurring in the location of ecotone and temperate forest species.

Climate-linked range shifts have been observed in the northern hardwood-boreal forest ecotone, where there has been a decrease in boreal and an increase in northern hardwood basal area [14,15]. (Basal area is the total cross-sectional area of all stems measured at breast height per unit land area.) The rate of temperature change due to global warming is projected to be much faster in temperate broadleaf and mixed forests than in temperate coniferous forests [16]. Temperature being considered the primary control on treeline formation and maintenance, global warming is also facilitating treeline advances to greater altitudes and latitudes [17,18]. The synergistic effects of warming temperatures and intensifying droughts have also triggered concerns. Tree mortality and range shifts can result from extreme water stress and lead to other ecological consequences including changes in carbon storage and dynamics, changes in microclimate, and changes in future production of important habitat and food sources [19,20].

Land Formation and History

On the geological time scale, climate has shaped the land on which temperate forests now grow. Between 110,000 and 10,000 years ago, during the Late Pleistocene age, glacial ice sheets left large parts of the Eurasian and North American continents as treeless, frozen tundra [21,22]. Below the tundra, latitudes and altitudes stretched the temperate forests, and beyond these, temperate, grassy plains. As the glaciers retreated, they left behind glacial till and lake-bottom sediments. As the zones of tundra followed the shrinking glaciers, the boreal and temperate forests encroached on the tundra belt until the matrix of ecoregions looked much like it does today. Since the end of the last ice age, human activity, including human-caused climate change, has become a primary factor in shaping temperate forests.

The three largest temperate forest regions, eastern North America, western and central Europe, and northeastern Asia, have played an important role in the development of human society. In North America, temperate forests were heavily impacted by the spread of European colonization in the 1700s. Much of the temperate forest was cleared, logged, and/or burned to make way for agricultural needs and human settlements [23,24]. In the 18th century, the temperate forests of Europe were the birthplace of new lines of scientific study and management, becoming what is now modern forestry [25,26]. These practices were applied in Europe and North America, where temperate forests provided the wood for fuel, building materials, and raw material in industrial processing during the Industrial Revolution of the 19th and 20th centuries [27,28]. With the spread and development of forestry practices, large amounts of temperate forest were logged for the production of timber, pulp, paper goods, and other forest products. During the 19th and into the 20th century, much of the agricultural land in the temperate range in North America was abandoned and the temperate forest grew back, peaking in its coverage in the mid-20th century [29]. Today, it is once again under threat from continuous loss and fragmentation primarily through human developments.

In Europe, many native temperate deciduous forests have been extirpated and replaced with planted monocultures, many of which are not representative of the native vegetation [30]. Most of the remaining European temperate forests display a simplified structure and composition resulting from centuries of tree harvesting and forest grazing. In other places, land use change and conversion continue to affect forest ecosystems. For example, the temperate forests of Asia, which cover parts of China, Japan, and Korea, were largely unaffected by glaciation through the Pleistocene [31,32]. Because of this, Asian temperate forests have higher levels of biodiversity than their counterparts in Europe or North America and are home to species with ancestries that extend much further back in geologic history. However, more recent heavy logging and fires have reduced forested land area and caused species such as the red-crowned crane (*Grus japonensis*) and the red panda (*Ailurus fulgens*) to be listed as endangered [33]. Collectively, temperate forests across the globe may have experienced more severe impacts over broader areas than any other large forest biome [34].

Structure and Composition

In contrast to grasslands or desert ecosystems, temperate forests have a more complex structure composed of different layers of vegetation. Temperate forest layers typically include a layer of mosses and lichens on the forest floor (and often on tree trunks and limbs), herbaceous and shrub layers, a subcanopy of trees, and a taller dominant canopy tree layer [35]. The complexity of the subcanopy and canopy layers can vary and often depend on the light requirements of the dominant tree species. Trees are the dominant life form of these ecosystems and can be either deciduous or evergreen. Their height or the tallest layer determines the volume of the ecosystem and therefore the amount of biomass it contains, and is a measure of the ecosystem's productivity.

Temperate forests undergo different stages in ecosystem succession as shown in Figure 26.2, allowing for the coexistence of plant species with different life history strategies. These adaptations are often linked to specific growing conditions associated with different stages of forest growth. As a result, plant community composition and forest structure change gradually over time [36,37]. For example, in northeastern temperate forests in North America, fast growing tree/shrub species that require high light availability such as birch (*Betula* spp.) and poplar (*Populus* spp.) are often the first to establish following a disturbance. As trees grow and the canopy begins to close, conditions become more favorable for more long-lived, shade-tolerant species such as maple (*Acer* spp.), oak (*Quercus* spp.), beech (*Fagus* spp.), and hemlock (*Tsuga* spp.). These later-successional trees become dominant or codominant as shown in Figure 26.2 and eventually outcompete early successional species, reaching heights of 35–40 m [38]. This process takes place over hundreds of years and can be reset at any time by a disturbance of natural or human origin. Indeed, within a relatively large and healthy naturally occurring forest area, multiple successional stages are present at once.

Disturbance regimes, which refer to the type, intensity, and frequency of disturbances in a particular region, have a substantial impact on the composition and age structure of vegetation [39–41]. In temperate forests, eolic events such as hurricanes, tornadoes, and thunderstorm downbursts are a major contributor to forest dynamics. Fire was once an important natural regulator. With land use changes and suppression of fire by humans, many temperate forests have become increasingly dominated by late

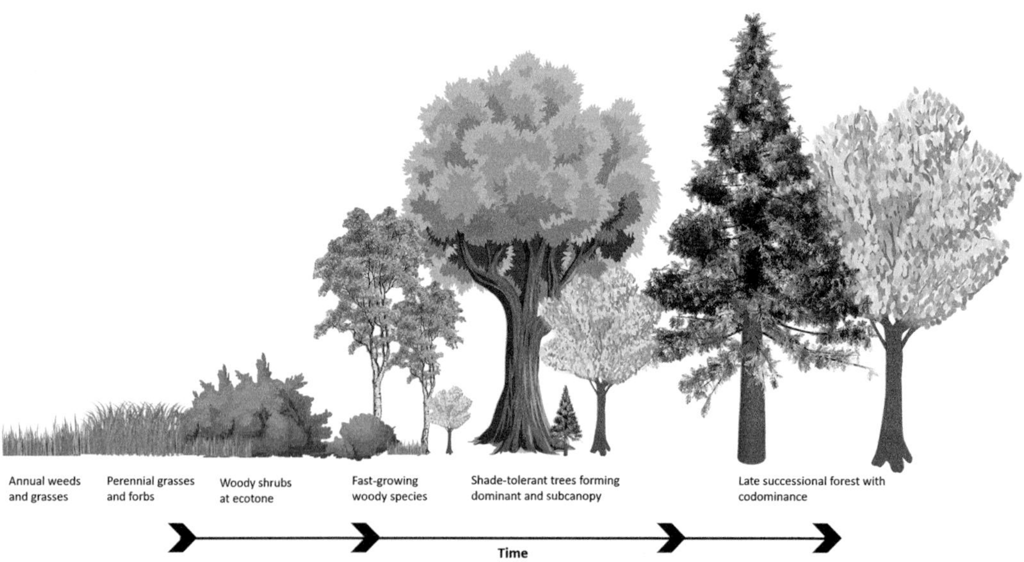

FIGURE 26.2 General stages of secondary succession in temperate forests. (Note: succession is typically not unidirectional, and multiple stages of succession are often present at the same time in large, intact temperate forests.)

successional species [42]. Other natural disturbances include pest/disease outbreaks, flood or drought, and snow and landslides.

Temperate Deciduous Forests

The temperate deciduous forest canopy is composed mostly of broadleaved angiosperm tree species with the inclusion of some coniferous tree species. Temperate deciduous broadleaved forests are often defined by species associations. Association names are most useful for distinguishing broad differences in forest type and are often representative of variation in soils, topography, and climate. For example, in eastern North America, some common associations are oak–hickory and beech–maple [43]. The temperate deciduous broadleaved forest that extends across East Asia (from 30° to 60°N) displays the greatest diversity of vegetation, a result of being less severely glaciated during the last ice age. In North America, the Florida peninsula and Gulf of Mexico coastal region was the southern limit for temperate forest refuge, while in Europe, the east to west mountain range of the Alps limited southern migration. Other locations of deciduous broadleaved forests include areas around the eastern Black Sea, mountainous regions in Iran and Caspian Sea, and a narrow strip of South America including southern Chile and Argentina [35,44–46].

The composition of deciduous temperate forest depends on latitude. In North America, birches (*Betula* spp.), aspen (*Populus* spp.), and maples (*Acer* spp.) are common, along with needle-leaved conifers such as pines (*Pinus* spp.). Coniferous species tend to decline in abundance further south, but they remain in localized areas that exhibit drier, more nutrient-poor conditions and higher altitudes. In Europe, forests at higher elevations are composed mainly of conifers (*Picea* spp. and *Abies* spp.) and *Fagus* spp., as well as larch (*Larix decidua*).

In central temperate deciduous forests, broadleaved deciduous species in Europe include oaks (*Quercus* spp.), European beech (*Fagus sylvatica*), and hornbeam (*Carpinus betulus*) [with *Fraxinus* spp. and *Castanea* spp. often just as common as *Carpinus*]. Common trees in North America include sugar maple (*Acer saccharum*), American beech (*Fagus grandifolia*), basswoods (*Tilia* spp.), oaks, and hickories (*Carya* spp.). In Asia, Japanese beech (*Fagus crenata*), oaks, maples, ashes (*Fraxinus* spp.), and basswoods are most common. At lower latitudes, broadleaved deciduous species are joined by broadleaved evergreen angiosperm trees, such as oaks, beeches (*Nothofagus* spp.), eucalyptus (*Eucalyptus* spp.), and pines (*Podocarpus* spp.).

Plant Life

Deciduous leaves are the most distinctive feature of temperate deciduous broadleaved forests. Autumn changes in leaf color and senescence have become very important to these regions both economically and ecologically. "Leaf peeping" tourists travel to deciduous forest destinations to view displays of reds, oranges, and yellows and support local economies [47,48]. The dropping of leaves in the fall is also a phenological event or recurring biological cycle that changes the seasonal microhabitat, creating unique understory conditions for other plants and animals.

The onset of plant phenophases such as spring leaf flush and autumn leaf senescence is controlled by two major environmental signals: photoperiod (relative length of days/nights) and total degree days (relative measure of heating/cooling). However, with the onset of global climate change, recent studies recognize that temperature plays a more important role in many phenological events [49,50].

Plant hormonal signals to leaves cause chlorophyll, the compound that absorbs light for photosynthesis and makes the leaves look green, to degrade. Some nutrients from the leaves are partially recovered and moved (i.e., retranslocated) to stem and trunk tissues before the leaves fall off [51–54]. Buds are formed and go into dormancy until the following spring, when the accumulation of degree days over a temperature threshold triggers leaf growth. Canopy tree species differ in their timing of bud break and leaf expansion, with the development of the forest canopy taking place over a period of a month or more [55]. Once intact canopies are in full leaf flush and the growing season is underway, typically only

1%–5% of sunlight reaches the forest floor [56–58]. Therefore, the portion of the year when canopy trees are leafless creates a unique opportunity of increased light availability for understory vegetation.

With the beginning of the warm season, herbaceous plants, shrubs, and small trees begin growth before the canopy does. Because maximum incident solar radiation flux and zenith angle of the sun are larger in the spring months, there is greater transmission of light and a greater likelihood for understory plants to reach photosynthetic capacity before canopy leaf flushing [59,60]. In an evolutionary adaptation designed to maximize the amount of light received, some plants, known as spring ephemerals, complete most of their annual growth cycle in the few weeks between snowmelt and the closure of the tree canopy. Other more shade-tolerant species begin or continue to grow after canopy closure [61].

Animal Life

Many animals living in temperate deciduous forests are mast-eaters (nuts and acorns) or omnivores. Herbivores (e.g., deer, hare, and rodents) perform important tasks in the ecosystem, such as aiding in plant seed dispersal, but may cause stress when their populations are overly abundant. Small rodents, such as squirrels and chipmunks (*Tamias* spp.), and some birds, such as blue jays (*Cyanocitta cristata*), cache large seeds/nuts in the autumn for sustenance in the winter [62]. Seeds that animals fail to recover may grow in their buried location. Omnivores include raccoon, opossum, skunk, fox, and black bear in North America and badger, mink, and marten in Europe. Asian forest fauna also include primates such as macaques. In temperate forests around the world, humans have drastically reduced the populations of many large carnivores such as wolves in North America and Europe and tigers in Asia, and now smaller carnivores such as coyote (*Canis latrans*) and lynx (*Lynx* spp.) fulfill part of this niche [63,64].

Some animals have adapted unique behaviors to better suit them for the varied seasons. For example, some mammals store fat and hibernate or go into torpor during the cold months. Some birds take shelter in the trees by becoming cavity-nesters, while others are migratory and avoid the cold winters by flying to more southerly climes.

Soils and Nutrients

Temperate deciduous forest soils show variation with latitude and altitude, as well as with moisture. At the northern or higher extent of the range, where there is a greater presence of needle-leaf evergreen tree species, soils tend to be relatively acidic and nutrient poor. By contrast, soils are strongly leached and less productive in the southern or lower ranges. Temperate deciduous forest soils, however, tend to be deep and fertile and do not freeze year-round. Brown forest soils (alfisols, in the American soil taxonomy) develop under the nutrient-demanding broadleaf trees whose leaves bind the major nutrient bases [65,66]. This causes leaf litter to be less acidic and provide a rich humus layer. Along with the favorable summer growing season, this is one of the reasons why the range of this biome has historically been a favorite target for agricultural use.

Temperate Evergreen Forest

Temperate broadleaved evergreen forests occur at the warm, moist end of the latitudinal gradient of temperate forests and are characterized by long, hot, humid growing seasons, and relatively milder winters. Precipitation is high (1,000–1,750 mm), and the forest is dominated by broadleaved evergreen angiosperms. These forests occur in eastern Asia, coastal regions of New Zealand and Australia, western coastal regions of North America and parts of Chile, South Africa, and warmer European regions [35,44–46]. There are several forest variants that occur in regions of maritime and dry-season climates. Maritime broadleaved forests are dominated by tall, broadleaved evergreen angiosperms and narrow to broadleaved conifers, and their range exhibits very high precipitation (2,000–3,000 mm). Maritime needle-leaved forests occur only in a narrow band along the coastline of North America from California to Alaska and throughout coastal southern Europe and are known for needle- and scale-leaved evergreen conifers such as giant sequoias in North America or angiosperms such as evergreen oaks in

Europe. Dry-season broad-sclerophyll forests occur in areas with periodic drought, exhibiting more of a Mediterranean climate, with hot dry summer and cool wet winters.

Needle-leaved conifers can be interspersed within deciduous canopies or occur in patches. They tend to grow at higher altitudes near treeline or in drier regions. Conditions can be highly seasonal, with severe winters and warm, humid summers. Conifer tree species include fir (*Abies* spp.), spruce (*Picea* spp.), pine (*Pinus* spp.), cedar (*Thuja* spp.), hemlock (*Tsuga* spp.), and larch (*Larix decidua*), among others.

Plant Life

While broadleaved trees in the Northern Hemisphere are exposed to variable environmental stress, broadleaved trees in the Southern Hemisphere grow under weaker seasonal forces. Because of milder climates and less-extreme variability in the Southern Hemisphere, it is more advantageous for trees to expose leaves all year-round in order to maximize photosynthesis and carbon gain [67,68]. Temperate broadleaved evergreen tree leaves are generally thick with smooth margins as opposed to deciduous leaves, which are generally thin and lobed. These differences are in part due to higher levels of rainfall and humidity. Southern Hemisphere species maintain a waxy outer layer on leaves to repel water. In these mild temperate evergreen forests, species compositions are mainly from the families of *Lauraceae*, *Fagaceae*, *Theaceae*, *Magnoliaceae*, and *Hamamelidaceae*, all of which have a relatively low resistance to cold and freezing temperatures [69,70]. The very dry-adapted forests of Australia are also found in the Southern Hemisphere. Fire disturbance in this dry region has allowed for the domination of *Eucalyptus* spp., which has thick bark for fire resistance [71,72].

Narrow-leaved evergreen trees are also relatively common in many temperate evergreen forests. These species produce their seeds in compact structures called cones, which allow for greater protection of reproductive material from predation as well as some unique adaptations to environmental conditions [73]. Some evergreen conifers have flat, triangular scale-like leaves (e.g., *Cupressaceae* and some *Podocarpaceae*), while others produce thick, needle-like leaves, an adaptation that prevents desiccation and increases the leaf area for photosynthesis [74,75]. Rates of area- or massed-based leaf photosynthesis in temperate needle-leaved evergreen trees are typically much lower than those for leaves of temperate deciduous tree species, except when amortized over the life of the leaf [76]. Temperate deciduous trees only keep their leaves for one growing season, while temperate evergreen forests keep their leaves for much longer.

The moist temperate coniferous forests dominated by tree species such as giant sequoias (*Sequoiadendron giganteum*), coastal redwoods (*Sequoia sempervirens*), Douglas firs (*Pseudotsuga menziesii*), and kauris (*Agathis australis*) sustain the highest levels of biomass in any terrestrial ecosystem [77]. These forests, rare and known for a rich variety of plant and animal species, grow in moist regions and are characterized by a higher abundance and diversity of epiphytes and vines.

In maritime climates, fog can provide a significant amount of the annual water available to vegetation [78]. Forming over adjacent oceans and water bodies, the moist air carries humidity and nutrients that can be intercepted by plants, reducing water stress caused by transpiration. Some studies suggest that some leaf shapes have evolved for more efficient collection of fog drip. It has also been hypothesized that fog in dry summer months is essential in supporting the sustained growth of large trees such as the redwoods of California [79,80].

Temperate coniferous forests at higher altitudes often have narrow tree leaf morphologies which allow heavy snow to slough off branches before breaking them [73,74]. These forests are dense, with trees growing close together to minimize wind damage; wind minimization is important because conifers often have very shallow root systems due to higher allocation of nutrients in the top layers of soil [81,82]. Evergreen trees in higher, drier areas allocate carbon to thick bark for greater protection from low-heat summer fires. Some conifers have adapted to take advantage of fire events, producing serotinous cones in which seeds are maintained in a dormant state until a fire occurs, when cones open and release the seeds for germination [83,84].

Because the canopy of these temperate forests is evergreen, the vegetation in the understory experiences low light availability. Understory plants have adapted to maximize their light/energy capture as well as their photosynthetic capacity under such conditions. Shaded plants tend to allocate more growth to leaves, displaying larger, thinner leaves which take up more horizontal area to maximize light capture [76,85]. Physiologically, shade leaves saturate at lower light levels and have lower light compensation points, which allows them to maintain positive rates of net photosynthesis at low light levels. Understory plants also adapt to respond quickly to environmental changes in order to make use of sunflecks or brief, unpredictable periods of high sunlight usually lasting only seconds [86–88]. Instantaneous adjustments in photosynthetic rates allow plants to be efficient with the limited light they receive to put toward carbon gain.

Animal Life

The fauna of the temperate evergreen forest is similar to that of the deciduous forest ecosystems, with many smaller mammals, large ungulates, and a few larger omnivorous mammals contributing to biological interactions. Many of these species, mostly birds and small rodents, have adapted skills to compensate for the scarcity of food in these nutrient-poor systems. The red squirrel (*Sciurus vulgaris*) removes cone scales to reach the nutrient-rich seeds of the coniferous trees [89,90]. They often hoard these seeds, many of which they do not recover, resulting in seed dispersal. In dry forests at higher latitudes, large ungulates such as moose (*Alces alces*) browse nutritious spring vegetation and wetland foliage in warmer seasons and on young tree shoots when the ground is frozen and covered with snow [91,92]. Some fauna in the coniferous evergreen forests located at higher latitudes have also adapted to the cold, snowy winters. For example, bears (*Ursus* spp.) are able to go into long periods of dormancy, maintaining both body temperature and metabolism without eating, drinking, defecating, or urinating [93]. In austral regions, temperate evergreen forests are also home to many endemic species such as Leadbeater's possum (*Gymnobelideus leadbeaten*), Parma wallaby (*Macropus parma*), and Albert's lyrebird (*Menura alberti*).

Soils and Nutrients

Soils of temperate evergreen forests are well leached and low in productivity. Ultisols represent older, unglaciated soils that have been weathered to a much greater degree than those of temperate deciduous forests. Ultisols are generally less fertile than most temperate deciduous forest soils and were further degraded under plantation and subsistence agriculture in both the colonial and postcolonial periods in North America [66,94].

Threats

Given that seasonality is a key feature of many temperate forest ecosystems and that temperature has a great influence on seasonal changes, global climate change is the most immediate threat to the structure and function of temperate forest ecosystems. For deciduous canopies, observational evidence has linked global climate change to differences in the timing of spring leaf growth and autumn leaf drop in temperate regions worldwide. Warm springs cause leaves to grow earlier, sometimes by up to 1 month (www.usanpn.org), lengthening the duration of the growing season in some temperate forests [95–97]. These changes in phenology can impact processes essential to survival, growth, and reproduction of all organisms as well as the complex interactions between species. Mutualistic interactions with pollinators may be especially vulnerable because of the potential for temporal mismatching as different species do not necessarily respond similarly or at the same rate to changes in environmental variation. Other evidence shows alterations in predator–prey and plant–herbivore interactions as a consequence of changing temperatures. In addition to the agricultural implications, climate-related shifts in phenology are having other economic impacts. In Japan, cherry blossom (*Prunus serrulata*) trees are blooming earlier than they did historically, causing concerns regarding the traditional festivals and tourism that surround the annual event [98]. Warmer temperatures have also been connected to poleward shifts in suitable conditions for species which could result in lowered productivity (e.g., maple syrup) or extirpation.

Rapid global climate change increases the vulnerability of some species to certain pathogens or fungi [99] while also increasing the infection severity and spread of those pests. For example, in the northwestern United States, heavy rains and wet weather during warm periods are optimal conditions for sudden oak death infection [100]. In many cases, humans accidentally introduce these diseases. Chestnut blight (*Cryphonectria parasitica*), a fungus introduced to America from Europe in 1904, is one of the best-known examples. In four decades, the chestnut blight largely eliminated American chestnut (*Castanea dentata*), one of the most important and widespread tree species across the eastern temperate forest of North America, from the continent [101,102].

Other introductions to temperate forests include non-native plant invaders that have, in some cases, taken over large portions of forested land, changing the identity of the landscape [103–105]. Invasive species can potentially reduce biodiversity, extirpate native species, and negatively impact ecosystem structure and services, thereby impacting the economy and human health. Species such as garlic mustard (*Alliaria petiolata*), native to Europe and introduced to forests of North America, exude chemicals that may disturb natural plant–plant and plant–soil processes and interactions [106]. Invasive plants may be unpalatable or even poisonous to native fauna. Examples include the Japanese knotweed (*Fallopia japonica*) [107] in Europe and North America and St. John's Wort (*Hypericum perforatum*) in Australia [108]. Negative effects of invasive plants on human populations include economic losses [109] and even disease as in the case of Japanese barberry (*Berberis thunbergii*), a shrub that, in its introduced North American environment, creates favorable habitat for vectors of Lyme disease (*Borrelia burgdorferi*) [110].

And yet, temperate forest ecosystems face threats beyond those exacerbated by climate change. The spatial coincidence of high human population density and associated impacts with temperate forests leave these ecosystems particularly susceptible to threats related to human development and pollution. Historically, farming has been one of the main reasons for deforestation in temperate biomes. Today, an increasing number of roads, residential and other developments have left forests highly fragmented, raising concerns over the healthy functioning of these ecosystems [111]. Increases in impervious surfaces and modern agricultural and industrial practices are raising nitrogen deposition on a global scale, affecting forest communities sensitive to nutrient fluxes. Many North American temperate forests are sensitive to nitrogen saturation, a phenomenon which occurs when atmospheric and microbial supply of nitrogen exceeds biotic demand. Excess nitrogen can lead to decreases in biodiversity by altering species competition, increasing susceptibility to fungal infection, inhibiting essential mutualistic associations with mycorrhizae, and facilitating species invasions. Furthermore, some research suggests that chronic excess nitrogen may decrease the survival of common temperate tree species and alter the species composition, structure, and function of future temperate forests [112].

Conclusions

Temperate forests are ecologically distinct biomes and provide important regional and global ecosystem services. The species in these ecosystems have evolved unique physiological traits to function under a variety of seasonal stresses. Being closely related to variation in environmental conditions, temperate forests may act as an ecological indicator of global climate change, changes in nutrient cycling, and other environmental alterations caused by human activities. Temperate forests have been and are of important use to industry and tourism. As they continue to play an important role in regional cultures and economies, sustainable use of these resources and proper protection and management of temperate forests will be important in ensuring the health and productivity of the biome.

References

1. Thomas, S.C.; MacLellan, J. Boreal and temperate forests, in forests and forest plants. In *Encyclopedia of Life Support Systems (EOLSS); UNESCO*; Eolss Publishers: Oxford, UK, 2004.
2. Chabot, B.F.; Mooney, H.A., Eds.; *Physiological Ecology of North American Plant Communities*; Chapman and Hall: New York, 1985.
3. Small, C.J.; McCarthy, B.C. Spatial and temporal variation in the response of understory vegetation to disturbance in a central Appalachian oak forest. *J. Torrey Bot. Soc.* **2002**, *129*(2), 136–153.
4. Gendron, F.C.; Comeau, P.G. Temporal variation in the understory photosynthetic photon flux density of a deciduous stand: the effects of canopy development, solar elevation and sky conditions. *Agric. Forest. Meteorol.* **2001**, *106*, 23–40.
5. Woodward, F.I. *Climate and Plant Distribution*; Cambridge University Press: New York, 1987.
6. Axelrod, D.I. Origin of deciduous and evergreen habits in temperate forests. *Evolution* **1966**, *20*, 1–15.
7. Food and Agriculture Organization of the United Nations. *State of the World's Forests*; FAO: Rome, 2009.
8. Vankat, J.L. *Boreal and Temperate Forests*; John Wiley & Sons Ltd: Chichester, England, 2002.
9. Reich, P.B.; Bolstad, P. Temperate forests: Evergreen and deciduous. In Terrestrial Global Productivity; Mooney, H., Roy, J., Saugier, B., Eds.; Academic Press: San Diego, CA, 2001.
10. Graumlich, L.J. A 1000-year record of temperature and precipitation in the Sierra Nevada. *Quarter. Res.* **1993**, *39*, 249.
11. Elsevier. Coniferous forests. In *Ecosystems of the World*; Vol. 6; Elsevier: Amsterdam, the Netherlands, 2005.
12. Ovington, J.D. *Temperate Broad-Leaved Evergreen Forests*; Elsevier: Amsterdam, the Netherlands, 1983.
13. Ohte, N; Tokuchi, N. Hydrology and biogeochemistry of temperate forests. In *Forest Hydrology and Biogeochemistry: Synthesis of Past Research and Future Directions*; Levia, D.F.; Carlyle, D.; Tanaka, T., Eds.; Springer Science+Business Media: New York, 2011.
14. Beckage, B.; Osborne, B.; Gavin, D.G.; Pucko, C.; Sic- cama, T.; Perkns, T. A rapid upward shift of a forest ecotone during 40 years of warming in the Green Mountains of Vermont. *Proc. Nat. Acad. Sci.* **2008**, *105* (11), 4197–4202.
15. Sittaro, F.; Paquette, A.; Messier, C.; Nock, C. Tree range expansion in eastern North America fails to keep pace with climate warming at northern range limits. *Global Change Biol.* **2017**, *23* (8), 3292–3301.
16. Loarie, S.R.; Duffy, P.B.; Hamilton, H.; Asner, G.P.; Field, C.B.; Ackerly, D.D. The velocity of climate change. *Nature* **2009**, *462* (24), 1052–1055.
17. Harsch, M.A.; Hulme, P.E.; McGlone, M.S.; Duncan, R.P. Are treelines advancing? A global meta-analysis of treeline response to climate warming. *Ecol. Lett.* **2009**, *12*, 1040–1049.
18. Mayor, J.R.; Sanders, N.J.; Classen, A.T.;Bardgett, R.D.; Clement, J.C.; Fajardo, A.; Lavorel, S.; Sundqvist, M.K.; Bahn, M.; Chisholm, C.; Cieraad, E.; Gedalof, Z.; Grigulis, K.; Kudo, G.; Oberski, D.L.; Wardle, D.A. Elevation alters ecosystem properties across temperate treelines globally. *Nature* **2017**, *542*, 91–95.
19. Adams, H.D.; Guardiola-Claramonte, M.; Barron-Gafford, G.A.; Villegas, J.C.; Breshears, D.D.; Zou, C.B.; Troch, P.A.; Huxman, T.E. Temperature sensitivity of drought-induced tree mortality portends increased regional die-off under global-change-type drought. *Proc. Nat. Acad. Sci.* **2009**, *106* (17), 7063–7066.
20. Neaumann, M.; Mues, V.; Moreno, A.; Hasenauer, H.; Seidl, R. Climate variability drive recent tree mortality in Europe. *Global Change Biol.* **2017**, *23* (11), 4788–4797.
21. Denton, G.H.; Hughes, T.J. The Arctic ice sheet: an outrageous hypothesis. In *The Last Great Ice Sheets*; Denton, G.H., Hughes, T.J., Eds.; John Wiley: New York, 1981.

22. Cox, C.B.; Moore, P.D. *Biogeography: An Ecological and Evolutionary Approach*; John Wiley: New York, 2010.
23. Foster, D.R. Land use history (1730–1990) and vegetation dynamics in central New England. *J. Ecol.* **1995**, *76*, 135–151.
24. Foster, D.R.; Motzkin, G.; Slate, B. Land-use history as long-term broad-scale disturbance: regional forest dynamics in central New England. *Ecosystems* **1998**, *1L*, 96–119.
25. Ferrow, B.E. *A Brief History of Forestry in Europe, the United States and Other Countries*; University Press: Cambridge, MA, 1991.
26. Volin, V.C.; Buongiorno, J. Effects of alternative management regimes on forest stand structure, species composition, and income: a model for the Italian Dolomites. *Forest Ecol. Manag.* **1996**, *87* (1–3), 107–125.
27. Perlin, J.A. *Forest Journey: The Role of Wood in the Development of Civilization*; Harvard University Press: London, UK, 1991.
28. Lehtinen, A.A.; Donnor-Amnell, J.; Saether, B., Eds. *Politics of Forests: Northern Forest-Industrial Regimes in the Age of Globalization*; Ashgate Publishing: Great Britain, 2004.
29. USDA Forest Service. *U.S. Forest Facts and Historical Trends*; FS-696; USDA Forest Service: Washington, DC, 2001.
30. Zerbe, S. Restoration of natural broad-leaved woodland in Central Europe on sites with coniferous forest plantations. *Forest Ecol. Manag.* **2001**, *167* (1–3), 27–42.
31. Grichuk, V.P. Late Pleistocene vegetation history. In *Late Quaternary Environments of Soviet Union*; Vwlichko, A.A., Ed.; University of Minnesota Press: Minneapolis, MN, 1984.
32. Ehlers, J; Gibbard, P.L. The extent and chronology of Cenozoic global glaciations. *Quarter. Int.* **2007**, *164–165*, 6–20.
33. www.iucnredlist.org (accessed March 2012).
34. Williams, M. *Americans and Their Forests*; Cambridge University Press: Cambridge, MA, 1989.
35. White, M.A. Deciduous Forests. Encyclopedia.com 2001 (accessed January 2012).
36. Thompson, J.N. Treefalls and colonization patterns of temperate forest herbs. *Am. Midland Nat.* **1980**, *104*, 176–184.
37. Rees, M.; Condit, R.; Crawley, M.; Pacala, S.; Tilman, D. Long-term studies of vegetation dynamics. *Science* **2001**, *293* (5530), 650–655.
38. Allaby, M. *Temperate Forests. Biomes of the Earth Series*; Chelsea House: New York, 2006.
39. Oliver C.D.; Larson, B.C. Forest stand dynamics. In Biological Resource Management Series; McGraw Hill: New York, 1990.
40. Schelhaas, M.; Nabuurs, G.; Schuck, A. Natural disturbances in the European forests in the 19th and 20th centuries. *Global Change Biol.* **2003**, *9* (11), 1620–1633.
41. Lorimer, C.G.; White, A.S. Scale and frequency of natural disturbances in the northeastern U.S. implications for early successional forest habitats and regional age distributions. *Forest Ecol. Manag.* **2003**, *185*, 41–64.
42. Aber, J.D.; Nadelhoffer, K.J.; Steudler, P.; Melillo, J.M. Nitrogen saturation in northern forest ecosystems. *BioScience* **1989**, *39*, 378–386.
43. Braum, E.L. *Deciduous Forests of Eastern North America*; Hafner Publishing Co.: New York, 1972.
44. Röhig, E.; Ulrich, B. Temperate deciduous forests. In *Ecosystems of the World*; Röhig, E.; Ulrich, B. Eds.; Elsevier: Amsterdam, the Netherlands, 1991.
45. Archibold, O.W. *Ecology of World Vegetation*; Chapman and Hall: London UK, 1995.
46. Sayre, A.P. *Temperate Deciduous Forest*; Twenty-First Century Books: New York, 1994.
47. Beckley, T.M. The nestedness of forest dependence: a conceptual framework and empirical exploration. *Soc. Nat. Resour.: An Int. J.* **1998**, *11* (2), 101–120.
48. Norris, G.A. The Economical Impact of Climate Change on the New England Region. New England Regional Assessment, 2008.

49. Mahall, B.E.; Bormann, F.H. A quanititative description of the vegetative phenology of herbs in a northern hard- wood forest. *Bot. Gaz.* **1978**, *139*, 467–481.
50. Korner, C.; Basler, D. Phenology under global warming. *Science* **2010**, *327* (5972), 1461–1462.
51. Havrenek, W.M.; Tranquillini, W. Physiological processes during winter dormancy and their ecological significant. In *Ecophysiology of Coniferous Forests*; Smight, W.K., Hinckley T.M. Eds.; Springer-Verlag: Berlin, 1995.
52. Hutchison, B.A.; Matt, D.R. The distribution of solar radiation within a deciduous forest. *Ecol. Monogr.* **1977**, *47*, 185–207.
53. Reifsnyder, W.E.; Furnival, C.M.; Horowitz, J.L. Spatial and temporal distribution of solar radiation beneath forest canopies. *Agric. Meteorol.* **1971**, *9*, 21–38.
54. Hoch, W.A.; Singsaas, E.L.; McCown, B.H. Resorption protections. Anthocyanins facilitate nutrient recovery in autumn by shielding leaves from potentially damaging light levels. *Plant Physiol.* **2003**, *113* (3), 1296–1305.
55. Dreiss, L.M.; Volin, J.C. Influence of leaf phenology and site nitrogen on invasive species establishment in temperate deciduous forest understories. *Forest Ecol. Manag.* **2013**, *296*, 1–8.
56. Gill, D.S.; Amthor, J.S.; Bormann, F.H. Leaf phenology, photosynthesis and the persistence of saplings and shrubs in a mature northern hardwood forest. *Tree Physiol.* **1998**, *18*, 281–289.
57. Sefcik, L.T.; Zak, D.R.; Ellsworth, D.S. Photosynthetic responses to understory shade and elevated carbon dioxide concentration in four northern hardwood tree species. *Tree Physiol.* **2006**, *26*, 1589–1599.
58. Hattenschwiler, S.; Korner, C. Tree seedling responses to in situ CO -enrichment differ among species and depend on understorey light availability. *Global Change Biol.* **2000**, *6*, 213–226.
59. Richardson, A.D.; O'Keefe, J. Phenological differences between understory and overstory: a case study using the long-term Harvard Forest Records Book. In *Phenology of Ecosystem Processes: Applications in Global Change Research*; Noormets, A., Eds.; Springer: New York, 2009.
60. Augspurger, C.K. Early spring leaf out enhances growth and survival of saplings in a temperate deciduous forest. *Oecologia* **2008**, *156*, 281–286.
61. Kudo, G.; Ida, T.Y.; Tani, T. Linkages between phenology, pollination, photosynthesis and reproduction in deciduous forest understory plants. *Ecology* **2008**, *89*, 321–331.
62. Li, H.; Zhang, Z. Effects of mast seeding and rodent abundance on seed predation and dispersal by rodents in Prunus armeniaca (Rosaceae). *Forest Ecol. Manag.* **2008**, *242*, 511–517.
63. Nowak, R.M. Evolution and taxonomy of coyotes and related *Canis*. In *Coyotes: Biology, Behavior, and Management*; Bekoff, M., Ed.; Academic Press: New York, 1987.
64. Thurber, J.M.; Peterson, R.O. Changes in body size associated with range expansion in the coyote (*Canis latrans*). *J. Mammal.* **1991**, *72* (4), 750–755.
65. Ulrichm, B. Stability, elasticity and resilience of terrestrial ecosystems with respect to matter balance. In *Potentials and Limitations of Ecosystem Analysis*; Schulze, E.D., Zwölfer, H. Eds.; Springer Verlag: Berlin, 1987.
66. Hendrick, R.L. Forest Types and Classification. In *The Forests Handbook, Volume 1: An Overview of Forest Science*; Evans, J., Ed.; Blackwell Science Ltd: Oxford, UK, 2008.
67. Chabot, B.F.; Hicks, D.J. The ecology of leaf life spans. *Annu. Rev. Ecol. Syst.* **1982**, *13*, 229–259.
68. Hallik, L.; Niinements, U.; Wright, I.J. Are species shade and drought tolerance reflected in leaf-level structural and functional differentiation in Northern Hemisphere temperate woody flora? *New Phytol.* **2009**, *184*, 257–274.
69. Lieth, H.; Box, E.O.; Peet, R.K.; Masuzawa, T.; Yamada, I.; Fujiwara, K.; Maycock, P.F. *Handbook of Vegetation Science*; Kluwer Academic Publishers: the Netherlands, 1995.
70. Bannister, P. Godley review: a touch of frost? Cold hardiness of plants in the southern hemisphere. *New Zealand J. Bot.* **2007**, *45* (1), 1–33.
71. Beard, J.S. Temperate forests of the Southern Hemisphere. *Vegio.* **1990**, *89* (1), 7–10.

72. Fire in ecosystems of southwest Western Australia: impacts and management. *Symposium Proceedings* (Vol. 1): 225–268, Perth, Australia, 2002.
73. Leslie, A.B. Predation and protection in the macroevolutionary history of conifer cones. *Proc. R. Soc. B* **2011**, *278*, 3003–3008.
74. Sprugel, D.G. The relationship of evergreenness, crown architecture, and leaf size. *Am. Nat.* **1989**, *133* (4), 465–479.
75. Brodribb, T.J.; Field, T.S.; Sack, L. Viewing leaf structure and evolution from a hydraulic perspective. *Func. Plant Biol.* **2010**, *37*, 499–498.
76. Busing, R.T.; Fujimori, T. Biomass, productions and woody detritus in an old coast redwood (*Sequoia sempervirens*) forest. *Plant Ecol.* **2005**, *177*, 177–188.
77. Ewing, H.A.; Weathers, K.C.; Templer, P.H.; Dawson, T.E.; Firestone, M.K.; Elliott, A.M.; Boukili, V.K. Fog water and ecosystem function: heterogeneity in a California redwood forest. *Ecosystems* **2009**, *12*, 417–433.
78. Canadell, J.; Jackson, R.B.; Ehleringer, J.R.; Mooney, H.A.; Sala, O.E.; Schulze, E.D. Maximum rooting depth of vegetation types at the global scale. *Oecologia* **1996**, *108* (4), 583–595.
79. Burgess, S.S.O.; Dawson, T.E. The contribution of fog to the water relations of Sequoia sempervirens (D. Don): foliar uptake and prevention of dehydration. *Plant, Cell Environ.* **2004**, *27* (8), 1023–1034.
80. Simonin, K.A.; Santiago, L.S.; Dawson, T.E. Fog interception by Sequoia sempervirens (D. Don) crowns decouples physiology from soil water deficit. *Plant, Cell Environ.* **2009**, *32*, 882–892.
81. Enright, N.J.; Marsula, R.; Lamont, B.B.; Wissel, C. The ecological significance of canopy seed storage in fire-prone environments: a model for non-sprouting shrubs. *J. Ecol.* **1998**, *86*, 946–959.
82. Echenwalder, J.E. *Conifers of the World*; Timber Press, Inc.: Portland, OR, 2009.
83. Givnish, T.J. Adaptation to sun and shade: a whole plant perspective. *Aust. J. Plant Physiol.* **1988**, *15*, 63–92.
84. Midgley, J.; Bond, W. Pushing back in time: the role of fire in plant evolution. *New Phytol.* **2011**, *191* (1), 5–7.
85. Boardman, N.K. Comparative photosynthesis of sun and shade plants. *Annu. Rev. Plant Physiol.* **1977**, *28*, 355–377.
86. Chazdon, R.L.; Pearcy, R.W. The importance of sunflecks for understory plants. *BioScience* **1991**, *41* (11), 760–766.
87. Brantley, S.T.; Young, D.R. Contribution of sunflecks is minimal in expanding shrub thickets compared to temperate forest. *Ecology* **2009**, *90*, 1021–1029.
88. Lertzman, K.P.; Sutherland, G.D.; Inselberg, A.; Saunders, S.C. Canopy gaps and the landscape mosaic in a coastal temperate rain forest. *Ecology* **1996**, *77*, 1254–1270.
89. Miyaki, M. Seed dispersal of the Korean pine, *Pinus koraiensis*, by the red squirrel, *Sciurus rulgaris*. *Ecol. Res.* **1987**, *2*(2), 147–157.
90. Wauters, L.A.; Gurnell, J.; Martinoli, A.; Tosi, G. Does interspecific competition with introduced grey squirrels affect foraging and food choice of Eurasian red squirrels? *Animal Behav.* **2001**, *61* (6), 1079–1091.
91. Hjeljord, O.; Sundstol, F.; Haagenrud, H. The nutritional value of browse to moose. *J. Wildlife Manag.* **1982**, *46* (2), 333–343.
92. Wam, H.K.; Hjeljord, O. Moose summer and winter diets along a large scale gradient of forage availability in southern Norway. *Eur. J. Wildlife Res.* **2010**, *56*, 745–755.
93. Hellgren, E.C. Physiology of hibernation in bears. *Int. Assoc. Bear Res. Manag.* **1995**, *10*, 467–477.
94. Mbagwu, J.S.C.; Lal, R.; Scott, T.W. Effects of desurfacing alfisols and ultisols in southern Nigeria: I. Crop performance. *Soil Sci. Soc. Am. J.* **1984**, *48*, 828–833.
95. Menzel, A.; Fabian, P. Growing season extended in Europe. *Nature* **1999**, *397*, 659.

96. Cleland, E.E.; Chuine, I.; Menzel, A.; Mooney, H.A.; Schwartz, M.D. Shifting plant phenology in response to global change. *Trends Ecol. Evol.* **2007**, *22* (7), 357–365.
97. Badeck, F-W.; Bondeau, A.; Bottcher, K.; Doktor, D.; Lucht, W.; Schaber, J.; Sitch, S. Responses of spring phenology to climate change. *New Phytol.* **2004**, *162* (2), 295–309.
98. Aono, Y.; Kazui, K. Phenological data series of cherry tree flowering in Kyoto, Japan, and its application to reconstructions of springtime temperatures since the 9th century. *Int. J. Climatol.* **2008**, *28* (7), 905–914.
99. Ayres, M.P.; Lombardero, M.J. Assessing the consequences of global change for forest disturbance from herbivores and pathogens. *Sci. Total Environ.* **2000**, *262*, 263–286.
100. Kelly, M.; Meentemeyer, R.K. Landscape dynamics of the spread of Sudden Oak Death. *Photogrammetric Eng. Rem. S.* **2002**, *68* (10), 1001–1009.
101. Anagnostakis, S.L. Chestnut blight: the classical problem of an introduced pathogen. *Mycologia* **1987**, *79* (1), 23–37.
102. Paillet, F.L. Chestnut: history and ecology of a transformed species. *J. Biogeogr.* **2002**, *29*, 1517–1530.
103. Walker, L.R.; Smith, S.D. Impacts of invasive plants on community and ecosystem properties. In *Assessment and Management of Plant Invasions*; Luken, J.O., Thieret, J.W., Eds.; Springer: New York, 1997.
104. Kohli, R.K.; Jose, S.; Singh, H.P.; Batish, D.R. *Invasive Plants and Forest Ecosystems*; CRC Press: Boca Raton, FL, 2009.
105. Daehler, C.C. Performance comparisons of co-occuring native and alien invasive plants: implications for conservation and restoration. *Annu. Rev. Ecol. Evol. Syst.* **2003**, *34*, 183–211.
106. Stinson, K.A.; Campbell, S.A.; Powell, J.R.; Wolfe, B.E.; Callaway, R.M.; Thelen, G.C.; Hallet, S.G.; Prati, D.; Klironomos, J.N. Invasive plant suppresses the growth of native tree seedlings by disrupting belowground mutualisms. *PLoS Biol.* **2006**, *4* (5), e140.
107. Krebs, C.; Gerber, E.; Mathies, D.; Schaffner, U. Herbivore resistance of invasive species and their hybrids. *Oecologia* **2011**, *167* (4), 1041–1052.
108. Southwell, I.A.; Bourke, C.A. Seasonal variation in hypericin content of *Hypericum perforatum* (St. John's Wort). *Phytochemistry* **2001**, *56* (5), 437–441.
109. Pimentel, D.; Lach, L.; Zuniga, R.; Morrison, D. Environmental and economic costs associated with non-indigenous species in the United States. *BioScience* **2000**, *50* (1), 53–65.
110. Williams, S.C.; Ward, J.S.; Worthley, T.E.; Stafford, K.C. Managing Japanese barberry (Ranunculales: Berberidacea) infestations reduces blacklegged tick (Acari: Ixodidae) abundance and infection prevalence with *Borrelia burgdorferi* (Spirochaetales: Spirochaetaceae). *Environ. Entomol.* **2009**, *38* (4), 977–984.
111. Foley, J.A.; DeFries, R.; Asner, G.P.; Barford, C.; Bonan, G.; Carpenter, S.R.; Chapin, F.S.; Coe, M.T.; Daily, G.C.; Gibbs, H.K.; Hellowski, J.H.; Holloway, T.; Howard, E.A.; Kucharik, C.A.; Monfreda, C.; Patz, J.A.; Prentice, I.C.; Ramankutty, N.; Snyder, P.K. Global consequences of land use. *Science* **2005**, *309* (5734), 570–574.
112. Simkin, S.M.; Allen, E.B.; Bowman, W.D.; Clark, C.M.; Belnap, J.; Brooks, M.L.; Cade, B.S.; Collins, S.L.; Geiser, L.H.; Gilliam, F.S.; Jovan, S.E.; Pardo, L.H.; Schulz, B.K.; Stevens, C.J.; Suding, K.N., Throop, H.L.; Waller, D.M. A continental analysis of ecosystem vulnerability to atmospheric nitrogen deposition. *Proc. Natl. Acad. Sci. Unit. States Am.* **2016**, *113*, 4086–4091.

27

Forests: Tropical Rain

Introduction 227
What Is (and What Is Not) a Tropical Rain Forest? 227
Where Do Tropical Rain Forests Occur? 227
South and Central America • Africa • Asia • New Guinea and Australia • Madagascar • Island Rain Forests
The Importance of Tropical Rain Forests 230
Threats to Tropical Rain Forests 231
Conservation of Tropical Rain Forests 232
Conclusions 232
References 233

Richard T. Corlett
Chinese Academy of Sciences

Introduction

Tropical rain forests cover only 6% of the Earth's total land area but have attracted more public attention than any other terrestrial ecosystem. The key roles these forests play in the leading environmental concerns of the 21st century—biodiversity loss and climate change will ensure that this attention will continue. The public image of tropical rain forests is oversimplified, however, and a major aim of this entry is to introduce the diversity of rain forests that exists today. This is followed by a consideration of their importance, the threats to their survival, and the options for their conservation.

What Is (and What Is Not) a Tropical Rain Forest?

Tropical rain forests are the tall, dense, evergreen forests that form the natural vegetation of those areas of the lowland tropics that are hot and wet all (or almost all) of the time.[1] Tropical rain forests extend into areas with a short (1- to 3-month) dry season, particularly on deep, water-retaining soils or where the dry season is moderated by fog or lower temperatures, but with a progressively longer dry season, the proportion of deciduous trees increases and rain forests are eventually replaced by tropical dry forests (tropical deciduous forests) of various types. The forest also changes with an increase in altitude on high mountains and, at around 1,000 m above sea level, tropical lowland rain forest is replaced by tropical montane forest, which is shorter and generally less species rich. An alternative term, tropical moist forest, has often been used in a broader sense to include semi-evergreen and montane forests as well as "true" tropical rain forests, but neither term is used consistently.

Where Do Tropical Rain Forests Occur?

On a more symmetrical planet, tropical rain forests would form a broad band 5–10 north and south of the equator, interrupted only by the oceans. On our complex Earth, however, interactions between wind direction and mountain ranges, along with a variety of other factors, cause gaps in this band,

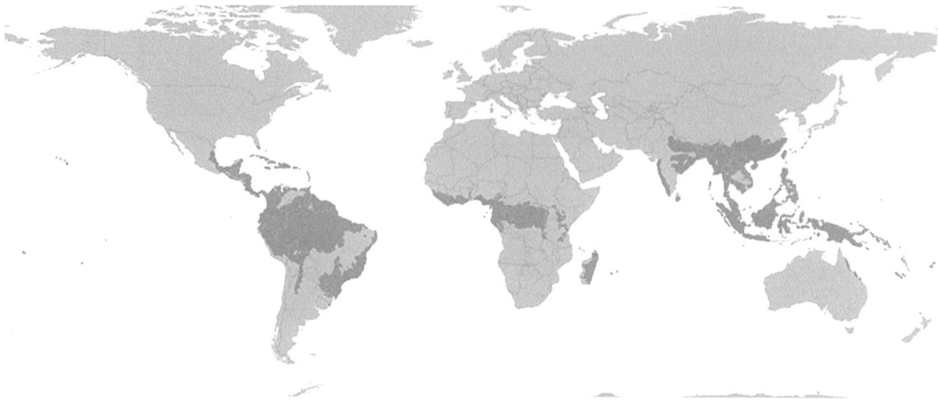

FIGURE 27.1 Map showing the global extent of tropical rain forests. (Adapted from Wikimedia Commons.)

most notably in East Africa, as well as places where the tropical rain forest extends north or south of the expected latitude (Figure 27.1).[1] More recently, huge areas of tropical rain forest have been converted to crops and other land uses, creating further fragmentation.

Tropical rain forests occur in five major regions and as isolated areas on many small oceanic islands (Figure 27.1).[1] Exchanges of species between the major regions have been difficult for tens of millions of years, so they have very few species in common and many genera and even families are confined to only one or two regions. In general, the differences between rain forest regions are greater for animals than plants, reflecting both the greater age of plant lineages and the greater dispersal ability of many plants. Asian and African tropical rain forests share most of their plant and animal families, as well as numerous genera, because they have been connected by land since Africa collided with Eurasia 20 million years ago, but they share few species, because the land link has been too dry for forest for most of that period. Asia and New Guinea have never had a land connection but have been close enough for at least 20 million years to exchange species able to cross relatively short water gaps. South America was an island until between 15 and 3 million years ago, when it became connected to the north through the Isthmus of Panama, resulting in an influx of northern animal. Madagascar, in contrast, has been an island for 90 million years and was reached only by species able to make substantial sea crossings. Oceanic islands, by definition, have never been connected to other land masses and have acquired their entire flora and fauna across the water by dispersal.

South and Central America

The tropical rain forests of South and Central America—the Neotropics—are the most distinctive, reflecting their long period of isolation.[1] Moreover, for many groups of organisms, including trees, birds, bats, and butterflies, the richest Neotropical sites support more species than the richest sites in the other regions. There are three main blocks of rain forest in the Neotropics. The largest block—indeed, the largest block of tropical rain forest in the world—is centered on the Amazon basin and stretches from the foothills of the Andes to the Atlantic Ocean. Most of this block is in Brazil, but there are also large areas in Bolivia, Peru, Ecuador, Colombia, Venezuela, Guyana, Surinam, and French Guiana. A second block extended as a narrow band along the Atlantic coast of Brazil and was separated from the Amazon block by a broad swath of drier vegetation types. Most of this block has now been cleared, leaving only small, isolated fragments. The third block extended from the Pacific coasts of Ecuador and Columbia through Central America to southern Mexico and has also been badly fragmented. Characteristic organisms of Neotropical rain forests include the following: the epiphyte family Bromeliaceae, the New World Monkeys (Platyrrhini), the New World fruit bats (Phyllostomidae), the sloths (Pilosa), the coatis and other members of the raccoon family (Procyonidae), the agoutis and other caviomorpha rodents (Dasyproctidae, Agoutidae), the antbirds and other suboscine passerines (Tyranni), the hummingbirds (Trochilidae), and the leafcutter ants (Attini). Neotropical rain forests are also characterized by the absence of any really large mammals and of such characteristic Old World groups as the apes, hornbills, and honeybees (although these were recently introduced).

Africa

The second largest block of tropical rain forest is in Central Africa, centered on the Congo River basin and extending over much of the Democratic Republic of the Congo, Republic of the Congo, Gabon, and Cameroon. A separate block, now heavily fragmented, extended along the coast of West Africa, from Ghana, through the Ivory Coast and Liberia to eastern Sierra Leone. There are also small patches of tropical rain forest in East Africa that have been isolated from the larger blocks for millions of years.

The tropical rain forests of Africa share many families and genera with Asia but also support an endemic family of fruit-eating birds, the turacos (Musophagidae).[1] Diversity is relatively low in many groups of organisms, particularly plants, as a result of the gradual drying of the climate over the last 30 million years. The termites, however, are exceptionally diverse, and these forests are notable for the abundance and diversity of large, ground-living herbivorous mammals, including the lowland gorilla, mandrill, African elephant, okapi, and many species in the cattle family, including the bongo and forest buffalo.

Asia

The tropical rain forests of Asia consist of a relatively uniform block covering much of mainland South East Asia and the large islands of Borneo, Sumatra, and Java, and more or less distinctive outlying areas in the Philippines, Sulawesi, the Western Ghats of India, and the wet zone of Sri Lanka. Indonesia and Malaysia have the most tropical rain forests, but most countries in the Asian tropics have some. Most families in Asian rain forests are shared with Africa, but several mammal families are unique to Asia: the gibbons (Hylobatidae), tarsiers (Tarsiidae), tree shrews (Tupaiidae), and colugos (Cynocephalidae).[1] These are also the only rain forests with rhinoceroses, although both species are on the verge of extinction, and the only lowland rain forests with bears. There are also several small families of birds that are unique to the region. Botanically, the most distinctive feature is the dominance of the upper layers of the forests by the tree family Dipterocarpaceae. Dipterocarps dominate both the biomass and the ecology of these forests in a way that has no parallel in the other rain forest regions. The most dramatic manifestation of this is the tendency of the majority of dipterocarp species over large areas to flower and fruit together at irregular, multiyear intervals triggered by drought and low temperatures.[1] Many non-dipterocarps join in these mass flowering and fruiting events, but dipterocarps usually dominate.

New Guinea and Australia

The island of New Guinea supports the fourth largest block of tropical rain forest, divided politically between the Indonesian provinces of Papua and West Papua and the independent country of Papua New Guinea. There are also several, much smaller, areas of tropical rain forest in the northeast of neighboring Australia. Although the rain forests of Australia and New Guinea have many similarities, there are also differences that reflect their very different histories.[1] Rain forest covered much of northern Australia until around 15 million years ago, since then it has been progressively restricted by the drying of the continent while most of the rain forest in New Guinea is on land that was uplifted above sea level only 10–15 million years ago.[1] As a result, while the remnants of Australian rain forests have a very distinctive flora, including two woody climbers in an ancient family of flowering plants that is found nowhere else (Austrobaileyaceae), the younger rain forests in New Guinea are dominated by rain forest plants of Asian origin. The faunas are more similar, however, and have fewer elements of Asian origin. Rats and bats are the only native placental mammals, and there are no primates, deer, squirrels, or placental carnivores. Marsupials, including possums, wallabies, and tree kangaroos, occupy most mammalian niches, but there are no large carnivores. The distinctive bird fauna includes the flightless, fruit-eating cassowaries (Casuariidae), which are the largest of all rain forest birds, the birds of paradise (Paradisaeidae), the honeyeaters (Meliphagidae), and a number of smaller families that are found nowhere else. There are also representatives of many birds families found in Asia and Africa, but no bulbuls, barbets, woodpeckers, or babblers.

Madagascar

Until recently, there was a band of tropical rain forest along the east coast of this large and mostly dry island, 425 km (266 miles) from Africa. Most plant families and genera are shared with African rain forests, but the fauna is far more distinctive, with many groups showing evidence of a single colonization event followed by adaptive radiation into the many empty niches on the island.[1] The best known example is the lemurs (Lemuriformes), whose ancestors arrived from Africa 50 to 60 million years ago.[2] Apes, monkeys, and squirrels never reached Madagascar, so the lemurs have radiated into a great diversity of forms, the largest of which became extinct soon after people first arrived on the island 2,000 years ago. Similar radiations following single colonization events account for all the rodents, insectivores, and carnivores. There were also some spectacular bird radiations, with one group, the vangas, which arrived around 25 million years ago, now occupying foraging niches that are filled by a variety of different bird families elsewhere.[3] Frogs and reptiles show similar patterns, as do some groups of insects.

Island Rain Forests

Although the rain forests on each island usually support only a fraction of the species diversity found in the major rain forest regions, most species are confined to one island, so that the island rain forests together support a substantial fraction of the total global rain forest diversity.[1] The larger Caribbean islands were colonized by bats, rodents, and primates, like Madagascar, as well as the Neotropical sloths, although all the larger mammals are now extinct. They also had spectacular radiations in the *Anolis* lizards[4] and the *Eleutherodactylus* frogs.[5] More distant islands, such as Fiji, were reached by frogs, snakes, and lizards, but bats are the only mammals, and many bird families are missing. The most remote islands with tropical rain forest are in the Hawaiian archipelago, 3,765 km (2,340 miles) from the nearest major land mass.[1] Terrestrial mammals are represented by a single species of insectivorous bat, the birds by only 13 families, and there were no ants, termites, mosquitoes, cockroaches, or swallowtail butterflies until people introduced them. The birds and many insect groups are dominated by a few, highly diverse, evolutionary radiations originating from one or a few colonists. Well-studied examples include the Hawaiian honeycreepers[6] and moths (*Hyposmocoma*).[7] Many plants disperse better than most rain forest animals, so floral diversity declines more slowly with distance from the mainland, but the Hawaiian rain forests lack many widespread plant groups, while one-eighth of the native flora (126 species) is made up of the descendants of a single ancestral species of Campanulaceae (the bell flower family) that arrived in the archipelago 8 million years ago.[8]

The Importance of Tropical Rain Forests

Intact tropical rain forests have multiple values, with their relative importance depending on your particular perspective. Most values can be included under one of four general headings: biodiversity, carbon, products, and services. Tropical rain forests support more plant and animal species than any other vegetation type on Earth. An area of 500 hectares (1,250 acres) of rain forest at Garza Cocha in Ecuador supports more butterfly species (676) than the whole of North America,[9] while a mere 52 hectares

(128 acres) at Lambir in Malaysian Borneo contains as many tree species (1,175) as the entire temperate forest of the northern hemisphere (Asia, Europe, and North America; 1,166 species).[10] Similar patterns occur in a range of other organisms, giving rise to suggestions that tropical rain forests, which cover about 6% of Earth's land surface, support 50% or more of all non-marine species on Earth. Although widely repeated, this number is little more than a guess, and we do not have a complete species list for any area of tropical rain forest. Many commercial crop species (fruits, nuts, timber, spices, etc.) have close relatives in the rain forest that are potentially critical resources for breeding.

Concern for global warming is responsible for another way of looking at tropical rain forest: as a store of carbon compounds that are easily converted to and from the major greenhouse gas, carbon dioxide.[11] All organisms contain carbon, as do soils, but in tropical rain forests, most is in the large trees. Plants remove carbon dioxide from the atmosphere when they photosynthesize in sunlight, and return it through respiration and when a part or all of the plant dies. An undisturbed rain forest is a massive carbon store, more or less in equilibrium, with plant photosynthesis balanced by the respiration of plants, animals, and microbes. Disturbances, such as logging and firewood collection, release part of this store into the atmosphere. Conversion to another land use, such as a soybean field, releases an amount of carbon (as carbon dioxide) equal to the difference between the amount stored in the forest, including the soil, and the amount in the crop. The degradation and clearance of tropical forests is a source of carbon dioxide second in magnitude only to fossil fuels, but these very large emissions are more or less offset at present by an equally large uptake by the surviving forests.[11] Much of this uptake is in forests regrowing after logging or clearance, which make more than a third of all tropical forests, but even intact forests are currently taking up more carbon dioxide than they release.

The most valuable product from tropical rain forests in international trade is timber, but a huge variety of non-timber products (firewood, edible plants, resins, fibers, honey, medicines, meat, etc.) are also harvested and may be locally more valuable as sources of both income and employment.[1] Apart from carbon fixation, considered separately in the preceding text, the most valuable service performed by tropical rain forests is usually the maintenance of a reliable supply of clean water. Although this service is generally viewed as "free," the benefits of maintaining forest cover for this purpose are easily understood, and payments from hydropower companies and urban areas are a potential source of revenue for protecting the forest. Other services, such as the provision of wild pollinators for agriculture crops, are more local and the benefits less visible, although not necessarily less important.

Threats to Tropical Rain Forests

The fact that most tropical rain forest species cannot live outside the forest means that conversion to other land uses is the greatest threat. The world is currently losing about 0.5% of its tropical rain forests every year, with rates of loss varying hugely from place to place.[12] The highest rates are currently in Indonesia and parts of the Brazilian Amazon, while most of the Congo Basin in Central Africa still has relatively low rates. Until fairly recently, most clearance was by smallholders, but increasingly the lowland tropics are becoming dominated by the plantations and ranches of big landowners. This is particularly true in Indonesia, where a single crop, oil palm, has replaced vast areas of lowland rain forest in Sumatra and Borneo, and in parts of the Amazon region, where cattle ranching and soybeans are the major threats. Smallholders still dominate in Africa, but commercial agriculture and plantations are now expanding there as well. Deforestation leads to biodiversity losses long before all the forest has gone because the forest fragments that persist in many human-dominated landscapes are not large enough to support all the species present in continuous forest.

Much of the remaining tropical rain forest has been exploited in other ways through logging (i.e., timber harvest), hunting, and/or the harvesting of a wide range of non-timber forest products. Logging has the most visible impact, because the biggest trees are removed, but studies suggest that the impacts on biodiversity are much less than with clearance, and largely reversible if the logged forest is protected from additional impacts such as fire. The impact of overhunting is less visible because most rain forest animals are hard to see anyway, but can be more pervasive because so many processes within the forest, such as seed dispersal, depend on animals.[13] Dramatic evidence for this indirect impact comes from Sarawak, Malaysia, where hunting has led to a decline in tree diversity and other changes in the forest[14] and the Brazilian Atlantic forest where the loss of large fruit eaters has resulted in a reduction in fruit sizes.[15] Hunting may also lead to local, regional, and, eventually, global extinction of favored animal species, such as the critically endangered forest rhinoceroses in tropical Asia. Other widespread human impacts on tropical rain forests include fires (usually only after fragmentation or logging has opened up the canopy), invasive species (particularly on oceanic islands), and, increasingly, air pollution. Climate change is an additional threat over the next few decades. By 2100, the regions occupied by tropical rain forests today will be 2°C–5°C warmer, creating climates that do not exist today anywhere on Earth.[16] Changes in rainfall are also expected but more difficult to predict. The potential biological impacts of the expected changes in temperature and rainfall are currently poorly understood but could be severe.[17] One major concern is that higher temperatures and more frequent droughts will convert intact forests from a carbon sink, as they are now, into a significant source, thus accelerating climate change.[11]

Conservation of Tropical Rain Forests

Some rain forest areas are still protected by their remoteness from human populations, but these areas are shrinking, and eventually, the only unexploited forests that remain will be in protected areas (national parks, nature reserves, etc.). Existing protected areas vary hugely in their size and in the aims and effectiveness of their management, but there is ample evidence that well-managed protected areas can work.[1] In much of the tropics, particularly in Asia, there are no large, unexploited areas of tropical rain forest left for protection, and new protected areas will need to incorporate logged or hunted forests that have the potential for recovery. Many of the values of tropical rain forests can also persist in areas that are well managed for sustainable timber production and, potentially, for other forms of resource exploitation. Most logging and hunting in tropical rain forests are currently illegal and thus unregulated, so the control of these activities is another key to conservation both in and outside the protected areas. This requires effective enforcement, both on the ground and along the supply chain that links producers through local and regional markets to the consumers, often in other countries. Unfortunately, successful conservation management requires higher standards of governance than that currently exist in some tropical rain forest countries, where corrupt practices and inefficiency often limit the ability of governments to convert good intentions into effective action.[18]

Tropical rain forests are mostly in developing countries with many priorities other than biodiversity protection, so funding rain forest conservation is a major problem. The massive amount of carbon stored in intact rain forests is the basis for plans to financially reward countries that reduce their carbon emissions by reducing deforestation and forest degradation. These policies are collectively known by the acronym REDD+ (Reduced Emissions from Deforestation and Degradation).[18] REDD+ is a complex and controversial issue, but the potentially large amounts of money involved make it an attractive option for tropical forest conservation although conserving carbon and biodiversity are not necessarily the same thing. Where most rain forest has already been lost, restoring forest on marginal land of little agricultural value has both carbon and biodiversity benefits. A variety of reforestation techniques have been developed, ranging from promoting natural forest succession to actively planting the site with nursery-raised saplings.[19,20] So far, the areas restored have been tiny in comparison with the total area of rain forest lost each year, but these projects could be scaled up if funding becomes available from REDD+ or other sources.

Conclusions

The protection and sustainable use of the remaining tropical rain forests are critical for both biodiversity conservation and reducing greenhouse gas emissions, and can also contribute to the livelihoods of tropical people. The diversity of tropical rain forests will require diverse solutions, but well-managed protected areas and the control of illegal hunting and logging are likely to be effective everywhere. International concerns for the carbon dioxide emitted by the clearance and degradation of tropical rain forests may provide the funding to make this possible.

References

1. Corlett, R.T.; Primack, R.B. *Tropical Rainforests: An Ecological and Biogeographical Comparison*; 2nd Ed.; Wiley-Blackwell: Oxford, UK, 2011.
2. Kappeler, P.M.; Nunn, C.L.; Vining, A.Q.; Goodman, S.M. Evolutionary dynamics of sexual size dimorphism in non-volant mammals following their independent colonization of Madagascar. *Sci. Rep.* **2019**, *9*, 1454.
3. Jønsson, K.A.; Fabre, P.H.; Fritz, S.A.; Etienne, R.S.; Ricklefs, R.E.; Jørgensen, T.B.; Fjeldså, J.; Rahbek, C.; Ericson, P.G.P.; Woog, F.; Pasquet, E.; Irestedt, M. Ecological and evolutionary determinants for the adaptive radiation of the Madagascan vangas. *Proc. Nat. Acad. Sci.* **2012**, *109* (17), 6620–6625.
4. Losos, J.B. *Lizards in an Evolutionary Tree: Ecology and Adaptive Radiation of Anoles*; University of California Press: Berkeley, CA, 2009.
5. Dugo-Cota, A.; Carles-Vilà, A.R.; Gonzalez-Voyer, A. Ecomorphological convergence in Eleutherodactylus frogs: A case of replicate radiations in the Caribbean. *Ecol. Lett.* **2019**, doi:10.1111/ele.13246.
6. Lerner, H.R.L.; Meyer, M.; James, H.F.; Hofreiter, M.; Fleischer, R.C. Multilocus resolution of phylogeny and timescale in the extant adaptive radiation of Hawaiian honeycreepers. *Curr. Biol.* **2011**, *21* (21), 1838–1844.
7. Haines, W.P.; Schmitz, P.; Rubinoff, D. Ancient diversification of *Hyposmocoma* moths in Hawaii. *Nat. Commun.* **2014**, *5*, 3502.
8. Knox, E.B.; Li, C. The East Asian origin of the giant lobelias. *Am. J. Bot.* **2017**, *104* (6), 924–938.
9. DeVries, P.J. Butterflies. In *Encyclopedia of Biodiversity*; Levin, S.A., Ed.; Academic Press: San Diego, CA, 2001; 559–573.
10. Wright, S.J. Plant diversity in tropical forests: A review of mechanisms of species coexistence. *Oecologia* **2002**, *130*, 1–14.
11. Mitchard, E.T. The tropical forest carbon cycle and climate change. *Nature* **2018**, *559*, 527–534.
12. Keenan, R.J.; Reams, G.A.; Achard, F.; de Freitas, J.V.; Grainger, A.; Lindquist, E. Dynamics of global forest area; results from the FAO Global Forest Resources Assessment 2015. *For. Ecol. Manag.* **2015**, *352*, 9–20
13. Corlett, R.T. The importance of animals in the forest. In *Managing the Future of Southeast Asia's Valuable Tropical Rainforests: A Practitioner's Guide to Forest Genetics*; Wickneswari, R., Cannon, C., Eds.; Springer: Dordrecht, 2011; 83–92.
14. Harrison, R.D.; Tan, S.; Plotkin, J.B.; Slik, F.; Detto, M.; Brenes, T.; Itoh, A.; Davies, S.J. Consequences of defaunation for a tropical tree community. *Ecol. Lett.* **2013**, *16* (5), 687–694.
15. Galetti, M.; Guevara, R.; Cortes, M.C.; Fadini, R.; Von Matter, S.; Leite, A.B.; Labecca, F.; Ribeiro, T.; Carvalho, C.S.; Collevatti, R.G.; Pires, M.M; Guimaraes, P.R., Jr.; Brancalion, P.H.; Ribeiro, M.C.; Jordano, P. Functional extinction of birds drives rapid evolutionary changes in seed size. *Science* **2013**, *340* (6136), 1086–1090.
16. IPCC. Global Warming of 1.5°C, an IPCC special report on the impacts of global warming of 1.5°C above pre-industrial levels and related global greenhouse gas emission pathways, in the context of strengthening the global response to the threat of climate change, sustainable development, and efforts to eradicate poverty; Intergovernmental Panel on Climate Change: Geneva, Switzerland, 2018.
17. Scheffers, B.R.; De Meester, L.; Bridge, T.C.L.; Hoffmann, A.A., Pandolfi, J.M., Corlett, R.T.; Butchart, S.H.M.; Pearce-Kelly, P.; Kovacs, K.M.; Dudgeon, D.; Pacifici, M.; Rondinini, C.; Bickford, D.; Watson, J.E.M. The broad footprint of climate change from genes to biomes to people. *Science* **2016**, *3454* (6313), aaf7671.

18. Peh, K.S.H. Governance and conservation in the tropical developing world. In *Conservation Biology: Voices from the Tropics*; Sodhi, N. S., Gibon, L., Raven, P.H., Eds.; Wiley: Oxford, UK, 2013; 216–225.
19. Phelps, J.; Webb, E.L.; Adams, W.A. Biodiversity cobenefits in forest carbon emissions reduction policy. *Nat. Clim. Change* **2012**, *2*, 497–503.
20. Lamb, D. Regreening the Bare Hills: Tropical Forest Restoration in the Asia-Pacific Region; Springer: Heidelberg, 2011.

28

Savannas and Grasslands

Niti B. Mishra and
Kenneth R. Young
University of Texas at Austin

Introduction 235
Structural Properties and Geographic Distribution 236
Determinants of Vegetation Structure 237
Plant Available Moisture/Nutrients 237
Fire 239
Herbivores 239
Paradigms/Models of Tree-Grass Interaction in Savannas 241
Biodiversity and Productivity 242
Ecological Concerns and Sustainable Use 242
Conclusion 243
References 244

Introduction

Savannas and grasslands are important ecosystems spread over more than 20% of the Earth's terrestrial surface. While savannas are mostly tropical ecosystems that feature the unique co-existence of dominant woody and herbaceous life forms, grasslands are found in both tropical and temperate climates and are primarily dominated by grasses (Figure 28.1).[1] The two biomes are distinct but share some similar ecological processes and biotic assemblages. Due to their large geographical extent, these ecosystems contribute to more than 30% of global net primary productivity and also significantly influence the

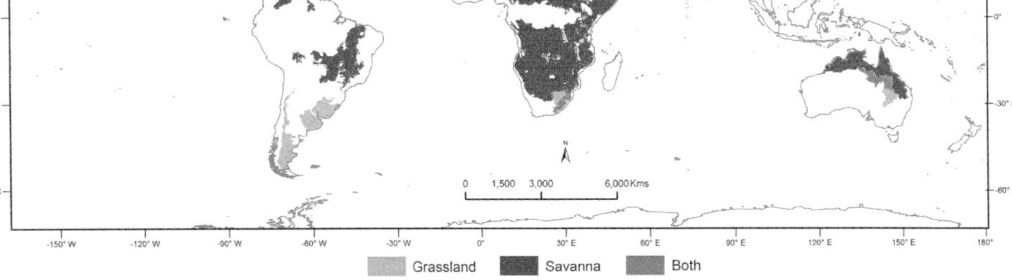

FIGURE 28.1 World-wide distribution of natural savannas and grasslands.

global land-atmosphere energy balance and carbon and nutrient cycles.[2,3] Savannas and grasslands support most of the world's livestock; pasturelands and protected wildlife areas are the basis of socio-economic activities such as pastoralism, tourism, and conservation[4,5] With a high variation in the ratio of woody to herbaceous life forms, the resulting spatial and structural heterogeneity of land cover is a defining characteristic of savanna systems. Due to their nutrient rich soil, temperate grasslands are productive, especially in North America where this biome houses a major crop and cereal-producing area. Changes in savanna and grassland occurrence and vegetation structure are primarily the result of the interaction of four factors: plant available moisture (PAM), plant available nutrient (PAN), grazing pressure, and fire. The complex interaction of these factors produces variations in vegetation structure at both local and regional scales. This high variation has made it difficult for studies to provide an over-arching and simple definition of savannas; it furthermore has led many ecologists to consider savannas as ecologically unstable mixtures of woody and herbaceous life forms.

Structural Properties and Geographic Distribution

Savannas have often been described in the ecological literature as landscapes with continuous grass cover and scattered trees/shrubs. However, the definition of the term "savanna" is debated given that the tree-grass ratio within these systems varies greatly from nearly treeless grasslands to areas with up to 80% woody canopy cover.[6] The woody cover component in savannas consists of trees/shrubs of evergreen or deciduous phenology and may include broad- or fine-leafed species. The herbaceous component can be either perennial or annual plant types of varying height. Since landscapes in various ecosystems fit this vegetation structural criterion, savannas in different parts of the world have been called rangelands, open forest, parkland, or simply grassland. The variation of the tree to grass ratio and its interpretation have also impacted the estimation of total area under savanna ecosystems. For example, while Atjay (1979) estimated it to be 22.5 million km^2, Alson (1983) found it to be 33.7 million km^2.

Based on climatic setting and disturbance regime, the world's savannas can be grouped into four broad categories:

- *Tropical and subtropical savannas* cover an eighth of the global terrestrial area (~1600 million ha) and represent the majority of savanna located in tropical and subtropical latitudes of Africa, Australia, South America, India, and South East Asia (Figure 28.1). African tropical savannas are scattered across a wide variety of soil conditions with annual rainfall ranging from 200 to 1800 mm; they form a semi-circle around the western central rainforest, bordered by deserts to the north and south.[7–9] In African savanna, fine-leaved woody vegetation mainly dominated by *Acacia* species occurs on nutrient-rich soils in low-lying semi-arid areas in Botswana, South Africa, and Namibia. On the contrary, broad-leaved savanna species (e.g., *Combretum* spp., *Lonchocarpus nelsii,* and *Terminalia* spp.) can be seen in the sub-humid interior plateau on infertile and highly weathered soils.[10] The Guinea-type savanna woodlands in humid climates replace moist semi-deciduous vegetation and form a transition to evergreen moist forests, whereas northern Sudan-type savannas with scattered *Acacia* trees form a transition with Sahara desert vegetation.[10,11] The Miombo woodland savanna covers an additional 2.7 million km^2 area across central and southern Africa and is characterized by the presence of deciduous species of *Brachystegia, Isoberlinia,* and *Fulbbernardia,* forming a discontinuous canopy with an average height of 10 m.[12] In South America, tropical savanna comprises the biologically rich Brazilian *Cerrado* and the Colombian and Venezuelan *Llanos,* together covering over 2 million km^2 in area. Rich in fauna, the *Cerrado* bioregion contains a gradient from completely open savannas to closed canopy forest along environmental shifts in rainfall, topography, and human influence. Besides producing most of the beef cattle of Brazil, with the implementation of soil improvement techniques and irrigation, *Cerrado* areas are now also important producers of soy and grains.[3,13] The Australian tropical savanna is contiguous with little topographic variation and covers about 12% of the total tropical savanna area.[14] In northern

Australia, savanna open *Eucalyptus* spp. woodlands is the dominant land cover type. The woody canopy cover decreases from northern Australia toward the continent's interior, following the rainfall gradient.[15,16] In general, the woody vegetation in Australian tropical savannas has predominantly evergreen tree species that is different from African tropical savannas with mostly deciduous tree species.[17] Also, compared to the tropical savannas in other continents, Australian tropical savannas have lower population density and thus are ecologically less modified. However, due to increasing economic activities, particularly mining and agriculture, their future remains uncertain.[15,18]

- *Temperate grasslands* are known as prairies in North America (covering 5 million km^2), steppes in Eurasia, pampas in Argentina and Uruguay in South America, and the veld region in South Africa. These mid-latitude grasslands experience wet summers and dry winters; the climate can range from temperate to semi-arid to sub-humid. Compared to tropical savannas, temperate grasslands have lower biodiversity. Typical kinds of grasses include *Bouteloua dactyloides* (buffalo grass), *Lolium* (ryegrass), *Avena* (wild oats), and *Nassella pulchra* (purple needlegrass), which also are all part of the diet of herbivorous animals. Grass height depends on the amount of rainfall received from late spring to early summer, which ranges in total between 450 and 700 mm.[9] Often lightning from thunderstorms can ignite large grass fires in temperate grasslands, which keeps trees from overshadowing the grasses and acts to maximize biodiversity by creating landscape heterogeneity. Soils are deep and dark with fertile upper layers. A longitudinal rainfall gradient can be used to help distinguish between tall- and short-grass grasslands in North America. The fauna in the temperate grasslands varies from herbivores such as bison, pronghorn, and deer (in North America), the Asian camel and various species of horses and saiga antelope (in Eurasia).
- Tropical *flooded savannas* are characterized by strong seasonality, that is, a dry period which causes severe water shortage and a wet period during which extensive flooding occurs.[19] Examples of flooded savannas include Zambezian flooded savanna of Mozambique, Okavango Delta, Botswana, and the Pantanal flooded savannas of Bolivia, Brazil, Paraguay, and some of the *Llanos* of Venezuela.[20–22]
- *Montane savannas* are high-altitude savannas, often of human origin due to burning, and include areas such as Angolan scarp savanna, Campos Rupestres montane savanna (Southeastern Brazil), degraded montane savanna of Ethiopia, and woodlands and high-altitude grasslands (e.g., páramo) of the Andes mountains.[9] At present, montane savannas are the most widespread vegetation type on the mountain chains running through Africa from Lesotho and South Africa to eastern Sudan. They are also abundant in India, Nepal (Terai), and Sri Lanka.

Determinants of Vegetation Structure

Findings from various studies suggest that the ecosystem structure and function in savannas and grasslands can be understood as the result of the complex interplay of four main ecological factors: PAM, PAN, fire, and herbivory (Figure 28.2).

Plant Available Moisture/Nutrients

At regional to continental scales, PAM is the most important determinant of ecosystem structure and function in both savannas and grasslands. PAM can be quantified in terms of mean annual precipitation, number of days during which rainfall is greater than evapotranspiration, or the number of days during which water is available in the soil. In general, tree cover in savanna ecosystems increases with increasing rainfall; there is a corresponding decreasing in herbaceous cover (Figure 28.3).[10,23] Like savannas, grasslands are water-limited ecosystems, and their locations are associated with places having dry periods that occur at least once a year. Grasslands are found in areas with mean annual precipitation ranging between 150 and 1200 mm. As temperature increases, the forest-grassland and desert-grassland

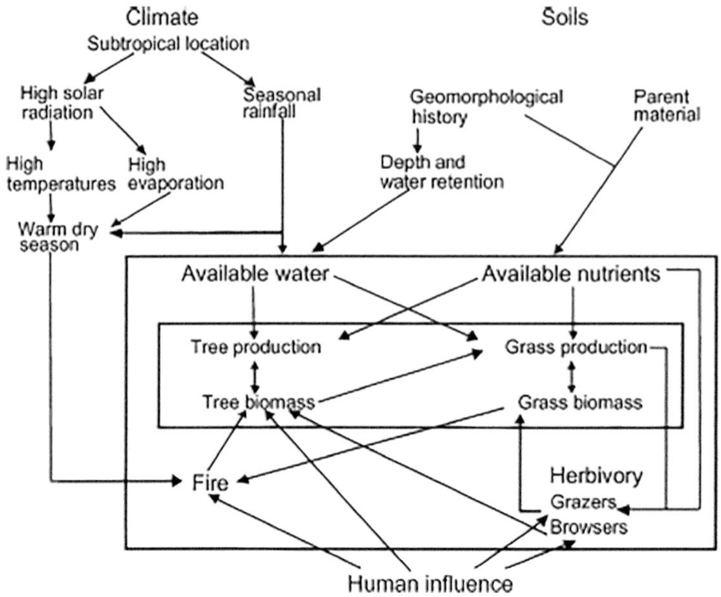

FIGURE 28.2 Interaction between four important ecological determinants of savanna structure (i.e., PAN, PAM, fire, and herbivore). This interaction is realized at any given savanna or grassland location by climatic condition and soil type.
Source: Reproduced from House et al.,[42] with permission from Blackwell Publishing.

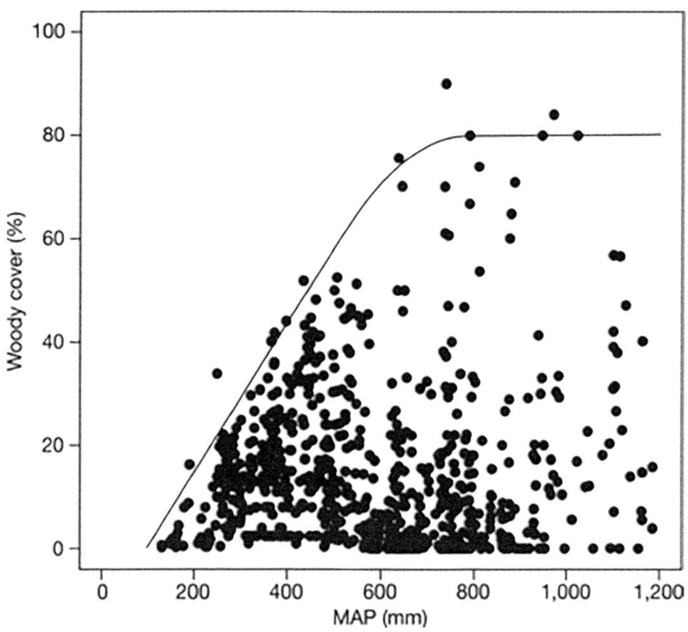

FIGURE 28.3 Woody cover in African savannas as a function of mean annual precipitation (MAP). Data from 854 sites across Africa suggest that below 650 mm both herbaceous and woody life form coexist and above 650 mm MAP values woodlands can dominate. Below 101 mm MAP value trees are generally absent.
Source: Reproduced from Sankaran et al.,[10] with permission from Nature Publishing Group.

boundaries occur at a higher amount of precipitation due to the increasing evaporation rate. For example, in North America where temperature isopleths run in an east-west direction and isohyets in a north-south direction, the grassland-forest boundary has a strong SE-NW orientation.[24] Unlike PAM, PAN acts as spatial a determinant of ecosystem structure in savannas and grasslands at a much finer spatial scale. This is because nutrient availability is a function of not only the quality of soil parent material, but also its interaction with PAM that facilitates the release of nutrients via mineralization.[1] Plant growth in savannas takes place during high periods of PAM when nutrients are mineralized. In turn, this creates changes in landscape mosaics and across topographic/edaphic gradients. The critical interaction of PAN and PAM has often been used to classify and compare savanna systems. Fine-scale changes in soil properties from site to site in a given landscape can thus result in different vegetation physiognomic patterns even within the same PAM regime.[7,15]

Fire

Fire is an important additional landscape scale determinant of the composition, ecology, and distribution of the savannas and grassland ecosystems; it controls the abundance of woody life forms below a precipitation determined boundary.[10] Savanna fires are predominantly surface fires that burn dry grasses and shrubs. These fires can spread rapidly due to the hot, windy, low-humidity conditions of the dry season. Early dry season fires are generally less intense than late dry season fires. The patch size of burned area can range from less than a hectare to several hundred square kilometers, creating a relatively fine-grained disturbance mosaic.[6] Since grasses and shrubs can grow rapidly post-fire, the same region can burn every year. Fire return intervals in savannas are commonly on the order of every one to three years and are further influenced by climatic factors and land-use practices.[25] With exceptions of sparsely populated areas where fire caused by lightening predominates, today most savanna fires are human-caused for reasons such as land clearing for farming, improving pasture for grazing, providing nutrients to crops, making fire breaks around human settlements, and fires due to accidents.[6,26,27]

In both savanna and grassland ecosystems, fire serves as a key ecological buffering mechanism that maintains woody plant densities.[28] The results of studies based on long-term fire exclusion plots in southern Africa and Australian savanna have shown that decreased fire frequencies can lead to bush encroachment, while increased fire frequencies can lead to conversion to grassland (Figure 28.4).[29–31] Changing fire frequency has shifted the balance between savanna and forest vegetation cover over much of the seasonal tropics and recent studies also are attempting to understand how plant traits (e.g., savanna versus forest trees) determine the dynamics of savanna-forest boundary.[32,33] Besides fire frequency, the timing of fire in relation to the reproductive phenology can also limit or support plant reproduction, resulting in altered species composition in savanna and grassland vegetation communities.[34] In addition, fire activity in these ecosystems has also been found to directly affect bird and insect species richness and their behavioral responses. For example, studies in the northern territory of Australia found that grass removal by fire allowed greater access to fallen grass seed, resulting in higher numbers of granivores attracted to the burned areas. Insectivorous birds showed similar patterns, while the species preferring tall dense grass or understory vegetation were least abundant after fire events, which affect the spatial extent of recovering vegetation.[35]

Herbivores

The importance of herbivore as an ecological factor varies among savanna and grassland regions. The two main kinds of herbivores influencing vegetation structure in these ecosystems are large native mammals (often ungulates) and the comparatively overlooked but important invertebrates such as termites, ants, and grasshoppers. In African savannas, mammals, both wild and domesticated, exert an unparalleled control on the ecosystem structure by acting as grazers and browsers consuming vegetation, by trampling soils and by destroying trees and shrubs.[36,37] Large herbivores can change vegetation structure and alter

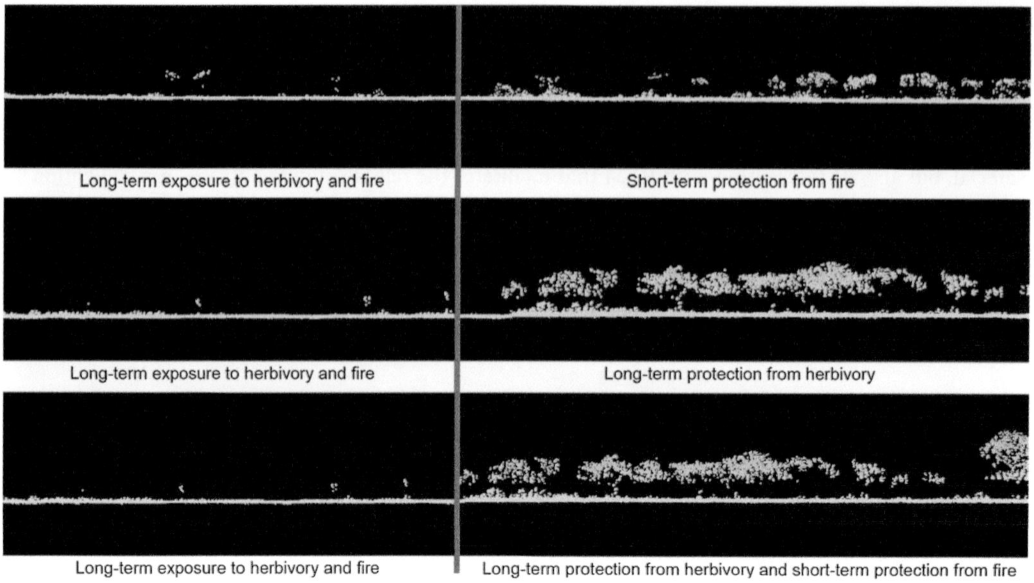

FIGURE 28.4 Cross-profile comparisons of the herbivore and fire experimental treatments and the control landscape. Profiles display the light detection and ranging point cloud of both ground surface and vegetation canopy returns. The solid vertical line marks the boundary between control and treatment sites.
Source: Reproduced from Levick et al.,[31] with permission from Elsevier Publishing.

species composition in savanna ecosystems (Figure 28.4). For example, in the African Serengeti system, after the eradication of rinderpest in the 1960s, the wildebeest population continued to increase until the 1980s, which transformed it in part into a woodland-dominated state due to the increased recruitment of sapling and adult trees.[38] There is an additional set of interactions with predators that control the behavior of some of the herbivores. Savanna structure and function are undoubtedly the result of all these interactive effects of herbivores and fire. More recent studies examining the impact of herbivore in southern African savanna found that areas under short-term (6 years) exclusion of herbivore contained 38–80% less bare ground compared to those sites that were exposed to herbivores. Additionally, areas under long-term exclusion from herbivores (>22 years) differed significantly in terms of their aboveground vegetation structure with up to 11-fold greater woody canopy cover in areas without herbivores.[39] Besides grazing, increased livestock pressure in farmlands in African and Australian savannas also impacted soil characteristics, resulting in losses of sensitive crust, soil compaction, erosion, and a related loss of nutrients.[40,41]

Although much smaller individually than the large and conspicuous mammalian grazers and browsers, insects are important herbivores in savanna systems, particularly in areas of low soil fertility. The key role of termites as ecosystem engineers in savannas is well established as they alter the mineral and organic composition of soil, and affect their hydrology, drainage, and infiltration rates as well as influencing decomposition and nutrient cy cling.[1,42] Grasshoppers are important grazing insects in the majority of tropical savannas. In Nylsvley, South Africa, an average grasshopper biomass of 2.3 kg ha^{-1} was estimated to consume 130 kg ha^{-1} yr^{-1} biomass in *Burkea* savanna and 406 kg ha^{-1} yr^{-1} in *Acacia* savanna, values that represented 40% and 73% of the total herbivore consumption, respectively.[43] While there are site-specific studies highlighting the role of insect herbivores in savannas, in general there is a lack of landscape and regional scale studies describing their abundance and role in these ecosystems.

The spatiotemporal interaction and dynamics of the four determinants of savannas discussed above (i.e., PAM, PAN, fire, and herbivore) have traditionally been used to explain the ecosystem structure and function in savannas as well as the much-debated issue of woody plant encroachment. However, more recent studies based on results from experimental plots (with manually controlled disturbance regimes) and outputs from deterministic models of savanna dynamics have emphasized the role of increased atmospheric CO_2 concentration that could be responsible for increasing the woody cover in savannas.[44,45] Recent studies argue that increasing atmospheric CO_2 concentration can influence the growth rate of juvenile tree saplings that can grow faster to reach fire-resistant height. The resulting increased tree recruitment is causing the conversion of savannas into woodlands.[44,46,47]

Paradigms/Models of Tree-Grass Interaction in Savannas

The coexistence of woody and herbaceous life forms in the same ecosystem type has always puzzled ecologists. Over the last few decades of savanna ecology research, several conceptual models have been proposed to explain how trees and grasses can coexist in savanna systems. These models can broadly be categorized as either competition based or demographic based. Savanna structure, under competition-based models, is expected to be the result of spatial and temporal niche separation in resource (i.e., water and nutrients) acquisition; in these cases, fire and herbivore effects are only considered as modifiers.[48–50] Mosaic patterns of soils and of topography would lead to complicated vegetation patterns. Alternatively, demographic-based models assume that tree-grass coexistence occurs in savannas because climatic variability and disturbances act as bottlenecks to limit successful tree establishment. Proposed by Walter (1971), the *root niche separation model* considers water as the primary limiting factor for which trees and grasses compete in savannas. While grasses are rooted only in topsoil, trees have roots in both topsoil and subsoil with exclusive access to deeper water; this separation in theory could lead to the stable coexistence of the two. The *phenological niche separation model* proposes the wet season as a potential axis for niche separation between savanna trees and grasses.[1,42] Deciduous savanna trees achieve full leaf expansion early in the start of rainy season, whereas peak leaf areas of grasses are achieved only after several months.[7] In this scenario, trees have exclusive access to resources during both the early and the late rainy season while grasses are more competitive during the proper rainy season. Under a *balanced competition model,* coexistence in savannas arises because trees, being the superior competitor for light and for soil resources, become self-limiting at a biomass insufficient to exclude grasses, which are inferior competitors.[1] In the absence of external perturbations, this model predicts a threshold with increasing rainfall above which woody plants dominate and below which grasses replace trees. The *hydrologically driven competition-colonization model* emphasizes the tradeoffs between the competitive ability and the colonization potential of trees and grasses, which changes in response to fluctuations in soil water stress caused by inter-annual rainfall variability, and how that further changes in response to fluctuations in rainfall. Model predictions on the balance between trees and grasses are sensitive to the magnitude and variance of those inter-annual fluctuations.[51]

Demographic bottleneck models integrate the effects of multiple drivers (e.g., fire, herbivore, and climatic fluctuations) and rather than emphasizing the post-disturbance competitive interaction, they highlight the direct effect of these disturbances on germination, mortality, and the demographic transition of trees in savannas.[52–56] There exist alternative viewpoints among researchers using demographic bottleneck models in regards the degree of control of savanna structure and functioning. The transitional "disequilibrium" system approach, for example, considers disturbances such as fire and grazing as not only modifiers but also maintainers of the unique savanna structure by buffering the system against shifts to alternate states.[52] Alternatively, another viewpoint is that the nonequilibrium dynamics are driven by geographic variations in rainfall in arid regions,[56] while disequilibrium dynamics are driven by variations in fire intensity in more mesic areas.[56] In arid and semi-arid savannas, due to spatiotemporal variability in rainfall, tree recruitment is pulsed in time with occasional heavy rainfall events, provided that the mature trees live long enough to bridge these events. However in mesic savannas, where trees can dominate canopy cover, frequent high-intensity fires prevent the seedlings from escaping the flames and thereby control tree density.[56] Further, browsing by herbivores may prevent the seedlings from establishing as mature trees.[28,52,56]

Recently, demographic models have gained favor over competition-based models, which were found insufficient in explaining long-term tree-grass coexistence in the spatially explicit modeling of interactions in savannas using field data.[28] Although the different models discussed here have been supported by empirical evidence from different sites around the Earth, no single model has been able to provide a generic mechanism that provides a single explanation for all cases. Furthermore, observational data collected during various studies in savanna ecosystems suggest that the savanna structure and function results from the interaction of all the processes discussed above, implying that holistic explanations will be most robust. More recent studies examining the correlation among selected environmental variables and global savanna distribution emphasize that climatic variables do not necessarily have a deterministic effect on savanna distribution as divergences from the limits of savanna exit globally that cannot be explained by environmental determinants.[57,58] Studies using dynamic global vegetation models to resolve savanna distribution in different continents emphasize that although climatic determinants are important, extensive savannas in some high rainfall areas suggest a decoupling of climate and vegetation.[59,60]

Biodiversity and Productivity

The spatial patterns of biodiversity in grasslands and savannas vary widely, with many areas with these biomes having relatively high levels of species richness and types of organisms.[61] For example, Pampas grasslands of South America have more than 400 species of grasses and tall grass prairies have more than 250 plant species.[62] In contrast to these areas, some grasslands of Patagonia have fewer than 30 species of plants.[63] Considering the large area of grasslands, the overall faunal diversity is low with a total of 477 species of birds and 245 species of mammals (i.e., equivalent to 5% of the total number of species in each group).[64] Even within climatically similar grasslands, the pattern of diversity of large mammals is significantly different. For example, in Africa and South America there are approximately 200 total species of large mammals, but the contribution of different orders is very different. For example, in Africa, the order *Artiodactyla* (Bovidae) has 80 species, whereas in South America there are fewer than 20.[65] The tropical savannas of Africa represent the world's greatest diversity of ungulates (over 40 different species). Up to 16 grazing and browsing species of antelopes can coexist in a given area in African savannas. Besides the above-ground diversity in faunal life, below-ground organisms are also an important component in savanna and grassland ecosystems. Grasslands in general have been found to have more biomass and higher productivity below ground compared to above.[66] The concentration of organic matter below ground provides the substrate for diverse bacteria, fungal, and nematode groups. More than 100 species of fungi and more than 200 species of nematodes have been reported for the short-grass steppe and tall-grass prairie, respectively[67] Tropical savanna and temperate grasslands account for 30% (i.e., 19.9 Gt yr^{-1}) and 9.1% (i.e., 5.6 Gt yr^{-1}) of the global net primary production (NPP), respectively, demonstrating that these are relatively productive ecosystems[68] In tropical savanna systems, although the relative productivity of trees and grasses are highly variable, the NPP of grass layers is 2–3 times that of the tree NPP.

Ecological Concerns and Sustainable Use

Savannas and grasslands are among the oldest ecosystems used by humans. Anthropogenic usage of the resources in these ecosystems is increasing and humans have influenced the ecological factors affecting savanna vegetation by controlling fire and by using it for landscape management. With burning, humans have either modified or totally cleared natural vegetation in savannas and grasslands for agricultural purposes. The introduction of animal husbandry systems has resulted in changes in grazing and browsing pressure and has undoubtedly modified the competitive balance between woody and herbaceous life forms in savannas. Compared to savannas, grasslands are even more highly altered ecosystems, with less than half remaining in an intact natural condition. Intensive mechanized agriculture activities have replaced more than 40% of the global temperate grasslands and another 15% have been converted due to urbanization.[69] The remaining area is dominated by grasses, but is degraded and vulnerable to desertification.

Increasing the human exploitation of natural resources in both grasslands and savannas has resulted in a variety of environmental concerns such as decreasing vegetation cover and changing species composition, loss of biodiversity, increased soil erosion resulting in nutrient loss and decreased productivity, and shifts in water balance and availability.[4,7] As a result of changing land use and a resultant increase in grazing/browsing pressure and altered fire regime in southern African and Australian savannas, large areas are experiencing the encroachment of woody species. Presumably human influences shift the competitive edge toward woody plant species.[70] Weed infestation in certain tropical savannas such as the Northern Territory of Australia is posing huge environmental and economic threat as they are changing the savanna ecosystem structure and resultant fire regime.[71,72] While certain weed species are toxic (e.g., *Argemone ochroleuca*), others chock waterways or are so dense that animals cannot move through them.[73,74] Changing vegetation structure of this sort may limit the grazing potential for local people, and hence may be endangering the livelihood of dependent ranching and pastoral communities.[40,75] Large parts of natural vegetation in Brazil's *Cerrado* have been cleared and modified to not only produce 70% of the beef cattle of that country, but with the implementation of soil modification techniques and irrigation, this ecoregion is now also important for grain and soy. There are environmental risks associated with monoculture plantation establishment and with burning that affects carbon, soil acidity, and friability.[3,13] In North America, the utilization of *Prairie* grasslands for grazing, cattle ranching, extensive agriculture, and urbanization has modified most of the temperate grasslands. Ecological issues in *Prairies* include woody plant expansion, biodiversity loss due to monospecific crop agriculture, gully erosion, and decreased productivity[73,74] Changing land use practices have also resulted in the invasion of non-native species, a problem encountered in savanna and grasslands worldwide.[76,77]

Besides human impacts, future conditions and sustainability in savanna and grassland systems will also be determined by impacts of global climate change, which may alter spatiotemporal patterns of rainfall and thus influence available moisture, nutrients, and land cover patterns. Studies predict that the increased concentration of atmospheric CO_2 will allow invasion and denser stands of C3 woody species in savannas as such species can potentially utilize higher CO_2 concentration more efficiently than C4 plants. Consequently, trees will have faster post-burn recovery rates and faster sapling growth, enabling them to dominate.[78–80]

Conclusion

The dynamics of the world's savannas and grasslands continue to be shaped by humans through changed ecological controls. Given the significant geographical extent of these ecosystems, any changes in the structural and functional properties of these ecosystems, either due to human influence or climatic variability, are likely to impact human well-being. Thus, increased knowledge about the interaction of ecological processes and resulting spatial patterns and their temporal dynamics will improve our understanding. This information is required for ecologically informed decision making in regards the future sustainability of these systems. Historically, savannas have been ignored in ecological research until recently when their carbon sequestration potential (a third that provided by tropical rainforests and temperate evergreen forests) and biodiversity importance became apparent. As a result, historical data on savannas are limited and most of modern savanna studies are short-term site studies involving only selected ecological aspects. Savanna research needs large-scale climate and bio-physical data along the broad environmental gradients of precipitation and water limitations. By managing fire activity in savannas, their productivity could be significantly increased. Savannas are suitable for agro-forestry and thus can contribute to carbon sequestration, greenhouse mitigation, and also provide more livelihood opportunities.

References

1. Scholes, R.J.; Archer, S.R. Tree-grass interactions in savannas. Annu. Rev. Ecol. Syst. **1997**, *28*, 517–544.
2. Flanagan, L.B.; Wever, L.A.; Carlson, P.J. Seasonal and interannual variation in carbon dioxide exchange and carbon balance in a northern temperate grassland. Global Change Biol. **2002**, *8* (7), 599–615.
3. Mistry, J. Savannas. Prog. Phys. Geogr. **2000**, *24* (4), 601–608.
4. Adeel, Z.; de Kalbermatten, G.; Assessment, M.E. *Ecosystems and Human Well-Being: Desertification Synthesis;* Island Press, 2005.
5. Stott, P. Recent trends in the ecology and management of the world's savanna formations. Prog. Phys. Geogr. **1991**, *15* (1), 18–28.
6. Roy, D.P.; Boschetti, L.; Giglio, L. Remote sensing of global savanna fire occurence, extent and properties. In *Ecosystem Function in Savannas: Measurement and Modeling at Landscape to Global Scales;* Hill, M.J., Hanan, N.P., Eds.; CRC Press, 2011.
7. Scholes, R.J.; Walker, B.H. *Cambridge Studies in Applied Ecology and Resource Management; An African Savanna: Synthesis of the Nylsvley Study*; Cambridge University Press: Cambridge, England, 1993; xii+306 p.
8. Backeus, I. Distribution and vegetation dynamics of humid savannas in Africa and Asia. J. Veg. Sci. **1992**, *3* (3), 345–356.
9. Archibold, O.W. *Ecology of World Vegetation;* Chapman and Hall Ltd.: London, 1995; xi+510 p.
10. Sankaran, M.; Hanan, N.P.; Scholes, R.J.; Ratnam, J.; Augustine, D.J.; Cade, B.S.; Gignoux, J.; Higgins, S.I.; Le Roux, X.; Ludwig, F.; Ardo, J.; Banyikwa, F.; Bronn, A.; Bucini, G.; Caylor, K.K.; Coughenour, M.B.; Diloufl, A.; Ekaya, W.; Feral, C.J.; February, E.C.; Frost, P.G.H.; Hiernaux, P.; Hrabar, H.; Metzger, K.L.; Prins, H.H.T.;Ringrose, S.; Sea, W.; Tews, J.; Worden, J.; Zambatis, N. Determinants of woody cover in African savannas. Nature **2005**, *438* (7069), 846–849.
11. Menaut, J.C.; Barbault, R.; Lavelle, P.; Lepage, M. African savannas: Biological systems of humification and mineralization; In *Ecology and Management of World's Savannas;* Tothill, J.C., Mott, J.J., Eds.; Australian Academy of Science: Canberra, 1985; 14–33.
12. Chidumayo, E.N. Climate and phenology of savanna vegetation in southern Africa. J. Veg. Sci. **2001**, *12* (3), 347–354.
13. Furley, P. Tropical savannas: Biomass, plant ecology, and the role of fire and soil on vegetation. Prog. Phys. Geogr. **2010**, *34* (4), 563–585.
14. Lehmann, C.E.; Prior, L.D.; Bowman, D.M. Decadal dynamics of tree cover in an Australian tropical savanna. Austral Ecol. 2009, *34* (6), 601–612.
15. Williams, R.J.; Duff, G.A.; Bowman, D.M.J.S.; Cook, G.D. Variation in the composition and structure of tropical savannas as a function of rainfall and soil texture along a large-scale climatic gradient in the Northern Territory, Australia. J. Biogeogr. **1996**, *23* (6), 747–756.
16. Bowman, D.; Walsh, A.; Milne, D. Forest expansion and grassland contraction within a Eucalyptus savanna matrix between 1941 and 1994 at Litchfield National Park in the Australian monsoon tropics. Global Ecol. Biogeogr. **2001**, *10* (5), 535–548.
17. Bowman, D.; Prior, L. Turner Review No. 10. Why do evergreen trees dominate the Australian seasonal tropics? Aust. J. Bot. **2005**, *53* (5), 379–399.
18. Woinarski, J.; Braithwaite, R.W. Conservation foci for Australian birds and mammals. Search 1990, *21* (65), 8.
19. Barreto, G.R.; Herrera, E.A. Foraging patterns of capybaras in a seasonally flooded savanna of Venezuela. J. Trop. Ecol. **1998**, *14* (1), 87–98.
20. Sarmiento, G.; Pinillos, M.; da Silva, M.P.; Acevedo, D. Effects of soil water regime and grazing on vegetation diversity and production in a hyperseasonal savanna in the Apure Llanos, Venezuela. J. Trop. Ecol. **2004**, *20* (2), 209–220.

21. Hoogesteijn, A.; Hoogesteijn, R. Cattle ranching and biodiversity conservation as allies in South America's flooded savannas. Great Plains Res. **2010**, *20* (1, Sp. Iss. SI), 37–50.
22. Mishra, N.B.; Crews, K.A.; Neuenschwander, A.L. Sensitivity of EVI-based harmonic regression to temporal resolution in the lower Okavango Delta. Int. J. Remote Sensing **2012**, *33* (24), 7703–7726.
23. Bucini, G.; Hanan, N.P. A continental-scale analysis of tree cover in African savannas. Global Ecol. Biogeogr. **2007**, *16* (5), 593–605.
24. Bailey, R.G.; *Ropes, L. Ecoregions: The Ecosystem Geography of the Oceans and Continents*; Springer: New York, 1998.
25. Van Wilgen, B.W.; Govender, N.; Biggs, H.C.; Ntsala, D.; Funda, X.N. Response of savanna fire regimes to changing fire-management policies in a large African National Park. Conserv. Biol. **2004**, *18* (6), 1533–1540.
26. Miranda, H.S.; Sato, M.N.; Neto, W.N.; Aires, F.S. Fires in the cerrado, the Brazilian savanna. In *Tropical Fire Ecology: Climate Change, Land Use, and Ecosystem Dynamics;* Springer-Verlag: Berlin, **2009;** 427–450.
27. Gregory, N.C.; Sensenig, R.L.; Wilcove, D.S. Effects of controlled fire and livestock grazing on bird communities in East African savannas. Conserv. Biol. **2010**, *24* (6), 1606–1616.
28. Jeltsch, F.; Weber, G.E.; Grimm, V. Ecological buffering mechanisms in savannas: A unifying theory of long-term tree-grass coexistence. Plant Ecol. **2000**, *150* (1–2), 161–171.
29. Gillson, L.; Ekblom, A. Resilience and thresholds in savannas: Nitrogen and fire as drivers and responders of vegetation transition. Ecosystems **2009**, *12* (7), 1189–1203.
30. Hoffmann, W.A.; Schroeder, W.; Jackson, R.B. Positive feedbacks of fire, climate, and vegetation and the conversion of tropical savanna. Geophys. Res. Lett. **2002**, *29* (22), 4.
31. Levick, S.R.; Asner, G.P.; Kennedy-Bowdoin, T.; Knapp, D.E. The relative influence of fire and herbivory on savanna three-dimensional vegetation structure. Biol. Conserv.**2009** *142* (8), 1693–1700.
32. Hoffmann, W.A.; Adasme, R.; Haridasan, M.; de Carvalho, M.T.; Geiger, E.L.; Pereira, M.A.; Gotsch, S.G.; Franco, A.C. Tree topkill, not mortality, governs the dynamics of savanna-forest boundaries under frequent fire in central Brazil. Ecology **2009**, *90* (5), 1326–1337.
33. Hoffmann, W.A.; Geiger, E.L.; Gotsch, S.G.; Rossatto, D.R.; Silva, L.C.; Lau, O.L.; Haridasan, M.; Franco, A.C. Ecological thresholds at the savanna-forest boundary: How plant traits, resources and fire govern the distribution of tropical biomes. Ecol. Lett. **2012**, *15* (7), 759–768.
34. Williams, R.J.; Myers, B.A.; Eamus, D.; Duff, G.A. Reproductive phenology of woody species in a north Australian tropical savanna. Biotropica **1999**, *31* (4), 626–636.
35. Woinarski, J.; Brock, C.; Fisher, A.; Milne, D.; Oliver, B. Response of birds and reptiles to fire regimes on pastoral land in the Victoria River District, Northern Territory. Rangeland J. **1999**, *21* (1), 24–38.
36. Hopcraft, J.G.C.; Olff, H.; Sinclair, A.R.E. Herbivores, resources and risks: Alternating regulation along primary environmental gradients in savannas. Trends Ecol. Evol. **2010** *25* (2), 119–128.
37. Pringle, R.M.; Young, T.P.; Rubenstein, D.I.; McCauley, D.J. Herbivore-initiated interaction cascades and their modulation by productivity in an African savanna. Proc. Natl. Acad. Sci. U. S. A. **2007**, *104* (1), 193–197.
38. Sinclair, A.R.E. Serengeti past and present. In *Serengeti II: Dynamics, Management, and Conservation of an Ecosystem*; University of Chicago Press: Chicago, USA, 1995; 3–30.
39. Asner, G.P.; Levick, S.R.; Kennedy-Bowdoin, T.; Knapp, D.E.; Emerson, R.; Jacobson, J.; Colgan, M.S.; Martin, R.E. Large-scale impacts of herbivores on the structural diversity of African savannas. Proc. Natl. Acad. Sci. **2009**, *106* (12), 4947–4952.
40. Dougill, A.J.; Thomas, D.S.G.; Heathwaite, A.L. Environmental change in the kalahari: Integrated land degradation studies for nonequilibrium dryland environments. Ann. Assoc. Am. Geogr. **1999**, *89* (3), 420–442.
41. Beyer, S.; Kinnear, A.; Hutley, L.B.; McGuinnessa, K.; Gibb, K. Assessing the relationship between fire and grazing on soil characteristics and mite communities in a semiarid savanna of northern Australia. Pedobiologia **2011**, *54* (3), 195–200.

42. House, J.I.; Archer, S.; Breshears, D.D.; Scholes, R.J. Conundrums in mixed woody-herbaceous plant systems. J. Biogeogr. **2003**, *30* (11), 1763–1777.
43. Gandar, M. Ecological Notes and Annotated Checklist of the Grasshoppers (Orthoptera-Acridoidea) of the Savanna Ecosystem Project Study Area, Nylsvley, SANSP Report 74; Cooperative Scientific Programmes: CSIR, Pretoria, South Africa, 1983.
44. Bond, W.J.; Midgley, G.F. Carbon dioxide and the uneasy interactions of trees and savannah grasses. Philos. Trans. Royal Soc. B Biol. Sci. **2012**, *367* (1588), 601–612.
45. Buitenwerf, R.; Bond, W.; Stevens, N.; Trollope, W.S.W. Increased tree densities in South African savannas: >50 years of data suggests CO_2 as a driver. Global Change Biol. **2012**, *18* (2), 675–684.
46. Bond, W.J.; Woodward, F.I.; Midgley, G.F. The global distribution of ecosystems in a world without fire. New Phytol. **2005**, *165* (2), 525–537.
47. Higgins, S.I.; Bond, W.J.; Trollope, W.S. Fire, resprouting and variability: A recipe for grass-tree coexistence in savanna. J. Ecol. **2000**, *88* (2), 213–229.
48. Walker, B.H.; Ludwig, D.; Holling, C.S.; Peterman, R.M. Stability of semi-arid savanna grazing systems. J. Ecol. **1981**, *69* (2), 473–498.
49. Belsky, A.J. Tree grass ratios in East-African savannas—A comparison of existing models. J. Biogeogr. **1990**, *17* (4–5), 483–489.
50. Eagleson, P.S.; Segarra, R.I. Water-limited equilibrium of savanna vegetation systems. Water Resour. Res. **1985**, *21* (10), 1483–1493.
51. Fernandez-Illescas, C.P.; Rodriguez-Iturbe, I. Hydrologically driven hierarchical competition-colonization models: The impact of interannual climate fluctuations. Ecol. Monogr. **2003**, 73 (2), 207–222.
52. Jeltsch, F.; Milton, S.J.; Dean, W.R.J.; van Rooyen, N. Tree spacing and coexistence in semiarid savannas. J. Ecol. **1996**, *84* (4), 583–595.
53. Holdo, R.M. Elephants, fire, and frost can determine community structure and composition in Kalahari woodlands. Ecol. Appl. **2007**, *17* (2), 558–568.
54. Zimmermann, J.; Higgins, S.I.; Grimm, V.; Hoffmann, J.; Linstadter, A. Grass mortality in semi-arid savanna: The role of fire, competition and self-shading. Perspect. Plant Ecol. Evol. Syst. **2010**, *12* (1), 1–8.
55. Sankaran, M.; Ratnam, J.; Hanan, N.P. Tree-grass coexistence in savannas revisited – insights from an examination of assumptions and mechanisms invoked in existing models. Ecol. Lett. **2004**, 7 (6), 480–490.
56. Higgins, S.I.; Bond, W.J.; Trollope, W.S.W. Fire, resprouting and variability: A recipe for grass-tree coexistence in savanna J. Ecol. **2000**, *88* (2), 213–229.
57. Lehmann, C.E.; Archibald, S.A.; Hoffmann, W.A.; Bond, W.J. Deciphering the distribution of the savanna biome. New Phytol. **2011**, *191* (1), 197–209.
58. Hanan, N.P.; Lehmann, C.E. *Tree-Grass Interactions in Savannas: Paradigms, Contradictions, and Conceptual Models*; CRC Press: Boca Raton, FL, 2011.
59. Hill, M.J.; Hanan, N.P. *Ecosystem Function in Savannas: Measurement and Modeling at Landscape to Global Scales*; Taylor & Francis: U.S., 2010; 39.
60. Murphy, B.P.; Bowman, D.M. What controls the distribution of tropical forest and savanna? Ecol. Lett. **2012**, *15* (7), 748–758.
61. Sala, O.E. Temperate grasslands. In *Global Biodiversity in a Changing Environment;* Springer: New York, 2001; 121–137.
62. Freeman, C. The flora of Konza Prairie: A historical review and contemporary patterns. In *Grassland Dynamics: LongTerm Ecological Research in Tallgrass Prairie;* Oxford University Press: New York, 1998; 69–80.
63. Golluscio, R.A.; Sala, O.E. Plant functional types and ecological strategies in Patagonian forbs. J. Veg. Sci. **1993**, *4* (6), 839–846.

64. Groombridge, B. *Global Biodiversity: Status of the Earth's Living Resources*; Chapman & Hall: Cambridge, England, 1992.
65. Vrba, E.S. Mammal evolution in the African Neogene and a new look at the Great American Interchange. In *Biological Relationships between Africa and South America;* Yale University Press: New Heaven, USA, 1993; 393–432.
66. Sims, P.L.; Singh, J. The structure and function of ten western North American grasslands: III. Net primary production, turnover and efficiencies of energy capture and water use. J. Ecol. **1978,** *66* (2), 573–597.
67. Knapp, A.K.; Briggs, J.M.; Hartnett, D.C.; Collins, S.L. *Grassland Dynamics: Long-Term Ecological Research in Tallgrass Prairie*; Oxford University Press: New York, 1998.
68. Roy, J.; Mooney, H.A.; Saugier, B. *Terrestrial Global Productivity;* Academic Press: London, 2001.
69. Heidenreich, B. *What Are Global Temperate Grasslands Worth? A Case for Their Protection;* Temperate Grasslands Conservation Initiative: Vancouver, BC, Canada, 2009; 21.
70. Van Auken, O. Shrub invasions of North American semiarid grasslands. Annu. Rev. Ecol. Syst. **2000,** *31,* 197–215.
71. Silva, J.F. Biodiversity and stability in tropical savannas, In *Biodiversity and Savanna Ecosystem Processes;* Springer: New York, 1996; 161–171.
72. Setterfield, S.A.; Rossiter-Rachor, N.A.; Hutley, L.B.; Douglas, M.M.; Williams, R.J. Biodiversity Research: Turning up the heat: The impacts of *Andropogon gayanus* (gamba grass) invasion on fire behaviour in northern Australian savannas. Divers. Distrib. **2010,** *16* (5), 854–861.
73. Briggs, J.M.; Knapp, A.K.; Blair, J.M.; Heisler, J.L.; Lett, M.S.; Hoch, G.A.; McCarron, J.K. An ecosystem in transition: Causes and consequences of the conversion of mesic grassland to shrubland. Bioscience **2005,** *55* (3), 243–254.
74. Derksen, D.A.; Anderson, R.L.; Blackshaw, R.E.; Maxwell, B. Weed dynamics and management strategies for cropping systems in the northern Great Plains. Agron. J. **2002,** *94* (2), 174–185.
75. Sinden, J.; Jones, R.; Hester, S.; Odom, D.; Kalisch, C.; James, R.; Cacho, O. *The Economic Impact of Weeds in Australia*, Report to the CRC for Australian Weed Management, Technical Series; University of Adelaide Press: Australia, 2004; 8.
76. Archer, S.; Vavra, M.; Laycock, W.; Pieper, R.D. Woody plant encroachment into southwestern grasslands and savannas: Rates, patterns and proximate causes. In *Ecological Implications of Livestock Herbivory in the West;* Society for Range Management: Denver, USA, 1994.
77. Fensham, R.; Fairfax, R.; Archer, S. Rainfall, land use and woody vegetation cover change in semi-arid Australian savanna. J. Ecol. **2005,** *93* (3), 596–606.
78. Bond, W.J.; Midgley, G.F. A proposed CO_2-controlled mechanism of woody plant invasion in grasslands and savannas. Global Change Biol. **2000,** *6* (8), 865–869.
79. Tews, J.; Jeltsch, F. Modelling the impact of climate change on woody plant population dynamics in South African savanna. BMC Ecol. **2004,** *4* (1), 17.
80. Tews, J.; Esther, A.; Milton, S.J.; Jeltsch, F. Linking a population model with an ecosystem model: Assessing the impact of land use and climate change on savanna shrub cover dynamics. Ecol. Model. **2006,** *195* (3–4), 219–228.

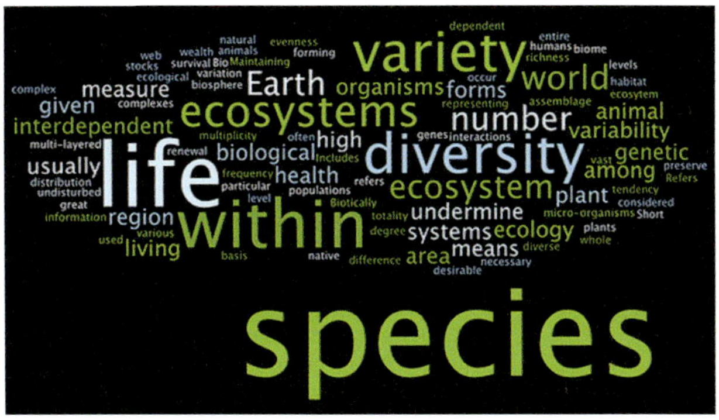

FIGURE 4.1 This word cloud results from a Google search for the definition of biodiversity. (http://thepimmgroup.org/1022/defining-biodiversity-toward-a-consensus/.)

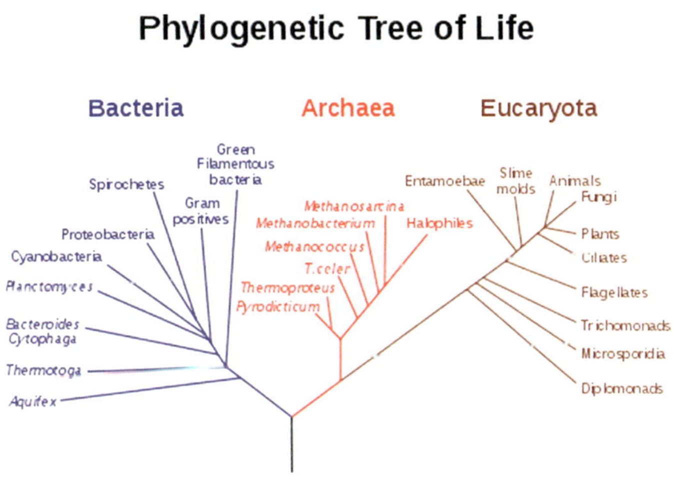

FIGURE 4.2 Phylogenetic tree based on nonribososmal (r)RNA sequences groups all life into three Domains Bacteria, Archaea, and Eucarya, which are above Kingdoms in the taxonomic classification system.
Source: Adapted from Woese, et al.[4]

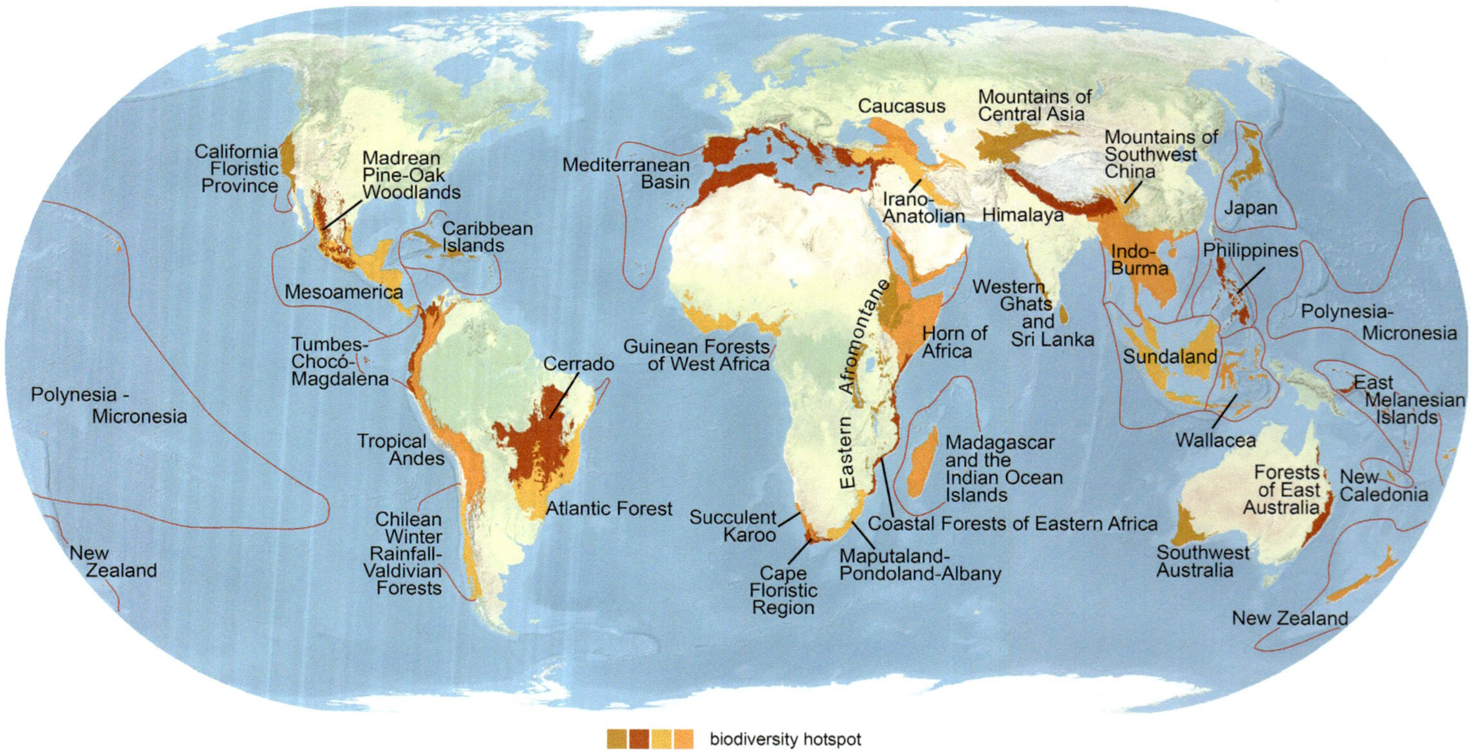

FIGURE 10.1 Map of global biodiversity hotspots.
Source: Courtesy of Conservation International.[7]

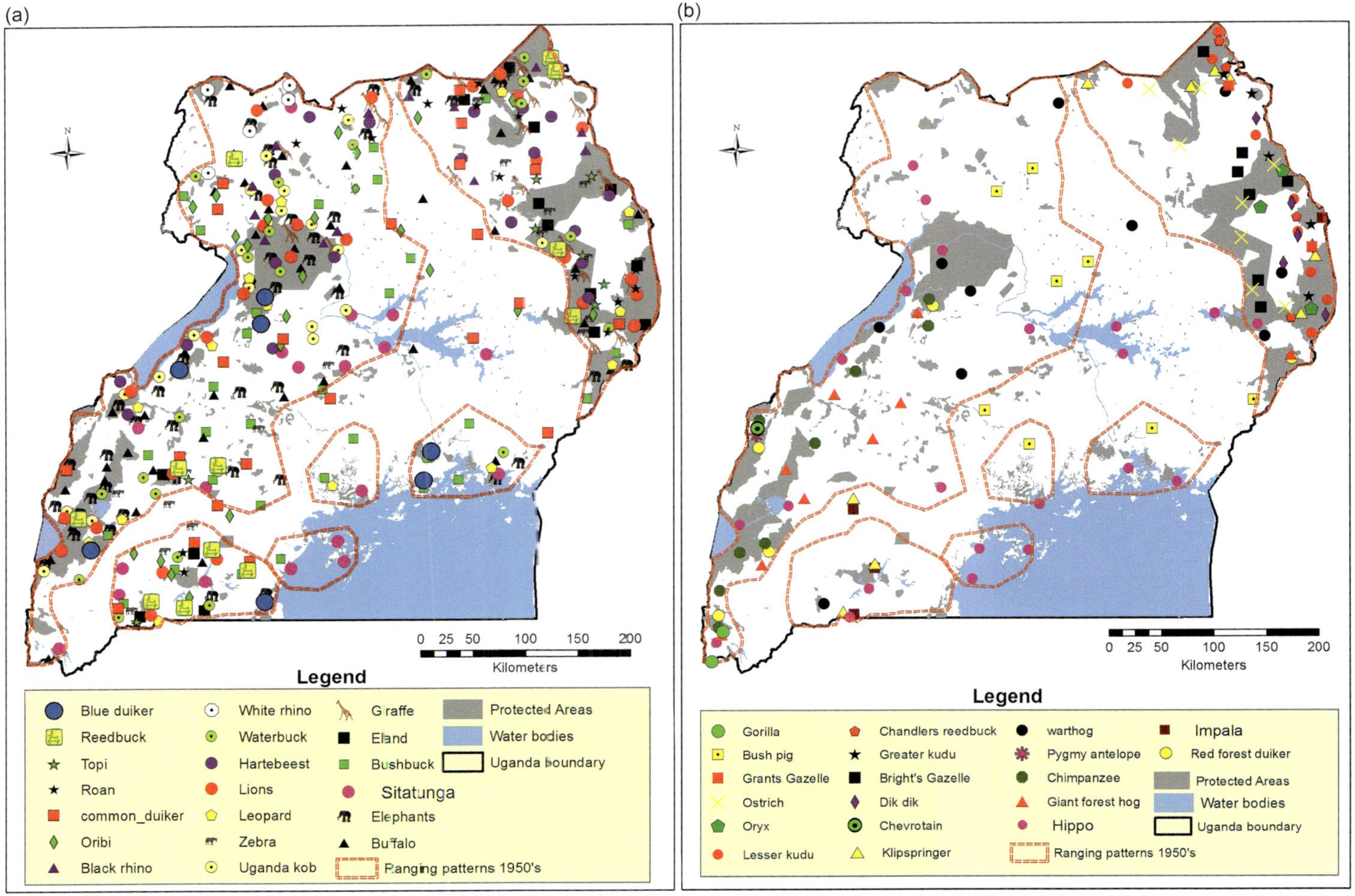

FIGURE 14.4 Distribution of wildlife in Uganda in the 1950s (a) and the current protected area system (b).

FIGURE 19.1 The Lesser long-nose bat, *Leptonycteris curasoae yerbabuenae*, feeding at flowers of *Agave palmeri palmeri* blossom during the night in the Sonoran Desert. *L. curasoae yerbabuenae* is also an essential pollinator of saguaro cactus *(Carnegiea gigantea)* that defines the Sonoran ecosystem.
Source: Photo credit Steven Buchmann.

FIGURE 19.2 A male *Habropoda pallida* emerging from a ground nest, which is a network of tunnels beneath the surface that house brood chambers. Ground nesting aids in the aeration of soil and nutrient cycling.
Source: Photo credit Steven Buchmann.

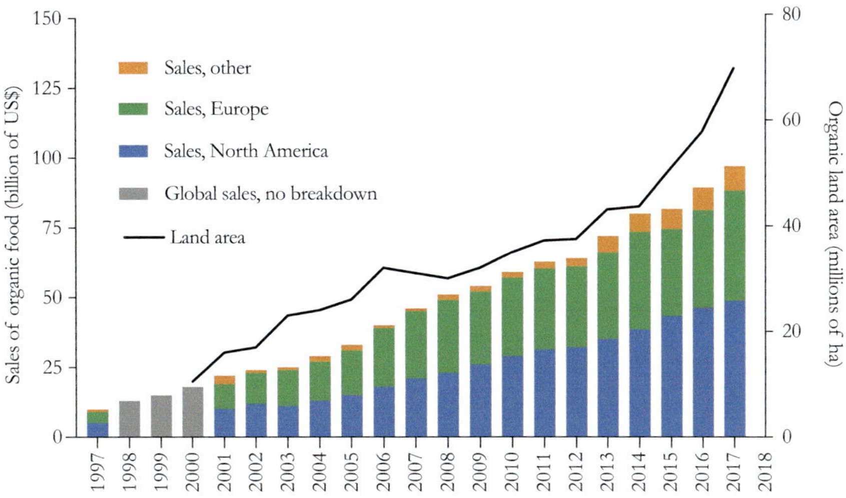

FIGURE 29.1 Annual global market for organic foods and land area of organic production. Sales of organic food in nominal billions of US dollars (bars) are broken down by contributions from North America (blue), Europe (green), and the rest of the world (orange). Total global land area under organic production (black line) is in millions of hectares [2,60–62]. European and North American contributions were not available for the years 1998 through 2000, shown as gray bars.

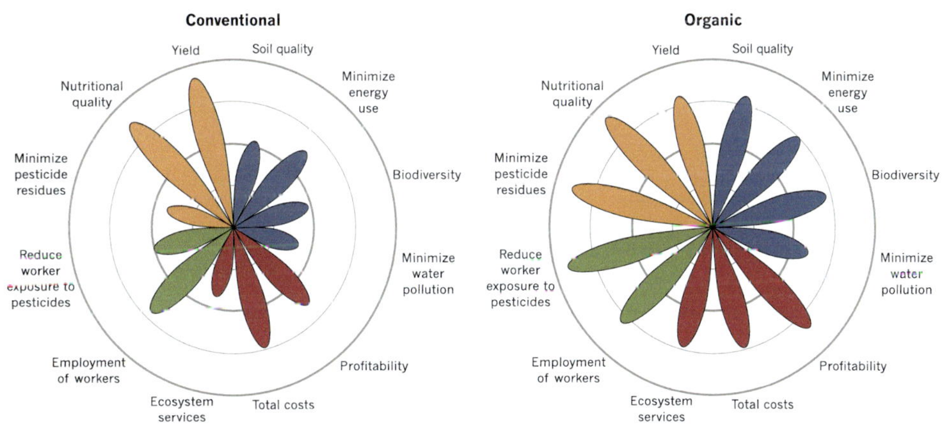

FIGURE 29.2 Assessment of organic farming relative to conventional farming in the four major areas of sustainability (as taken from Reganold & Wachter [2]). Lengths of the 12 flower petals are qualitatively based on the studies discussed in this entry and in Reganold & Wachter [2], and indicate the level of performance of specific sustainability metrics relative to the four circles representing 25%, 50%, 75%, and 100%. Orange petals represent areas of production; blue petals represent areas of environmental sustainability; red petals represent areas of economic sustainability; and green petals represent areas of well-being.

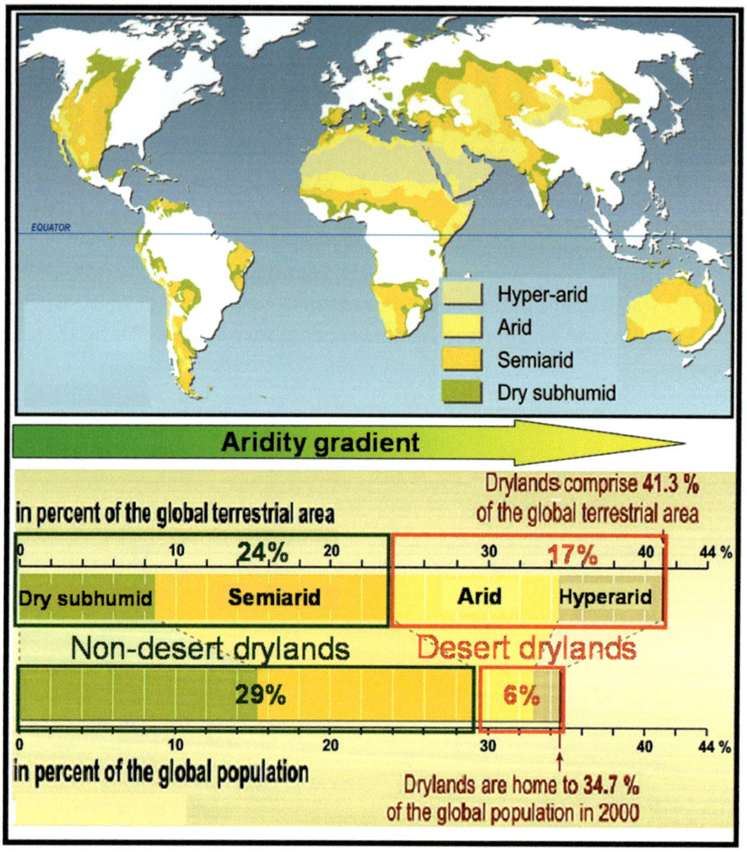

FIGURE 31.1 The global drylands. Geographic layout of the four sections of the aridity gradient, their percent share of global land, their human populations, and their percent share of global population.
Source: Adapted from Adeel et al.[1]

FIGURE 32.1 Tiger on abandoned logging road photographed by photo trap, Tembat Forest Reserve, Terengganu, Peninsular Malaysia.
Source: Photo by Rimba/Reuben Clements.

FIGURE 32.2 Some populations of the serotinous South African shrub *Protea repens* exhibit mostly closed cones (a), while others exhibit mostly open cones (b), revealing an underlying intraspecific genetic diversity that could enhance the resilience of *P. repens* to global environmental changes. (Source: Photos by Roland de Gouvenain.)

FIGURE 34.1 Representatives of important algal and plant lineages: (**a**) charophytes alga Coleochaete, a member of the Coleochaetales; (**b**) charophytes alga Chara, a member of the Charales; and (**c**) Anthoceros, a hornwort; (**d**) Marchantia, a liverwort.

FIGURE 37.1 Experiment design and setup in the Changbai Mountain, Northeast China.

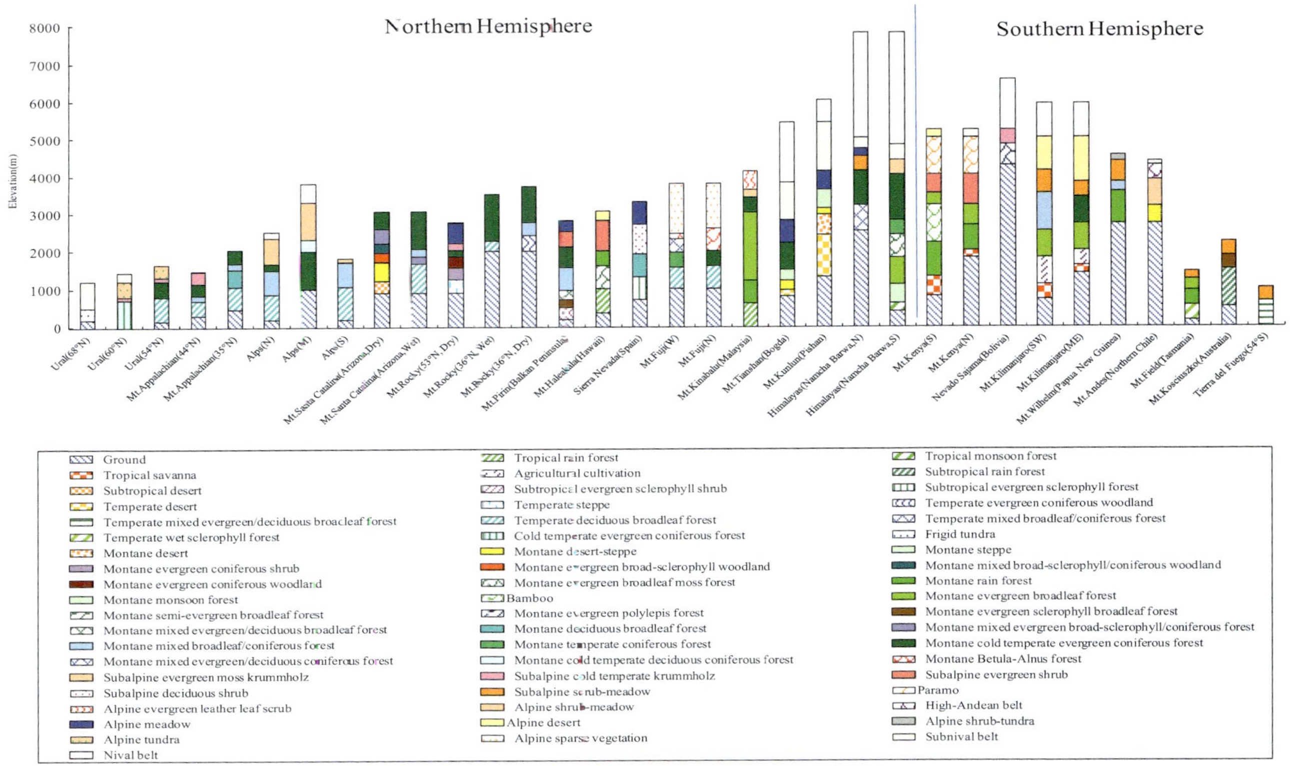

FIGURE 41.1 Representative spectra of altitudinal belts of the global main mountain ranges.

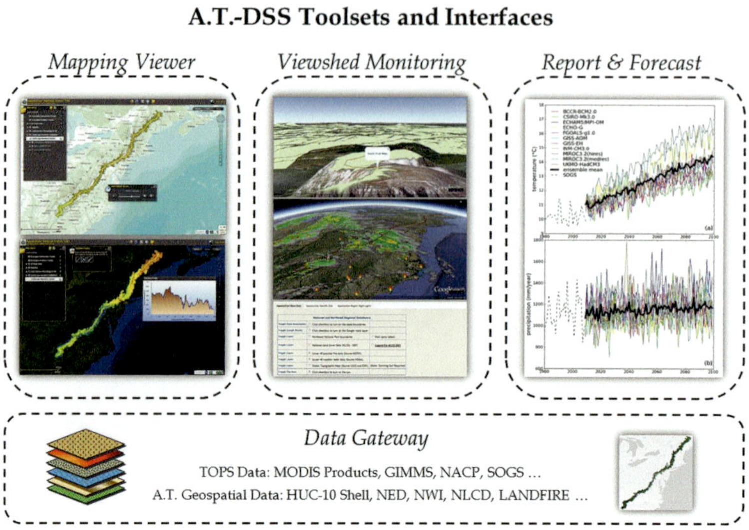

FIGURE 43.1 The toolsets and interfaces were organized by their primary functions, namely, the Mapping Viewer, Viewshed Monitoring, Report & Forecast, and Data Gateway modules (www.edc.uri/edu/ATMT-DSS).

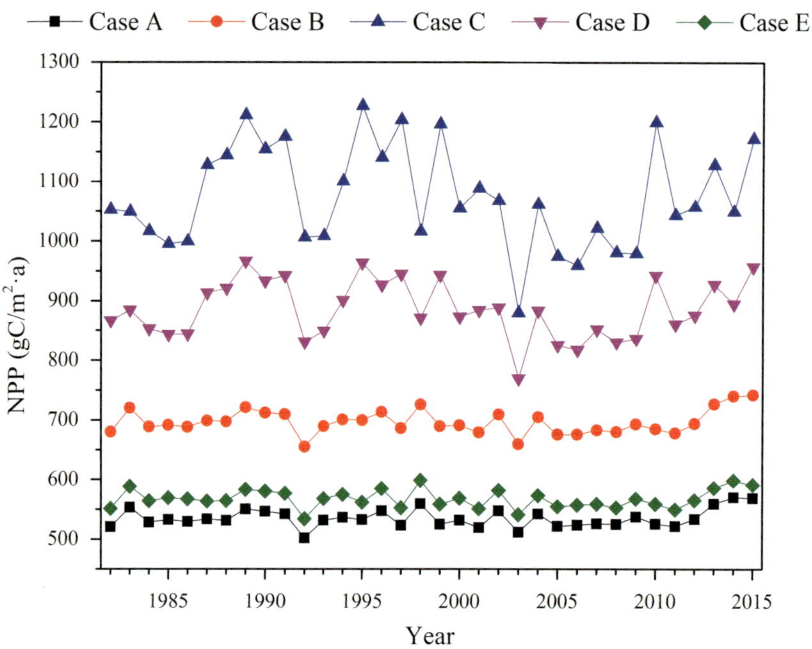

FIGURE 44.4 Annual NPP value for various cases from 1982 to 2015 in Northeast China.

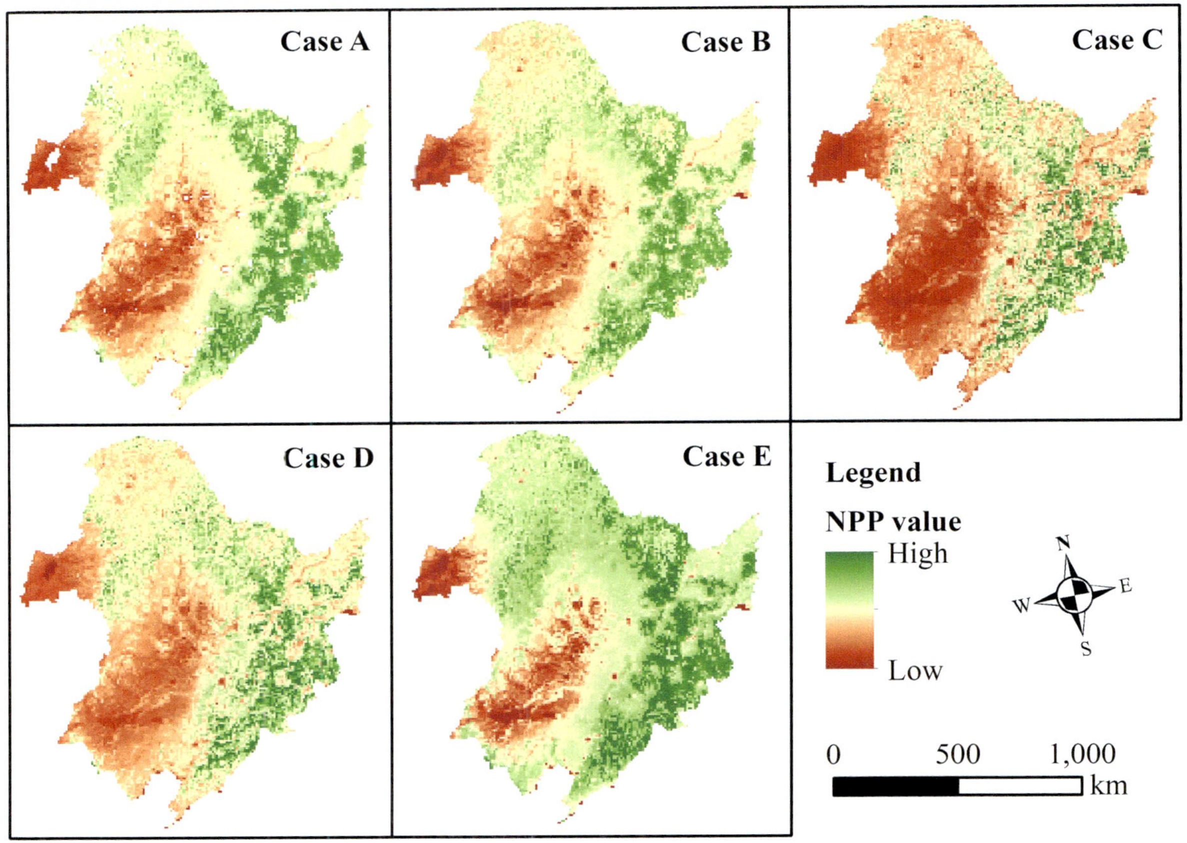

FIGURE 44.5 Spatial patterns of NPP estimations based on various FPAR cases in Northeast China.

FIGURE 45.1 (a) Balata leaf, (b) Populus ussuriensis leaf, (c) Apricot leaf, (d) Red twig dogwood leaf, and (e) Grape leaf; the main vein was parallel to the incident direction in the measurements.

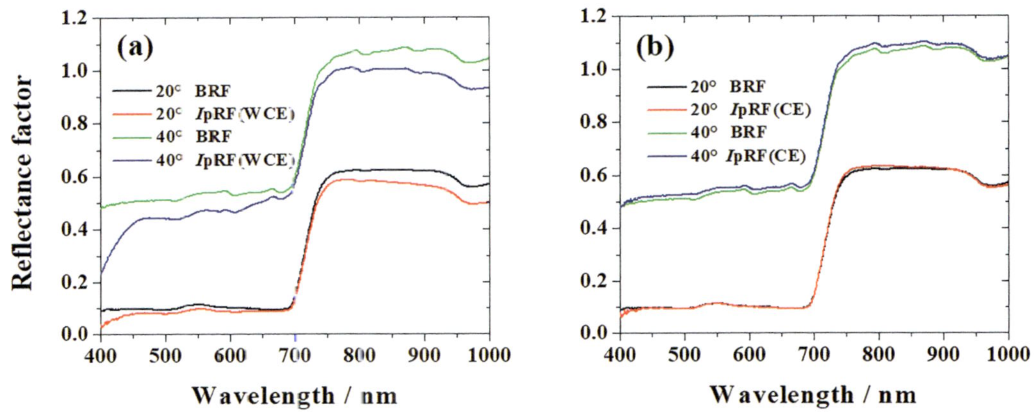

FIGURE 45.2 Comparison between the BRF and IpRF of the Balata leaf in the forward scattering directions in the principal plane: (a) without considering the extinction (WCE) of the polarizer, and (b) considering the extinction (CE) of the polarizer, the illumination zenith angle was 40°.

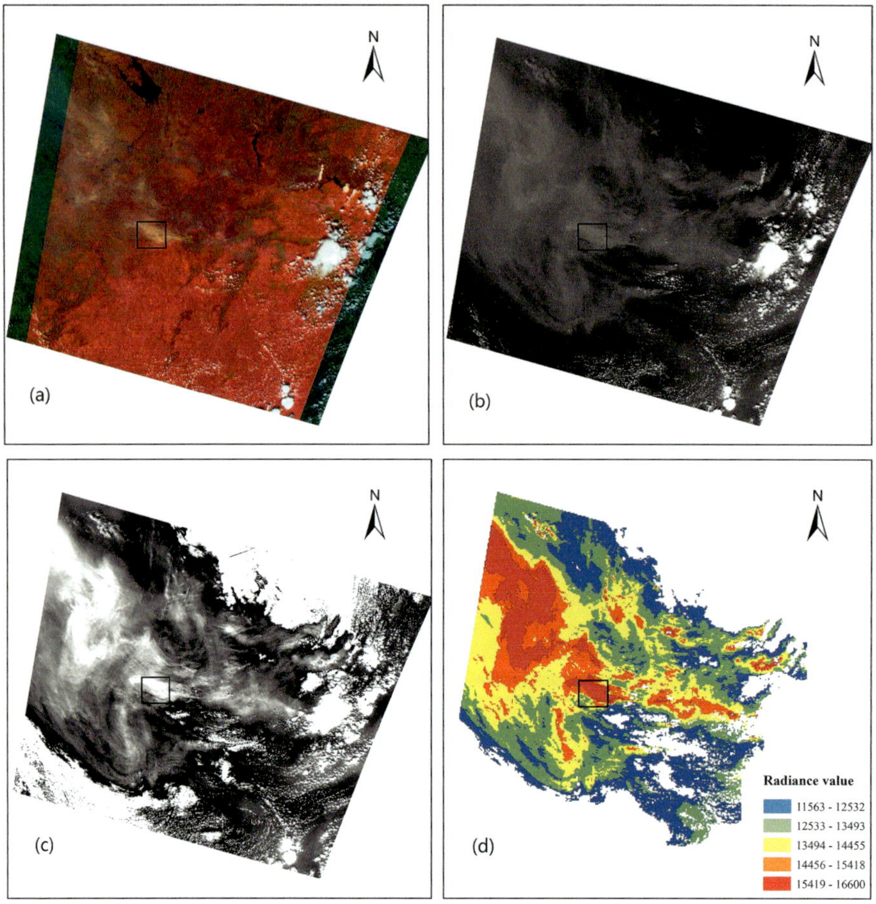

FIGURE 46.1 Example to identify smoke plumes on August 8, 2010. CAI image with band 431 (a), CAI image with band 1 (b), smoke plume areas identified by using thresholds of band 1 (c), and a subset of smoke plumes using ROI (d) (Guo et al. 2017).

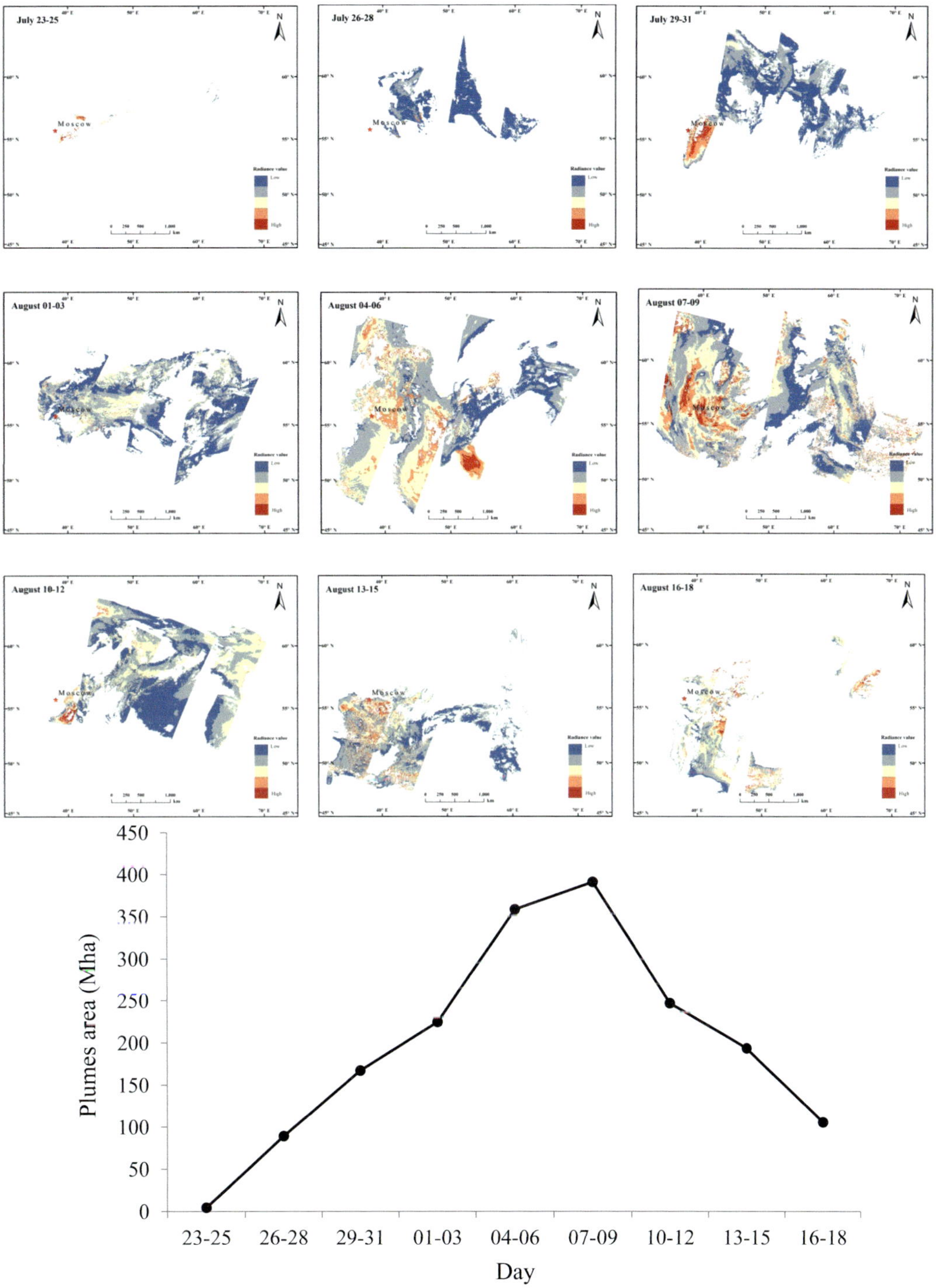

FIGURE 46.2 Three-day smoke plume areas in the study region. The preceding nine figures show the distribution of three-day smoke plumes, and the last figure shows the smoke plume area in million hectares (23–25 of *X*-axis means July 23–25 and 1–3 means August 1–3). Here, the three-day description means the sum of 3 days but without overlapping regions (Guo et al. 2017).

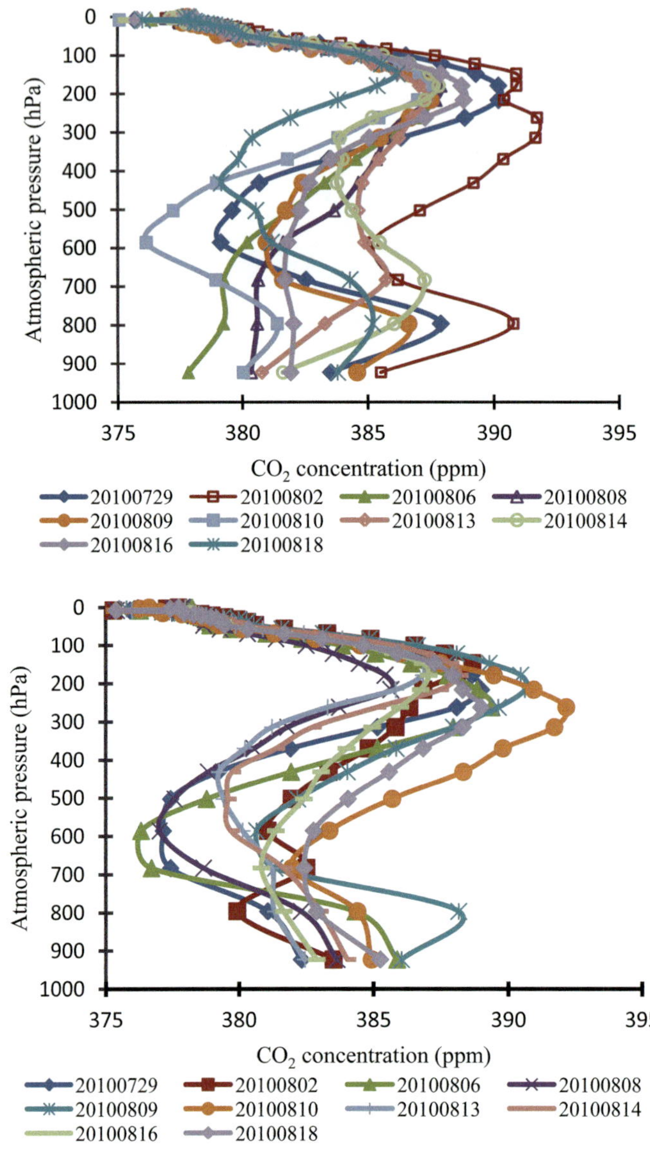

FIGURE 46.4 CO_2 concentration changes against atmospheric pressure. FTS TIR points were covered by smoke plumes (a) and in clear sky (b) (Guo et al. 2017).

III

Ecological Processes

29

Agriculture: Organic

John P. Reganold
Washington State University

Jonathan M. Wachter
Marin Agricultural Land Trust

Introduction 251
Organic Agricultural Practices 252
History and Extent of Organic Agriculture 253
Organic Certification 253
The Sustainability of Organic Agriculture 254
Conclusions 256
References 256

Introduction

During the past two decades, organic farming became one of the fastest growing segments of agriculture all over the world (Figure 29.1). Increasing demands and expanding supermarket involvement with organic foods in North America and Europe have created demand for more types of organic food, including tropical and out-of-season foods, and opened the market for growing and exporting organic foods in Asia, Africa, and Latin America [1]. In 2017, the US$97 billion global market for organic foods was dominated by the United States (47% of the global market) and the European Union

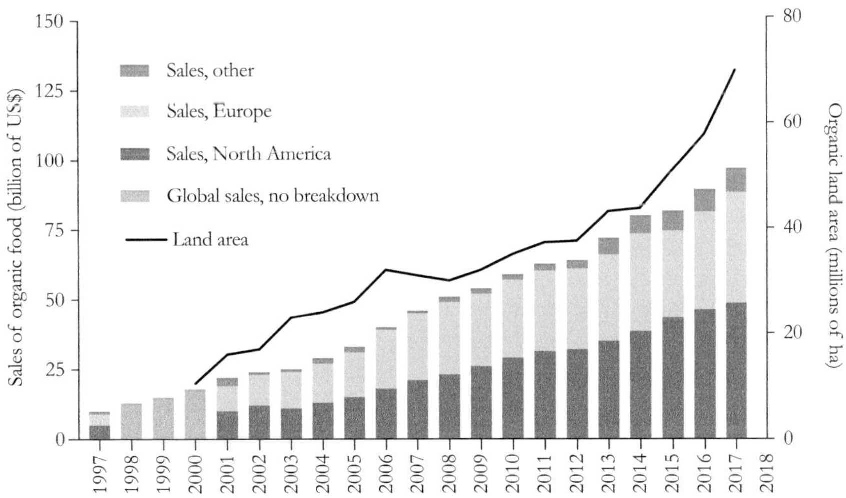

FIGURE 29.1 **(See color insert.)** Annual global market for organic foods and land area of organic production. Sales of organic food in nominal billions of US dollars (bars) are broken down by contributions from North America (blue), Europe (green), and the rest of the world (orange). Total global land area under organic production (black line) is in millions of hectares [2,60–62]. European and North American contributions were not available for the years 1998 through 2000, shown as gray bars.

(37%), followed by China (8%) [1]. As a result of organic agriculture being a quickly expanding global market force, more arable land, research funding, and research test sites are being devoted to organic farming worldwide.

Organic farming systems do not represent a return to the past. Organic agriculture, sometimes called biological or ecological agriculture, combines traditional conservation-minded farming methods with modern farming technologies. It virtually excludes such conventional inputs as synthetic fertilizers, pesticides, and pharmaceuticals. Instead, it uses naturally derived chemicals or products as defined by organic certification programs, and it emphasizes improving the soil with compost additions and animal and green manures, controlling pests naturally, rotating crops, and diversifying crops and livestock [2]. Organic farmers use modern equipment, certified seed, soil and water conservation practices, improved crop varieties, and the latest innovations in feeding and handling livestock. Organic farming systems range from strict closed-cycle systems that go beyond organic certification guidelines by limiting external inputs as much as possible, to more standard systems that simply follow organic certification guidelines.

Organic agriculture is an alternative to conventional agriculture that seeks socioeconomic and environmental sustainability on local and global scales. Organic farming systems have been shown to be energy-efficient, environmentally sound, productive, stable, and tending toward long-term sustainability [2,3]. This entry discusses organic agricultural practices, the history and extent of organic agriculture, organic certification, and the sustainability of organic agriculture.

Organic Agricultural Practices

Organic farming systems rely on ecologically based practices. A central component of organic farming systems is the rotation of crops—a planned succession of various crops grown on one field. Rotating crops often results in increased yields when compared to monocultures (growing the same crop on the same field year after year). In most cases, monocultures can be perpetuated only by adding large amounts of fertilizers and pesticides. Rotating crops provides better weed and insect control, lower disease pressure, more efficient nutrient cycling, and other benefits. Alternating two crops, such as corn and soybeans, is considered a simple rotation. More complex rotations require three or more crops and often a four- to seven-year (or more) cycle to complete. In growing a more diversified group of crops in rotation, a farmer is less affected by price fluctuations of one or two crops, and may experience more year-to-year financial stability. There are disadvantages, too, however. They include the need for more equipment to grow a number of different crops, the reduction of acreage planted with government-supported crops, and the need for more time and information to manage a greater number of crops.

Maintaining healthy soils by regularly adding crop residues, manures, and other organic materials to the soil is another central feature of organic farming. Organic matter improves soil structure, increases its water storage capacity, enhances fertility, promotes soil biological activity, and contributes to the tilth, or physical condition, of the soil. The better the tilth, the more easily the soil can be tilled or direct-seeded, and the easier it is for seedlings to emerge and for roots to extend downward. Water readily infiltrates soils with good tilth, thereby minimizing surface runoff and soil erosion. Organic materials also feed earthworms and soil microbes, which contribute to soil aeration and structure, and plant health.

The main sources of plant nutrients in organic farming systems are nitrogen-fixing green manures, composted animal manures and plant materials, and plant residues. A green manure crop is typically a grass or legume that is plowed into the soil or surface-mulched at the end of a growing season to enhance soil productivity and tilth. Green manures help to control weeds, insect pests, and soil erosion, while also providing forage for livestock and cover for wildlife. Fertility on organic farms is usually biologically mediated, depending on macro- and microorganisms to breakdown organic materials and make nutrients available to plants.

Weed management is often cited as a primary challenge for organic farmers. Crop rotations are designed to diversify resource competition against weeds, and to diversify timing and type of operations in which weeds can be managed. Dense green manures, cover crops, and pasture phases of a rotation can compete against weeds, as well as providing a heavy mulch to physically suppress emergence. Direct methods of weed management, such as shallow tillage, solarization, and hoeing, are also common.

Organic farmers use disease-resistant crop varieties and biological controls (such as natural predators or parasites that keep pest populations below injurious levels). They also select tillage methods, planting times, crop rotations, and plant-residue management practices to optimize the environment for beneficial insects that control pest species, or to deprive pests of a habitat. Organically certified pesticides, usually used as a last resort to control insects, diseases, and weeds, are applied when pests are most vulnerable or when any beneficial species and natural predators are least likely to be harmed.

History and Extent of Organic Agriculture

European organic agriculture emerged in 1924 when Rudolf Steiner held his course on biodynamic agriculture. In the 1930s and 1940s, organic agriculture was developed in Britain by Lady Eve Balfour and Sir Albert Howard, in Switzerland by Hans Müller, in the United States by J. I. Rodale, and in Japan by Masanobu Fukuoka. Since the beginning of the 1990s, development of organic agriculture in Europe has been supported by government subsidies. In many other countries of the world, organic agriculture was established because of the growing demand for organic products in Europe, the United States, and Japan. Today, organic farming is gaining increasing acceptance by the public at large.

Organic agriculture is practiced in almost all countries of the world, and its share of agricultural land and farms is growing everywhere. In 2017, 69.8 million hectares were managed organically worldwide [1]. The top three countries with the most organic agricultural land were Australia (35.6 million hectares), Argentina (3.4 million hectares), and China (3 million hectares). By region, Oceania held 51% of the world's organic land, followed by Europe (21%), Latin America (11%), and Asia (9%). In 2017, of the 2.9 million organic producers, India had the highest number with 835,000 producers, followed by Uganda with 210,352 and Mexico with 210,000.

Organic Certification

Organic certification provides verification that products are indeed produced according to certain standards. For consumers who want to buy organic foods, the organic certification standards ensure that they can be confident in knowing what they are buying. For farmers, these standards give guidelines for organic management practices and allow products to capture high-value markets and premium prices. For the organic products industry, these standards provide an important marketing tool to help boost trade in organic products.

"Certified organic" means that agricultural products have been grown and processed according to the specific standards of various national, state, or private certification organizations. Certifying agents review applications from farmers and processors for certification eligibility, and qualified inspectors conduct annual onsite inspections of their operations. Inspectors talk with producers and observe their production or processing practices to determine if they are in compliance with organic standards. Organic certification standards detail the methods, practices, and substances that can be used in producing and handling organic crops and livestock, as well as processed products. For example, the U.S. Department of Agriculture's organic standards went into effect in 2002, establishing national standards that food labeled "organic" must meet, whether it is grown in the United States or imported from other countries [4]. The European Union developed a regulatory framework to oversee organic practices across all of Europe beginning in 1991. The International Federation of Organic Agriculture Movements (IFOAM) established the IFOAM Accreditation Programme to provide international equivalency of certification bodies worldwide, effective in 2000 [5].

While requirements vary slightly between agencies, they all prohibit the use of synthetic pesticides and fertilizers, the use of growth hormones and antibiotics in livestock (with some exceptions for the sake of humane treatment), and the use of genetically modified organisms. Critics of organic certification argue that it encourages farmers to follow the minimum standards for organic certification and undervalue the agroecological elements of organic farming, that the cost of certification gives favor to large farms over small operations, and that national and global certification schemes do not account for important local and regional differences in farming systems [2,6].

The Sustainability of Organic Agriculture

Organic agriculture addresses many serious problems afflicting world food production: high energy costs, groundwater contamination, soil erosion, depletion of fossil resources, low farm incomes, and risks to human health and wildlife habitats. However, just because a farm is organic does not mean that it is sustainable. For any farm to be sustainable, whether it be organic or conventional, it must (i) produce adequate yields of good quality, (ii) enhance the natural-resource base and environment, (iii) be profitable, and (iv) contribute to the well-being of farmers, farm workers, and rural communities [7]. So, if an organic farm produces high yields of nutritious food and is environmentally friendly and energy-efficient, but is not profitable, then it is not sustainable. Likewise, a conventional farm that meets all the sustainability criteria but pollutes a nearby river with sediment because of soil erosion is not sustainable.

The performance of organic farming systems in the context of various sustainability metrics (Figure 29.2) indicates that they better balance multiple sustainability goals than their conventional counterparts. Although organic farming systems produce lower yields compared with conventional agriculture and thus have higher land use for similar yields, they are on average more profitable and environmentally friendly. Organic systems also deliver equal or more nutritious foods with less to no pesticide residues. In addition, initial evidence indicates that organic agriculture is better at enhancing the delivery of ecosystem services, other than yield, as well as some social sustainability benefits.

Several reviews and meta-analyses of numerous organic/conventional yield comparison studies show that yields from organic farms are on average 8%–25% lower than those from conventional farms [8–12]. However, with certain crops, growing conditions, and management practices, organic systems are able to close the yield gap between organic and conventional systems. For example, de Ponti et al. [10].

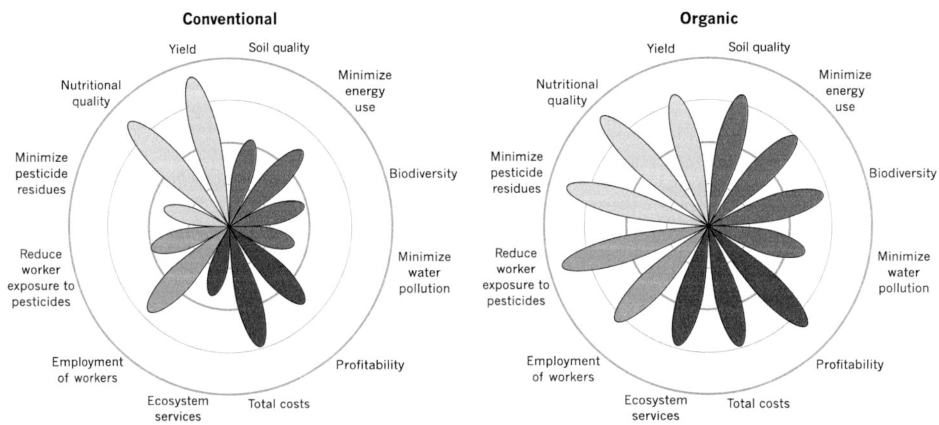

FIGURE 29.2 **(See color insert.)** Assessment of organic farming relative to conventional farming in the four major areas of sustainability (as taken from Reganold & Wachter [2]). Lengths of the 12 flower petals are qualitatively based on the studies discussed in this entry and in Reganold & Wachter [2], and indicate the level of performance of specific sustainability metrics relative to the four circles representing 25%, 50%, 75%, and 100%. Orange petals represent areas of production; blue petals represent areas of environmental sustainability; red petals represent areas of economic sustainability; and green petals represent areas of well-being.

found that the best yielding organically grown crops are rice (6% lower than conventional), soybeans (8% lower), corn (11% lower), and grass-clover (11% lower). Organic farming systems still require greater land use to produce similar yields as their conventional counterparts, a big concern by some critics given the task of feeding a growing world population without further agricultural expansion [13–15]. Conversely, some contend that the environmental advantages of organic agriculture far outweigh the lower yields, and that increasing research and breeding resources for organic systems would reduce the yield gap [9,10]. Others suggest that multifunctional farming systems, such as organic, coupled with more plant-based diets and reduced food waste, are necessary elements of a more sustainable food system [2,16].

Research has found that organic foods, compared with their conventional counterparts, have significantly less to no pesticide residues [17–20]. Thirteen of 16 reviews or meta-analyses of numerous studies in the scientific literature [20–32] have found some evidence of organic food being more nutritious (for instance, having higher concentrations of vitamin C, total antioxidants, and total omega-3 fatty acids) than conventionally grown food. The other studies [19,33,34] concluded that there were no consistent nutritional differences between organic and conventional foods. However, one of these three studies found that conventional chicken and pork had a 33% higher risk for contamination with antibiotic-resistant bacteria compared with organic alternatives [19].

Reviews and meta-analyses of comparison studies on environmental quality generally support the perception that organic farming systems are more environmentally friendly than conventional farming systems [35–49]. For example, such aggregate studies have found that organic farming systems have greater soil carbon levels, enhanced soil quality, and less soil erosion compared with conventional systems [35,37,38,41–44]. In addition, organic farms generally have more plant diversity, greater faunal diversity (insects, soil fauna and microbes, and birds), and often more habitat and landscape diversity than conventional farms [35–41,43–45,47]. Most functional groups, such as herbivores, pollinators, predators, and producers (plants), are more diverse in organic farming systems [43,45,47,49].

As organic agriculture uses virtually no synthetic pesticides, there is little to no risk of synthetic pesticide pollution of ground and surface waters [2,35]. With respect to nitrate and phosphorous leaching and greenhouse gas emissions, organic farming systems score better than conventional farming when expressed per production area; however, given the lower land-use efficiency of organic farming in developed countries, this positive effect is less pronounced or not present at all when it is expressed per unit product [14,35,38,43,46,48]. Organic systems have been shown to usually be more energy efficient—energy consumed per hectare and per product—than their conventional counterparts [14,38,41,44].

In the only meta-analysis of the financial performance of organic and conventional agriculture, Crowder and Reganold [50] examined a global dataset spanning 55 crops grown on five continents over a 40-year period. With actual price premiums (higher prices awarded to organic foods), organic agriculture was significantly more profitable (22%–35%) and had higher benefit/cost ratios (20%–24%) than conventional agriculture. Price premiums for organically grown crops averaged between 29% and 32%. Total costs (variable costs plus fixed costs = total costs) were not significantly different, and variable costs and fixed costs did not differ significantly between organic and conventional systems [50]. Labor costs (part of variable costs) were significantly higher (7%–13%) with organic farming systems.

Studies used in the economic meta-analysis [50] did not account for environmental costs (negative externalities) or ecosystem services (environmental benefits from good farming practices), which have been shown to favor organic agriculture. That is, environmental costs tend to be lower and benefits higher in organic agriculture. For example, studies generally show that conventional practices decrease the ability of farms to provide some economically significant ecosystem services relative to organic practices [51–53]. For consumers who believe that organic farming is more environmentally friendly, organic premiums may serve as a stand-in for the monetary value of such costs and benefits.

We know relatively little in how well organic, conventional, and other farming systems are performing in the social well-being area; for example, few organic/conventional comparison studies have been done examining farmer resilience, farmer wages and benefits, and labor conditions [2,3]. Available data indicate that both organic and conventional farming systems need to make significant progress to meet social sustainability goals [2,3,7]. However, organic farming has been shown to have some sociocultural strengths, such as increased social interactions between farmers and consumers [54,55], greater employment of farm workers and cooperation among farmers [56,57], and reduced exposure of farm workers to pesticides and other chemicals [58,59].

Conclusions

If we want to feed a growing world population, producing adequate yields is vital. But, as described above, sufficient productivity is only one of four main goals that must be met for agriculture to be sustainable. The other three are enhancing the natural-resource base and environment, making farming financially viable, and contributing to the well-being of farmers and their communities. Mainstream conventional farming systems have provided growing supplies of food and other products, but often at the expense of the other three main sustainability goals. The performance of organic farming systems in the context of sustainability metrics indicates that they better balance multiple sustainability goals than their conventional counterparts (Figure 29.2). Although organic farming systems produce lower yields compared with conventional agriculture, they are more profitable and environmentally friendly, and deliver equal or more nutritious foods with less to no pesticide residues. In addition, initial evidence indicates that organic agriculture is better at enhancing the delivery of ecosystem services, other than yield, as well as some social sustainability benefits.

Although organic agriculture has an untapped potential role in global food and ecosystem security, no one agricultural system alone will safely feed the planet. Rather, a blend of innovative farming systems, including agroforestry, integrated farming, conservation agriculture, mixed crop/livestock, and still undiscovered systems, will be needed for future global food and ecosystem security. Organic agriculture, now occupying about 1.4% of global agricultural land, can play a much larger role in feeding the world, especially a world facing potential resource limits and an unpredictable climate. The challenge facing policymakers is to create an enabling environment for scaling-up organic and other innovative farming systems to move toward truly sustainable production systems. To make this happen will require mobilizing the full arsenal of effective policies, scientific and socioeconomic advances, farmer ingenuity, and public engagement.

References

1. Willer, H.; Lernoud, J., Eds. *The World of Organic Agriculture: Statistics and Emerging Trends 2019*; Research Institute of Organic Agriculture (FiBL): Frick, and International Federation of Organic Agriculture Movements (IFOAM): Bonn, 2019.
2. Reganold, J.P.; Wachter, J.M. Organic agriculture in the twenty-first century. *Nat. Plants*. 2016, 2 (2), e15221. doi: 10.1038/NPLANTS.2015.221
3. Seufert, V.; Ramankutty, N. Many shades of gray—the context-dependent performance of organic agriculture. *Sci. Adv.* 2017, 3 (3), e1602638. doi: 10.1126/sciadv.1602638
4. www.ams.usda.gov/nop/ [accessed June 2019].
5. www.ifoam.org [accessed June 2019].
6. Schmid, O. Development of standards for organic farming. In *Organic Farming: An International History*; Lockeretz, W., Ed.; CABI Publishing: Oxfordshire, 2007; 152–174.
7. National Research Council. *Toward Sustainable Agricultural Systems in the 21st Century*. The National Academies Press: Washington, DC., 2010.
8. Stanhill, G. The comparative productivity of organic agriculture. *Agr. Ecosyst. Environ.* 1990, 30, 1–26.
9. Badgley, C.; Moghtader, J.; Eileen Quintero, E.; Zakem, E.; Chappell, M.J.; Avilés-Vázquez, K.; Samulon, A.; Perfecto, I. Organic agriculture and the global food supply. *Renew. Agr. Food Syst.* 2007, 22, 86–108.

10. de Ponti, T.; Rijk, B.; van Ittersum, M.K. The crop yield gap between organic and conventional agriculture. *Agr. Syst.* 2012, 108, 1–9.
11. Seufert, V.; Ramankutty, N.; Foley, J.A. Comparing the yields of organic and conventional agriculture. *Nature* 2012, 485 (7397), 229–232.
12. Ponisio, L.C.; M'Gonigle, L.K.; Mace, K.C.; Palomino, J.; de Valpine, P.; Kremen, C. Diversification practices reduce organic to conventional yield gap. *Proc. R. Soc. B* 2015, 282, 20141396.
13. Connor, D.J.; Mínguez, M.I. Evolution not revolution of farming systems will best feed and green the world. *Glob. Food Secur.* 2012, 1, 106–113.
14. Clark, M.; Tilman, D. Comparative analysis of environmental impacts of agricultural production systems, agricultural input efficiency, and food choice. *Environ. Res. Lett.* 2017, 12, 064016.
15. Kirchmann, H. Why organic farming is not the way forward. *Outlook Agr.* 2019, 48 (1), 22–27.
16. Kremen, C.; Miles, A. Ecosystem services in biologically diversified versus conventional farming systems: benefits, externalities, and trade-offs. *Ecol. Soc.* 2012, 17, 40.
17. Baker, B.P.; Benbrook, C.M.; Groth III, E.; Benbrook, K.L. Pesticide residues in conventional, integrated pest management (IPM)-grown and organic foods: insights from three U.S. data sets. *Food Addit. Contam.* 2002, 19 (5), 427–446.
18. Pussemier, L.; Larondelle, Y.; Van Peteghem, C.; Huyghebaert, A. Chemical safety of conventionally and organically produced foodstuffs: a tentative comparison under Belgian conditions. *Food Control* 2006, 17, 14–21.
19. Smith-Spangler, C.; Brandeau, M.L.; Hunter, G.E.; Bavinger, J.C.; Pearson, M.; Eschbach, P.J.; Sundaram, V.; Liu, H.; Schirmer, P.; Stave, C.; Olkin, I.; Bravata, D.M. Are organic foods safer or healthier than conventional alternatives? *Ann. Intern. Med.* 2012, 157 (5), 348–366.
20. Barański, M.; Srednicka-Tober, D.; Volakakis, N.; Seal, C.; Sanderson, R.; Stewart, G.B.; Benbrook, C.; Biavati, B.; Markellou, E.; Giotis, C.; Gromadzka-Ostrowska, J.; Rembiałkowska, E.; Skwarło-Sońta, K.; Tahvonen, R.; Janovská, D.; Niggli, U.; Nicot, P.; Leifert, C. Higher antioxidant and lower cadmium concentrations and lower incidence of pesticide residues in organically grown crops: a systematic literature review and meta-analyses. *Brit. J. Nutr.* 2014, 112, 794–811.
21. Soil Association. *Organic Farming, Food Quality and Human Health: A Review of the Evidence*; Soil Association: Bristol, 2000.
22. Brandt, K.; Mølgaard, J.P. Organic agriculture: does it enhance or reduce the nutritional value of plant foods? *J. Sci. Food Agr.* 2001, 81, 924–931.
23. Worthington, V. Nutritional quality of organic versus conventional fruits, vegetables, and grains. *J. Alter. Complem. Med.* 2001, 7, 161–173.
24. Williams, C. M. Nutritional quality of organic food: shades of grey or shades of green? *Proc. Nutr. Soc.* 2002, 61, 19–24.
25. Magkos, F., Arvaniti, F.; Zampelas, A. Organic food: nutritious food or food for thought? A review of the evidence. *Int. J. Food Sci. Nutr.* 2003, 54, 357–371.
26. Rembialkowska, E. Quality of plant products from organic agriculture. *J. Sci. Food Agr.* 2007, 87, 2757–2762.
27. Benbrook, C.; Zhao, X.; Yáñez, J.; Davies, N.; Andrews, P. *New Evidence Confirms the Nutritional Superiority of Plant-based Organic Foods*; The Organic Center: Washington, DC, 2008, http://organic-center.org
28. Hunter, D.; Foster, M.; McArthur, J.O.; Ojha, R.; Petocz, P.; Samman, S. Evaluation of the micronutrient composition of plant foods produced by organic and conventional agricultural methods. *Crit. Rev. Food Sci.* 2011, 51, 571–582.
29. Lairon, D. Nutritional quality and safety of organic food: a review. *Agron. Sustain. Dev.* 2010, 30 (1), 33–41.
30. Brandt, K.; Leifert, C.; Sanderson, R.; Seal, C.J. Agroecosystem management and nutritional quality of plant foods: the case of organic fruits and vegetables. *Critical Rev. Plant Sci.* 2011, 30 (1–2), 177–197.

31. Palupi, E.; Jayanegara, A.; Ploeger, A.; Kahl, J. Comparison of nutritional quality between conventional and organic dairy products: a meta-analysis. *J. Sci. Food. Agr.* 2012, 92, 2774–2781.
32. Średnicka-Tober, D.; Barański, M.; Seal, C.J.; Sanderson, R.; Benbrook, C.; Steinshamn, H.; Gromadzka-Ostrowska, J.; Rembiałkowska, E.; Skwarło-Sońta, K.; Eyre, M.; Cozzi, G.; Larsen, M.K.; Jordon, T.; Niggli, U.; Sakowski, T.; Calder, P.C.; Burdge, G.C.; Sotiraki,S.; Stefanakis, A.; Stergiadis, S.; Yolcu, H.; Chatzidimitriou, E.; Butler, G.; Stewart, G.; Leifert, C. Higher PUFA and *n*-3 PUFA, conjugated linoleic acid, α-tocopherol and iron, but lower iodine and selenium concentrations in organic milk: a systematic literature review and meta- and redundancy analyses. *Brit. J. Nutr.* 2016, 115, 1043–1060.
33. Bourn, D.; Prescott, J. A comparison of the nutritional value, sensory qualities, and food safety of organically and conventionally produced foods. *Critical Rev. Food Sci.* 2002, 42, 1–34.
34. Dangour A.D.; Dodhia, S.K.; Hayter, A.; Allen, E.; Lock, K.; Uauy, R. Nutritional quality of organic foods: a systematic review. *Am. J. Clin. Nutr.* 2009, 90, 680–685.
35. Alföldi, T.; Fliessbach, A.; Geier, U.; Kilcher, L.; Niggli, U.; Pfiffner, L.; Stolze, M.; Willer, H. Organic agriculture and the environment. In *Organic Agriculture, Environment, and Food Security*; Scialabba, N.E.-H., Hattam, C., Eds.; Food and Agriculture Organization: Rome, 2002, http://www.fao.org/docrep/005/y4137e/y4137e00.htm [accessed 15 September 2012].
36. Lotter, D.W. Organic agriculture. *J. Sustain. Agr.* 2003, 21 (4), 59–128.
37. Kaspercyk, N.; Knickel, K. Environmental impacts of organic farming. In *Organic Agriculture: A Global Perspective*; Kristiansen, P., Taji, A., Reganold, J., Eds; Comstock Publishing Associates: Ithaca, NY, 2006; 259–294.
38. Mondelaers, K.; Aertsens, J.; Van Huylenbroeck, G. A meta-analysis of the differences in environmental impacts between organic and conventional farming. *Brit. Food J.* 2009, 111 (10), 1098–1119.
39. Crowder, D.W.; Northfield, T.D.; Strand, M.R.; Snyder, W.E. Organic agriculture promotes evenness and natural pest control. *Nature* 2010, 466, 109–112.
40. Geiger, F.; Bengtsson, J.; Berendse, F.; Weisser, W.W.; Emmerson, M.; Morales, M.B.; Ceryngier, P.; Liira, J.; Tscharntke, T.; Winqvist, C.; Eggers, S.; Bommarco, R.; Pärt, T.; Bretagnolle, V.; Plantegenest, M.; Clement, L.W.; Dennis, C.; Palmer, C.; Oñate, J.J.; Guerrero, I.; Hawro, V.; Aavik, T.; Thies, C.; Flohre, A.; Hänke, S.; Fischer, C.; Goedhart, P.W.; Inchausti, P. Persistent negative effects of pesticides on biodiversity and biological control potential on European farmland. *Basic Appl. Ecol.* 2010, 11, 97–105.
41. Gomiero, T.; Pimentel, D.; Paoletti, M.G. Environmental impact of different agricultural management practices: conventional vs. organic agriculture. *Crit. Rev. Plant Sci.* 2011, 30 (1–2), 95–124.
42. Gattinger, A.; Muller, A.; Haeni, M.; Skinner, C.; Fliessbach, A.; Buchmann, N.; Mäder, P.; Stolzea, M.; Smith, P.; Scialabb, N.E.-H.; Niggli, U. Enhanced top soil carbon stocks under organic farming. *P. Natl. Acad. Sci. USA* 2012, 109 (44), 18226–18231.
43. Lynch, D.H.; Halberg, N.; Bhatta, G.D. Environmental impacts of organic agriculture in temperate regions. *CAB Rev.* 2012, 7, 1–17.
44. Tuomisto, H.L.; Hodge, I.D.; Riordan, P.; Macdonald, D.W. Does organic farming reduce environmental impacts?—a meta-analysis of European research. *J. Environ. Manage.* 2012, 112, 309–320.
45. Kennedy, C.M.; Lonsdorf, E.; Neel, M.C.; Williams, N.M.; Ricketts, T.H.; Winfree, R.; Bommarco, R.; Brittain, C.; Burley, A.L.; Cariveau, D.; Carvalheiro, L.G.; Chacoff, N.P.; Cunningham, S.A.; Danforth, B.N.; Dudenhöffer, J.H.; Elle, E.; Gaines, H.R.; Garibaldi, L.A.; Gratton, C.; Holzschuh, A.; Isaacs, R.; Javorek, S.K.; Jha, S.; Klein, A.M.; Krewenka, K.; Mandelik, Y.; Mayfield, M.M.; Morandin, L.; Neame, L.A.; Otieno, M.; Park, M.; Potts, S.G.; Rundlöf, M.; Saez, A.; Steffan-Dewenter, I.; Taki, H.; Viana, B.F.; Westphal, C.; Wilson, J.K.; Greenleaf, S.S.; Kremen, C. A global quantitative synthesis of local and landscape effects on wild bee pollinators in agroecosystems. *Ecol. Lett.* 2013, 16, 584–599.
46. Skinner, S.; Gattinger, A.; Muller, A.; Mäder, P.; Fließbach, A.; Stolze, M.; Ruser, R.; Niggli, R. Greenhouse gas fluxes from agricultural soils under organic and non-organic management—a global meta-analysis. *Sci. Total Environ.* 2014, 468–469, 553–563.

47. Tuck, S.; Winqvist, C.; Mota, F.; Ahnström, J.; Turnbull, L.A.; Bengtsson, J. Land-use intensity and the effects of organic farming on biodiversity: a hierarchical meta-analysis. *J. Appl. Ecol.* 2014, 51, 746–755.
48. Lee, K.S.; Choe, Y.C.; Park, S.H. Measuring the environmental effects of organic farming: a meta-analysis of structural variables in empirical research. *J. Environ. Manage.* 2015, 162, 263–274.
49. Lichtenberg, E.M.; Kennedy, C.M.; Kremen, C.; Batáry, P.; Berendse, F.; Bommarco, R.; Bosque-Pérez, N.A.; Carvalheiro, L.G.; Snyder, W.E.; Williams, N.M.; Winfree, R.; Klatt, B.K.; Åström, S.; Benjamin, F.; Brittain, C.; Chaplin-Kramer, R.; Clough, Y.; Danforth, B.; Diekötter, T.; Eigenbrode, S.D.; Ekroos, J.; Elle, E.; Freitas, B.M.; Fukuda, Y.; Gaines-Day, H.R.; Grab, H.; Gratton, C.; Holzschuh, A.; Isaacs, R.; Isaia, M.; Jha, S.; Jonason, D.; Jones, V.P.; Klein, A-M.; Krauss, J.; Letourneau, D.K.; Macfadyen, S.; Mallinger, R.E.; Martin, E.A.; Martinez, E.; Memmott, J.; Morandin, L.; Neame, L.; Otieno, M.; Park, M.G.; Pfiffner, L.; Pocock, M.J.O.; Ponce, C.; Potts, S.G.; Poveda, K.; Ramos, M.; Rosenheim, J.A.; Rundlöf, M.; Sardiñas, H.; Saunders, M.E.; Schon, N.L.; Sciligo, A.R.; Sidhu, C.S.; Steffan-Dewenter, I.; Tscharntke, T.; Veselý, M.; Weisser, W.W.; Wilson, J.K.; Crowder, D.W. A global synthesis of the effects of diversified farming systems on arthropod diversity within fields and across agricultural landscapes. *Glob. Change Biol.* 2017, 23, 4946–4957.
50. Crowder, D.W.; Reganold, J.P. Financial competitiveness of organic agriculture on a global scale. *Proc. Natl. Acad. Sci. USA* 2015, 112, 7611–7616.
51. Porter, J.; Costanza, R.; Sandhu, H.; Sigsgaard, L.; Wratten, S. The value of producing food, energy, and ecosystem services within an agro-ecosystem. *Ambio* 2009, 38, 186–193.
52. Sandhu, H.; Wratten, S.; Costanza, R.; Pretty, J.; Porter, J.R.; Reganold, J. Significance and value of non-traded ecosystem services on farmland. *PeerJ* 2015, 3, e762.
53. Östman, Ö.; Ekbom, B.; Bengtsson, J. Yield increase attributable to aphid predation by ground-living polyphagous natural enemies in spring barley in Sweden. *Ecol. Econ.* 2003, 45, 149–158.
54. MacRae, R.J.; Frick, B.; Martin, R.C. Economic and social impacts of organic production systems. *Can. J. Plant Sci.* 2007, 87 (5), 1037–1044.
55. Gruère, G.; Nagarajan, L.; King, E.O. The role of collective action in the marketing of underutilized plant species: lessons from a case study on minor millets in South India. *Food Policy* 2009, 34, 39–45.
56. Prihtanti, T. M., Hardyastuti, S., Hartono, S. Social-cultural functions of rice farming systems. *Asian J. Agr. Rural Dev.* 2014, 4, 341–351.
57. Mendoza, T.C. Evaluating the benefits of organic farming in rice agroecosystems in the Philippines. *J. Sustain. Agr.* 2004, 24, 93–115.
58. Eddleston, M.; Karalliedde, L.; Buckley, N.; Fernando, R.; Hutchinson, G.; Isbister, G.; Konradsen, F.; Murray, D.; Piola, J.C.; Senanayake, N.; Sheriff, R.; Singh, S.; Siwach, S.B.; Smit, L. Pesticide poisoning in the developing world—a minimum pesticides list. *Lancet* 2002, 360, 1163–1167.
59. Thundiyil, J. G.; Stober, J.; Besbelli, N.; Pronczuk, J. Acute pesticide poisoning: a proposed classification tool. *World Health Organization Bulletin* 2008, 86, 205–209.
60. Willer, H.; Lernoud, J., Eds. *The World of Organic Agriculture: Statistics and Emerging Trends 2016*; Research Institute of Organic Agriculture (FiBL): Frick, and International Federation of Organic Agriculture Movements (IFOAM): Bonn, 2016.
61. Willer, H.; Lernoud, J., Eds. *The World of Organic Agriculture: Statistics and Emerging Trends 2017*; Research Institute of Organic Agriculture (FiBL): Frick, and International Federation of Organic Agriculture Movements (IFOAM): Bonn, 2017.
62. Willer, H.; Lernoud, J., Eds. *The World of Organic Agriculture: Statistics and Emerging Trends 2018*; Research Institute of Organic Agriculture (FiBL): Frick, and International Federation of Organic Agriculture Movements (IFOAM): Bonn, 2018.

30

Crops and the Atmosphere: Trace Gas Exchanges

Jürgen Kreuzwieser and Heinz Rennenberg
University of Freiburg

Introduction 261
Impacts of Trace Gases on Crops 261
Factors Controlling Trace Gas Exchange 262
Plant Physiological Controls 263
Internal Transfer Resistances 264
Conclusion 265
References 265

Introduction

The atmosphere mainly consists of nitrogen (78% by volume), oxygen (21% by volume) and the noble gas argon (0.93% by volume) together making up >99.9% of the atmosphere's composition; the remainder is known as trace gases. Trace gases are typically present in the range of parts per trillion by volume (pptv) to parts per million by volume (ppmv); they include the greenhouse gases carbon dioxide (CO_2), methane (CH_4), nitrous oxide (N_2O), ethane, water vapor, and ozone (O_3); the air pollutants sulfur dioxide (SO_2), ammonia (NH_3), nitric oxide (NO), nitrogen dioxide (NO_2), peroxyacylnitrates (PAN), nitric acid (HNO_3), and carbon monoxide (CO); and a number of volatile organic compounds (VOCs) (Table 30.1). Among VOCs are isoprenoids (mainly isoprene and monoterpenes) and many oxygenated species such as alcohols, aldehydes, and organic acids.[2] Because of their reactivity, VOCs strongly affect the oxidation capacity of the troposphere and influence the concentration and distribution of several other trace gases, including CH_4 or CO.[3] On a regional scale, VOCs significantly contribute to the formation of tropospheric O_3.[4] The main source of VOCs (~90%) is natural emission by vegetation.[2]

Impacts of Trace Gases on Crops

The effects of atmospheric trace gases on crops are quite diverse and depend on the type of gas, its concentration, the duration of exposure, and the amount taken up, as well as a range of plant internal factors. Direct phytotoxic effects due to exposure to high concentrations of pollutants such as O_3 and SO_2 on crop plants include, among others, visible leaf injury; changes in chloroplast structure and cell membranes; disturbances of stomatal regulation, respiration, and photosynthesis; and reductions in growth and yield.[5] However, because sulfur (S) is an essential nutrient for plants, SO_2 absorbed by foliage may also be used as an additional source of sulfur in polluted areas, in addition to sulfate from the soil.[6] The same principle applies to nitrogen (N). Thus, effects of trace gases can be divided into phytotoxic effects caused by protons, organic compounds, SO_2, NO_2, NH_3, and O_3; and nutritional effects caused by S-and N-containing gases and CO_2.[7]

TABLE 30.1 Range of Ambient Concentrations of Different Trace Gases in the Atmosphere

Trace Gas	Ambient Concentrations
Ar	9,340 ppm
CO_2	379 ppm (2005)
CH_4	1774 ppb (2005)
N_2O	319 ppb (2005)
Isoprene	ppt–several ppb
Monoterpenes	ppt–several ppb
Alcohols	1–30 ppb
Carbonyls (aldehydes, ketones)	1–30 ppb
Alkanes	1–3 ppb
Alkenes	1–3 ppb
Esters	ppt–several ppb
Carbonic acids	0.1–16 ppb
SO_2	0.5–20 ppb
H_2S	0–0.2 ppb
NO_2	0.5–15 ppb
O_3	20–80 ppb

Source: Data from Forster, et al.,[1] Guenther, et al.,[2] Kesselmeier & Staudt,[3] Bender & Weigel,[5] and Herschbach, et al.[6]

Factors Controlling Trace Gas Exchange

Trace gases can be exchanged between the atmosphere and aboveground plant parts by 1) dry deposition as gases or aerosols; 2) wet deposition as dissolved compounds in rainwater or snow; or 3) interception of compounds dissolved in mist or cloud water[7] (Figure 30.1). The direction of the exchange (i.e., emission vs. deposition) and its velocity (i.e., the exchange rates) is controlled by the physicochemical conditions and by internal plant factors.

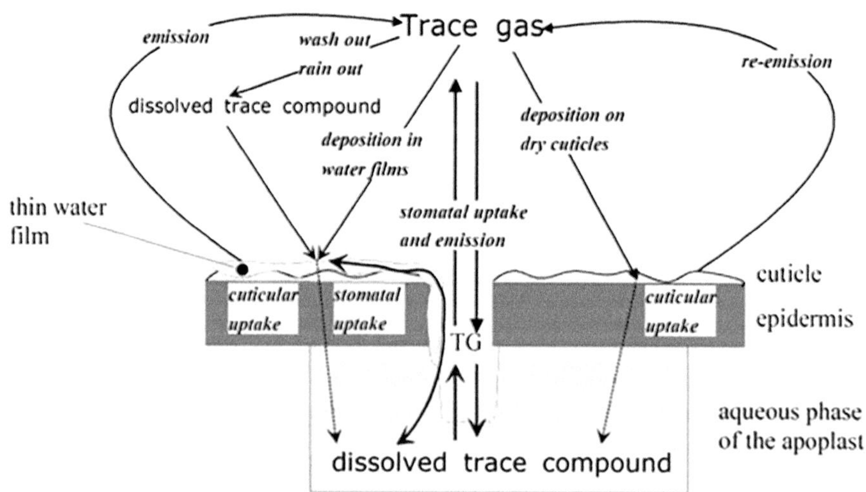

FIGURE 30.1 Main access routes for trace gases to enter a leaf: Uptake via the stomata and cuticular uptake.
Source: Redrawn from Rennenberg and Geβler.[8]

TABLE 30.2 Typical Range of Rates of Exchange for Different Trace Gases between Crops and the Atmosphere

Trace Gas	Range of Exchange[a] [μg g^{-1} (leaf d.wt.) h^{-1}]
SO_2	–300 to 0
H_2S	–2 to 2
O_3	–300 to 0
NO_2	–3 to 1
NH_3	–3 to 3
Isoprene	0 to 0.1
Monoterpenes	0 to 20
Oxygenated VOCs	–50 to 50

[a] Negative values indicate deposition and positive values indicate emission of respective trace gas.

The gradient in the gas concentration between substomatal cavities and the atmosphere is the driving force for gas exchange. The gas flux is a diffusive (passive) process and can be described by Fick's law of diffusion. Accordingly, the net flux of a trace gas is zero when the substomatal concentration is equal to the concentrations in the surrounding ambient air. This concentration is referred to as the compensation point for the particular gas. When trace gas concentrations outside the leaves are higher than those in the substomatal cavities, a net flux into the leaves will take place (deposition), and vice versa. Therefore, crops may act both as a source (if ambient concentrations are lower than substomatal concentrations) or as a sink of a specific gas (Table 30.2). This dual behavior has been observed for a variety of gases (SO_2, H_2S, NO_2, NH_3, organic acids, and aldehydes). Compensation points for pollutants such as NH_3, e.g., range between 1 and 6 parts per billion by volume (ppbv) (see Husted et al.[9] and references therein). They depend mainly on the plant species or cultivar, development stage, temperature, and status of N nutrition of the plants. Generally, compensation points increase with increasing availability of nutrients in the soil, which suggests that this is one mechanism by which plants cope with excess nutrient supply.[8]

Plant Physiological Controls

The existence of a compensation point depends on the capacity of a plant to produce the trace gas to be exchanged, or to consume it. Therefore, compensation points do not exist for compounds that cannot be produced (e.g., O_3) or consumed (e.g., isoprene). It is evident that for each of the many gases exchanged between crop plants and the atmosphere, specific metabolic pathways exist, not all of which are understood yet. As an example, some details on the exchange of nitrogen compounds are presented (Figure 30.2).

Both NO_2 and NH_3 can be taken up by aboveground parts of plants, mainly via the stomata of leaves.[7] In the aqueous phase of the apoplast, NO_2 is either disproportionated yielding equal amounts of NO_2^- and NO_3^-, or it reacts with apoplastic ascorbate.[10] Because disproportionation of NO_2 in water is slow at atmospheric NO_2 concentrations, the reaction with ascorbate may be of more importance. Upon conversion to either NO_3^- or NO_2^-, these anions are transported to the cytoplasm, where they are reduced by the assimilatory nitrate reduction pathway yielding NO_4^+ and the amino acid glutamine. Atmospheric NH_3 dissolves in the aqueous phase of the apoplastic space to yield NO_4^+, which is then taken up into the cytoplasm.

Both NH_3 and NO_2 can also be emitted by plants. NH_3 may be released from cellular NO_4^+ pools when plants are exposed to excess nitrogen in the soil. In addition, it can be released from drying water films at the leaf surface. During this process, the remaining NH_4^+ concentrations on the surface will increase. By contrast, the chemical source of NO_2 emitted by the leaves is largely unknown; it has been proposed that nitrate reductase may be involved in the reduction of NO_2^- to NO_2.[8]

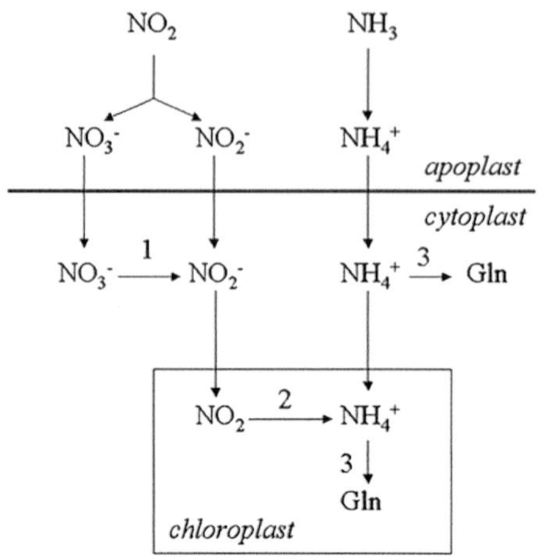

FIGURE 30.2 Main processes involved in the assimilation of atmospheric NH_3 and NO_2 taken up by the leaves. 1: nitrate reductase, 2: nitrite reductase, 3: glutamine synthetase (cytoplasmatic and chloroplastic isoforms).
Source: Redrawn from Rennenberg, and Geβler.[8]

The rate of trace gas emission does not necessarily depend on the actual rate of production. Some volatile compounds are produced and then stored in particular pools. For example, some plant reservoirs contain high amounts of monoterpenes, which can be emitted throughout the day and night independent of light-dependent biosynthesis.[3]

Through their effect on biochemical pathways, biotic and abiotic factors (e.g., stress factors) and the developmental stage of plants influence the rate of trace gas emission. For example, stress caused by wounding, chilling, iron deficiency, O_2 deficiency, or induction of oxidative stress caused by exposure to O_3 or SO_2 can lead to the production of the VOCs hexenal, hexanal, formaldehyde, formate, ethene, ethane, ethanol, and acetaldehyde.[3]

Internal Transfer Resistances

The transfer of gases in and out of plants is often described by a resistance analogy. Along the path from the sites of production (or consumption) to the atmosphere, a series exists of mainly internal plant resistances. Because biosynthesis and consumption of volatile compounds usually take place in the cytoplasm or in other compartments of the cell, the gas must pass across the bordering membranes. The diffusive flux through these membranes is determined by the molecular size and the lipophilic character of the particular compound. Polar molecules such as organic and inorganic acids are not likely to be dissolved in the lipophilic membranes; therefore, the diffusive flux should be slow. Carrier proteins can facilitate the transport of these polar compounds across membranes from the cytoplasm into the apoplastic space, similar to the transport of organic acids.[11]

The compounds are transferred from the liquid phase in the apoplastic space to the gaseous phase in the substomatal cavity, or vice versa. The volatilization into the internal leaf air space depends on 1) chemical properties of both the aqueous phase and the individual compound to be emitted (e.g., its solubility in the apoplastic solution); and 2) physical factors such as the ambient concentration of the gas, its vapor pressure, and temperature.[11] A reduction of the apoplastic pH reduces the resistance for inorganic and organic acids, because these compounds become protonated and thereby more volatile. When present in the gaseous phase of the apoplastic space, trace gases can escape from the leaves through either the cuticle or the stomata. However, because of their polarity, the lipophilic cuticle constitutes a strong barrier, and therefore diffusion through the stomata is the main pathway. However, nonstomatal emission and deposition have also been observed in air pollutants such as NO, NO_2, and SO_2, but at much lower rates than stomatal exchange. For this reason, factors that influence the stomatal aperture exert a strong influence on the rate of gas exchange between plants and the atmosphere. The control by the stomata of emission and deposition of NO_2, SO_2, O_3, PAN, and other trace gases was observed in many studies, including in investigations of crop plants. Thus, concentration and time of exposure are not the only factors determining the effect of an air pollutant on vegetation. Plants usually close their stomata during hot, dry conditions when, for instance, O_3 levels are high. This may provide some protection for the plants from O_3 injury. Alternatively, in northern Europe where O_3 concentrations are lower than in southern and central Europe, the potential O_3 uptake at a given O_3 concentration may be higher because of higher air humidity, leading to high rates of stomatal O_3 uptake.

Conclusion

Trace gases influence not only natural ecosystems but also agricultural crops. Future studies should focus on the impact of different combinations of air pollutants (e.g., increased nitrogen input combined with elevated O_3 concentrations) on plants and should include aspects of global climate change (e.g., higher temperatures, droughts, and increased frequency of heavy rainfalls and droughts in combination with elevated CO_2 concentrations). Future management practices should allow for optimum plant growth, while simultaneously reducing the loss of pollutants from the plant–soil system.[12]

References

1. Forster, P.; Ramaswamy, V.; Artaxo, P.; Berntsen, T.; Betts, R.; Fahey, D.W.; Haywood, J.; Lean, J.; Lowe, D.C.; Myhre, G.; Nganga, J.; Prinn, R.; Raga, G.; Schulz, M.; Van Dorland, R. Changes in Atmospheric Constituents and in Radiative Forcing. In: *Climate Change 2007: The Physical Science Basis. Contribution of Working Group I to the Fourth Assessment Report of the Intergovernmental Panel on Climate Change.* Solomon, S.; Qin, D.; Manning, M.; Chen, Z.; Marquis, M.; Averyt, K.B.; Tignor, M.; Miller, H.L. Eds.; Cambridge University Press: Cambridge and New York, 2007; 137–153 pp.
2. Guenther, A.B.; Hewitt, C.N.; Erickson, D.; Fall, R.; Geron, C.; Graedel, T.; Harley, P.; Klinger, L.; Lerdau, M.; McKay, W.A.; Pierce, T.; Scholes, B.; Steinbrecher, R.; Tallamraju, R.; Taylor, J.; Zimmerman, P. A. Global model of natural volatile organic compound emissions. J. Geophys. Res. **1995**, *100*, 8873–8892.
3. Kesselmeier, J.; Staudt, M. Biogenic volatile organic compounds (VOC): An overview on emission, physiology and ecology. J. Atmos. Chem. **1999**, *33*, 23–88.
4. Ashworth, K.; Wild, O.; Hewitt, C.N. Impacts of biofuel cultivation on mortality and crop yields. Nat. Clim. Change **2013**, *3*, 492–496.
5. Bender, J.; Weigel, H.-J. Changes in atmospheric chemistry and crop health: A review. Agron. Sustain. Dev. **2011**, *31*, 81–89.
6. Herschbach, C.; De Kok, L.J.; Rennenberg, H. Net uptake of sulfate and its transport to the shoot in tobacco plants fumigated with H_2S or SO_2. Plant Soil. **1995**, *175*, 75–84.
7. Wellburn, A.R. Why are atmospheric oxides of nitrogen usually phytotoxic and not alternative fertilizers? New Phytol. **1990**, *115*, 395–429.
8. Rennenberg, H.; Geßler, A. Consequences of N Deposition to Forest Ecosystems—Recent Results and Future Research Needs. In *Forest Growth Responses to the Pollution Climate of the 21st Century;* Sheppard, L.J.; Neil Cape, J., Eds.; Kluwer Academic Publication: Dordrecht, the Netherlands, 1999; 47–64.
9. Husted, S.; Schjoerring, J.K.; Nielsen, K.H.; Nemitz, E.; Sutton, M.A. Stomatal compensation points for ammonia in oilseed rape plants under field conditions. Agr. Forest Meteorol. **2000**, 105, 371–383.

10. Ramge, P.; Badeck, F.W.; Ploechl, M.; Kohlmaier, G.H. Apoplastic antioxidants as decisive elimination factors within the uptake of nitrogen dioxide into leaf tissues. New Phytol. **1993**, *125*, 771–785.
11. Gabriel, R.; Schafer, L.; Gerlach, C.; Rausch, T.; Kesselmeier, J. Factors controlling the emissions of volatile organic acids from leaves of *Quercus ilex* L. (Holm oak). Atmos. Environ. **1999**, *33*, 1347–1355.
12. Rogasik, J.; Schroetter, S.; Schnug, E. Impact of air pollutants on agriculture. Phyton **2002**, *42*, 171–182.

31

Desertification

Introduction 267
What Is Desertification? 267
Desertification and Land Degradation, Drylands and Non-Drylands 268
How LDD Is Caused? 269
Who Causes LDD? 269
Why Land Users Cause LDD? 271
Where and How Much LDD There Is? 272
Why LDD Is of Concern? 273
What Has Been Done and What Can Be Done? 275
Desertification and Climate, and Climate Change 275
Conclusions 276
References 277
Bibliography 278

Uriel N. Safriel
Hebrew University of Jerusalem

Introduction

"Desertification" is a human-driven process, which impinges on a significant Earth's natural capital, land, by reducing its productivity. Desertification is probably as old as the pastoral and farming livelihoods, but has attracted attention only in the 20th century, such that as it turned into the new millennium, desertification has been ranked among the greatest contemporary environmental problems.[1] Though literally the term links with deserts, desertification hardly occurs in deserts. Rather, desertification affects non-desert areas, all around the globe, and is of major concern, as it is a driver of poverty, mainly in the rural areas of the developing world, but at the same time it puts at risk food security at the global scale. This entry explains what desertification is, and describes how it occurs, where, by whom, and why. It then elaborates on its implications to people across scales, from local to global. This is followed by scanning through the measures taken by people and the international community to address this threat to human survival and global sustainability. The entry closes with the effect of global climate change on desertification, currently and in the future.

What Is Desertification?

Deserts—areas of poor vegetation cover and of the driest climate on earth—were molded by climatic processes, such as scarce rainfall and intense evaporation. These jointly constrain plant productivity such that much of the desert appears lifeless. As the world's deserts were formed by natural climatic processes over long periods during which they have grown and shrunk, the term "desertification" can grammatically describe desert formation. However, the term was apparently coined first in a 1949 publication,[2] not for describing a climate-driven outward spreading of deserts, but for a human-driven decline of land productivity to levels reminiscent to those of deserts, while climate remained a non-desert one. The term came into widespread international use only as of the 1970s, then being defined and redefined ever since. All definitions, however, relate to a land attribute critical to all people, everywhere and always—the marketed biological products, food and fiber, which are derived from the land's biological productivity. Thus, desertification is best expressed by a persistent reduction, relative to the potential, of the land's biological productivity of an economic value.[1,3]

Desertification and Land Degradation, Drylands and Non-Drylands

The decline in biological productivity of economic value is driven by biophysical processes of "land degradation," which can operate on lands used by people anywhere on earth, but mostly affect lands whose productivity is constrained by water. Water is one of the raw materials required for the biological production process, but atmospheric CO_2 and soil minerals, as well as light are required too. Each of these can limit productivity, even if all the others are available in abundance. Land degradation occurs in areas of low water and high evaporation, thus making soil water the limiting factor. This water scarcity exists in all lands exposed to a potential annual evaporative loss that is at least about one and one-half times greater than the mean annual gain through rainfall. These are "drylands," and depending on their degree of aridity, they are classified into desert drylands (hyper arid and arid drylands) and non-desert ones (semiarid and dry subhumid drylands) (Figure 31.1).

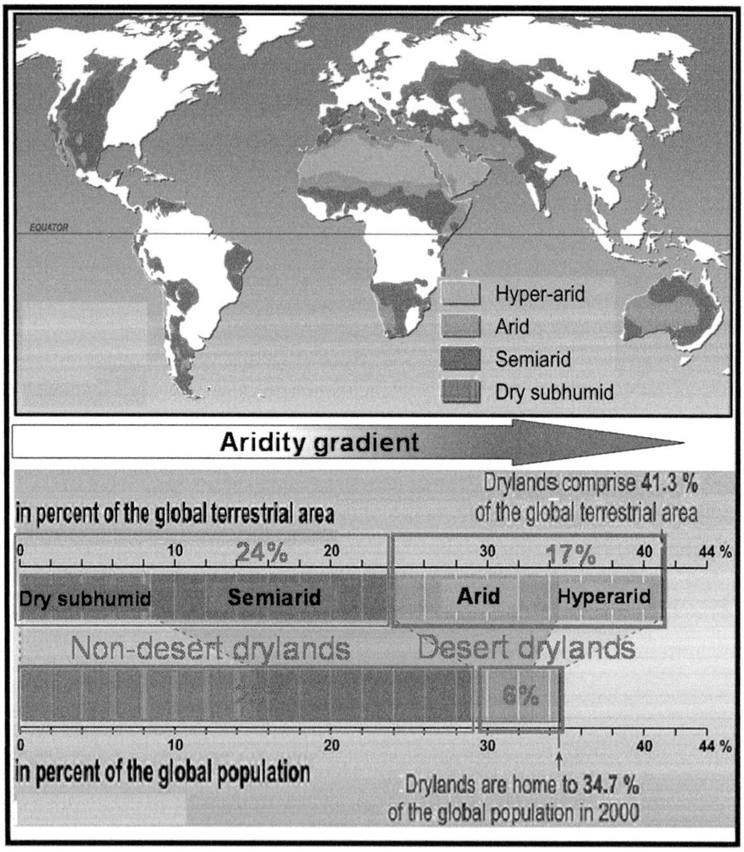

FIGURE 31.1 **(See color insert.)** The global drylands. Geographic layout of the four sections of the aridity gradient, their percent share of global land, their human populations, and their percent share of global population.
Source: Adapted from Adeel et al.[1]

The United Nations Convention to Combat Desertification (UNCCD) confined its mandate (to "combat desertification") to degradation of drylands only, excluding the most arid ones, the hyper arid desert drylands. Also, the stakeholders of the UNCCD habitually use the two terms, "land degradation" and "desertification" interchangeably or jointly. However, there is an increasing recent tendency to address "land degradation" as a process, while regarding "desertification" as a terminal, extreme state of this process; and, to use both terms for processes impinging on biological productivity in non-dryland areas too.[3,4] These tendencies have driven an increasing usage of an acronym for Land Degradation and Desertification, namely, LDD.

How LDD Is Caused?

Biological productivity depends on the land's soil, which provides trees, forage and crop plants physical anchor, and raw materials for the production process. The loose and aerated texture of the soil is due to its organic matter that binds the soil's mineral particles together. The binding creates aggregates with air spaces between them that retain water in which the plant mineral nutrients are dissolved. This particulate texture of soil, however, makes it vulnerable to the mechanical force of wind, that blows away the soil's fine components what often leaves sandy areas and dunes, and of rainwater that carries soils away. These soil erosion processes reduce fertility as most nutrients are confined to the uppermost soil layers, which are directly exposed to the eroding forces. Since the rate of soil formation is extremely slow, for all practical purposes soil is nonrenewable and hence topsoil loss is one of the major causes of persistent productivity decline. It is plant cover that provides for soil protection, by mitigating wind force and reducing the impact of raindrops that would otherwise destroy the delicate soil texture. Breakage of the aggregates results in soil crusting, leading to land surface sealing that reduces soil water absorption. Rather than infiltrating and enriching soil moisture, rainwater turn into surface runoff that erodes the soil or leaching away dissolved soil minerals and its organic compounds. These losses directly reduce fertility and indirectly the water holding capacity of the soil, what also slows down the soil microbial activity essential in nutrient recycling. Thus, due to soil erosion plants may lose their physical support as well as their production resources. Another major driver of declining productivity is soil salinization. Almost all types of water, including rainwater, contain some dissolved salts, which are left in the top soils after soil moisture evaporates or is transpired by the plants. This accumulated salinity reduces plants' ability to suck moisture from the soil, and the resulting soil sodicity (predominance of sodium ion in soil moisture) limits water absorption by soils. To conclude, while climatically driven water scarcity makes dryland productivity low compared to most other lands,[1] drylands are vulnerable to soil erosion and salinization (Figure 31.2), the major physical and chemical processes, respectively, leading to LDD, whose ultimate expression is productivity loss.

Who Causes LDD?

Depending on whether the biological products of economic value are derived directly from wild plants (e.g., woody plants used for timber and firewood, and other edible wild species); or indirectly from free-ranging livestock feeding on wild plants; or from cultivated crop plants; lands and their productivity are managed by their users through interventions expressed in natural resource exploitation rates, livestock stocking rates, and soil fertility levels, respectively. These interventions sustain peoples, societies, and their prosperity, yet they often pose potential threats to the land's productivity, which could ultimately impinge on the land users themselves, and on others. This is because while all productivity is derived from plants, their exploitative removal, either through grazing or harvesting, exposes the land to degradation processes once a threshold of plant cover is removed and/or one of regeneration relative to exploitation rate is crossed.

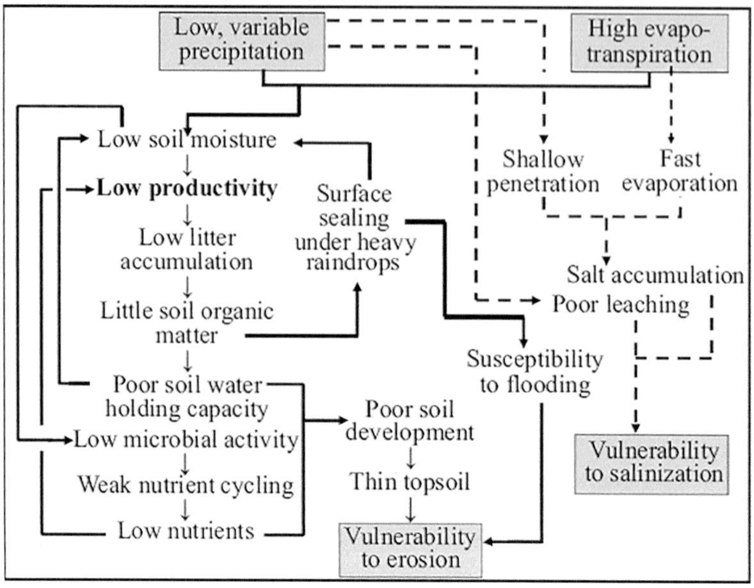

FIGURE 31.2 The dryland syndrome—climate-driven natural processes driving soil biophysical processes leading to inherent low productivity, to erosion vulnerability (solid arrows, two negative feedback loops), and to salinization vulnerability (broken arrows).
Source: Adapted from Fig. 2 in p. 232, Safriel.[31]

Regarding the pastoral livelihood, high livestock densities maintained throughout a long period reduce the overall soil protective cover; exploit vegetation faster than it can regenerate, and export nutrients (through selling livestock and their products) faster than their natural renewability. These encourage non-forage plants to dominate the range, promote encroachment of bushes that obstruct livestock foraging movement, and reduce soil physical resilience to livestock trampling, which compacts soil surface, reduces rainfall infiltration thus increasing risks of powerful surface runoff that erodes the soil and changes the local topography. Individually or combined, these processes lead to a persistent rangeland productivity loss (Figure 31.3). As to the farming livelihood, long fallow periods expose the croplands to erosion forces, but in short fallow periods, a high number of crops cultivated each year and a large amount of nutrients exported in the harvested agricultural products, individually or combined are of a potential to cross the threshold of land degradation. This is through nutrient loss that is faster than its renewability and a decline in soil organic matter, which reduces soil permeability, and the water holding capacity leading to increased erosion vulnerability. Irrigation, used mainly in drylands for increasing the productivity of otherwise rainfed agriculture, has dramatically increased not only the economic productivity but also soil salinity. Dryland water is scarce and hence not used for leaching soil salinity and when used, nutrients are leached too. Furthermore, much of dryland water is brackish, and irrigation systems devoid of drainage cause water logging, leading to further soil salinization. The irrigation-driven increased economic productivity amplifies nutrient export through harvest, which needs to be balanced by fertilizers. This combination of irrigation and fertilization could generate immediate economic gains, often succeeded by LDD (Figure 31.3).

Practitioners of both livelihoods, but more often of the pastoral one, also exploit the woody plants of natural or planted woodlands and of rangelands, for timber and fuel wood, or remove all vegetation cover of woodlands or rangelands, to be replaced by expanding cultivation. These processes, too, expose large swaths of land to erosion processes and eventual LDD. To conclude, human land uses directed at increasing the inherently low dryland productivity of economic value are of a potential that

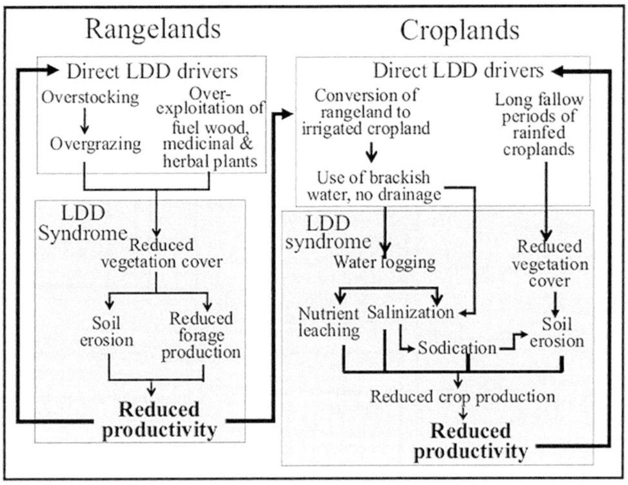

FIGURE 31.3 Actions of land users that constitute the proximate, direct drivers (top boxes) of the land degradation and desertification (LDD) syndrome (bottom boxes) and their feedbacks and linkage (thick arrows), in each of the two prevailing dryland livelihoods. For further explanation see text.

is often materialized to bring about LDD. The persistent decline in productivity, rather than its aspired increase, attests to a preceding shift from grazing, cropping, and exploitation, to overgrazing, overcropping, and overexploitation—activities of land users that constitute the direct, biophysical drivers of the LDD processes.[5]

Why Land Users Cause LDD?

It can be expected that land users are well adapted to their local environment and hence possess the knowledge for using their lands sustainably, i.e., without degrading it. Since land degradation does occur in areas used by experienced users, factors beyond their control may be involved. Indeed, indirect drivers of change, or underlying causes of LDD that affect the direct, proximate causes that change non-degraded to a degraded land can be invoked (e.g., Figure 31.3). These can be population (or demographic), economic, technological, policy and institutional, cultural, and political drivers.[6] Each of these (Figure 31.4) can affect the direct LDD biophysical drivers alone, jointly, linearly or interactively, across different temporal and spatial scales.[7] However, in spite of an increasing research and political interest in these LDD drivers, there is not enough field evidence, hence also agreement, regarding the role and relative significance of each of the variables that structure the network of pathways to LDD, and therefore the LDD "syndrome" is currently illustrated by paradigms that can be perceived as structured programs for further research. An "LDD paradigm" (Figure 31.5) integrates elements of the Desertification Paradigm,[8] the Dryland Livelihood Paradigm[9] and the Dryland Development Paradigm.[10] Under this paradigm LDD is driven by mutually interactive biophysical processes within the land's "Ecological System, and by mutually interactive socio-economic and policy processes within a "Human System," integrated into an interactive "Human-Environment System" that oscillates between land's sustainable use and LDD. The LDD paradigm attends the effect of LDD on human wellbeing at different spatial scales—effects that often further exacerbate rather than mitigate the LDD state. These are mainly poverty at the local scale, and migrations, refugees, conflicts and foreign interventions at the regional and even global scales. In short, land users are driven to overexploit land, mostly by economic factors that in themselves can be driven by demographic variables, social and cultural factors, and by economic and social policies and political processes molded at diverse regional and global processes on which they have no control.

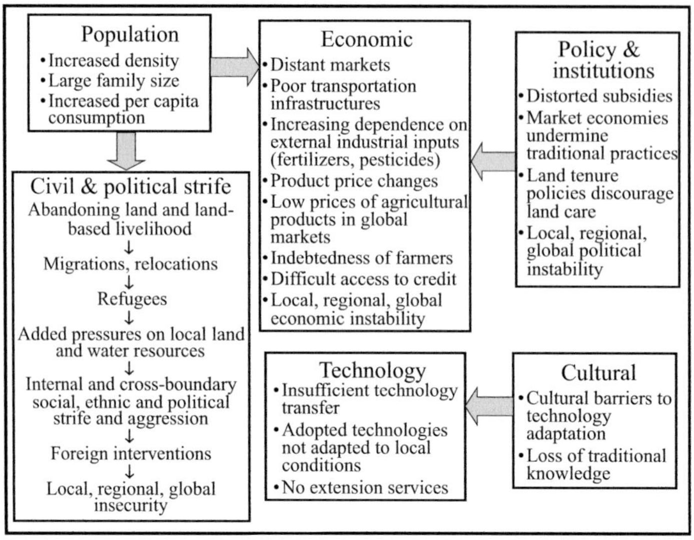

FIGURE 31.4 Underlying socioeconomic and policy indirect land degradation and desertification drivers and their interactions.
Source: Adapted from Fig. 3, p. 173 in Reynolds et al.[7]

Where and How Much LDD There Is?

Since biological productivity expressed in market-traded products of economic value is regularly monitored a detected decline in the production of crops or in livestock-derived products can be used to delineate LDD-affected areas and quantify their spatial extent. However, the quantities of marketable agricultural products per unit area do not necessarily reflect land productivity, as these much depend on the land users' management decisions, driven mainly by economic considerations. Therefore, instead of monitoring marketed land products scientists turned to monitoring land and soil attributes serving as indicators of productivity. But since different studies use different indicators and measurement methods, it is not surprising that five global assessments during the last four decades resulted in LDD estimates ranging from 15% to 63% of global land and 4% to 74% of its subset of global drylands.[11] A critical review of these studies suggested (with "medium certainty") that 10–20% of the global drylands have been degraded by year 2000,[8] while an earlier study qualified 12% of non-dryland as degraded.[12] The latter study qualified 5.7% and 3.8% of Asian and African lands as degraded, respectively, and 1.9%, 1.7%, 1.2%, and 0.8% of lands in South America, Europe, North America, and Australasia as degraded, respectively; which combined to classify 15.1% of global land as degraded.[13]

The development of frequent and accessible time series of Earth monitoring from space enabled assessing not just the amount of degraded land but the more recent rates of ongoing LDD. Satellite-born sensors of radiation reflected and emitted from the Earth's surface, and especially from its vegetation cover, are used for following the course of biological productivity changes within given time periods. Most studies attend selected regions but one[14] addresses the dynamics of biological productivity of vegetation cover at a global scale. This 23-year (1981–2003) analysis of remote sensing images revealed persistently declining overall productivity (i.e., not just the subset of direct economic value) at 24% of the global land, on which 1.5 billion people reside. It mainly occurred in Africa south of the equator, Southeast Asia and south China, north-central Australia, the Argentinean Pampas and swaths of the Siberian and North American taiga. The study also found that land degradation in China is more severe in forests than in croplands. However, measures taken to control the productivity–rainfall interaction need to be refined in making such assessment fully reliable.[15] Nevertheless, the available studies demonstrate that as much as a quarter of the global land is already degraded, and active LDD is still in force, which points at the vulnerability of non-degraded lands around the globe, especially in drylands.

Why LDD Is of Concern?

LDD reduces the capacity of earth to provide food; hence, it is of concern not only to the direct land users but to humanity at large. Furthermore, even the LDD's local effect on the well-being of the direct land users is implicated with generating externalities of a global concern, as the local LDD-driven poverty leads to within-country, cross-boundary, and overseas migrations of refugees, which often trigger conflicts of variable severity and foreign and international interventions (Figures 31.4 and 31.5). This component of the LDD paradigm is widely alluded to, though hard data on the causation and the dimension of each of these processes at the global scale are scarce. Regarding poverty, countries' statistics of poor rural land users do not distinguish between LDD-driven poverty and poverty not driven by LDD; it adds the poverty of land users with that of rural people who do not live off the land, as well as that of the country's dryland and non-dryland inhabitants. Thus, only localized statistics (e.g., 68–84% poverty

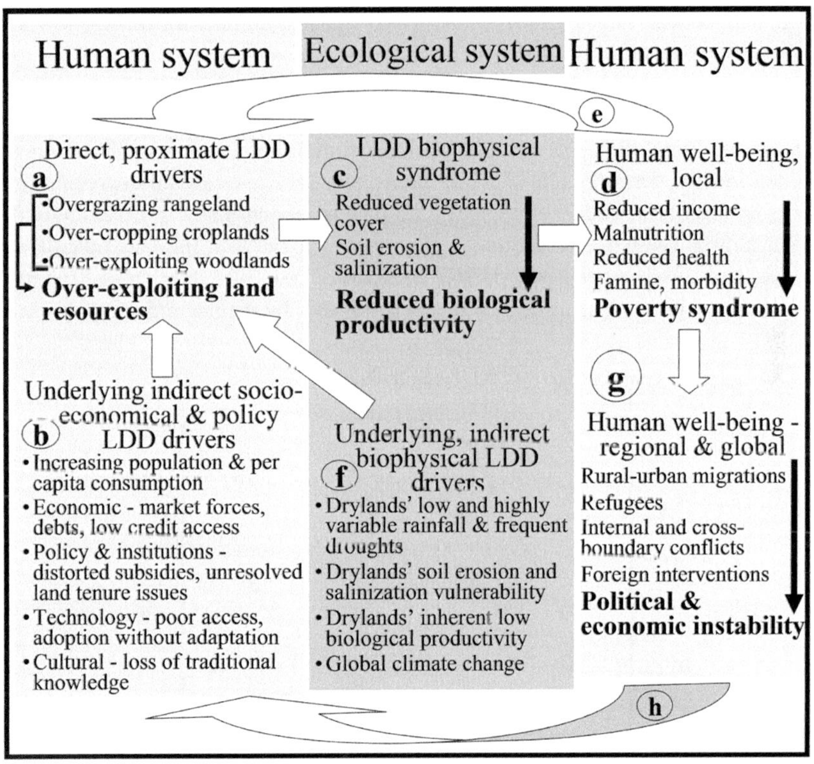

FIGURE 31.5 The land degradation and desertification (LDD) paradigm—interactions and feedback loops: Direct proximate LDD drivers (a) (detailed in Fig. 3), driven by underlying socioeconomic and policy drivers (b) (detailed in Fig. 4) generate the LDD syndrome (c) (detailed in Fig. 3); the reduced human well-being impinges on the direct drivers thus further exacerbating LDD (e); underlying indirect biophysical drivers also drive people to further exacerbate LDD (f) (detailed in Fig. 2); the locally affected human well-being (d) also has regional and global repercussions (g), and human well-being at all levels impinges on the underlying LDD drivers (h). Vertical arrows indicate the linear effects of variables from top to bottom of lists, while bullets indicate the independent effects of each variable.

in the drylands in northern Kenya that is higher than in other, non-dryland parts of the country) suggest a prevalence of poverty in drylands of developing countries, and also has led to a rough estimate of about nearly one billion people in poor rural drylands.[16] A direct, quantifiable linkage between poverty and LDD at the global scale is so far provided only by the observation that the poverty indicators—infant mortality and malnutrition of under-5-year-old children track the trajectory of desertification severity across the global aridity gradient (Figure 31.6). Similar to the LDD-poverty nexus, there is no hard data on the linkage among poverty, migrations, refugees, and conflicts, and evidence for LDD being instrumental is circumstantial. Most cited are migrations following severe droughts, while only a few cases present a direct linkage, for example, the Tambacounda region of Senegal's 1960s migrations driven by soil erosion-triggered decline of crop production—to the capital, other cities, other African countries, and Europe.[17] Social scientists also disagree on the role of environmental degradation (LDD included) in generating conflicts and violence, and concluded that empirical evidence is still insufficient to demonstrate a tangible effect.[18] However, comprehensive research of the Darfur conflict[19] asserted that LDD in both croplands and rangelands has been one of the major factors in that crisis but it has been also intertwined with a range of other socioeconomic and political issues. Even if LDD does not always generate local poverty of regional and global repercussions, any local loss of land productivity undermines food security not only locally, but also at the global scale. It has been estimated that for feeding the global population by year 2050, 8.9 million km^2 of good cultivable land would be required, on top of the 15 million km^2 already under cultivation.[20] However, only 4.4 million km^2 of cultivable land will be available.[21] Therefore, LDD which practically converts cultivable to non-cultivable land could dramatically aggravate the already fragile global food security. LDD is of concern not only due to loss of crop and forage production but also due to loss of other benefits that people derive from the land, such as water regulation and purification, local climate amelioration, pollination, biodiversity conservation, and others. These benefits, called "ecosystem services," are an expression of the land's functionality as an ecological system. Adopting a holistic approach, LDD is increasingly described as a persistent, substantial reduction in a bundle of services provided to humans by cultivated, range, or natural ecosystems.[1,22] The persistent decline in the provision of one of these dryland ecosystem services, carbon sequestration (the uptake of atmospheric CO_2 by plants, and transforming it to plant

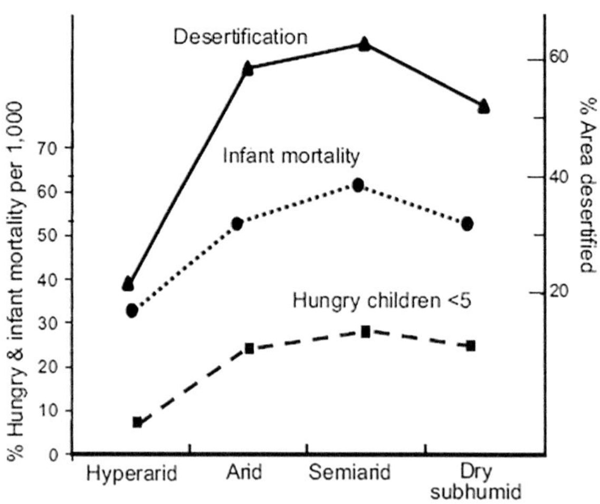

FIGURE 31.6 Trends of land degradation and desertification and poverty indicators across the aridity gradient of the global drylands. Land degradation and desertification (LDD) data are of GLASOD,[12] with the "low degradation" category excluded; data of infant mortality per 1000 births and percentage of hungry children under the age of 5, drylands of OECD countries excluded, are from the Millennium Ecosystem Assessment.[8]
Source: Adapted from Fig. 4 of Safriel.[32]

biomass and litter) is of global concern, too. All dryland vegetation and all dryland soils store 14% and 20% of the global organic carbon, respectively.[8] LDD-associated soil erosion and loss of plant cover and plant productivity deplete this land's storage of organic carbon. This locally removed organic carbon is eventually oxidized and emitted back to the atmosphere, thus increasing atmospheric CO_2 concentration and exacerbating climate change, globally. Indeed, drylands' LDD-driven emissions are estimated to comprise about 4% of the total global emissions from all sources combined (in year 2000),[1] and added LDD would further amplify global climate change.

What Has Been Done and What Can Be Done?

That most lands used in drylands are not yet affected by LDD suggests that drylands can be used sustainably. Many cases have been documented, mostly in east and west African drylands, in which rural communities adapted to rapid population growth, globalization, market development, and technological change and avoided LDD. These apparently detrimental changes triggered innovative actions: investments in more intensive yet sustainable land use, which helped reaping benefits of the enlarged market opportunity; exploitation of local bioclimatic and social comparative advantages; access to technologies that increased land and labor productivity faster than population growth; and improved access to growing markets.[23] Avoiding LDD can be also achieved by adding income generating "alternative" land-independent livelihoods like tourism and aquaculture,[24] or generating supplementary income from seasonal migration to urban centers.[17]

Nevertheless, it is developing countries that share most of the global drylands,[8] as well as probably the most rural poverty and LDD risks. Noting these facts and being aware of the global threats of dryland LDD, the international community negotiated and adopted the UNCCD, with the intention of coordinating and promoting assistance of industrial (developed) to developing countries in addressing their LDD problems. Entering into force in 1996, the 195 country parties of the UNCCD together with the Secretariat and the subsidiary bodies of the Convention succeeded in raising awareness, encouraging knowledge and research, and promoting capacity-building and partnerships in projects on the ground, as well as installing mechanisms for assessing success. A major thrust of the international community is to drive the political process of setting a sustainable development goal—striving for Land Degradation Neutral World—and its associated target of achieving Zero Net Land Degradation (ZNLD) by 2030, initiated at the United Nations Conference on Sustainable Development convened in June 2012 in Rio De Janeiro. The rationale behind this target is that completely halting further decline of the productive, non-degraded land within a relatively short time is unattainable. However, the amount of non-degraded land that becomes degraded within a time period can be offset by a similar amount of already degraded land, whose productivity is restored within the same time period. Once this offset is attained, the net rate of LDD is zero. Setting such goals and targets can help shape expectations and create the conditions for all stakeholders to assess progress and take appropriate action in addressing LDD.

As to reducing LDD for addressing the ZNLD target, "sustainable land management" (SLM) is widely prescribed. In practice, any land that supports the user does not lead LDD to qualify as SLM. More specifically, practices of land use need to adaptively maintain the resilience of the used ecosystems to all LDD drivers, such that the thresholds in ecosystem functions will not be crossed and the service provision will be sustained.[25] Avoiding degradation may be easier and cheaper than restoring already degraded land but, as land is often degraded inadvertently while restoration is planned, its success prospects are relatively high. The first stage is removal of the LDD causes, to be followed by selecting from the rich literature of ecological restoration[26] the toolbox appropriate for the local state and the set restoration target.

Desertification and Climate, and Climate Change

Drylands are not only exposed to climate-driven water scarcity but also to high between-year variability. This includes relatively short periods (<5 years) during which rainfall and hence productivity remain far below the long-term average. These periods are termed "droughts," and since their associated productivity decline is reversible, this productivity decline does not qualify as desertification. It can be expected that a drought-driven productivity decline would persist after the drought terminate, provided that it has been severe and exceptionally long. Such events, if ever occurred, have not been reliably documented. On the other hand, drylands in which human-induced LDD processes have been already initiated are likely to be pushed by a drought to cross the desertification threshold, such that productivity decline would persist after the drought terminates. Such cases are rare, too. Even the reduced productivity during the Sahel droughts of 1968–1973 and 1982–1984 of the 20th century, which took a heavy toll on properties and lives and aroused the world's concern that culminated in adopting the UNCCD, was restored once the droughts terminated.[27] Nevertheless, though droughts alone do not cause LDD, they do exacerbate LDD and its effect on human well-being, and hence droughts are listed among the underlying, indirect biophysical LDD drivers (Figure 31.5). The role of drought in desertification is laid out in the full name of the UNCCD—a convention *"to combat desertification, in those countries experiencing serious droughts..."* and in its objective, that is not only to *"combat desertification"* but also to *"mitigate the effects of droughts."*[28] In spite of the difference between "combating" and "mitigating," an emerging and widely used acronym for the UNCCD's subject matter is Desertification, Land Degradation, and Drought, or DLDD.

Coining the DLDD acronym is partly justified given that the frequency of droughts and their severity are projected to increase, driven by the ongoing global warming and climate change. Indeed, climate change (and poverty) is projected to intensify desertification by 2050 in all the four Millennium Ecosystem Assessment's scenarios of global governance.[1] Climate change's effect on global aridity is already evident. Comparison of the climatic conditions, precipitation, and temperatures that prevailed in 1931–1960 and used for setting the drylands' boundaries, with those prevailing in the period of 1961–1990, demonstrated that the spatial extent of the African hyper-arid and arid drylands increased by 51 and 3 million ha, respectively, which comprise 1.7% and 0.1% of their area extent in 1931–1960, respectively. More significant is the "loss" of 25 million ha of African non-drylands of 1931–1960 that turned into dry subhumid drylands during 1961–1990.[29] Furthermore, rainfall decreases and evapotranspiration increases are projected to intensify this trend in many drylands by mid-century (IPCC 2008).[30] To conclude, though current LDD is driven by humans at the local scale, climate change, apparently human-driven, too, also leads to local productivity persistent decline, that is LDD, but one in which the local land users are not directly involved.

Conclusions

Desertification is a persistent reduction in land's biological productivity of an economic value, as well as in other services provided by the land's ecosystems. It is an extreme, terminal state of the land degradation process, and LDD can occur anywhere but the focus is on drylands; 40% of global land whose inherent low productivity is constrained by water. LDD comprises a syndrome of dynamically interactive biophysical processes, directly driven by land management practices of land users, themselves driven by underlying socioeconomic and policy interlinked processes that cross temporal and spatial scales, as well as by the inherent vulnerability of dryland soils to LDD, and by the effects of droughts.

Though the cumulative area of degraded land is only about a quarter of the globe's productive land, the process is still ongoing apparently at an increasing rate. LDD is therefore of global concern—the poverty it inflicts on the local land users is a source of migrations, refugees, and conflicts across boundaries and scales; and the local loss of productive land risks the already fragile global food security. In addition, LDD and global climate changes are interlinked; climate change currently increases dryness and hence LDD vulnerability, and LDD reduces the land's carbon storage locally thus increasing emissions and warming, globally. The good news is that the UNCCD succeeded in raising awareness to the LDD plight, such that the international community seems ready to commit itself to significantly reducing further degradation and investing in restoring already degraded lands. Knowledge, institutions and instruments are available, and the political will and resources need to be secured.

References

1. Adeel, Z.; Safriel, U.; Niemeijer, D.; White, R. *Ecosystems and Human Well-Being: Desertification Synthesis*; The Millennium Ecosystem Assessment; World Resources Institute: Washington, DC, 2005.
2. Aubreville, A. Climats, Forest, et Desertification de l'Afrique Tropicale. Societe de Editions 3/ Geographiques, Maritime et Coloniales, Paris, 1949.
3. Vogt, J.V.; Safriel, U.; Von Maltitz, G.; Sokona, Y.; Zougmore, R.; Bastin, G.; Hill, J. Monitoring and assessment of land degradation and desertification: towards new conceptual and integrated approaches. Land Degrad. Dev. **2011**, 22, 150–165.
4. Safriel, U.N. Global Land Degradation: State, Risks and Prospects Under Global Change. In *Soil, Society and Global Change*; Bigas, H., Gudbransson, G.I., Monanarella, L.; Arnalds, O., Eds.; European Communities: Italy, 2009; 46–52.
5. http://dsd-consortium.jrc.ec.europa.eu/documents/WG1_White-Paper_Draft-2_20090818.pdf (accessed March 2012). White Paper of DSD Working Group 1, August 2009, Integrated Methods for Monitoring and Assessing Desertification/Land Degradation Processes and Drivers.
6. Geist, H.J.; Lambin, E.F. Dynamic causal patterns of desertification. Bioscience **2004**, *54*, 817–829.
7. Reynolds, J.F.; Grainger, A.; Stafford Smith, D.M.; Bastin, G.; Garcia-Barrios, L.; Fernandez, R.J.; Janssen, M.A.; Jurgens, N.; Scholes, R.J.; Veldkamp, A.; Verstraete, M.M.; Von Maltitz, G.; Zdruli, P. Scientific concepts for an integrated analysis of desertification. Land Degrad. Dev. **2011**, *22*, 166–183.
8. Safriel, U.N.; Adeel, Z. Dryalnd Systems. In *Ecosystems and Human Well-Being: Current State and Trends*; Hassan, R., Scholes, R., Ash, N., Eds.; Island Press: Washington, 2005; 623–662.
9. Safriel, U.N. Alternative livelihoods for attaining sustainability and security in drylands. In *Coping with Global Environmental Change, Disasters and Security*; Brauch, H.G., Oswald Spring, U., Mesjasz, C., Grin, J., Kameri-Mbote, P., Chourou, B., Dunay, P., Birkmann, J. Eds.; Vol 5, Hexagon Series on Human and Environmental Security and Peace, edited by J. Springer: Berlin, 2011; 835–852.
10. Reynolds, J.F.; Stafford Smith, D.M.; Lambin, E.F.; Turner, B.L. II,; Mortimore, M.; Batterbury, S.P.J.; Downing, T.E.; Dowlatabadi, H.; Fernandez, R.J., Herrick, J.E., Huber-Sannwald, E.; Jiang, H.; Leemans, R,; Lynam, T.; Maestre, F.T.; Ayarza, M., Walker, B. Global desertification: building a science for dryland development. Science **2007**, *316,* 847–851.
11. Safriel, U.N. The assessment of global trends in land degradation. In *Climate and Land Degradation*; Sivakumar M.V.K., Ndiaugui, N., Eds.; Springer: Berlin, 2007; 1–38.
12. http://www.isric.org/projects/global-assessment-humaninduced-soil-degradation-glasod (accessed March 2012) GLASOD, Global Assessment of Soil Degradation, International Soil Reference and Information Centre, Wageningen, Netherlands, and United Nations Environment Programme, Nairobi, Kenya; 1990.
13. Middleton, N.; Thomas, D. Eds., *World Atlas of Desertification*; Arnold: London, 1997; 18 p.
14. Bai, Z.; Dent, D.; Olsson, L.; Schaepman, M.E. Proxy global assessment of land degradation. Soil Use Manag. **2008**, *24*, 223–243.
15. Wessels, K. Comments on proxy global assessment of land degradation, by Bai et al. 2008. Soil Use Manag. **2009**, *25*, 91–92.
16. Dobie, P. Poverty and the drylands. UNDP, Nairobi. September 2001. http://arabstates.undp.org/contents/file/Poverty-and-the-Drylands-Challenge-Paper.pdf (accessed March 2012).
17. Leighton, M. Desertification and migration. In *Governing Global Desertification: Linking Environmental Degradation, Poverty and Participation;* Johnson, P.-M., Mayrand, K., Paquin, M., Eds.; Ashgate Publishing: Aldershot, England, 2006; 43–58.
18. Safriel, U.; Adeel, Z. Development paths of drylands—is sustainability achievable? Sustain. Sci. J. **2008**, *3* (1), 117–123.

19. Sudan Post-Conflict Environmental Assessment: United Nations Environment Programme; Nairobi, Kenya, 2007. http://postconflict.unep.ch/publications/UNEP_Sudan.pdf (accessed March 2012).
20. Fischer, G.; Shah, M.; van Velthuizen, H.' Nachtergaele, F. O. Global agroecological assessment for agriculture in the 21st century. 2001, http://www.iiasa.ac.at/Research/LUC/SAEZ/index.html (accessed March 2012).
21. Tilman, D.; Fargione, J.; Wolff, B.; D'Antonio, C.; Dobson, A.; Howarth, R.; Schindler, D.; Schlesinger, W. H.; Simberloff, D.; Swackhamers, D. Forecasting agriculturally driven global environmental change. Science **2001**, *292*, 281–284.
22. Verstratete, M. M.; Scholes, R.J.; Stafford Smith, M. Climate and desertification: looking at an old problem through new lenses. Front. Ecol. Environ. **2009**, *7* (8), 421–428.
23. Nkonia, E.; Winslow, M.; Reed, S.; Mortimore, M.; Mirzagaev, A. Monitoring and assessing the influence of social, economic and policy factors on sustainable land management in drylands. Land Degrad. Dev. **2011**, *22*, 240–247.
24. Adeel, Z.; Safriel, U. Achieving sustainability by introducing alternative livelihoods. Sustain. Sci. J. **2008**, *3* (1), 125–133.
25. Cowie, A.L.; Penman, T.D.; Gorssen, L.; Winslow, M.D.; Lehmann, J.; Tyrrell, T.D.; Twomlow, S.; Wilkes, A.; Lal, R.; Jones, J.W.; Paulsch, A.; Kellner, K.; Akhtar-Schuster, M. Towards sustainable land management in the drylands: scientific connections in monitoring and assessing dryland degradation, climate change and biodiversity. Land Degrad. Dev. **2011**, *22*, 248–260.
26. Bainbridge, D.A. A. Guide for Desert and Dryland Restoration; Island Press: Washington, 2007.
27. Prince, S.D.; Brown De Colstoun, E, Kravitz, L,L. Evidence from rain use efficiencies does not indicate extensive Sahelian desertification. Clim. Change **1988**, *4*, 359–458.
28. United Nations Convention to Combat Desertification (UNCCD) http://www.unccd.int.
29. Hulme, M.; Marsh, R; Jones, P.D. Global change in a humidity index between 1931–1960 and 1961–1990. Clim. Res. **1992**, 2, 1–22.
30. Parry, M.L.; Canziani, O.F.; Palutikof J.P., van der Linden, P. J.; Hanson, C.E. Eds.; Climate Change 2007: Impacts, Adaptation and Vulnerability. Contribution of Working Group II to the Fourth Assessment Report of the Intergovernmental Panel on Climate Change. Cambridge University Press: Cambridge, UK and New York, NY, USA., 2007.
31. Safriel, U.N. Dryland development, desertification and security in the Mediterranean. In *Desertification in the Mediterranean Region. A Security Issue*. Kepner, W.G., Rubio, J.L., Mouat, D.A., Pedrazzini, F., Eds.; NATO Security through Science Series, Volume 3, Springer Publishers: Germany, 2006; 227–250.
32. Safriel, U.N. Deserts and Desertification: Challenges but also Opportunities. Land Degrad. Dev. **2009**, *20*, 353–366.

Bibliography

Allington, G.R.H.; Valone, T.J. Reversal of desertification: the role of physical and chemical soil properties. J. Arid Environ. **2010**, *74*, 973–977.

Briassoulis, H. *Policy Integration for Complex Environmental Problems. The Example of Mediterranean Desertification*; Ashgate: Burlington, USA, 2005.

Corell, E. *The Negotiable Desert. Expert Knowledge in the Negotiations of the Convention to Combat Desertification*; Linkoping University, Lindoping: Sweden, 1999; 286 pp.

Fantechi, R.; Peter, D.; Balabanis, P.; Rubio, J.L. *Desertification in a European Context: Physical and Socio-Economic Aspects*; European Commission: Luxembourg, 1955.

Grainger, A. *The Threatening Desert: Controlling Desertification*; Earthscan and UNEP: Nairobi, 1990.

Safriel, U.N. The concept of sustainability in dryland ecosystems. In *Arid Lands Management – Toward Ecological Sustainability*; Hoekstra, T. W., Shachak, M., Eds.; University of Illinois Press: Urbana, 1999; 117–140.

Schlesinger, W.H.; Reynolds, J.F.; Cunningham, G.L.; Huenneke, L.F.; Jarrell, W.M.; Virginia O.A.; Whitford, W.G. Biological feedbacks in global desertification. Science **1990**, *247*, 1043–1048.

Swift, J.; Leach, M.; Mearns, R. *Desertification. Narratives, Winners & Losers*. James Currey: Oxford, UK, 1996.

Tiffen, M.; Mortimore, M.; Gichuki, F. *More People, Less Erosion: Environmental Recovery in Kenya*; J. Wiley: Chichester, England, 1994.

Thomas, D.S.G.; Middleton, N. *Desertification: Exploring the Myth*; J. Wiley: Chichester, 1994.

Williams, M.A.J.; Balling, R.C. *Interactions of Desertification and Climate*; Arnold, Ahalsted Press: 1996; 372.

Xu, D.Y.; Kang, X.W.; Zhuang, D.F; Pan, J.J. Multi-scale quantitative assessment of the relative roles of climate change and human activities in desertification – A case study of the Ordos Plateau, China. J. Arid Environ. **2010**, 74, 498–507.

32

Ecological and Evolutionary Processes

Roland C. de Gouvenain
Rhode Island College

Gopalasamy Reuben Clements
Sunway University

Introduction 281
Natural Resources Are Components of Ecosystems 281
Ecosystems and Their Resources Are Shaped by Evolutionary Changes 282
Natural Resource Overexploitation Jeopardizes Ecosystem Health 282
Ecologically Sustainable Natural Resource Management Can Protect Evolutionary Processes 283
Evolutionary Conservation Strategies Are Needed to Insure Maintenance of Genetic Biodiversity in the Context of Climate Change 284
Conclusion 285
References 285

Introduction

Natural resources, whether renewable like forests or fisheries, or non-renewable like crude oil or minerals, have provided generations of human beings with food, shelter, and spiritual or aesthetical enjoyment. Unfortunately, these resources are too often being destroyed or extracted at unsustainable rates. Only when we begin responsibly managing our resources within an ecosystem framework that considers long-term social good will, we achieve some equilibrium between extraction and conservation.

Natural Resources Are Components of Ecosystems

Biotic (living) natural resources such as forests, wildlife, and fisheries are vital components of the different ecosystems on our planet; for instance, sphagnum moss is a component of the arctic tundra or boreal ecosystems, and gray reef sharks are components of tropical coral reef ecosystems. Abiotic (non-living) natural resources (for instance, the atmosphere, water, and soils) influence the health of those ecosystems. Soils, which took thousands of years to develop under the activity of microbes, fungi, and invertebrates, are also products and components of ecosystems; they sustained the advent of agriculture 10,000 years ago and support today's agricultural productivity [1,2]. Located deep underground where no life is found today, abiotic natural resources such as fossil fuels are nonetheless products of ancient forest or marine ecosystems now fossilized [1]. Thus, past and present ecosystems are the source of many of our natural resources today [2,3].

Ecosystems and Their Resources Are Shaped by Evolutionary Changes

All living natural resources such as forests, fisheries, and wildlife are not only shaped by ecosystems, but they also influence how these ecosystems function and evolve. From the beginning of life on Earth 3.5 million years ago, ecosystems and their biotic and abiotic natural resources have been shaped by evolutionary changes in the world's biota (all the living organisms that inhabit the Earth). In fact, evolutionary forces were already in motion between 3 and 2 billion years ago, when immense colonies of single-celled *Cyanobacteria* were slowly transforming the early atmosphere of the Earth from a reducing one (low in oxygen gas) to an oxidizing one (high in oxygen gas) as a result of their photosynthetic activity [4]. This atmospheric transformation itself would, about 2.5 billion years ago, foster the evolution of more complex organisms that relied on aerobic respiration to harvest the energy contained in their food. These primitive eukaryotes would eventually (around 550 million years ago) evolve into animals, plants, and fungi [4,5]. Approximately 350 million years ago, during the tropical climate of the Carboniferous period, the Earth's biota would similarly shape its ecosystems when newly evolved tree-sized vascular terrestrial plants produced vast *Carboniferous forests*. These ancient forests were then fossilized over millions of years into thick coal deposits that fueled the industrial revolution and still power today's electricity production [1,5]. Just as an oxygen-rich atmosphere was the product of photosynthesis by Cyanobacteria colonies, and just as today's coal deposits are the product of the Carboniferous forests, today's ecosystems are the product of their component organisms that have themselves evolved over millions of years in response to abiotic and biotic sets of natural selection factors relevant to each ecosystem.

Natural Resource Overexploitation Jeopardizes Ecosystem Health

The current natural resource management is driven by the exponential growth of the human population (around 7.7 billion people as of 2019) and socioeconomic goals that seek to maximize resource extraction with little attention to the perils of overexploitation [1,6]. Technological advances have accelerated the rate at which resources are being harvested, and this has resulted in irreversible alterations to ecosystems, especially those that are not managed according to sustainability guidelines, such as forest concessions that do not carry out reduced-impact logging. By increasing the amount and rate of extraction of biotic and abiotic natural resources, human populations change (often irreversibly) the species composition and ecology of natural ecosystems. This in turn influences the natural selection processes that shape these ecosystems, in addition to impacting the amount and location of natural resources available for future generations.

From the cold-adapted microorganisms in the permafrost soils of Siberia to the salt-adapted trees of tropical mangroves, many plant, animal, and fungi species are pushed to the limit of their ecological tolerance because of the loss or transformation of their natural environment by humans. Perhaps the most well-known example is that of polar bears faced with an ever-shrinking Arctic Sea ice under the influence of global climate change, but many more examples abound, including the loss of Asian tigers impacted by the logging of Asian tropical dipterocarp forests and their replacement with oil palm or rubber plantations, the disappearance of migratory bird species due to loss of stop-over wetlands, the collapse of oceanic fisheries due to overfishing, and the poaching of charismatic large mammals in increasingly logged rain forests [7–13]. Although some cases of rapid evolutionary adaptive response of plants and insects to environmental degradation or to climate change have been documented [14,15], most studies report negative impacts (including species extinction) from the overharvesting of biotic resources or from the impacts of abiotic resource extraction [11,13,15].

By reducing the biodiversity (the diversity of species and their genetic diversity) of the Earth, we are not only losing plant, animal, and fungi species that could potentially provide future generations with critical resources, including yet undiscovered medicinal compounds, but also losing the ecological services these ecosystems provide us with (such as the storage of clean water in forest watersheds).

Furthermore, we are destroying the very species and ecological processes that have made these ecosystems productive and resilient to natural or human-induced environmental changes [15,16]. Losing walleye pollock (a prey species for predatory marine fish, birds, and mammals) because of climate change affects the other marine species in the Bering Sea food web, and degrades the health and productivity of the oceanic ecosystems that support economically valuable fisheries [17].

As species and genetic diversities are lost, the raw material upon which natural selection operates is lost as well. Loss of intertidal species (for instance, mangrove trees and seagrass beds) to development reduces the ability of coastal ecosystems to filter polluted effluents or to protect shorelines from storm flooding [13], which can hinder future evolutionary adaptation of these coastal ecosystems to climate and sea-level changes. Worldwide, nearly 30% of the currently fished species are considered collapsed (>90% decline), and this decline occurred faster in species-poor ecosystems than in species-rich ones, suggesting that maintaining ecosystem biodiversity is the key to future ecosystem resilience in the face of human impact [13,16]. Monoculture plantations of cash crops, such as oil palm and rubber, are rapidly transforming tropical forests into biological deserts [7,18]. Southeast Asia has the highest rate of tropical deforestation, losing around 1.0% of its forests per year [19], and some countries like the Philippines and Singapore have nearly no forest cover left. Indonesia's forest ecosystems are impacted by even higher rates of deforestation (2.6% per year), and two-thirds of its original forests are now replaced with oil palm plantations. In large part as a result of this rapid deforestation, more than 25% of the mammals, more than 20% of the birds, and more than 15% of the native plants species of Southeast Asia are now extinct [11].

Ecologically Sustainable Natural Resource Management Can Protect Evolutionary Processes

To be ecologically sustainable, management of natural resources must protect the ecological integrity of ecosystems, including their biodiversity, so that the evolutionary processes that generated these ecosystems can be maintained in the future. This will increase the likelihood that these ecosystems will be able to evolve in response to changing global conditions [16,20], especially because it is difficult to predict with certainty the characteristics of future terrestrial and aquatic environments [21]. For example, maintaining habitat connectivity (through wildlife corridors, for instance) can help ensure a beneficial exchange of genetic material among populations isolated by habitat conversion [9,22,23]. Conserving the health of soil ecosystems by protecting them from erosion and maintaining a diversity of nutrient-cycling microbes can ensure that future generations will enjoy sustained agricultural productivity [24].

Perhaps the single-most important indicator of ecosystem health is biodiversity, and the genetic diversity it manifests is what natural selection can operate on to allow the natural processes of evolution to keep these ecosystems healthy, resilient, and productive [16,20]. Not only can productive fisheries not exist in species-poor oceans, or economically viable wood supplies not be maintained in unsustainably logged landscapes, but species-poor ecosystems are also much more likely to collapse in the face of natural or human-induced environmental change than species-rich ones [16]. Terrestrial and marine species diversity enhances ecosystem processes such as nutrient cycling and primary production and increases the resilience of ecosystems to disturbance and therefore the capacity of these ecosystems to provide services to future human generations [13]. Sustainable resource extraction is not only common sense for long-term economic benefits to humans, but it is also a management practice that can give the ecosystems of the world a chance to adapt to and persist [20]. For instance, sustainably logged Malaysian tropical forests can still support reasonable densities of tiger populations [25], and, as long as forest structure and biodiversity are maintained, even old logging roads are traveled regularly by the big cats (Figure 32.1).

To be sustainable, natural resource management should be both ecologically responsible; that is, it should respect the "economy of nature" as defined by the German biologist Ernst Haeckel, and socially ethical, as suggested by the American ecologist Aldo Leopold [26]. Societal decisions regarding both the

FIGURE 32.1 (See color insert.) Tiger on abandoned logging road photographed by photo trap, Tembat Forest Reserve, Terengganu, Peninsular Malaysia. (Source: Photo by Rimba/Reuben Clements.)

exploitation and the conservation of natural resources should be based on ecological knowledge and research, because ecological sustainability is the key to long-term success [23,27], and on social justice, because extracting and conserving resources involve trade-offs that have ecological and social costs [28–30]. This is especially urgent, given the fact that most of the world's biodiversity "hotspots" (areas with both high biodiversity and high rates of natural habitat loss) [10] are also areas of relatively high human population density and growth [31].

To assess these trade-offs and make socially fair decisions regarding the extraction and conservation of natural resources, other points of view besides the dominant market-oriented western perspective should be solicited and appraised, including the traditional ecological knowledge of the world's rural cultures [32]. Giving resource management more legitimacy by involving local human communities in the decision process concerning the extraction and conservation of natural resources is likely to yield management actions that are not only ecologically but socially sustainable as well [28]. For instance, conservation of the Tampolo coastal rain forest in Madagascar has enjoyed support from local communities by mixing traditional covenant ceremonials, agricultural development, ecotourism, native tree species planting, and environmental education. In Thailand, a successful community-based conservation approach has even resulted in the recovery of ungulate populations that were previously subjected to poaching pressure [33]. Community-based conservation approaches may involve negotiating trade-offs and compromises that are complex and thus require more time to achieve than the traditional "top-down" conservation approach. However, the process itself may not only empower local communities to own and manage their resources as caretakers, but may also enhance the educational benefit for all parties involved and allow stakeholders to embrace this necessary commitment to long-term resource conservation for the sake of future human generations.

Evolutionary Conservation Strategies Are Needed to Insure Maintenance of Genetic Biodiversity in the Context of Climate Change

Because genetic diversity across populations of the same species [34,35] (Figure 32.2) can increase the chance that a species will evolve and adapt to a changing environment [36], genetic diversity is a major component of biodiversity and the basis for adaptation in a changing world [37] and should thus become a major focus of biodiversity conservation strategies and planning. Because assessments of intraspecific genetic variability are rare, more easily obtained surrogate measures such as habitat and ecotype diversity could be used to estimate that variability [37]. Today, the aim of evolutionary conservation is to protect not only hotspots of threatened species biodiversity but also hotspots of

FIGURE 32.2 **(See color insert.)** Some populations of the serotinous South African shrub *Protea repens* exhibit mostly closed cones (a), while others exhibit mostly open cones (b), revealing an underlying intraspecific genetic diversity that could enhance the resilience of *P. repens* to global environmental changes. (Source: Photos by Roland de Gouvenain.)

threatened evolutionary history, using several indices such as *Expected Phylogenic Diversity Loss*, *Heightened Evolutionary Distinctiveness and Global Endangerment* (HEDGE), and *Biogeographically Weighted Evolutionary Distinctiveness* (BED) [38]. Evolutionary conservation biologists argue that, in a rapidly changing environment driven by anthropogenic activities, protecting evolutionary history may be the most urgent goal of a new conservation strategy aimed at promoting the resilience of threatened ecosystems [38].

Conclusion

Unsustainable extraction of natural resources, pollution from the use of these resources, and anthropologically driven climate change have modified nearly the entire set of ecosystems found on Earth. But as long as ecological processes are not irremediably impacted, and species and genetic diversity is maintained, intrinsic properties of these ecosystems, including evolutionary histories, can allow those ecosystems to be resilient to global environmental changes [9]. It behooves us, as de facto caretakers of the Earth's environment, to conserve ecosystems and their specific and genetic diversity so that evolutionary processes can allow those ecosystems to persist and provide future generations with an ecologically and socially livable world.

References

1. Chiras, D.D.; Reganold, J.P. *Natural Resource Conservation: Management for a Sustainable Future*; Pearson Prentice Hall: Upper Saddle River, NJ, 2005; p. 644.

2. Holechek, J.L.; Cole, R.A.; Fisher, J.T.; Valdez, R. *Natural Resources: Ecology, Economics, and Policy*; Pearson Prentice Hall: Upper Saddle River, NJ, 2003; pp. 716–718.
3. Weddell, B.J. *Conserving Living Natural Resources in the Context of a Changing World*; Cambridge University Press: Cambridge, MA, 2002; p. 426.
4. Knoll, A. *Life on a Young Planet*; Princeton University Press: Princeton, NJ, 2003; p. 296.
5. Futuyma, D.J. *Evolution*; 2nd Ed.; Sinauer: Sunderland, 2009.
6. Rutherford, M.B. Resource management. In *Encyclopedia of Environmental Ethics and Philosophy*; Callicott, J.B.; Frodeman, R., Eds.; Macmillan – Gale: Farmington Hills, 2009; pp. 198–201.
7. Aziz, S.A.; Laurance, W.F.; Clements, R. Forests reserved for rubber. *Front. Ecol. Env.* 2010, *8*, 178.
8. Clements, R.; Rayan, D.M.; Ahmad Zafir, A.W.; Venkataraman, A.; Alfred, R.; Payne, J.; Ambu, L.N.; Sharma, D.S.K. Trio under threat: can we secure the future of rhinos, elephants and tigers in Malaysia? *Biodivers. Conserv.* 2010, *19*, 1115–1136.
9. Gardner, T.A.; Barlow, J.; Chazdon, R.; Ewers, R.M.; Harvey, C.A.; Peres, C.A.; Sodhi, N.S. Prospects for tropical forest biodiversity in a human-modified world. *Ecol. Lett.* 2009, *12*, 561–582.
10. Myers, N.; Mittermeier, R.A.; Mittermeier, C.G.; da Fonseca, G.A.B.; Kent, J. Biodiversity hotspots for conservation priorities. *Nature* 2000, *403*, 853–858.
11. Sodhi, N.S.; Koh, L.P.; Brook, B.W.; Ng, P.K.L. Southeast Asian biodiversity: an impending disaster. *Trends Ecol. Evol.* 2004, *19*, 654–660.
12. Venter, O.; Meijaard, E.; Possingham, H.; Dennis, R.; Sheil, D.; Wich, S.; Hovani, L.; Wilson, K. Carbon payments as a safeguard for threatened tropical mammals. *Conserv. Lett.* 2009, *2*, 123–129.
13. Worm, B.; Barbier, E.B.; Beaumont, N.; Duffy, J.E.; Folke, C.; Halpern, B.S.; Jackson, J.B.C.; Lotze, H.K.; Micheli, F.; Palumbi, S.R.; Sala, E.; Selkoe, K.A.; Stachowicz, J.J.; Watson, R. Impacts of biodiversity loss on ocean ecosystem services. *Science* 2006, *314*, 787–790.
14. Bone, E.; Farres, A. Trends and rates of microevolution in plants. *Genetica* 2001, *112–113*, 165–182.
15. Parmesan, C. Ecological and evolutionary responses to recent climate change. *Ann. Rev. Ecol. Evol. Syst.* 2006, *37*, 637–669.
16. Elmqvist, T.; Folke, C.; Nyström, M.; Peterson, G.; Bengts-son, J.; Walker, B.; Norberg, J. Response diversity, ecosystem change, and resilience. *Front. Ecol. Env.* 2003, *1*, 488–494.
17. Walther, G.-R.; Post, E.; Convey, P.; Menzel, A.; Parmesan, C.; Beebee, T.J.C.; Fromentin, J.-M.; Hoegh-Guldberg, O.; Bairlein, F. Ecological responses to recent climate change. *Nature* 2002, *416*, 389–395.
18. Koh, L.P.; Wilcove, D.S. Is oil palm agriculture really destroying tropical biodiversity? *Conserv. Lett.* 2008, *1*, 60–64.
19. Miettinen, J.; Shi, C.; Liew, S.C. Deforestation rates in insular Southeast Asia between 2000 and 2010. *Glob. Change Biol.* 2011, *17*, 2261–2270.
20. Fazey, I.; Gamarra, J.G.P.; Fischer, J.; Reed, M.S.; Stringer, L.C.; Christie, M. Adaptation strategies for reducing vulnerability to future environmental change. *Front. Ecol. Environ.* 2010, *8*, 414–422.
21. Lawler, J.J.; Tear, T.H.; Pyke, C.; Shaw, M.R.; Gonzalez, P.; Kareiva, P.; Hansen, L.; Hannah, L.; Klausmeyer, K.; Aldous, A.; Bienz, C.; Pearsall, S. Resource management in a changing and uncertain climate. *Front. Ecol. Environ.* 2010, *8*, 35–43.
22. Chave, J.; Wiegand, K.; Levin, S. Spatial and biological aspects of reserve design. *Environ. Model. Assess.* 2002, *7*, 115–122.
23. Sodhi, N.S.; Koh, L.P.; Clements, R.; Wanger, T.C.; Hill, J.K.; Hamer, K.C.; Clough, Y.; Tscharntke, T.; Posa, M.R.C.; Lee, T.M. Conserving Southeast Asian forest biodiversity in human-modified landscapes. *Biol. Conserv.* 2010, *143*, 2375–2384.
24. Stringer, L. Can the UN Convention to Combat Desertification guide sustainable use of the world's soils? *Front. Ecol. Environ.* 2008, *3*, 138–144.
25. Rayan, D.M.; Mohamad, S.W. The importance of selectively logged forests for tiger Panthera tigris conservation: a population density estimate in Peninsular Malaysia. *Oryx* 2009, *43*, 48–51.

26. Rozzi, R. The reciprocal links between evolutionary-ecological sciences and environmental ethics. *BioScience* 1999, *49*, 911–921.
27. Palmer, M.A.; Bernhardt, E.S.; Chornesky, E.A.; Collins, S.L.; Dobson, A.P.; Duke, C.S.; Gold, B.D.; Jacobson, R.B.; Kingsland, S.E.; Kranz, R.H.; Mappin, M.J.; Martinez, M.L.; Micheli, F.; Morse, J.L.; Pace, M.L.; Pascual, M.; Palumbi, S.S.; Reichman, O.J.; Townsend, A.R.; Turner; M.G. Ecological science and sustainability for the 21st century. *Front. Ecol. Environ.* 2005, *3*, 4–11.
28. Brechin, S.R.; Wilshusen, P.R.; Fortwangler, C.L, West, P.C. *Contested Nature: Promoting International Biodiversity with Social Justice in the Twenty-first Century*; State University of New York Press: Albany, NY, 2003; p. 321.
29. Lugo, A.E. Management of tropical biodiversity. *Ecol. Appl.* 1995, *5*, 956–961.
30. Peluso, N.L. *Rich Forests, Poor People: Resource Control and Resistance in Java*; University of California Press: Berkeley, CA, 1992; p. 321.
31. Cincotta, R.P.; Wisnewski, J.; Engelman, R. Human population in the biodiversity hotspots. *Nature* 2000, *404*, 990–992.
32. Menzies, C.R. *Traditional Ecological Knowledge and Natural Resource Management*; University of Nebraska Press: Lincoln, NE, 2006; p. 274
33. Steinmetz, R.; Chutipong, W.; Seuaturien, N.; Chirngsaard, E.; Khaengkhetkarn, M. Population recovery patterns of Southeast Asian ungulates after poaching. *Biol. Conserv.* 2010, *143*, 42–51.
34. Akman, M.; Carlson, J.E.; Holsinger, K.E.; Latimer, A.M. Transcriptome sequencing reveals population differentiation in gene expression linked to functional traits and environmental gradients in the South African shrub *Protea repens*. *New Phytol.* 2016, *210*, 295–309.
35. de Gouvenain, R.C.; Midgley, J.J.; Merow, C. Serotiny in the South African shrub *Protea repens* is associated with gradients of precipitation, temperature, and fire intensity. *Plant Ecol.* 2019, *220*, 97–109.
36. Eizaguirre, C., Baltazar-Soares, M. Evolutionary conservation—evaluating the adaptive potential of species. *Evol. Appl.* 2014, *7*, 963–967.
37. Graudal, L.; Aravanopoulos, F.; Bennadji, Z.; Changtragoon, S.; Fady, B.; Kjær, E.D.; Loo, J.; Ramamonjisoa, L.; Vendramin, G.G. Global to local genetic diversity indicators of evolutionary potential in tree species within and outside forests. *Forest Ecol. Manag.* 2014, *333*, 35–51.
38. Veron, S.; Clergeau, P., Pavoine, S. Loss and conservation of evolutionary history in the Mediterranean Basin. *BMC Ecol.* 2016, 16–43.

33

Ecology: Functions, Patterns, and Evolution

Fernando Valladares
Centro de Ciencias Medioambientales

Introduction 289
Exploring the Mechanisms: From Physiology to Ecophysiology and Functional Ecology 289
The Search of General Patterns: Comparative Ecology 290
Experimental Ecology 290
Evolutionary Ecology 291
References 291

Introduction

Ecology is the study of living organisms and their environment in an attempt to explain and predict. While natural history involves the accumulation of detailed data with emphasis on the autecology of each species, the objectives of ecology in general and of functional ecology in particular are to develop predictive theories and to assemble the data to develop general models. Functional ecology has three basic components: 1) constructing trait matrices through screening of various plant and animal species, 2) exploring empirical relationships among these traits, and 3) determining the relationships between traits and environments.[1] Studies in functional ecology encompass a wide range of approaches, from individuals to populations; from mechanistically detailed to deliberately simplified, black-box simulations; and from deductive to inductive.[2] Functional ecology is concerned with the links between structure and function, the existence of general patterns among species, and the evolutionary connections among these patterns. And functional ecology is, above all, timely and pertinent, because the environmental degradation associated with human development is rapidly destroying the very systems that ecologists seek to understand. If we are to anticipate the extent and repercussions of global change in natural habitats, we first need to understand how organisms and ecosystems function.

Exploring the Mechanisms: From Physiology to Ecophysiology and Functional Ecology

The study of plant functions has largely followed a reductionistic approach aimed at explaining the functions in terms of the principles of physics and chemistry. However, plant physiology, which is focused on molecules, organelles, and cells, has not been able to provide reliable predictions of the responses of vegetation to changes in their environment, due to the multiple interactions involved in the responses of plants and the hierarchical nature of plant organization.[3] The realization of the fact that the responses of higher organizational levels are not predictable from the dynamics of those of smaller scales led to the advent of plant ecophysiology, which is focused on organs—a more relevant scale of organization to address questions regarding plant performance. Ecophysiologists have primarily focused on the structural and functional properties of leaves. Leaves display wide variation in morphology and physiology, including differences in specific mass, carbon and nitrogen investments, stomatal densities, optical properties, and hydraulic and photosynthetic characteristics. The emphasis of ecophysiological studies on leaves is due to the profound implications of the interactions between leaf structure and function for the performance of plants in natural habitats.[4,5] However, the traditional ecophysiological approach proved to be insufficient in predicting plant distribution and responses to changing environments, which led to the development of functional plant ecology.[6] Functional ecology is centered on whole plants as the unit of analysis, encompassing a range of scales of organization from organs to whole organism architecture, and is based on a much broader conception of plant functions than that formulated by the earlier practitioners of ecophysiology.[7]

The Search of General Patterns: Comparative Ecology

Since elucidation of the range of possible functional responses of plants is not possible with the use of model organisms, such as those typically used in plant physiology, functional ecology arises as an essentially comparative science. Ideally, functional ecology deals with traits measured on a large number of species in order to minimize the influence of the peculiarities of the autecology of each species. Two main approaches have been followed to find general ecological patterns in nature: 1) screening, that is, the design of bioassays for a trait or a set of traits measured simultaneously on a large number of species, as in the classic study by Grime and Hunt[8] of the relative growth rate of 132 species of British flora, and 2) empiricism, or the search for quantitative relationships between measurable dependent and independent variables (e.g., correlations among pairs of traits or traits and environments) producing quantitative models using traits and not species, as in the general revision of leaf traits by Reich et al.[4] Models using traits are more general than those based on species and can be more easily transferred to different floras.[1] The question arises as to which trait must be measured. A possible answer can be obtained by analyzing the basic functions that organisms perform (i.e., resource acquisition, the ability to tolerate environmental extremes, and the ability to compete with neighbors) and then either looking for traits which measure these functions or carrying out direct bioassays of them. The most common limitations of this kind of study are the difficulties in finding unambiguous linkages of a trait to a specific function, and the so-called phylogenetic constraints that are due to the fact that phylogenetic proximity among species can influence their functional similarities.

Individual plants demonstrate important degrees of phenotypic variation, which must be considered in comparative studies. No two individuals of the same species exhibit the same final shape or functional features, regardless of how similar the genotypes of two individuals may be.[9] Part of this variation is due to phenotypic plasticity, that is, the capacity of a given genotype to render different phenotypes under different environmental conditions, and part is due to other reasons (ontogenetic stage, developmental instability). Plants of the same chronological age can be ontogenetically different, so interpretation of differences in phenotypic traits will depend on whether comparisons are made as a function of age, size, or developmental stage.

Experimental Ecology

Broad-scale comparisons must be driven by hypotheses and must be based on robust statistical designs. However, the finding of statistically significant patterns is no guarantee of underlying cause-and-effect relationships, which must be tested experimentally. Gradient analysis, where functional responses are examined along a clearly defined environmental gradient, is a powerful approach to exploring the relationships between plant function and environment, but it is prone to spurious relationships when there is a hidden factor covarying with the factor defining the gradient.[3] Inferences from gradient analysis and broad-scale comparisons are statistical in nature and must be confirmed experimentally.

Experiments are designed to test hypotheses, but sufficient knowledge must be available to specify more than a trivial hypothesis before thorough experimentation can be undertaken.[2] Unfortunately, this is not the case for many natural systems. Although experiments are a common practice in physiology and ecophysiology, where mechanisms can be explored and hypotheses tested under relatively well-controlled conditions, they are less common in functional ecology studies, especially when hypotheses must be tested in natural habitats. Experimental design should meet a number of standards that are not easy—and in certain cases not possible—to meet in field ecology. The controls should be randomly intermixed with the treatments, both in space and time, and both the control and the experiments must be replicated enough. Replicates are not easy to find in natural scenarios, and when they can be found they are frequently not truly independent of one another, leading to a weak experimental design due to pseudoreplication. In addition, multiple causality and indirect effects, which significantly complicate the interpretation of results, are commonplace in ecology. However, experimental ecology is burgeoning despite all this adversity, because experiments are irreplaceable elements in achieving relevant progress in our understanding of ecological processes, as has been revealed by a number of experimental manipulations of natural populations.[10]

Evolutionary Ecology

Functional ecology eventually leads to evolutionary ecology. Trends in functional traits across species and mechanisms linking cause and effect contribute to our understanding of evolutionary processes, especially when they are considered on a time scale long enough to allow for changes in gene frequencies, the essence of evolution. Whereas functional ecology is interested in the immediate influence of environment on a given trait, evolutionary ecology is aimed at understanding why some individuals have left the most offspring in response to long-term consistent patterns of environmental conditions. Functional ecology, focused on an ecological time scale (*now* time), asks questions of "how?" and is concerned with the proximate factors influencing an event. Evolutionary ecology is focused on an evolutionary time scale (geological time), asking questions of "why?" and concerned with the ultimate factors influencing an event. Neither is more correct than the other, and they are not mutually exclusive because ecological events can always be profitably considered within an evolutionary framework, and vice versa.[10]

There are five agents of evolution: natural selection (differential reproductive success of individuals within a population); genetic drift (random sampling bias in small populations); gene flow (migration movements of individuals among and between populations with different gene frequencies); meiotic drive (segregation distortion of certain alleles that do not follow the Mendelian lottery of meiosis and recombination); and mutation. Of these agents, only natural selection is directed, resulting in conformity between organisms and their environments.[10] Darwin's theory of natural selection is a fundamental unifying theory, and the many studies that have been carried out over the last century in support of it demonstrate the power of the rigorous application of the genetic theory of natural selection to population biology.

References

1. Keddy, P.A. A pragmatic approach to functional ecology. Funct. Ecol. **1992**, *6*, 62–626.
2. Mentis, M.T. Hypothetico-deductive and inductive approaches in ecology. Funct. Ecol. **1988**, *2*, 5–14.
3. Duarte, C. Methods in Comparative Functional Ecology. In *Handbook of Functional Plant Ecology*; Pugnaire, F.I., Valladares, F., Eds.; Marcel Dekker: New York, 1999; 1–8.
4. Reich, P.B.; Ellsworth, D.S.; Walters, M.B.; Vose, J.M.; Gresham, C.; Volin, J.C.; Bowman, W.D. Generality of leaf trait relationships: A test across six biomes. Ecology **1999**, *80*, 1955–1969.

5. Niinemets, U. Global-scale climatic controls of leaf dry mass per area, density, and thickness in trees and shrubs. Ecology **2001**, *82*, 453–469.
6. Pugnaire, F.I.; Valladares, F. *Handbook of Functional Plant Ecology*; Marcel Dekker: New York, 1999.
7. Valladares, F.; Pearcy, R.W. The functional ecology of shoot architecture in sun and shade plants of *Heteromeles arbutifolia* M. Roem., a Californian chaparral shrub. Oecologia **1998**, *114*, 1–10.
8. Grime, J.P.; Hunt, R. Relative growth rate: Its range and adaptive significance in a local flora. J. Ecol. **1975**, *63*, 393–422.
9. Coleman, J.S.; McConnaughay; Ackerley, D.D. Interpreting phenotypic variation in plants. Trends Ecol. Evol. **1994**, *9*, 187–191.
10. Pianka, E.R. *Evolutionary Ecology*; Benjamin-Cummings: San Francisco, 2000.

34

Land Plants: Origin and Evolution

Elizabeth R. Waters
San Diego State University

Introduction 293
Adaptations of Land Plants 293
When Did Land Plants Evolve? 294
Which Algae Are the Sister Group to Land Plants and How Are the Earliest Land Plants Related to the Rest of the Embryophytes? 294
Evolutionary Relationships among the "Bryophytes" 295
Conclusions 296
References 296

Introduction

The origin of land plants, an event that occurred at least 475 million years ago (MYA), is one of the most important events in biotic history. Once established on land, plants (embryophytes) transformed the terrestrial environment, and by doing so created numerous new niches for plants and other groups of organisms to exploit. It is now known that plants evolved once from fresh-water algae. The first land plants faced numerous challenges that included but were not limited to desiccation, temperature stress, and increased exposure to UV radiation. These stresses or challenges were the selective forces that led to a number of innovations or adaptations.

Adaptations of Land Plants

The land plant radiation is marked by a long list of morphological and molecular innovations and adaptations.[1,2] The morphological adaptations include a waxy cuticle and desiccation-resistant haploid spores. One of the major innovations in land plants is a complex body plan including an alternation of generations between saprophytes and gametophytes and the presence of multicellular sporangia, antheridia, and archegonia.[2,3] In recent years, considerable effort has been made to understand the molecular and biochemical adaptations associated with land plants. It is clear that there are many of these adaptations, including an increased complexity in phenolics and other secondary plant compounds.[2] Analysis of the complete genome of the moss *Physcomitrella patens* has revealed that the complexity of gene families associated with desiccation and other stress responses, photoreceptors, and transport proteins occurred early in land plant evolution.[4]

When Did Land Plants Evolve?

There is considerable fossil evidence that land plants originated in the Ordovician at least 475 MYA.[1,3,5] Wellman et al.[6] reported the presence of spore-containing plant fragments in Ordovician rocks from Oman. This finding is important because for the first-time spores (microfossils) that suggested an Ordovician origin of plants were associated with macroscopic plant fossils. In fact, the spore wall ultra-structure of these new fossils shows some similarity to extant liverworts. These data provide conclusive evidence that land plants existed at least 475 MYA and suggest that liverworts may be the most basal (earliest diverging) group of land plants (for more details, see further).

The importance of the origin of land plants in evolutionary history and the scarcity of early land plant fossils have motivated some scientists to estimate the time of origin of land plants using a molecular clock analysis of DNA and amino acids sequences. This work, however, has not been without controversy. In 2001, Heckman et al. proposed a date of at least 700 MYA for the origin of land plants.[7] This date has not been confirmed by either other molecular clock analyses or by fossil evidence. In contrast, using another set of molecular data, Sanderson[8] estimated the origin of land plants to be between 483 and 425 MYA. These dates are similar to that from the fossil evidence. The disparity between these two molecular age estimates (700 vs. ~475 MYA) is most likely due to methodological differences in these two studies. Heckman et al.[7] used proteins that are encoded in the nucleus and calibrated their tree with divergence dates between animal lineages. Sanderson[8] used plastid-coding genes, utilized methods that did not rely on a clock, and used a calibration point within the plant lineage.

In short, determining the exact time of past evolutionary events is always problematic, and the origin of land plants is particularly problematic due to the scarcity of molecular sequence data from a number of different algal and bryophyte lineages and the lack of many early plant megafossils. However, at this time the best-supported date for the origin of land plants is at least 475 MYA.

Which Algae Are the Sister Group to Land Plants and How Are the Earliest Land Plants Related to the Rest of the Embryophytes?

In the past ten years, numerous phylogenetic analyses of the embryophytes have been published. In fact, there are so many analyses that they cannot all be cited here. Within this literature, there are many controversies and many areas where there is general agreement. First, there is considerable morphological and molecular evidence that land plants or embryophytes (Embryobionta), consisting of the liverworts, mosses, and hornworts ("Bryophytes") and the tracheophytes (vascular plants), are monophyletic.[1,9] There is also general agreement that the closest extant algal relatives to land plants are the charophyte algae and that the charophyte algae with the embryophytes form the streptophytes (Strep-tobionta), which are a monophyletic group.[10–13] In summary, this wealth of data indicates that land plants evolved once from a fresh-water charophyte alga (Figure 34.1).

Due to the evolutionary relationship of the charophyte algae to the land plants, our understanding of which groups comprise the charophyte algae and how they are related to each other is highly relevant to our understanding of the origin of land plants. Recent progress has been made in some parts of the charophycean phylogenetic tree; however, there is still much that is currently equivocal. The charophyte algae sensu Mattox and Stewart[14] have traditionally contained five orders of freshwater algae: Charales, Coleochaetales, Zygnematales, Klebsor- midiales, and Chlorokybales.[15] However, it is now clear that the Mesostigmatales also belong within the charophyte algae.[10] There is also considerable evidence that *Mesostigma* and *Chlorokybus* represent the earliest diverging lineages within the Charophytes[10,11,13] (Figure 34.2). One active area of research is the determination of which charophyte lineage might be the sister group to land plants. Morphological analysis has suggested a close relationship between the Coleochaetales, the Charales, and land plants.[1,15] However, analysis of molecular data has been somewhat contradictory.[16–19] Some molecular phylogenies also support *Coleochatate* as the sister group to embryophytes,[17] while others support either the Zygnematales alone[16,18] or the Zygnematales plus

FIGURE 34.1 **(See color insert.)** Representatives of important algal and plant lineages: (**a**) charophytes alga Coleochaete, a member of the Coleochaetales; (**b**) charophytes alga Chara, a member of the Charales; and (**c**) Anthoceros, a hornwort; (**d**) Marchantia, a liverwort.

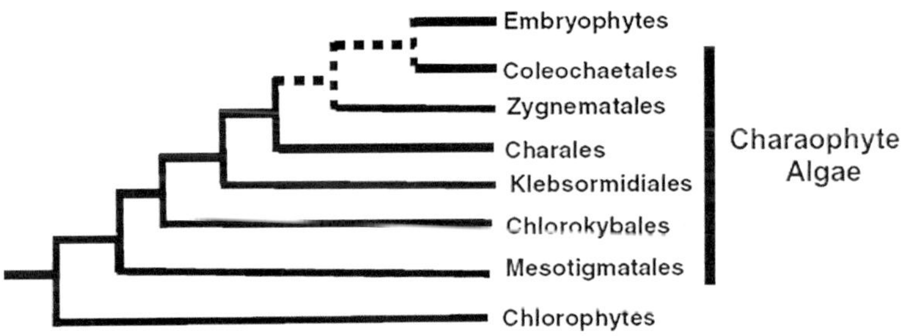

FIGURE 34.2 Relationships of the charophyte algal lineages and land plants based on data and analysis found in (Sanderson,[8] Leliaert, et al.,[11] and Graham, et al.[12]). Phylogenetic relationships that are not strongly supported in all or most analyses are depicted by a dotted line.

Coleochaetate[19] as sister group to the embryophytes. All of these studies have included a large number of species and a wide range of DNA sequences. It is evident that the determination of which group(s) of charophytes is (are) the sister group to land plants is a very difficult problem. These evolutionary events occurred almost 500 MYA and the branching patterns among these groups may be short and difficulty to determine based on data from extant taxa. It is hoped that complete genome sequences of key taxa will soon become available and that these data will help shed light on the identity of the charophyte algae group that gave rise to land plants.

Evolutionary Relationships among the "Bryophytes"

Determining the relationships among the extant land plant lineages and the identification of the lineage that branched off first in the evolution of land plants is crucial to our understanding of the transition from water to land. Thus, knowledge of the relationships among the basal land plant groups, the

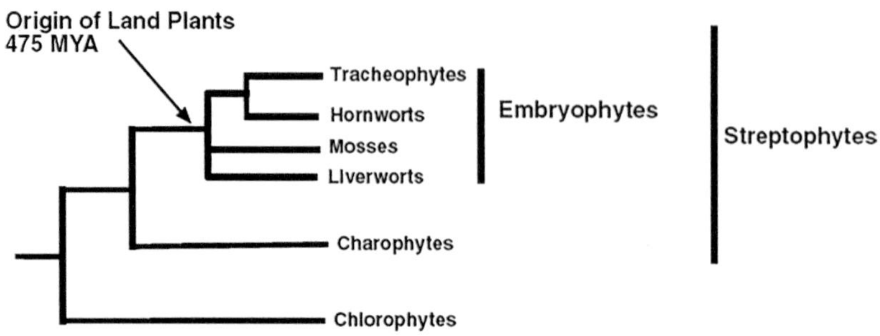

FIGURE 34.3 Relationships of the Bryophyte (Liverwort, Hornwort and Moss) lineages to charophyte green algae and the tracheophytes based on data and analysis found in (Heckman, et al.,[7] Becker & Marin,[13] and Mattox & Stewart[14]).

liverworts, hornworts, and mosses and the relationships of these groups to both the charophyte algae and the tracheophytes (vascular plants) is of great importance. There is clear agreement that all of the bryophytes are members of the embryophyte clade and that they are basal to the tracheophytes (vascular plants). However, it is also clear that the "Bryophytes" are a grade and are not a mono- phyletic group (Figure 34.3).

There is now considerable evidence that the liverworts are the earliest lineages of extant land plants.[9,20–22] It is also now clear that the hornworts, and not the mosses as had been previously reported, are the sister group to the tracheophytes. Interestingly, the hornworts are the least diverse of the three nonvascular or bryophyte groups with just 14 genera and 200–240 species.[9] Phylogenies now place the mosses basal to both the hornworts and the tracheophytes (Figure 34.3).

Conclusions

In summary, the embryophytes (Embryobionta) or land plants are a monophyletic group composed of the liverworts, mosses, hornworts, and vascular plants (tracheophytes). The origin of the land plants occurred at least 475 MYA, and their ancestor was a charophycean freshwater green alga, either a member of the Coleochaetales or the Zygnematales. The early evolution of the land plants is marked by a large number of important adaptations, including a desiccation-resistant cuticle, a complex body plan, the alternation of generations between sporophyte and gametophyte phases, the diversity of secondary plant compounds, and the increasing complexity of many gene and protein families.[4]

References

1. Kenrick, P.; Crane, P.R. *The Origin and Early Diversification of Land Plants: A Cladistic Study;* Smithsonian Institution Press: Washington, DC, 1997; 441 pp.
2. Waters, E.R. Molecular adaptation and the origin of land plants. Mol. Phylogenet. Evol. **2003**, *29,* 456–463.
3. Graham, L.E.; Cook, M.E.; Busse, J.S. The origin of plants: Body plan changes contributing to a major evolutionary radiation. Proc. Natl. Acad. Sci. **2000**, *97,* 4535–4540.
4. Rensing, S.A.; Lang, D.; Zimmer, A.D.; Terry, A.; Salamov, A.; Shapiro, H.; Nishiyama, T.; Perroud, P.F.; Lindquist, E.A.; Kamisugi, Y.; Tanahashi, T.; Sakakibara, K.; Fujita, T.; Oishi, K.; Shin, I.T.; Kuroki, Y.; Toyoda, A.; Suzuki, Y.; Hashimoto, S.; Yamaguchi, K.; Sugano, S.; Kohara, Y.; Fujiyama, A.; Anterola, A.; Aoki, S.; Ashton, N.; Barbazuk, W.B.; Barker, E.; Bennetzen, J.L.; Blankenship, R.; Cho, S.H.; Dutcher, S.K.; Estelle, M.; Fawcett, J.A.; Gundlach, H.; Hanada, K.; Heyl, A.; Hicks, K.A.; Hughes, J.; Lohr, M.; Mayer, K.; Melkozernov, A.; Murata, T.; Nelson, D.R.; Pils, B.; Prigge, M.; Reiss, B.; Renner, T.; Rombauts, S.; Rushton, P.J.; Sanderfoot, A.;

Schween, G.; Shiu, S.H.; Stueber, K.; Theodoulou, F.L.; Tu, H.; Van de Peer, Y.; Verrier, P.J.; Waters, E.; Wood, A.; Yang, L.; Cove, D.; Cuming, A.C.; Hasebe, M.; Lucas, S.; Mishler, B.D.; Reski, R.; Grigoriev, I.V.; Quatrano, R.S.; Boore, J.L. The Physcomitrella genome reveals evolutionary insights into the conquest of land by plants. Science **2008**, *319*, 64–69.

5. Gensel, P.G. The earliest land plants. Annu. Rev. Ecol. Evol. Syst. **2008**, *39*, 459–477.
6. Wellman, C.H.; Osterloff, P.L.; and Mohiuddin, U. Fragments of the earliest land plants. Nature **2003**, *425*, 282–285.
7. Heckman, D.S.; Geiser, D.M.; Eidell, B.R.; Stauffer, R.L.; Kardos, N.L.; Hedges, S.B. Molecular evidence for the early colonization of land by fungi and plants. Science **2001**, *293*, 1129–1133.
8. Sanderson, M.J.; Molecular data from 27 proteins do not support a precambrian origin of land plants. Amer. J. Bot. **2003**, *90*, 954–956.
9. Shaw, J.A.; Szoneyi, P.; Shaw, B. Bryophyte diversity and evolution: Windows into the early evolution of land plants. Am. J. Bot. **2010**, *98*, 352–369.
10. Lewis, L.A.; McCourt, R.M. Green algae and the origin of land plants. Am. J. Bot. **2004**, *91*, 15351556.
11. Leliaert, F.; Smith, D.R.; Moreau, H.; Herron, M.D.; Verbruggen, H.; Delwiche, C.F.; De Clerck, O. Phylogeny and molecular evolution of the green algae. Crit. Rev. Plant Sci. **2012**, *31*, 1–46.
12. Graham, J.; Wilcox, L.; Graham, L. *Algae;* 2nd ed.; Pearson: New York, 2008; 720.
13. Becker, B.; Marin, B. Streptophyte algae and the origin of embryophytes. Annu. Bot. **2009**, *103*, 999–1004.
14. Mattox, K.R.; Stewart, K.D. Classification of the green algae: A concept based on comparative cytology, In *Sys- tematics of the Green Algae*, Irvine, G. E. G. and John, D.M. Eds.; Academic Press: London, 1984; 29–72 p.
15. Graham, L.E. *Origin of Land Plants;* John Wiley and Sons Inc.: New York, 1993; 287.
16. Chang, Y.; Graham S.W. Inferring the higher-order phylogeny of mosses (Bryophyta) and relatives using a large, multigene plastid data set. Am. J. Bot. **2011**, *98*, 839–849.
17. Finet, C.; Timme, R.E.; Delwiche, C.F.; Marletaz, F. Multigene phylogeny of the green lineage reveals the origin and diversification of land plants. Curr. Biol. **2010**, *20*, 2217–2222.
18. Timme, R.E., Bachvaroff, T.R.; Delwiche, C.F. Broad phy- logenomic sampling and the sister lineage of land plants. PLoS One, **2012**, *7*, e29696.
19. Wodniok, S.; Brinkmann, H.; Glockner, G.; Heidel, A.J.; Philippe, H.; Melkonian, M.; Becker, B. Origin of land plants: do conjugating green algae hold the key? BMC Evol. Biol. **2011**, *104*, 11.
20. Qiu, Y.-L., Phylogeny and evolution of charophytic algae and land plants. J. Syst. Evol. **2008**, *46*, 287–306.
21. Qiu,Y-L.; Cho, Y., Cox, J.C.; Palmer, J.D. The gain of three mitochondrial introns identifies liverworts as the earliest land plants. Nature **1998**, *394*, 671–674.
22. Qiu, Y-L; Li, L.B.; Wang, B.; Chen, Z.D.; Knoop, V.; Groth-Malonek, M.; Dombrovska, O. The deepest divergences in land plants inferred from phylogenomic evidence. PNAS **2006**, *103*, 15511–15516.

35

Leaves: Elevated CO_2 Levels

G. Brett Runion
National Soil Dynamics Laboratory

Seth G. Pritchard
College of Charleston

Stephen A. Prior
National Soil Dynamics Laboratory

Introduction 299
Molecular and Cellular Scale 299
Whole Leaf Scale 301
Whole Plant Scale 301
Canopy and Ecosystem Scale 301
Conclusion 303
References 303

Introduction

The concentration of carbon dioxide (CO_2) in the atmosphere is increasing primarily because of fossil fuel use and land use change and is expected to double preindustrial levels during this century.[1] Leaves are the sites for the transfer of carbon from atmosphere to biosphere; thus, this increase has significant implications for plants. Carbon dioxide, along with sunlight and water, drive photosynthesis and it is the capture of carbon by leaves that enables plants to grow. This carbon capture by leaves forms the basis of agriculture and forestry and provides humans and other organisms with the food, clothing, and shelter they need to survive. This entry will provide a brief summary of the effects of the increasing atmospheric CO_2 on plant leaves at various scales from molecular to plant community.

Molecular and Cellular Scale

Photosynthesis is stimulated by elevated atmospheric CO_2 in nearly all C_3 plant species, typically on the order of 20–60% for plants grown with 700 ppm compared to those grown at 365 ppm.[2] The increase in photosynthesis is commonly linked to 1) faster C fixation by the ribulose 1,5-bisphosphate carboxylase/oxygenase (rubisco) enzyme; 2) suppression of photorespiration caused by an increase in CO_2:O_2 within the chloroplast stroma; 3) decreased oxidative damage to the photosynthetic apparatus; and 4) amelioration of water stress by enhanced water-use efficiency (WUE; the ratio of carbon fixed to water transpired).

Predicting the photosynthetic response of C_4 species (such as maize, sugarcane, sorghum, and many other grasses) to elevated CO_2 is somewhat more difficult. Because C_4 plants possess a CO_2-concentrating mechanism, they are far less sensitive to changes in atmospheric CO_2 concentration compared to C_3 plants. Stimulation of photosynthesis in C_4 plants, therefore, is only expected to occur during periods of water stress.[3] When exposed to drought, both C_3 and C_4 plants benefit from exposure to elevated CO_2 because they are able to maintain sufficient leaf CO_2 concentrations to maintain photosynthesis while keeping stomatal conductance values low to preserve water (i.e., WUE is increased).

Increased rates of carbon assimilation by leaves are accompanied by many biochemical and molecular adjustments.[4] For example, the stimulatory effect of elevated CO_2 on photosynthesis is often downregulated to some extent as a result of a reduction in rubisco content and maximum carboxylation capacity.[5] Alterations in expression patterns of genes coding for proteins involved in growth and development, the Calvin cycle, photorespiration, photosystem I and II subunits, carbon metabolism and utilization, respiration, aquaporins, antioxidants, and metabolism of secondary metabolites important for defense have all been reported. Whether specific leaf genes are upregulated or downregulated can vary depending on species and environmental conditions. Clearly, more work is needed at the transcriptomic, proteomic, and metabolomic levels to understand how leaves and whole plants will respond to global climate changes. Not only does research at this level of biological organization have the potential to explain higher level plant and canopy responses to elevated CO_2, these types of data may help guide researchers to select crop and tree species best able to exploit a future high CO_2 world.

Regardless of the nature of changes in gene expression occurring in leaves, increased photosynthesis in plants grown with elevated CO_2 nearly always results in a greater supply of carbohydrates to fuel additional leaf and whole plant growth (discussed further). However, there are feedbacks at the organelle level that have the potential to dampen higher photosynthesis and modify changes in gene expression caused by elevated CO_2. For example, an overproduction of carbohydrates from higher photosynthesis sometimes results in an excessive accumulation of leaf starch that may negatively affect the internal organization of chloroplasts (Figure 35.1).[6] Excessive accumulation of leaf starch occurs when carbohydrate production exceeds the plant's ability to transport, metabolize, or store available carbohydrates. In general, plants that receive adequate levels of water and nutrients have enough resources to build sufficient sinks and excessive accumulation of leaf starch is avoided. However, carbohydrates accumulate when plants are sink-limited as commonly occurs with low levels of nutrients and/or under water stress.[6]

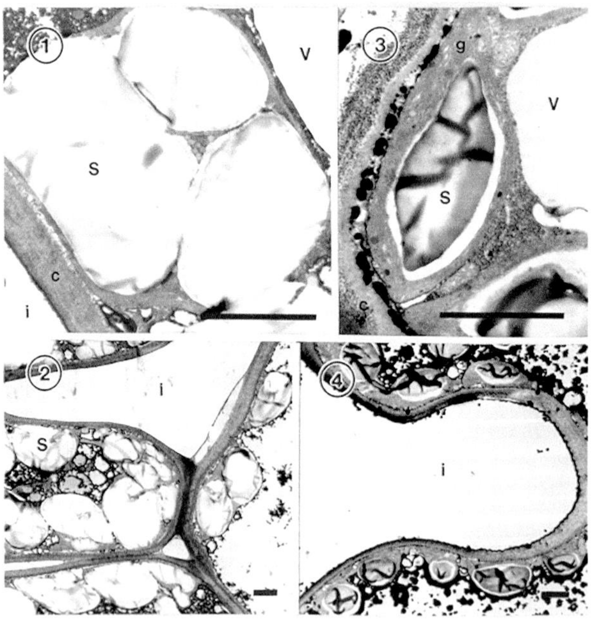

FIGURE 35.1 Electron micrograph of mesophyll chloroplasts from plants grown for 12 months under low nitrogen ($4\,g\,m^{-2}$), water stress (cyclically to −1.5 MPa predawn xylem pressure potential) and either elevated (720 μmol mol^{-1}; 1, 2) or ambient (365 μmol mo–$^{-1}$; 3, 4) as described by Prichard et al.[6] Note the much smaller starch inclusions in the ambient than the elevated CO_2 treatment. S, starch grains; g, granum; c, cell wall, V, vacuole, I, intercellular space. Scale bar = 2 μm.

Whole Leaf Scale

Increasing levels of atmospheric CO_2 have been shown to affect leaf growth, development, anatomy, and longevity. Plants grown under high CO_2 often exhibit increased area per leaf,[7] which might indicate a higher leaf-level photosynthetic capacity. Leaf mesophyll and vascular tissue cross-sectional areas can increase leading to increases in leaf thickness,[7] as can an extra layer of palisade cells.[8] Changes in leaf size are more commonly caused by increased cell expansion rather than increased cell division. Leaf stomatal density has been shown to decrease with elevated CO_2 in some, but not all, cases.[7] Although decreased stomatal density and partial stomatal closure can increase WUE, it can also lessen the amount of CO_2 diffusing from the atmosphere into the leaf interior (thereby reducing photosynthesis and influencing gene expression).

Frequently, leaf chemistry also changes under high CO_2. Leaves can have lower nitrogen (N) concentration,[9,10] higher carbon to nitrogen ratios,[9,10] and (as mentioned earlier) higher levels of nonstructural carbohydrates.[6,9] These changes can affect interactions of leaves with pathogens and insect pests while still attached to the plant and decompositional processes following leaf fall.[11] Changes in epicuticular wax on high CO_2-grown leaves have also been observed.[12,13] These changes could affect water relations, susceptibility to pests and diseases, and surface properties that are important in chemical applications.

Whole Plant Scale

Plants grown under elevated CO_2 generally have more leaves per plant and higher total leaf area per plant,[7] which usually, but not always, results in greater leaf area index.[14,15] The increase in leaf number and leaf area, coupled with increases in leaf-level photosynthesis, generally leads to an increase in plant biomass, including leaf biomass. Summaries have consistently shown that biomass response to atmospheric CO_2 enrichment varies between plants with a C_3 (33–40% increase) versus a C_4 (10–15% increase) photosynthetic pathway.[16] This assumes that other resources (water and nutrients) are not limiting.

Plants will allocate energy and photosynthate to those organs associated with acquiring the most limiting resource. Thus, when CO_2 is elevated, plants tend to allocate more to belowground growth; roots often show the largest response to elevated CO_2.[17] So, in addition to C_3 versus C_4 differences, plants also vary in their response to CO_2 depending on phenology or growth habit. Plants with large storage organs for the extra carbohydrates produced (i.e., tubers or woody root systems) tend to have a greater response to elevated CO_2, and experience less photosynthetic down- regulation, than those without such organs (i.e., small, fibrous root systems).[18]

One benefit of high CO_2 is thought to be a decrease in drought stress. This would seem logical given larger root systems being able to supply more water coupled with a decrease in transpiration. However, it is important to note that while an increase in leaf-level photosynthesis generally indicates an increase in whole plant growth, a decrease in leaf-level transpiration does not necessarily translate to a reduction in whole plant water use. Plants grown under elevated CO_2 are generally larger (assuming adequate supply of other resources) and larger plants can use more water on a whole plant basis. Such feedbacks on leaf-level responses that are manifested at the whole plant level are further modified at the canopy scale.[19]

Canopy and Ecosystem Scale

Although molecular, cellular, and whole leaf response to CO_2 enrichment ultimately control a plant's C budget and growth, it is difficult to extrapolate from such fine-scale responses to an entire crop or forest canopy because of feedbacks that operate at these broader scales. For example, the relationship between stomatal conductance and transpiration commonly observed at the leaf or whole- plant level sometimes disappears at the canopy level. This is because, as transpiration decreases (due to partial stomatal closure induced by high CO_2 in and around leaves), the air in and directly above the canopy can become dryer, thereby steepening the water vapor diffusion gradient between the very wet leaf interior and canopy air (resulting in faster transpiration). This feedback, coupled with greater total leaf area index in CO_2 enriched canopies, likely explains the observation that stand-level transpiration is reduced far less than would be predicted based on leaf- level responses.

There is another potentially important canopy-scale feedback that is linked to partial closure of stomates induced by higher atmospheric CO_2 levels.[20–21] Reduced stomatal conductance and transpiration by individual leaves can slow canopy transpiration, thereby decreasing transpirational cooling and increasing the temperature of individual leaves in addition to the entire plant canopy (Figure 35.2).[3] Higher canopy temperatures can drive faster transpiration and can modify the physiological responses of individual cells, leaves, and whole plants discussed earlier. Increased canopy temperatures caused by less cooling in CO_2-enriched environments, combined with increases in air temperatures caused by global warming, could be especially important for field crops. Wheat and rice yields, for example, commonly decrease by ~5% for each one degree Celsius increase in air temperature.[2]

Not all plants respond the same to elevated CO_2, therefore shifts in competitive interactions among plants growing in mixed stands, as is the case with unmanaged systems, are inevitable. Clearly, there

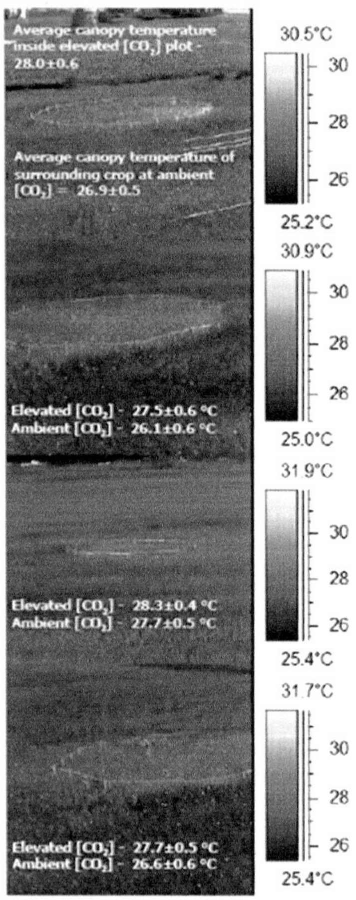

FIGURE 35.2 Infrared image showing that maize growing inside an elevated [CO_2] plot was warmer than maize growing under ambient [CO_2] outside the plot at 15.30 h on 15 July 15, 2004. At that time, the average canopy temperature inside the four elevated [CO_2] plots at SoyFACE was 27.9°C+0.2°C, significantly higher than the canopy temperature under ambient [CO_2] outside the plots (26.8°C+0.3°C).

Source: This figure was reproduced with permission from the author and is copyrighted © by the Royal Society, London.[3]

will be winners and losers as atmospheric CO_2 concentration continues to rise, which could substantially complicate interpretation of leaf, cellular, and whole-plant responses to elevated CO_2. For instance, we grew an assemblage of species common to Southeastern U.S. longleaf pine savannahs under ambient and elevated CO_2 and predicted that broadleaf plants would have a greater response to CO_2 than would pines.[22] Total plant biomass and gross primary productivity (GPP = total photosynthesis or photosynthesis per unit land area) were greatly increased under high CO_2, and there were winners and losers, but not as we expected. The oaks had no response to twice ambient levels of CO_2 and the pines out-competed all other species. The pines had begun to reach canopy closure after only three years of CO_2 exposure and the understory species, wiregrass in particular, were being shaded out indicating that elevated CO_2 had substantially altered ecosystem structure. These sorts of ecosystem-level responses to elevated CO_2 could cancel out leaf-level responses in some cases or enhance those same responses in others.

Despite the canopy-scale feedbacks discussed earlier, most researchers have found stimulation of both GPP and net primary productivity (NPP = the amount of carbon incorporated into plant biomass or GPP-autotrophic respiration) following exposure to elevated level of atmospheric CO_2.[23–25]

Conclusion

Leaves represent the first contact between atmosphere and biosphere. An entire biological chain of events begins with atmospheric CO_2 entering leaves, being processed via photosynthesis into starch that ultimately feeds the planet. Rising levels of atmospheric CO_2 impact leaves at various scales which build from effects on the enzyme rubisco, to changes in leaf cells which can increase leaf size and alter leaf function. These changes usually lead to larger, leafier plants which increase plant yield and ecosystem productivity. Still, leaf responses to elevated atmospheric CO_2 need to be further elucidated if we are to accurately understand and predict the development and growth of crops under future environmental conditions. Future research will need to address the following:

1. How alterations in leaf structure are related to both cellular and higher-level growth processes
2. Whether increased cell division is driven by greater rates of cell expansion or by molecular cues
3. The role of ultrastructural, anatomical, and morphological leaf adjustments in photosynthesis
4. How CO_2-induced changes in leaf structure and leaf function will alter the degree of impact of multiple plant stresses (abiotic and biotic)
5. How CO_2-induced shifts in leaf quality and/or changes in leaf-surface features (e.g., epicuticular waxes, trichomes) will alter herbivore and pathogen attacks
6. How elevated CO_2 affects mechanisms of herbicide tolerance in plant;
7. How changes in leaf anatomy and physiology impact interactions with insect and disease pests
8. How changes in leaf chemistry impact decomposition and the residence time of leaf-derived carbon in the soil

References

1. Keeling, C.D.; Whorf, T.P. Atmospheric CO_2 records from sites in the SIO air sampling network. In *Trends: A Compendium of Data on Global Change;* Carbon Dioxide Information Analysis Center: Oak Ridge National Laboratory, Oak Ridge, TN, 2001; 14–21 pp.
2. Pritchard, S.G.; Amthor, J.S. *Crops and Environmental Change*; Haworth Press: Binghamton, NY, 2005.
3. Leakey, A.D.B. Rising atmospheric carbon dioxide concentration and the future of C_4 crops for food and fuel. Proc. R. Soc., London, Ser. B, **2009**, *276*, 1333–2343.
4. Ahuja, I.; de Vos, R.C.H.; Bones, A.M.; Hall, R.D. Plant molecular stress responses face climate change. Trends Plant Sci. **2010**, *15*, 664–674.
5. Ainsworth, E.A.; Rogers, A. The response of photosynthesis and stomatal conductance to rising [CO_2]: Mechanisms and environmental interactions. Plant, Cell, Environ. **2007**, *30*, 258–270.

6. Pritchard, S.; Peterson, C.; Runion, G.B.; Prior, S.; Rogers, H. Atmospheric CO_2 concentration, N availability, and water status affect patterns of ergastic substance deposition in longleaf pine (*Pinus palustris* Mill.) foliage. Trees **1997**, *11*, 494–503.
7. Pritchard, S.G.; Rogers, H.H.; Prior, S.A.; Peterson, C.M. Elevated CO_2 and plant structure: A review. Glob. Chang. Biol. **1999**, *5*, 807–837.
8. Thomas, J.F.; Harvey, C.H. Leaf anatomy of four species grown under continuous CO_2 enrichment. Bot. Gaz. **1983**, *144*, 303–309.
9. Runion, G.B.; Entry, J.A.; Prior, S.A.; Mitchell, R.J.; Rogers, H.H. Tissue chemistry and carbon allocation in seedlings of *Pinus palustris* subjected to elevated atmospheric CO_2 and water stress. Tree Physiol. **1999**, *19*, 329–335.
10. O'Neill, E.G.; Norby, R.J. Litter quality and decomposition rates of foliar litter produced under CO_2 enrichment. In *Carbon Dioxide and Terrestrial Ecosystems*. Koch, G.W.; Mooney, H.A. Eds.; Academic Press: New York, 1996; 87–103 pp.
11. Torbert, H.A.; Prior, S.A.; Rogers, H.H.; Wood, C.W. Elevated atmospheric CO_2 effects on agro-ecosystems: Residue decomposition processes and soil C storage. Plant Soil **2000**, *224*, 59–73.
12. Graham, E.A.; Nobel, P.S. Long-term effects of a doubled atmospheric CO_2 concentration on the CAM species Agave deserti. J. Exp. Bot. **1996**, *47*, 61–69.
13. Prior, S.A.; Pritchard, S.G.; Runion, G.B.; Rogers, H.H.; Mitchell, R.J. Influence of atmospheric CO_2 enrichment, soil N, and water stress on needle surface wax formation in *Pinus palustris* (Pinaceae). Am. J. Bot. **1997**, *84*, 1070–1077.
14. Drake, B.G.; Gonzalez-Meler, M.A. More efficient plants: A consequence of rising atmospheric CO_2? Annu. Rev. Plant Physiol. Plant Mol. Biol. **1997**, *48*, 609–639.
15. Kimball, B.A.; Kobayashi, K.; Bindi, M. Responses of agricultural crops to free-air CO_2 enrichment. Adv. Agron. **2002**, *77*, 293–368.
16. Kimball, B.A. Carbon dioxide and agricultural yield: An assemblage and analysis of 430 prior observations. Agron. J. **1983**, *75*, 779–788.
17. Rogers, H.H.; Runion, G.B.; Krupa, S.V. Plant responses to atmospheric CO_2 enrichment with emphasis on roots and the rhizosphere. Environ. Pollut. **1994**, *83*, 155–189.
18. Runion, G.B.; Finegan, H.M.; Prior, S.A.; Rogers, H.H.; Gjerstad, D.H. Effects of elevated atmospheric CO_2 on non-native plants: Comparison of two important southeastern ornamentals. Environ. Control Biol. **2011**, *49(3)*, 107–117.
19. Runion, G.B.; Mitchell, R.J.; Green, T.H.; Prior, S.A.; Rogers, H.H.; Gjerstad, D.H. Longleaf pine photosynthetic response to soil resource availability and elevated atmospheric carbon dioxide. J. Environ. Qual. **1999**, *28*, 880–887.
20. Rogers, H.H.; Thomas, J.F.; Bingham, G.E. Response of agronomic and forest species to elevated atmospheric carbon dioxide. Science **1983**, *220*, 428–429.
21. Rogers, H.H.; Dahlman, R.C. Crop responses to CO_2 enrichment. Vegetatio **1993**, *104/105*, 117–131.
22. Runion, G.B.; Davis, M.A.; Pritchard, S.G.; Prior, S.A.; Mitchell, R.J.; Torbert, H.A.; Rogers, H.H.; Dute, R.R. Effects of elevated atmospheric carbon dioxide on biomass and carbon accumulation in a model regenerating longleaf pine community. J. Environ. Qual. **2006**, *35*, 1478–1486.
23. Ellsworth, D.S.; Oren, R.; Huang, C.; Phillips, N.; Hendrey, G.R. Leaf and canopy responses to elevated CO_2 in a pine forest under free-air CO_2 enrichment. Oecologia **1995**, *104*, 139–146.
24. DeLucia, E.H.; Hamilton, J.G.; Naidu, S.L.; Thomas, R.B.; Andrews, J.A.; Finzi, A.; Lavine, M.; Matamala, R.; Mohan, J.E.; Hendrey, G.R.; Schlesinger, W.H. Net primary production of a forest ecosystem with experimental CO_2 enrichment. Science **1999**, *284*, 1177–1179.
25. Witting, V.E.; Bernacchi, C.J.; Zhu, X.G.; Calfapietra, C.; Ceulemans, R.; Deangelis, P.; Gielen, B.; Miglietta, F.; Morgan, P.B.; Long, S.P. Gross primary production is stimulated for three *Populus* species grown under free-air CO_2 enrichment from planting through canopy closure. Glob. Chang. Biol. **2005**, *11*, 644–656.

36

Net Ecosystem Production (NEP)

Introduction 305
Measurement of NEP 306
Terrestrial versus Aquatic Ecosystems 307
Temporal and Spatial Patterns 307
Acknowledgments 313
References 313

Mark E. Harmon
Oregon State University

Introduction

Net ecosystem production (NEP) quantifies the balance of ecosystem primary production minus ecosystem respiration. First used by Woodwell & Whittaker,[1] NEP provides an index of ecosystem metabolism and determines, in part, whether there is a net gain or loss of carbon. In some ways it is analogous to the energy balance of an organism. It also parallels the relationship between gross primary production, autotrophic respiration, and net primary production (NPP). The sign of NEP can be either positive or negative. When NEP is zero, the ecosystem is in carbon balance. NEP is expressed as the mass of carbon per area per time and can be expressed in terms of gross primary production (GPP) or NPP:

$$\mathrm{NEP} = \mathrm{GPP} - \mathrm{Re}$$

$$\mathrm{NEP} = \mathrm{NPP} - \mathrm{Rh}$$

where Re is ecosystem respiration and Rh is heterotrophic respiration resulting from the metabolism of consumers and decomposers. The two equations are equivalent because

$$\mathrm{GPP} = \mathrm{NPP} + \mathrm{Ra}$$

and

$$\mathrm{Re} = \mathrm{Ra} + \mathrm{Rh}$$

Here Ra is autotrophic respiration (i.e., respiration from primary producers).

NEP is parallel to NPP in that

$$\mathrm{NPP} = \mathrm{GPP} - \mathrm{Ra}$$

Related terms include net ecosystem exchange (NEE), ΔC, and net ecosystem carbon balance (NECB). NEP is referenced to the ecosystem, whereas NEE is referenced to the atmosphere. This means that when

NEP is positive NEE is negative and vice versa. As typically measured in terrestrial ecosystems (e.g., eddy covariance) NEE involves vertical exchanges of carbon dioxide with the atmosphere outside the ecosystem. While most consider NEE is solely due to photosynthesis and respiration, Chapin et al.[2] include carbon dioxide losses associated with combustion also; moreover they include flows in any direction. In contrast to NEE, NEP may not involve exchanges outside the ecosystem itself. For example, respired carbon could be absorbed by leaf stomata before leaving the ecosystem. ΔC is the net change in carbon stored in a fixed area over a time period, typically a year. In some cases NEP and ΔC have that same value, but conceptually differ in that carbon can enter and leave ecosystems by means other than photosynthesis or respiration.[2,3] This is most evident in aquatic systems with flowing water where carbon is constantly being imported or exported independent of the physiologic processes of photosynthesis and respiration.[3] For many terrestrial systems there can be extended periods when carbon losses via erosion, leaching, combustion, and harvest can be very low, and this gives the appearance that NEP and AC are the same.[4] However, over very long periods or broad expanses these other losses can be substantial. NECB was offered by Chapin et al.[2] to address this discrepancy and includes other processes that either add or remove carbon from ecosystems:

$$\text{NECB} = \text{NEB} - \text{E} - \text{S} - \text{F} - \text{H} - \text{V}$$

where erosion and particle transport (E), solution (S), fire (F), harvest (H), and volatiles (V) are some of the largely nonphysiologic processes influencing carbon balances. An alternative to specifying the processes moving carbon in and out of ecosystems is to specify the form of carbon moving, as was done by Chapin et al.[2] While many of these flows are out of the ecosystem, and hence the signs in the equation are negative, in some cases erosion and particle transport as well as solution transport can add carbon to ecosystems (particularly aquatic ecosystems). Net Biome Production (NBP) is used to indicate the overall carbon balance of an ecosystem[5] or landscape;[2] however, ecosystems need not achieve a carbon balance at the landscape or biome level and so a term that indicates the biological level that this biological property emerges (i.e., the ecosystem) is preferred by some.[4]

NEP is also related to the concepts of autotrophy and heterotrophy, particularly in aquatic ecosystems. A system that exhibits autotrophy is one in which NEP is positive, that GPP > Re. In contrast, a system that exhibits heterotrophy is one in which NEP is negative and GPP < Re. This is most useful in aquatic ecosystems because it indicates whether an ecosystem is fixing enough carbon or energy to support itself, or whether is it being subsidized by inputs from outside the system.

Measurement of NEP

Until recently direct measurement of NEP or related terms was challenging. Traditional terrestrial methods involved first estimating NPP and then either measuring respiration losses or using a combination of decomposition rates and stores of detritus and soil to estimate Rh.[1] Accurately estimating NPP and Rh presents numerous methodological challenges, particularly belowground. As measuring GPP directly is difficult, most traditional estimates of NEP are done on an NPP basis. The advent of eddy covariance, a method that simultaneously measures carbon dioxide concentrations and vertical air movements, has allowed more direct estimates of NEP.[6] However, there are methodological issues with this technique as it tends to measure carbon exchanges better during the day than at night, the latter being when respiration is the primary process; moreover it assumes strictly vertical exchanges whereas advection (i.e., horizontal transport) is common at night. Both these limitations are likely to lead to an underestimation of Re, which could result in a positively biased estimate of NEP. Considerable effort has been made to eliminate these problems by locating measurements in areas with minimal topographic slope and using night time periods with sufficient turbulent mixing based on a minimum wind velocity.[7]

In flowing aquatic systems estimating NEP may not involve the measurement of carbon dioxide as in terrestrial systems. Instead, oxygen concentration and water velocity can indicate the rate at which photosynthesis versus respiration is occurring.[8] These net changes in oxygen can be converted to a net carbon balance. A confounding factor is the addition of oxygen-rich or -depleted waters from hyporheic or groundwater flows; the former would lead to an overestimate of GPP and the latter would overestimate Re.[9]

Terrestrial versus Aquatic Ecosystems

There are significant differences in NEP between terrestrial and aquatic systems related to the transport of carbon, which influence how this variable is examined. In most terrestrial systems the input of carbon by processes other than photosynthesis is minimal. While losses of carbon via particle and solution transport occur in terrestrial ecosystems, for many locations they are small terms relative to photosynthesis and respiration. In aquatic ecosystems, particulate and solution transport can be very significant. This is one reason that NEP can often approximate a net carbon balance in many terrestrial ecosystems, but it often does not in aquatic ecosystems. While it is rare for terrestrial ecosystems to have a semipermanent negative NEP, in aquatic ecosystems with significant outside (allochthonous) carbon input it is possible for NEP to remain negative for extensive periods. The converse is also true, in that aquatic systems with primarily autochthonous input NEP may remain positive for extensive periods. This explains why in terrestrial ecosystems the focus of NEP studies is often on long-term temporal changes,[10] whereas in flowing aquatic ecosystems the focus is on broad-scale spatial changes.[11] Despite some aquatic ecosystems having a positive NEP, over broad spatial extents inland waters appear to have a negative value of NEP primarily due to the input of carbon from upland terrestrial ecosystems.[12]

Temporal and Spatial Patterns

In terrestrial ecosystems there are several classic temporal patterns of NEP at a local level (e.g., a small plot of land), which are distinguished by the amount of biological legacies initially present. In the case of primary succession on freshly created surfaces, there is no legacy carbon and NEP can remain positive for extensive periods, but gradually approaches zero (Figure 36.1A). This is because Rh generally lags NPP, but with sufficient time the amount of carbon produced is theoretically equal to the amount dying and respiring. In the case of secondary succession, considerable legacy carbon is initially present and respiration losses can far exceed NPP. This means that NEP is initially negative after disturbance, but eventually becomes positive as legacy carbon respires away and NPP increases (Figure 36.1B). With enough time NEP in secondary succession will approach zero as in primary succession. In terms of carbon accumulation, primary and secondary succession are quite different, with the former steadily accumulating carbon to a steady-state and the latter first declining and then increasing toward a steady-state value (Figure 36.1C). Given that secondary succession can start with variable legacies (i.e., those left by clear-cut timber harvest differ from those left by fire or windstorms), there can be a considerable range in the magnitude and duration of the negative and positive phases of NEP during terrestrial succession. Although many factors control NEP temporal dynamics, the shapes of the underlying NPP and Rh curves are a logical focus. Aside from the maximum level of NPP reached in an ecosystem, the time to reach that maximum is important in determining the NPP time curve. The primary control on the shape of the Rh time curve is the relative rate of respiration (i.e., rate constant) and the amount of biological legacy starting the succession. The effect of the NPP and Rh time curves on NEP varies with primary versus secondary succession (Figure 36.2). In primary succession there is no or very little biological legacy, which means NPP is the primary controlling time curve. As the maximum NPP increases, so does the peak of NEP attained. More interesting is the effect of the time to reach the maximum NPP, with a decreasing, broader NEP peak, as the time to reach the maximum NPP increases (Figure 36.2A). This is largely due to the fact that Rh has more time, so to speak, to catch up to NPP when the increase in NPP is slower. In secondary succession the Rh time curve appears to have the greatest influence on the temporal pattern of NEP. Reducing the biological legacy causes NEP to move from

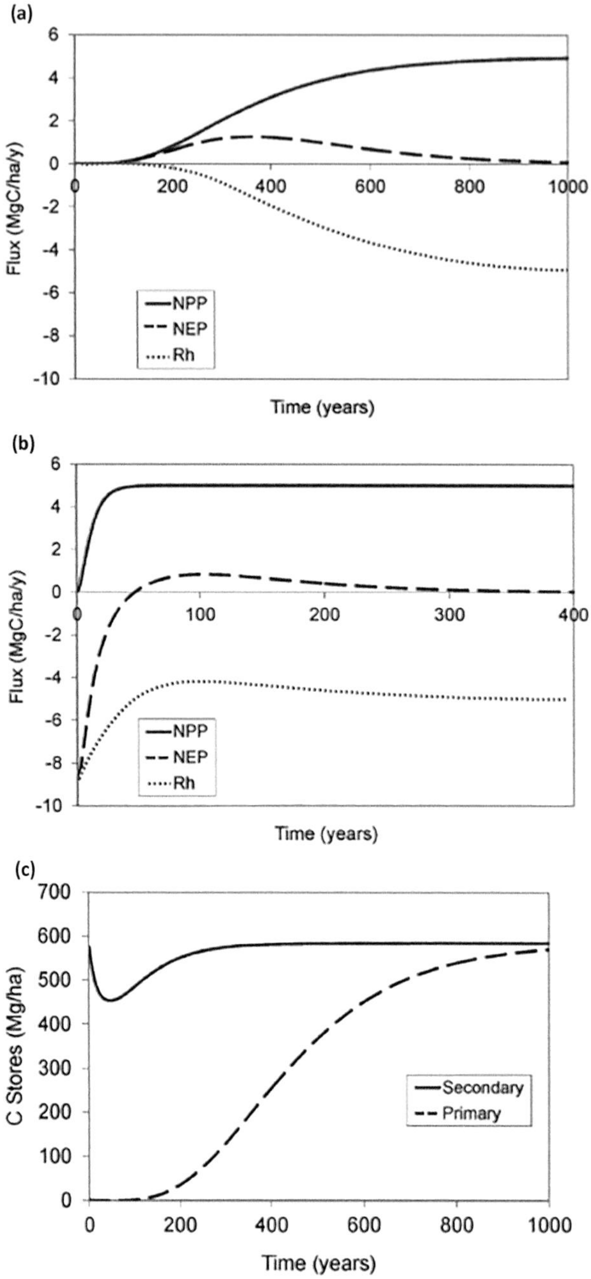

FIGURE 36.1 Temporal trends in net primary production (NPP), heterotrophic respiration (Rh), net ecosystem production (NEP), and carbon stores predicted by the classic model. (**a**) Trend of fluxes for primary succession assuming the NPP maximum is reached in 800 years, (**b**) trend in fluxes for secondary succession assuming the NPP maximum is reached in 25 years and no removal of carbon by the disturbance, and (**c**) trends in total carbon stores.

negative to positive faster and for the positive peak to be higher (Figure 36.2B). In contrast increasing the time to reach maximum NPP has little effect on the time to move from negative to positive NEP, although it does increase the amount of carbon lost during the negative phase and the amount gained in the positive phase (Figure 36.2C).

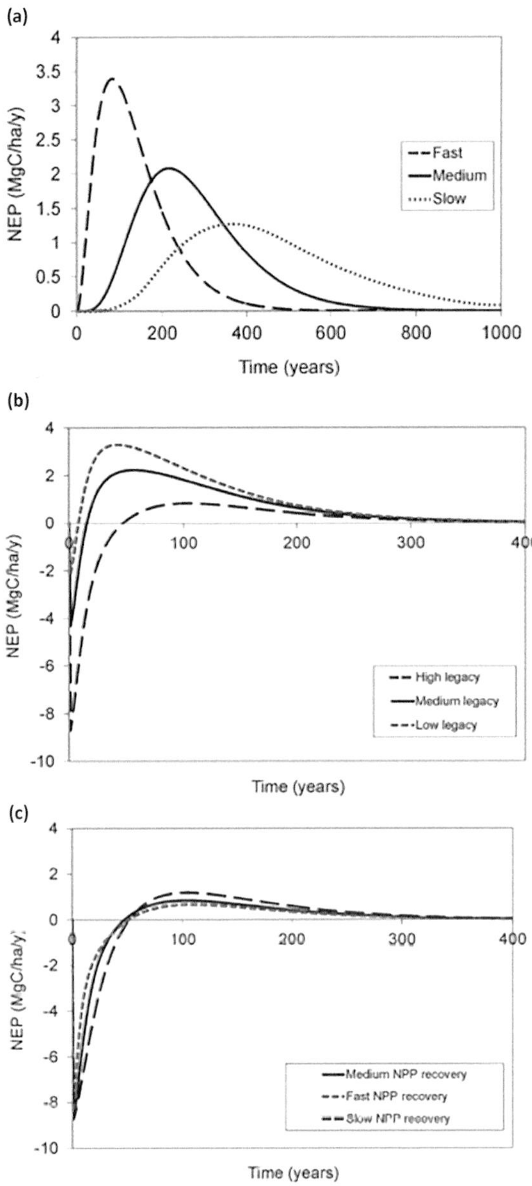

FIGURE 36.2 Potential variability in trends in net primary production (NPP), heterotrophic respiration (Rh), and net ecosystem production (NEP) during primary and secondary succession. (**a**) Effect of varying time to reach maximum NPP in a primary succession. Fast, medium, and slow times to reach maximum NPP were 12, 25, and 50 years, respectively. (**b**) Effect of varying the size of the biological legacy in a secondary succession. Low, medium, and high legacies represented 25, 50, and 100% of the maximum legacy possible. (**c**) Effect of varying the time to reach maximum NPP in a secondary succession. Fast, medium, and slow times to reach maximum NPP were 12, 25, and 50 years, respectively, and the maximum possible legacy started the succession.

While NEP may approach zero with sufficient time in terrestrial ecosystems it is unlikely that it remains zero for an extended period of time. A more useful concept is that it averages zero, but given variation in climate, small scale disturbances, and other factors it is rarely exactly zero as described by classical theory.[13] Some have questioned whether NEP ever approaches zero in forests (Figure 36.3A).[10]

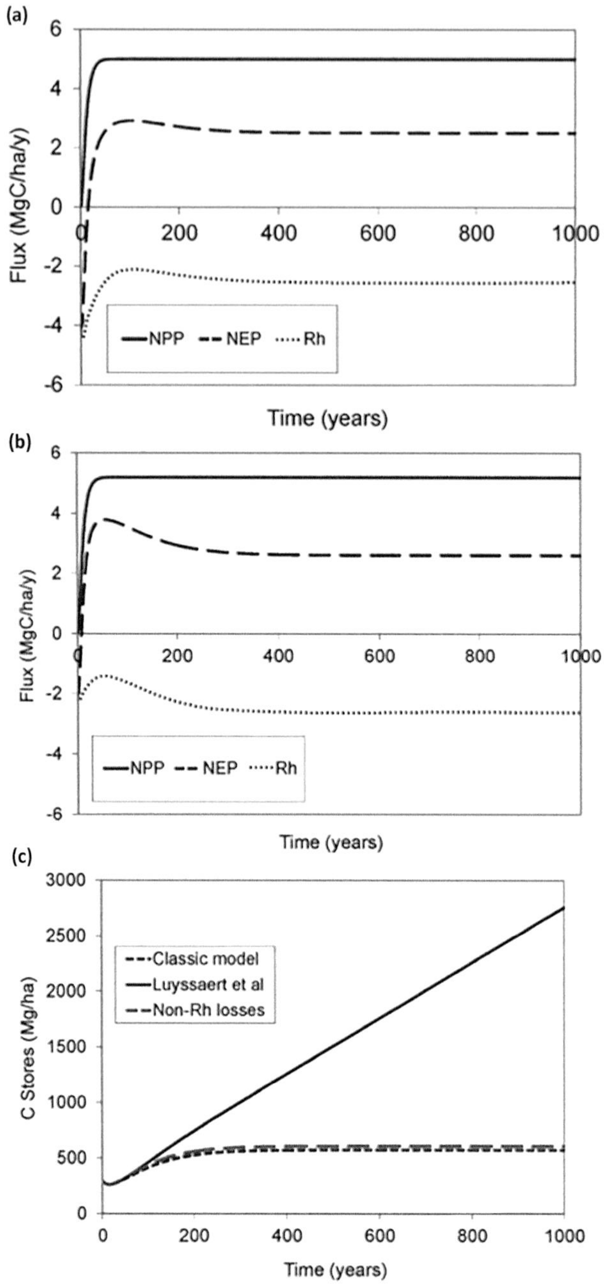

FIGURE 36.3 The possible consequences of a semipermanent, positive net ecosystem production (NEP) on terrestrial carbon stores. (**a**) The time trend in net primary production (NPP), heterotrophic respiration (Rh), and NEP predicted by Luyssaert et al.[10] (**b**) An alternative explanation that invokes nonrespiration losses. (**c**) Implied carbon stores of (**a**) and (**b**) along with that predicted by the classic model.

Given that NEP is not the same as NECB, NEP could remain positive for extended periods and carbon need not accumulate in terrestrial ecosystems; for example if there is a loss that is not associated with respiration such as erosion, significant solution losses, fire, or harvest (Figure 36.3B). Because these processes remove carbon that has been produced, it is no longer available as substrates for respiration and cannot contribute to NEP; hence NPP > Rh. Without these losses an extended period of positive and constant NEP would mean ecosystem carbon stores would increase indefinitely in a straight line manner (Figure 36.3C). This might be associated with forms of carbon that cannot be respired such as forms of soil organic matter.[14] However, classical studies of soil carbon increases over time using chronosequences indicate that while soil carbon accumulation can take long periods of time, the overall accumulation curve is not a straight line, but decreases in slope over time.[15,16] If there are major losses not associated with Rh, then it is possible that carbon accumulation is quite similar to the classical curve (Figure 36.3C), albeit for different reasons than originally proposed. Additional research is required to measure long-term trends in NEP and to document the specific mechanisms responsible for extensive periods of constant, positive NEP.

Another aspect of temporal changes in NEP at the local level in terrestrial system being explored is whether there is only a single negative and positive phase after major disturbances. The classic theory assumes homogeneity of producers and the material being respired. In addition, temporal lags other than those associated with NPP are assumed to be minimal. In forest ecosystems in particular, forms of carbon have vastly different longevities in life and after they die. For example, leaves might remain alive for a year, whereas tree stems might remain alive for centuries. Some disturbances create forms of carbon, such as standing dead trees, which do not immediately begin to decompose. The heterogeneity of carbon forms and lags associated with respiration losses could lead to a situation in which NEP has multiple negative and positive phases (Figure 36.4). Determining if this is the case is important because if there are multiple negative phases, one cannot assume a priori that once NEP is positive ecosystems will accumulate carbon in the future. Most of the temporal trends of NEP described above for terrestrial ecosystems consider periods of decades to hundreds of years. Over longer periods of time there is evidence that NPP can decline, as elements such as phosphorus become depleted due to extensive weathering.[17] This suggests that without replenishment of fresh minerals for weathering that NEP might become negative, as carbon created by earlier periods of higher production is decomposed away. However, this is likely a very slow rate of carbon loss and may be difficult to measure over the short term. Temporal patterns of NEP in aquatic ecosystems appear to have not been studied to the degree they have in upland ecosystems. This may stem from the fact that distinct disturbances in aquatic systems are hard to identify;[18] one could argue that aquatic systems are continuously disturbed and this

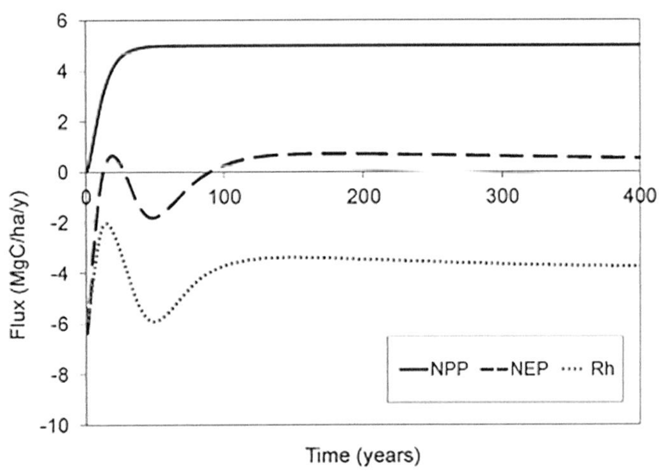

FIGURE 36.4 A hypothetical situation in which net ecosystem production (NEP) can have multiple negative and positive phases. In this system net primary production (NPP) is allocated between live and dead parts with life spans that differ by an order of magnitude. Dead parts have a fast and slow phase of decomposition with a temporal lag before newly created dead parts can decompose at maximum rates. Dead plant parts are also assumed to eventually form soil carbon with a very slow rate of decomposition.

process maintains their productivity. Nonetheless, over very long periods, some aquatic systems may undergo a temporal pattern similar to terrestrial primary succession as these ecosystems move from an oligotrophic to a eutrophic condition with both production and decomposition increasing over time. In aquatic systems that receive large amount of their carbon in dissolved forms from adjoining upland systems NEP may remain negative regardless of system age. In contrast, those receiving particulate carbon inputs from adjacent uplands may have an extensive period of positive NEP if particulates are deposited in anaerobic sediments that greatly slow decomposition. For those with high rates of particulate carbon decomposition such as flowing waters, an extensive period of negative NEP may occur. It is also possible that riparian streams that typically have a negative NEP due to high allochthonous input have a period of positive NEP after disturbance of their associated forests. This might be caused by the simultaneous increase in light levels as well as temporary reduction in carbon inputs from the adjoining upland ecosystem. If so, then this would contrast from the typical initial negative phase observed in disturbed terrestrial ecosystems. When NEP is considered over a broader spatial extent than a local one (e.g., multiple areas with asynchronous timing of disturbances), the temporal pattern of distinct negative and positive phases of NEP become less prominent. This is because the negative phase of some locations is being offset to some degree by the positive phases elsewhere. In an area in which local ecosystems are perfectly counterbalancing each other, NEP would remain constant. While theoretically possible, NEP would more likely be relatively constant until the disturbance regime influencing the broad area was changed. An underappreciated feature of NEP is that at the local scale disturbances lead to major, systematic temporal trends in NEP, whereas at broader scales it is changes in the nature of the disturbance regime in terms of frequency and severity that cause NEP to change systematically over time.[19] Another feature is that local regions need not reach a period of zero NEP for a broader region to approach zero. As long as the timing of disturbances in local regions is asynchronous the modulating of effect of one location against another can cause NEP to approach zero. The presence of losses other than respiration in a broad area can cause NEP at this scale to remain positive for an extended period without the overall system gaining carbon (Figure 36.5). Given these different temporal behaviors of NEP at different spatial scales, it is critical to clearly identify the spatial scale being considered when NEP is being examined.

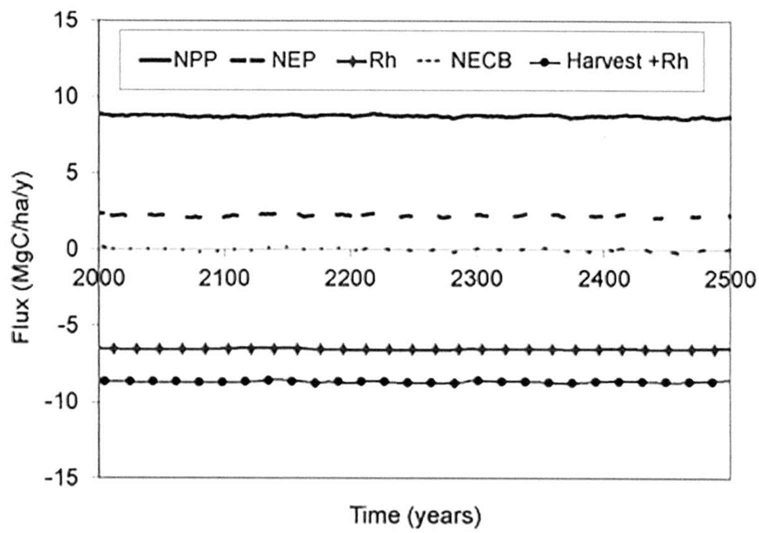

FIGURE 36.5 Landscape level trends in net ecosystem production (NEP), net primary production (NPP), heterotrophic respiration (Rh), and NECB for a forest being clear-cut harvested every 50 years. The difference between NEP and Net Ecosystem Carbon Balance (NECB) is the amount of carbon harvested annually.

Acknowledgments

The author thanks Drs. Walter Dodds and Stanley Gregory for their perspectives on aquatic ecosystems. This work was supported by the Kaye and Ward Richardson Endowment.

References

1. Woodwell, G.M.; Whittaker, R.H. Primary production in terrestrial communities. Am. Zool. **1968**, *8*, 19–30.
2. Chapin, F.S. III; Woodwell, G.M.; Randerson, J.T.; Rastetter, E.B.; Lovett, G.M.; Baldocchi, D.D.; Clark, D.A.; Harmon, M.E.; Schimel, D.S.; Valentini, R.; Wirth, C.; Aber, J.D.; Cole, J.J.; Goulden, M.L.; Harden, J.W.; Heimann, M.; Howarth, R.W.; Matson, P.A.; McGuire, A.D.; Melillo, J.M.; Mooney, H.A.; Neff, J.C.; Houghton, R.A.; Pace, M.L.; Ryan, M.G.; Running, S.W.; Sala, O.E.; Schlesinger, W.H.; Schulze. E.-D. Reconciling carbon-cycle concepts, terminology, and methods. Ecosystems **2006**, *9*, 1041–1050.
3. Lovett, G. M.; Cole, J.J.; Pace. M.L. Is net ecosystem production equal to ecosystem carbon storage? Ecosystems **2006**, *9*, 152–155.
4. Randerson, J.T.; Chapin, F.S. III; Harden, J.; Neff, J.C.; Harmon. M.E. Net ecosystem production: a comprehensive measure of net carbon accumulation by ecosystems. Ecol. Appl. **2002**, *12*, 937–947.
5. Schulze, E.-D.; Wirth, C.; Heimann, M. Climate change: Managing forests after Kyoto. Science **2000**, *289*, 2058–2059.
6. Goulden, M.L.; Munger, J.W.; Fan, S.-M.; Daube, B.C.; Wofsy. S.C. Measurements of carbon sequestration by long-term eddy covariance: methods and a critical evaluation of accuracy. Global Change Biol. **1996**, *2*, 169–182.
7. Paw, U,K.T.; Baldocchi, D.D.; Meyers, T.P.; Tilden, P.; Wilson, K.B. Kell B. Correction of eddy-covariance measurements incorporating both advective effects and density fluxes. Bound.-Lay. Meteorol. **2000**, *97*, 487–511.
8. Fellows, C.S.; Valet, H.M.; Dahm, C.N. Whole-stream metabolism in two montane streams: Contributions of the hyporheic zone. Limnol. Oceanogr. **2001**, *46*, 523–531.
9. Hall, R.O.; Tank, J.L. Correcting whole steam estimates of metabolism for groundwater inputs. Limnol. Oceanogr.: Methods **2005**, *3*, 222–229.
10. Luyssaert, S.; Schulze, E.-D.; Börner, A.; Knohl, A.; Hessenmöller, D.; Law, B.E.; Ciais, P.; Grace, J. Old-growth forests as global carbon sinks. Nature **2008**, *455*, 213–215.
11. Vannote, R.L.; Minshall, G.W.; Cummins, K.W.; Sedel, J.R.; Cushing, C.E. The river continuum concept. Can. J. Fish. Aquat. Sci. **1980**, *37*, 130–137.
12. Cole, J.J., Praire, Y.T.; Caraco, N.F.; McDowell, W.H.; Tranvik, L.J.; Striegl, R.G.; Duarte, C.M.; Kortelainnen, P.; Downing, J.A.; Middleburg, J.J.; Melack, J. Plumbing the global carbon cycle: Integrating inland waters into the terrestrial carbon budget. Ecosystems **2007**, *10*, 171–184.
13. Sierra C.A, Harmon, M.E.; Moreno, F.H.; Orrego, S.A.; del Valle, J.I. Spatial and temporal variation of net ecosystem production in a primary tropical forest in the Porce Region, Colombia. Global Change Biol. **2007**, *13*, 1–16.
14. Schmidt, M.W.I.; Torn, M.S.; Abiven, S.; Dittmar, T.; Guggenberger, G.; Janssens, I.A.; Kleber, M.; Kögel-Knabner, I.; Lehmann, J.; Manning, D.A.C.; Nannipieri, P.; Rasse, D.P.; Weiner, S.; Trumbore, S.E. Persistence of soil organic matter as an ecosystem property. Nature **2011**, *478*, 49–56.
15. Olson, J.S. Rates of Succession and Soil Changes on Southern Lake Michigan Sand Dunes Botanical Gazette **1958**, *119*, 125–170.
16. Sollins, P.; Spycher, G.; Topik, C. Processes of soil organic matter accretion at a mudflow chronosequence, Mt. Shasta, California. Ecology **1983**, *64*, 1273–1282.

17. Wardle, D.A.; Walker, L.R.; Bardgett, R.D. Ecosystem properties and forest decline in contrasting long-term chronosequences. Science **2004**, *305*, 509–513.
18. Resh, V.H.; Brown, A.V.; Covich, A.P.; Gurtz, M.E.; Li, H.W.; Minshall, G.W.; Reice, S.R.; Sheldon, A.L.; Wallace, J.B.; Wissmar, R.C. The role of disturbance in stream ecology. J. North Am. Benthol. Soc. **1988**, *7*, 433–455.
19. Smithwick, E.A.H.; Harmon, M.E.; Domingo, J.B. Changing temporal patterns of forest carbon stores and net ecosystem carbon balance: The stand to landscape transformation. Landscape Ecol. **2007**, *22*, 77–94.

37

Nitrogen Deposition and *Deyeuxia angustifolia* Encroachment in Alpine Tundra

Shengwei Zong
and Kunwei Wang
Northeast Normal University

Introduction 315
Design of the Experiment 316
Responses of Two Species to Environmental Change 317
Respective Roles of Elevated Temperature and Nitrogen Deposition....317
Future Changes in Alpine Tundra Vegetation in the Changbai Mountain 317
Conclusion 319
Acknowledgments.... 319
References 319

Introduction

Alpine ecosystems are characterized by strong winds, short growing seasons, high solar radiation, and low temperatures [1]. Due to the harsh environment, these systems are commonly considered to be at low risk of plant encroachments that originate from low elevations. However, upward expansion of plant species to higher elevations has occurred in many mountainous areas in the last decade. The upward trend is found to be accelerating, and the expansion is greater for herbaceous species, which are fast-growing, often with broad altitudinal and ecological niches.

Upward expansion of plant species from low elevations has been attributed to three main causes, i.e., climate change induced by elevated greenhouse gases [2], increased nitrogen deposition [3], and increased human disturbances [4]. Studies suggested that warmer climates make it possible for plant species to move from low to high elevations [5]. Rising temperatures have caused an amelioration of environmental conditions, such as longer growing season length, increased energy supply, and reduction of some other ecological constraints, which open alpine terrain for "invaders" from lower elevations [6,7]. However, few studies have been conducted to assess the effect of climate warming on the upward expansion of plant species and the interaction between encroaching plants and native plants in alpine regions [8], in particular in the Changbai Mountain of China. Besides climate warming, another explanation for plants expanding to high elevations is an increase in available nutrients, especially nitrogen resources. It has been proven that global terrestrial ecosystems have experienced increased atmospheric nitrogen deposition over the past decades [9]. Alpine regions commonly received higher nitrogen deposition because of more precipitation received than surrounding areas at lower elevations [10]. As alpine vegetation has developed under poor nutrient conditions, its sensitivity to increased nitrogen may result in changes to species composition [11]. However, it has been difficult to distinguish between the effects of climate warming and nitrogen deposition on the upward expansion of plant species. It is unclear which one is more important in driving the upward expansion of plants to alpine tundra.

Changbai Mountain Nature Reserve has one of the two alpine tundra areas in China. Comparing current conditions with historical data, we found that *Deyeuxia angustifolia* (Komarov) Y. L. Chang in birch forests on an adjacent mountain area had encroached the tree line and expanded into the alpine tundra. The widely distributed *D. angustifolia* patches have severe impacts on the alpine tundra ecosystem and lead to extirpation of typical tundra plants such as *Rhododendron chrysanthum* [12]. *D. angustifolia*, a narrow-leaf small reed, is characterized as a hydro-mesophyte perennial herb, with life traits that include fast growth, clonal and sexual reproduction, a short juvenile period, high seed production, and anemochory. These traits ensure they can overcome stressful abiotic conditions and survive in the harsh alpine tundra environment.

Design of the Experiment

According to the plant traits of *D. angustifolia* and *R. chrysanthum*, a hypothesis was proposed that nitrogen deposition rather than climate warming promotes *D. angustifolia* expansion within the alpine tundra. To test this hypothesis, a multi-factorial experiment was conducted in the Changbai Mountain study site in which warmth and nutrients were added to *D. angustifolia* communities. Treatments of increasing temperature and adding nitrogen load were conducted, whereas treatments, i.e., controls were not conducted on *D. angustifolia* communities (Figure 37.1). The widely used open-top chamber (OTC), a passive warming apparatus, was applied to simulate elevated atmospheric temperature as a surrogate of climate warming. To simulate atmospheric nitrogen deposition, external nitrogen was applied in equal monthly doses during the growing season with a solution of NH_4NO_3 in deionized distilled water. During the experiment, growth of both *D. angustifolia* and *R. chrysanthum* was observed.

FIGURE 37.1 **(See color insert.)** Experiment design and setup in the Changbai Mountain, Northeast China.

Responses of Two Species to Environmental Change

Low temperatures and poor nutrients are two common major constraints to plant growth in alpine habitats [13]. However, *R. chrysanthum*, the typical alpine tundra shrub, showed negligible response to the simulated environmental changes. This may be because evergreen shrubs in alpine tundra can be affected negatively by nutrient addition [14]. Compared with *R. chrysanthum*, *D. angustifolia* could effectively occupy the aboveground space by increasing tillers and growing rapidly by efficiently using nitrogen. These findings are consistent with reported studies. Aerts and Caluwe [15] found a difference in the nitrogen economy between evergreen and deciduous species, which can lead to deciduous species replacing evergreen species. Klanderud and Totland [16] found a shift in dominance hierarchies from shrub to graminoids and forbs under nutrient addition.

It was surprised to see that *D. angustifolia* responded negatively to warmer conditions since experiments across tundra sites worldwide have shown that responses of plant species, especially graminoids, to simulated increases in temperature and nutrient supply are generally positive [10,17–20]. In the alpine tundra of the Changbai Mountain, the combined effect of temperature and moisture is more influential for plant growth than temperature alone. *D. angustifolia* is a hydro-mesophyte perennial herb, so water deficiency would limit its establishment and growth. OTC treatment increased the air temperature while reducing soil moisture due to high evaporation [21]. Therefore, the negative response of *D. angustifolia* to OTC treatment was really because of the coupled effect of higher temperatures and less water.

Respective Roles of Elevated Temperature and Nitrogen Deposition

Although studies suggested a high correlation between climate warming and the upward expansion of plant species [4,10], climate warming does not explain the expansion of *D. angustifolia* in the Changbai Mountain. Growth of *D. angustifolia* showed a negative response to warmer conditions. Unlike other upwardly expanding plants [5], the effect of elevated temperature on *D. angustifolia* expansion is not decisive. Furthermore, *D. angustifolia* is commonly distributed at low elevations on all slope aspects in the alpine tundra of the Changbai Mountain, however, the upward expansion phenomenon occurred only on the western aspect of the slope. If climate warming is promoting the *D. angustifolia* expansion, the phenomenon could be expected on other slope aspects, at least on the southern slope with similar habitat conditions, since variations in climatic conditions on the four slope aspects were not found [7].

Nutrient supply as a driving growth mechanism is a reasonable explanation of *D. angustifolia* encroachment [22]. Alpine tundra of the Changbai Mountain has received an increasingly high load of atmospheric nitrogen deposition as reported [23]. Sufficient nitrogen supply is conducive to *D. angustifolia* encroachment. Additional nitrogen could accumulate on the western slope aspect because it is windward and winds prevail during the growing season [11], which bring more atmospheric nitrogen deposition by precipitation. Once established, *D. angustifolia* could effectively expand by utilizing nitrogen. Moreover, other plant species in the community that can tolerate lower nutrient levels are unable to compete successfully against this encroaching species [24]. Alatalo et al. [25] also found that for some herbaceous plant species with high nitrogen use efficiency, climate warming may not be the decisive factor in the process of moving upward to high elevations.

Future Changes in Alpine Tundra Vegetation in the Changbai Mountain

Biodiversity is highly correlated to changes in the abundance of a dominant species because the dominant species may strongly influence biotic conditions and is often an important driver of ecosystem function and community dynamics [24]. The native dominant species, *R. chrysanthum*, was suppressed as *D. angustifolia* expanded. Continued *D. angustifolia* expansion would change the biodiversity of tundra vegetation and may eventually result in a replacement of native biota, especially if the increased supply of nitrogen continues. This prediction is consistent with the findings of Wolf et al. [26] who studied the invasion of sweet clover (*Melilotus*) in Montane Grasslands, Rocky Mountain National Park, USA.

It can be anticipated that increases in nitrogen will have a greater effect on long-term vegetation changes and community biodiversity in the alpine tundra than increases in temperature. *D. angustifolia* encroachment has occurred during past decades, despite increased temperatures. Therefore, when the current nitrogen level (23 kg N/ha per year) increases to 46 kg N/ha per year in 2050 [27], *D. angustifolia* populations would significantly increase (Figure 37.2).

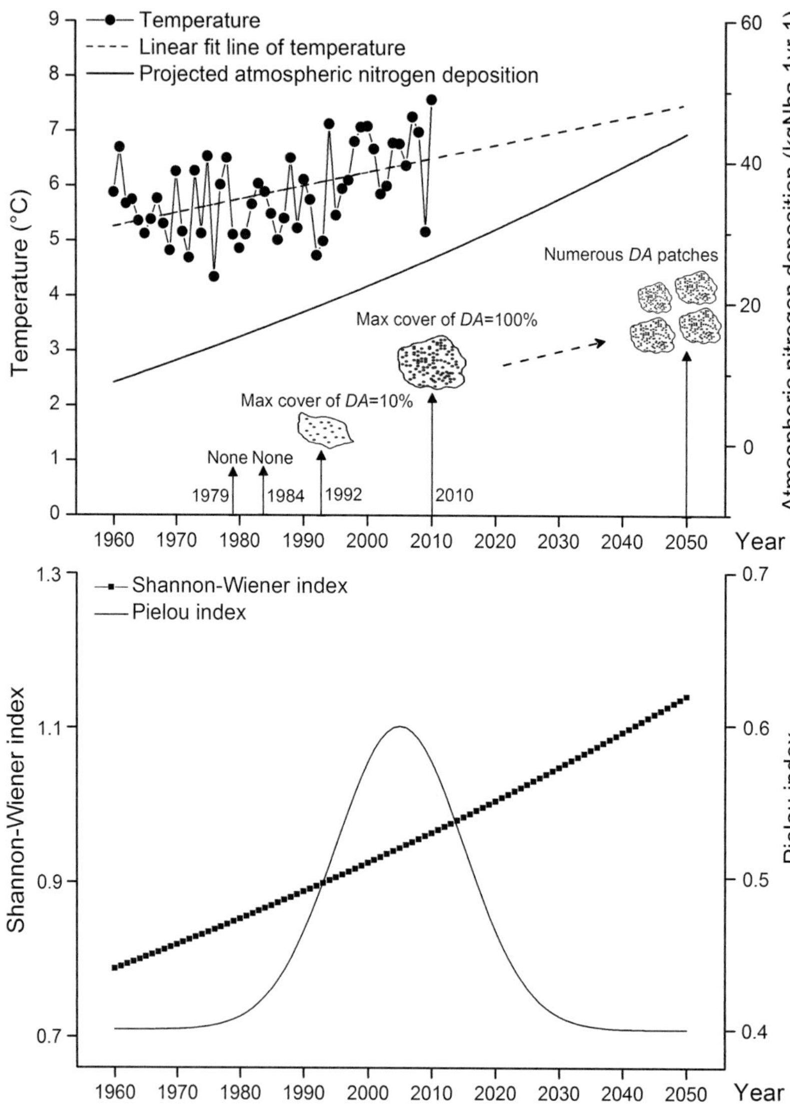

FIGURE 37.2 Theoretical prediction of long-term changes in the Shannon-Wiener diversity index and Pielou index for plant communities in the alpine tundra of the Changbai Mountain, based on changes in temperature, atmospheric nitrogen deposition, and the corresponding changes in distribution of *D. angustifolia* (DA).

Conclusion

Temperature increase had a negative effect on the growth of *D. angustifolia*, but nitrogen addition dramatically stimulated *D. angustifolia* expansion. Compared to *R. chrysanthum*, *D. angustifolia* could effectively occupy the aboveground space by efficiently using nitrogen. The native alpine shrub, *R. chrysanthum*, showed negligible responses to the simulated environmental changes, and its growth was suppressed by *D. angustifolia*, especially in the high encroachment community under nitrogen treatment. Nutrient perturbation may be more influential than temperature perturbation in promoting *D. angustifolia* encroachment upon the nutrient- and species-poor alpine tundra ecosystem in the Changbai Mountain.

Acknowledgments

This study was supported by National Natural Science Foundation of China (No. 41971124 and 41501089).

References

1. Körner, C., *Alpine Plant Life: Functional Plant Ecology of High Mountain Ecosystems; with 47 Tables*, 2003, Springer Science & Business Media, Berlin, Heidelberg.
2. Walther, G.R., S. Beißner, and C.A. Burga, Trends in the upward shift of alpine plants. *Journal of Vegetation Science*, 2005. 16(5): 541–548.
3. Thuiller, W., D.M. Richardson, and G.F. Midgley, Will climate change promote alien plant invasions?, in *Biological Invasions*, 2008, Springer, Berlin, Heidelberg. pp. 197–211.
4. Heer, C. and C. Körner, High elevation pioneer plants are sensitive to mineral nutrient addition. *Basic and Applied Ecology*, 2002. 3(1): 39–47.
5. Klanderud, K. and H.J.B. Birks, Recent increases in species richness and shifts in altitudinal distributions of Norwegian mountain plants. *The Holocene*, 2003. 13(1): 1–6.
6. Parolo, G. and G. Rossi, Upward migration of vascular plants following a climate warming trend in the Alps. *Basic and Applied Ecology*, 2008. 9(2): 100–107.
7. McDougall, K.L., et al., Plant invasions in mountains: Global lessons for better management. *Mountain Research and Development*, 2011. 31(4): 380–388.
8. Harte, J. and R. Shaw, Shifting dominance within a montane vegetation community: Results of a climate-warming experiment. *Science*, 1995. 267(5199): 876–880.
9. Phoenix, G.K., et al., Atmospheric nitrogen deposition in world biodiversity hotspots: The need for a greater global perspective in assessing N deposition impacts. *Global Change Biology*, 2006. 12(3): 470–476.
10. Huang, D., R.A. Haack, and R. Zhang, Does global warming increase establishment rates of invasive alien species? A centurial time series analysis. *PLoS One*, 2011. 6(9): e24733.
11. Fowler, D., et al., The influence of altitude on rainfall composition at great dun fell. *Atmospheric Environment (1967)*, 1988. 22(7): 1355–1362.
12. Curtis, C., et al., Nitrogen saturation in UK moorlands: The critical role of bryophytes and lichens in determining retention of atmospheric N deposition. *Journal of Applied Ecology*, 2005. 42(3): 507–517.
13. Zong, S., Z. Wu, and H. Du, Study on climate change in alpine tundra of the Changbai mountain in growing season in recent 52 years. *Arid Zone Research*, 2013. 30: 41–49.
14. Robinson, C.H., et al., Plant community responses to simulated environmental change at a high arctic polar semi-desert. *Ecology*, 1998. 79(3): 856–866.
15. Aerts, R. and de Caluwe, H., Nitrogen use efficiency of Carex species in relation to nitrogen supply. *Ecology*, 1994. 75(8): 2362–2372.
16. Klanderud, K. and Totland, Ø., Simulated climate change altered dominance hierarchies and diversity of an alpine biodiversity hotspot. *Ecology*, 2005. 86(8): 2047–2372.

17. Bowman, W.D., Accumulation and use of nitrogen and phosphorus following fertilization in two alpine tundra communities. *Oikos*, 1994. 70(2): 261–270.
18. Seastedt, T. and L. Vaccaro, Plant species richness, productivity, and nitrogen and phosphorus limitations across a snowpack gradient in alpine tundra, Colorado, USA. *Arctic, Antarctic, and Alpine Research*, 2001. 33(1): 100–106.
19. Elmendorf, S.C., et al., Global assessment of experimental climate warming on tundra vegetation: Heterogeneity over space and time. *Ecology Letters*, 2012. 15(2): 164–175.
20. Elmendorf, S.C., et al., Plot-scale evidence of tundra vegetation change and links to recent summer warming. *Nature Climate Change*, 2012. 2(6): 453.
21. Bokhorst, S., et al., Climate change effects on soil arthropod communities from the Falkland Islands and the Maritime Antarctic. *Soil Biology and Biochemistry*, 2008. 40(7): 1547–1556.
22. Hu, Y., et al., Responses of soil available nitrogen of natural forest and secondary forest to simulated N deposition in Changbai Mountain. *Journal of Northeast Forestry University*, 2009. 37(5): 36–42.
23. Liu, X. and H. Zhou, Analysis on the regional precipitation characteristics of western slope in Changbai Mountain. *Jilin Meteorology*, 2009. 3: 25–26.
24. Brooks, M.L., Effects of increased soil nitrogen on the dominance of alien annual plants in the Mojave Desert. *Journal of Applied Ecology*, 2003. 40(2): 344–353.
25. Alatalo, J.M., et al., Vascular plant abundance and diversity in an alpine heath under observed and simulated global change. *Scientific reports*. 2015. 5: 10197, doi:10.1038/srep10197.
26. Wolf, J.J., et al., Invasion by sweet clover (Melilotus) in montane grasslands, Rocky Mountain National Park. *Annals of the Association of American Geographers*. 2003. 93(3): 531–543.
27. Lü, C.Q., et al., Ecological effects of increased nitrogen deposition in terrestrial ecosystems. *Chinese Journal of Plant Ecology*. 2007. 31(2): 205–218 (In Chinese).

38

Nitrogen Fixation: Biological

Introduction 321
Symbiotic and Free-Living Nitrogen Fixers 321
Biochemistry of Nitrogen Fixation 322
Cell-Free Nitrogen Fixation 323
Purification of Nitrogenase 323
Requirement for Adenosine Triphosphate 323
Substrates Other than N_2 323
Control of Nitrogenase 323
Tertiary Structure of Nitrogenase 324
Associative Nitrogen Fixation 324
Genetics of the Nitrogenase Systems 324
References 324

Robert H. Burris
University of Wisconsin

Introduction

For centuries farmers practiced mixed cropping of leguminous and non-leguminous plants without knowing the basis of the empirically observed benefit from the practice. The first clear recognition that leguminous plants could utilize nitrogen (N_2) from the air is attributed to Boussingault's report in 1838.[1] Skepticism remained until Hellriegel and Wilfarth in Germany in 1886 reported definitive experiments demonstrating that pea plants (*Pisum sativum*) in association with bacteria borne in their root nodules were capable of utilizing N_2.[1]

Symbiotic and Free-Living Nitrogen Fixers

The symbiotic process of N_2 fixation by leguminous plants plus associated root nodule bacteria (*Rhizobium* species) is of greatest practical importance to agriculture, but there also are a number of free-living bacteria capable of N_2 fixation. The first of these recognized was the anaerobic organism *Clostridium pasteurianum* as reported by Winogradsky in 1893.[2] In 1901 Beijerinck recorded that the aerobic bacterium *Azotobacter chroococcum* also fixed N_2.[2] These early observations were followed by practical applications, and the practice of inoculation of leguminous seeds with cultures of the rhizobia became widespread. The bacteria were grown commercially and farmers applied these bacteria to leguminous seeds at the time of planting. There is a specificity between the plants and bacteria, and this gave rise to the recognition of cross-inoculation groups, i.e., groups of plants all of which are nodulated by the same species of rhizobia. For example, *Rhizobium leguminosarum* infects garden peas, sweet peas (*Lathyrus odoratus*), common vetch (*Vicia faba*), hairy vetch (*Vicia villosa*), and lentil (*Lens culinaris*), whereas *Rhizobium japonicum* is rather specific for soybeans (*Glycine max*). Not all strains of root nodule bacteria that infect a plant are equally effective in N_2 fixation, and this is referred to as strain variation. Because of strain variation, commercial inoculants often are a mixture of effective strains. Inoculants are grown as large batches of effective organisms in liquid culture, and then these cultures are mixed on a peat base. The finely ground peat retains moisture, and the microorganisms remain viable for long periods. The peat culture is mixed with seeds before planting, and the viable organisms are in proximity when the seeds germinate and produce roots. The bacteria induce curling of root hairs and invade the plant through the root hairs. After invasion they induce the production of root nodules and proliferate in the nodules. The bacteria undergo morphological and metabolic changes in the nodule and are referred to as bacteroides. Bacteroides in the nodule are capable of fixing N_2 from the air.

Claims of fixation of N_2 by cultures of the rhizobia outside the host plant in general could not be verified. In 1975 three laboratory groups reported success in achieving such fixation, by furnishing succinic acid as the substrate for growth and by culturing the organisms under a very low pressure of oxygen.[3]

The rhizobia function at a low oxygen concentration in the nodule. The nitrogenase enzyme system is labile to oxygen, but there must be sufficient oxygen to support the generation of adenosine triphosphate (ATP). Hemoglobin in nodules achieves this necessary balance in the oxygen level, keeping it adequate to support the action of cytochrome oxidase which generates ATP.

There are well over 10,000 legume plant species that fix N_2, and about 200 non-legumes. Only one of these nonlegumes has been reported to be nodulated by a *Rhizobium* species such as nodulate the legumes. This is *Parasponia andersonii*, a tree of the elm group. The other non-leguminous plants, such as the alder (e.g., *Alnus glutinosa*), fix nitrogen in association with actinomycetes. It proved extremely difficult to culture these actinomycetes, but since methodology to isolate and culture these organisms was developed in 1977,[3] information on them has expanded substantially. Some of the actinorhizal associations have found extensive use, particularly in tropical and low rainfall areas.

Biochemistry of Nitrogen Fixation

Early studies on biological nitrogen fixation emphasized practical applications of the process and the development of effective inoculants. Since 1925 there has been interest in the biochemistry of the N_2 fixation process. Wilson[1] utilized nodulated red clover plants (*Trifolium pratense*) to demonstrate that half-maximum fixation occurred at a pN_2 (partial pressure of N_2) of about 0.05 atmosphere (approximately 50 millibar). High levels of O_2 inhibited the fixation non-competitively. While growing clover plants under controlled atmospheres, Wilson[1] made the unexpected observation that fixation was inhibited by hydrogen gas (H_2). H_2 is a specific and competitive inhibitor of N_2 fixation. The H_2 inhibition of N_2 fixation results from an ordered, sequential process in which the H_2 must bind to the nitrogenase enzyme before the N_2.

There was speculation about the first or key product of N_2 fixation. Winogradsky[2] suggested that ammonia (NH_3) was the first product. Virtanen, [2] on the other hand, supported hydroxylamine as the key intermediate based on his claim that leguminous plants excreted aspartic acid, and that the hydroxylamine formed by fixation reacted with oxalacetic acid to form an oxime that then was converted to aspartic acid. Other investigators failed to confirm Virtanen's work. The use of the stable isotope ^{15}N as a tracer resolved the argument. ^{15}N-enriched N_2 was supplied to a culture of *Azotobacter vinelandii* fixing N_2 vigorously, and among the amino acids recovered from the hydrolysate of the cells, the highest enrichment in ^{15}N among the isolated amino acids was in glutamic acid[2] rather than in aspartic acid, which would have been supportive of Virtanen's hydroxylamine hypothesis. If NH_3 were the key product of fixation, it could logically be incorporated into glutamic acid by the action of known enzymes. Similar experiments were performed with species of bacteria from the genera *Clostridium, Chromatium, Chlorobium,* and *Rhodospirillum,* as well as the blue-green alga *Nostoc muscorum* and excised soybean nodules. In all cases the highest concentration of ^{15}N among the isolated amino acids was in glutamic acid.

Cell-Free Nitrogen Fixation

As the metabolism of nitrogenous compounds in intact microorganisms can be rather complex, it was important to verify the path of fixation in cell-free preparations where reactions can be defined more specifically. ^{15}N as a tracer furnished a sensitive tool for seeking reliable cell-free fixation. Carnahan and coworkers in 1960[3] developed a method for recovering active preparations consistently from *C. pasteurianum*. Such cell-free preparations supplied ^{15}N-enriched N_2 yielded NH_3 with over 50 atom % ^{15}N excess, a very convincing verification that NH_3 is the "key" intermediate in biological nitrogen fixation.

Purification of Nitrogenase

Mortenson[3] demonstrated that the cell-free enzyme system consisted of two proteins. When the two proteins were purified, it turned out that one component was a molybdenum–iron (MoFe) protein and the other was an iron protein. In purification on a chromatographic column, the MoFe protein was eluted before the Fe protein, so the MoFe protein often was referred to as protein 1 or component 1, and the Fe protein as protein 2 or component 2. A number of other terminologies were used but abandoned. Descriptive terms are dinitrogenase for the MoFe protein, dinitrogenase reductase for the Fe protein, and nitrogenase for the complex of the two components. The primary function (it also has other functions) of the Fe protein is to transfer electrons to the MoFe protein, hence the designation dinitrogenase reductase.

Requirement for Adenosine Triphosphate

In 1960 it was reported that ATP was inhibitory to N_2 fixation in cell-free preparations of nitrogenase, but in 1962 McNary[3] demonstrated that ATP is an absolute requirement for fixation by such preparations. The N–N bond is stable, so N_2 fixation is energy demanding whether it is accomplished chemically at high temperature and pressure, or whether it is accomplished by a biological system. A minimum of 16 ATPs are required for the reduction of one N_2 to $2NH_3$. Sixteen ATP are required under ideal conditions, and under most conditions in nature the requirement is 20 to 30 ATP.

Substrates Other than N_2

Nitrogenase is a versatile enzyme and is capable of reducing substrates other than N_2.[3] It was recognized early that H_2 is a by-product of the nitrogenase reaction. The next substrate to be recognized was nitrous oxide (N_2O) which is reduced to N_2. Cyanide and methyl isocyanide, and azide also can be reduced by the enzyme system. Nitrogenase can reduce acetylene to ethylene. Other substrates demonstrated include diazine, diazarine, cyanamide, and cyclopropene. Acetylene reduction to ethylene has become a widely used measure of nitrogenase activity. The substrate is easy to prepare, the ethylene formed as a product can be measured readily and accurately by gas chromatography, and the simple method has been applied to studies in the glasshouse and field.

Control of Nitrogenase

As nitrogen fixation is energy demanding, it is advantageous for the fixing organisms to turn off the system if they have access to a source of fixed nitrogen. Ludden[3] found that *Rhodospirillum rubrum* turned its nitrogenase off when supplied fixed nitrogen such as NH_3, or when it was placed in the dark and thus was deprived of light energy. When NH_3 was exhausted or when light was restored, it turned nitrogenase back on. Inactivation is achieved by ADP ribosylation of arginine 100 of one of the two equivalent subunits of dinitrogenase reductase, and reactivation is achieved by removal of this ADP-ribosyl group.

Tertiary Structure of Nitrogenase

The tertiary structure of the nitrogenase components was established in 1992.[2] This aids one in visualizing how the dinitrogenase and dinitrogenase reductase fit together and how the active Fe and MoFe sites are aligned.

Associative Nitrogen Fixation

The leguminous plants and their associated rhizobia are of great practical importance in agriculture, but the non-leguminous actinorhizal systems often fix N_2 vigorously and occupy important niches in ecosystems. Certain non-legumes can benefit from association with bacteria that occupy their root surface or multiply within the plant without forming nodules.[3,4] In Brazil it has been demonstrated that sugarcane can derive nitrogen from fixation by *Acetobacter diazotrophicus* carried inside the cane. *A. diazotrophicus* is a remarkable bacterium, as it can grow in a 25% sucrose solution and fix N_2 at a pH as low as 3.0.

Genetics of the Nitrogenase Systems

The genetics of biological N_2-fixing systems has occupied many investigators in recent years, as genetic manipulations may offer ways to improve the process or adapt it to new crops. Progress in our knowledge of the genetics of N_2-fixing organisms has led in recent years to a substantial reorganization of classification of these organisms.

References

1. Wilson, P.W. *The Biochemistry of Symbiotic NitrogenFixation;* The University of Wisconsin Press: Madison, WI, 1940; 302 pp.
2. Triplett, E.W. *Prokaryotic Nitrogen Fixation; Horizon*; ScientificPress: Wymondham, England, 2000; 800 pp.
3. Stacey, G.; Burris R.H.; Evans, H.J., Eds.; *Biological Nitrogen Fixation;* Chapman and Hall: New York, 1992; 943 pp.
4. Postgate, J.R. *The Fundamentals of Nitrogen Fixation;* Cambridge University Press: Cambridge, 1982; 252 pp.

39

Ozone: Crop Yield and Quality Effects

Håkan Pleijel
Göteborg University

Discovery of Ozone Effects on Crops 325
Open-Top Chambers 325
Crop Loss Assessment Programs 326
"Critical Levels" of Ozone 326
Ozone Uptake 326
Effects of Ozone on Protein Contents 326
Conclusion 328
References 329

Discovery of Ozone Effects on Crops

Systematic studies of ozone effects on crop yield and quality started around 1950 in California, U.S.A.,[1] where the problems caused by ozone and other photochemical oxidants were first discovered. At that time, the studies were mainly of an observational nature, and they were based on visible injury appearing on the plant leaves. Controlled experiments in the laboratory began later; however, initially these yielded largely qualitative data which were not suitable as a basis to estimate the magnitude of potential effects in the field.

But, it should be noted here that in certain crops, such as spinach and tobacco, leaf injury is of large economic importance, apart from potential reductions in absolute yield. In some crops visible leaf injury, expressed as necrotic spots on the leaves, is the most pronounced effect by ozone and other pollutants, while in others a reduction in the leaf life span, the so-called premature senescence, is most important, as it reduces the length of the growth period. The latter occurs in wheat, in which characteristic leaf injury at moderate ozone exposure is generally lacking, while early senescence in response to ozone is pronounced.[2]

Open-Top Chambers

With the introduction of the open-top chamber (OTC) as an exposure system, [3] which permits semi-controlled exposure of plants to gaseous pollutants under ecologically realistic conditions in the field, an important step was taken towards a quantitative understanding of the impacts of ozone and other pollutants. An OTC is a transparent plastic cylinder through which air (ambient air, filtered air, or air enriched with various levels of pollutant gases) is ventilated using a fan. OTCs can be mounted in plots of a field grown crop and kept in place throughout the growing season. Although the system alters the microclimate of the plants to a certain extent, [4] the ecological realism is much larger than in a closed chamber system in the laboratory.

Crop Loss Assessment Programs

In the U.S.A., a large experimental program, the National Crop Loss Assessment Network (NCLAN), involving the use of OTCs of crops, was executed during the 1980s. The NCLAN results showed that ambient levels of ozone had the potential to reduce the yield of a number of crops, including soy bean, wheat, and alfalfa.[5] In addition, economic estimates of crop losses for the U.S.A. indicated an annual loss for the farmers in the range of a few billion U.S. dollars.

A similar program, the European Open-top Chamber Programme (EOTC), was organized in Western Europe beginning in the second half of the 1980s until the early 1990s. Negative effects of ozone levels typical of wide areas of Europe were observed in beans and spring wheat.[6] In some, but not all, experiments with pastures the main effect of ozone was on the species composition, with clover being replaced by grasses. This effect acts to reduce the protein content and thus the fodder quality of the yield. Towards the end of the 1990s a European research program, Changing Climate and Potential Impacts on Potato Yield and Quality (CHIP), studied the effects of ozone on potato. It was concluded that current ozone levels in Europe can cause significant effects on potato, but that the sensitivity is less than in wheat.

Based on the experiences of both research programs in North America and Europe, it can be concluded that soybean, wheat, tomato, pulses, and watermelon can be classified as highly sensitive to ozone. Barley and certain fruits are rather insensitive, while potato, rapeseed, sugar beet, maize, and rice are intermediate with respect to ozone sensitivity. There exists a certain degree of intraspecific variation in ozone sensitivity, such as in different bean varieties, but a recent European study resulted in a rather uniform ozone response pattern for different cultivars of wheat.

"Critical Levels" of Ozone

The relationship between grain yield and ozone exposure was most consistent in spring wheat across cultivars and experimental sites, and, considering the pronounced sensitivity observed for this crop, the combined data were used as a basis to derive critical levels for ozone effects under the Convention on Long Range Transboundary Air Pollution of the United Nations Economic Commission for Europe (UNECE). For this purpose, the ozone exposure index AOT40 (the accumulated exposure over a concentration threshold of 40 nmol mol^{-1} ozone based on hourly averages) was used.[7]

Ozone Uptake

An important recent development in the field of crop loss assessments is the consideration of ozone uptake by the plant leaves, in place of a statistical index to characterize the concentration of ozone in the air surrounding the plants, such as the AOT40 exposure index. Key to this method is the quantification of ozone diffusion through the stomata of leaves.

The stomatal conductance of the plant leaves varies with a number of factors, such as soil and air moisture contents, temperature, solar radiation, phenology, and the levels of other pollutants, including carbon dioxide. Stomatal conductance models can now be used to establish relationships between yield loss and ozone uptake. An example is given in Figure 39.1 showing the relationship between relative yield in two Swedish wheat cultivars and the calculated, cumulated uptake of ozone by flag leaves, taking into account an uptake rate threshold of 5 nmol $m^{-2} s^{-1}$ (CUO^5).[8] This threshold provided the best correlation between effect and exposure, and it may represent the biochemical defense capacity of the plants. However, it remains to be shown that the two thresholds are directly related to each other.

Effects of Ozone on Protein Contents

Much less attention has been paid to the influences of ozone on crop quality as compared to efforts made to understand the relationship with yield. The most discussed quality aspect has been the protein concentration or content, for instance, of wheat grains. The combined results of sixteen OTC experiments performed

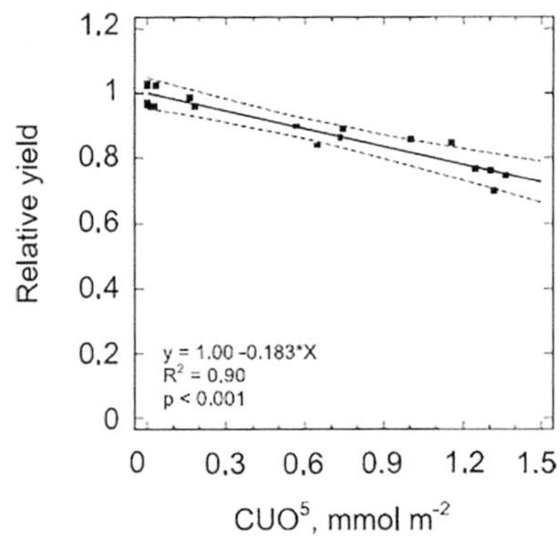

FIGURE 39.1 Relationship between the relative yield of field grown spring wheat and the cumulative ozone uptake (CUO) by the flag leaves based on stomatal conductance modeling. An ozone uptake rate threshold of 5 nmol $m^{-2}s^{-1}$ was used, since this threshold resulted in the best correlation between relative yield and ozone uptake. Five OTC experiments performed in Sweden were included.
Source: Adapted from Danielsson, Pihl Karlsson, et al.[8]

in four different European countries (Finland, Sweden, Denmark, and Switzerland) were used to show the relationship between grain yield and grain quality (Figure 39.2), and grain yield and the off take of grain protein per unit ground area (Fig. 3), at different ozone or carbon dioxide levels.[9] From Fig. 2, it can be inferred that, relative to control (always the OTC treatment with non filtered air), the grain concentration of protein was higher and yield was lower in elevated ozone, which is the opposite result of situations with elevated carbon dioxide or in chambers where the ozone had been reduced by using charcoal filters. This situation could be attributed to a so-called growth dilution effect by elevated carbon dioxide and air

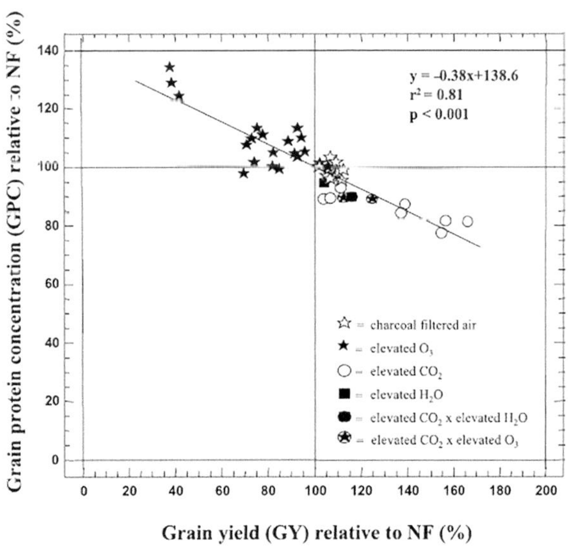

FIGURE 39.2 Relationship between grain protein concentration and grain yield in field grown spring wheat based on data from OTC experiments performed in Sweden, Finland, Denmark, and Switzerland. The experiments included ozone treatments, carbon dioxide treatments, and a few irrigation treatments. Relative scales were used for both axes: The effects were related to the OTC treatment with non filtered air (NF) without extra ozone or carbon dioxide fumigation in each of the sixteen experiments included.
Source: Adopted from Pleijel, Mortenson, et al.[9]

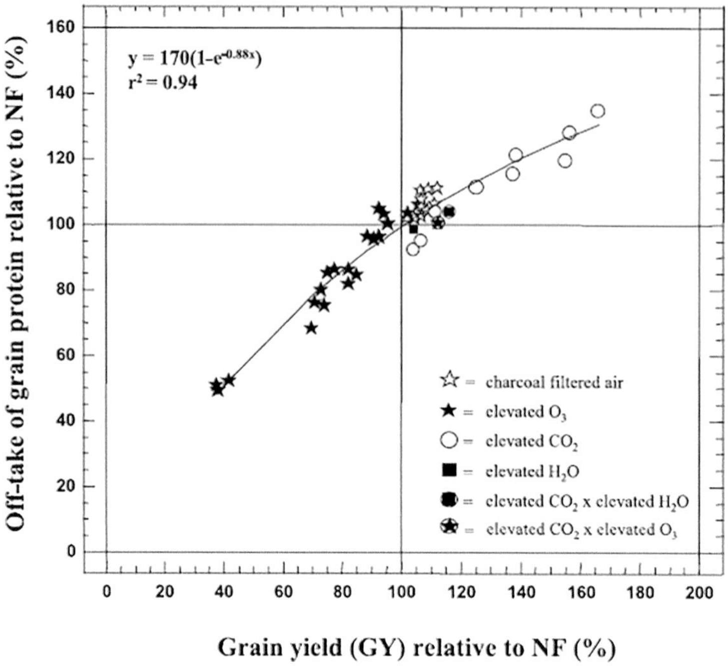

FIGURE 39.3 Relationship between grain protein off take per unit ground area and grain yield in field grown spring wheat based on data from OTC experiments performed in Sweden, Finland, Denmark, and Switzerland. The experiments included ozone treatments, carbon dioxide treatments, and a few irrigation treatments. Relative scales were used for both axes: The effects were related to the OTC treatment with non-filtered air (NF) without ozone or carbon dioxide fumigation in each of the sixteen experiments included.
Source: Adopted from Pleijel, Mortenson, et al.[9]

filtration stimulating the carbohydrate accumulation relatively more than the uptake of nitrogen at a given nitrogen fertilizer application rate, while in situations of elevated ozone, carbohydrate accumulation was negatively affected more strongly than protein accumulation.

However, Figure 39.3 shows that the protein yield per unit ground area was negatively affected by ozone and enhanced by elevated carbon dioxide, although the latter effect tended to saturate at a level determined by the availability of nitrogen in the soil.

Conclusion

There is strong evidence that current ozone concentrations over large areas in the industrialized world are high enough to cause yield loss in several important agricultural crops, but it remains a challenge to quantify these effects exactly. New approaches based on the uptake of the pollutant by plants, rather than the concentration in the air surrounding the plant, are promising.

The protein concentration of the yield tends to increase with increasing ozone in some crops in connection with declining yield. On the other hand, the protein yield per unit ground area tends to decrease. In future research, the understanding of additional quality effects of air pollutant exposure should be given more attention.

In the developing world, such as many countries in Asia, there is a great risk for crop losses due to a number of air pollutants for which emission rates are increasing dramatically.[10] A loss in food production in these countries represents a much larger problem than in the industrialized world. More research should be devoted to this problem in the decades to come.

References

1. Middleton, J.T.; Kendrick, J.B.; Schalm, H.W. Injury to herbaceous plants by smog or air pollution. *Plant Dis. Rep.* **1950**, *34*, 245–252.
2. Pleijel, H.; Ojanperä, K.; Danielsson, H.; Sild, E.; Gelang, J.; Wallin, G.; Skärby, L.; Selldén, G. Effects of ozone on leaf senescence in spring wheat—Possible consequences for grain yield. Phyton (Austria) **1997**, *37*, 227–232.
3. Heagle, A.S.; Body, D.; Heck, W.W. An open-top filed chamber to assess the impact of air pollution on plants. J. Environ. Qual. **1973**, *2*, 365–368.
4. Unsworth, M.H. Air pollution and vegetation, hypothesis, field exposure, and experiment. Proc. R. Soc. Edinb. **1991**, *97B*, 139–153.
5. Heck, W.W.; Taylor, O.C.; Tingey, D.T. *Assessment of Crop Loss from Air Pollutants*; Elsevier Applied Science: London, 1988.
6. Jäger, H.J.; Unsworth, M.H.; De Temmerman, L.; Mathy, P. *Effects of Air Pollution on Agricultural Crops in Europe. Results of the European Open-Top Chamber Project*; Air Pollution Research Report Series of the Environmental Research Programme of the Commission of the European Communities, Directorate-General for Science, Research and Development, 1992; 46.
7. Fuhrer, J.; Skärby, L.; Ashmore, M. Critical levels for ozone effects on vegetation in Europe. Environ. Pollut. **1997**, *97*, 91–106.
8. Danielsson, H.; Pihl Karlsson, G.; Karlsson, P.E.; Pleijel, H. Ozone uptake modeling and flux-response relationships—Assessments of ozone-induced yield loss in spring wheat. Atmos. Environ. **2003**, *37*, 475–485.
9. Pleijel, H.; Mortensen, L.; Fuhrer, J.; Ojanperä, K.; Daniels-son, H. Grain protein accumulation in relation to grain yield of spring wheat (*Triticum aestivum* L.) grown in open-top chambers with different concentrations of ozone, carbon dioxide and water availability. Agric. Ecosyst. Environ. **1998**, *72*, 265–270.
10. Emberson, L.D.; Ashmore, M.R.; Murray, F.; Kuylensti-erna, J.C.I.; Percy, K.E.; Izuta, T.; Zheng, Y.; Shimizu, H.; Sheu, B.H.; Liu, C.P.; Agrawal, M.; Wahid, A.; Abdel-Latif, N.M.; van Tienhoven, M.; de Bauer, L.I.; Domingos, M. Impacts of air pollutants on vegetation in developing countries. Water Air Soil Pollut. **2001**, *130*, 107–118.

40

Global Plant Breeding

Introduction 331
World Population and Food Needs 331
The Need for Global Plant Breeding 332
Accomplishments of the Global Plant-Breeding System 333
Benefits of Global Plant Breeding 334
New Directions in Global Plant Breeding 334
Conclusion 334
Acknowledgment 335
References 335

Manjit S. Kang
Kansas State University

Introduction

Global plant breeding generally refers to the development of improved, high-yielding crop cultivars through organizations with international mandate, such as the Consultative Group on International Agricultural Research (CGIAR). The CGIAR consortium, which started with four centers in 1971 with the intended mission of achieving sustainable food security and reducing poverty in developing countries through scientific research and research-related activities in the fields of agriculture, forestry, fisheries, policy, and environment, now has 15 research centers around the globe [1]. Thirteen of these research centers are located in less developed countries. The CGIAR system is co-sponsored by the World Bank, the Food and Agriculture Organization of the United Nations (FAO), the United Nations Development Program, and the United Nations Environment Program. Prior to the establishment of the CGIAR system, private organizations, such as the Rockefeller Foundation and the Ford Foundation, supported international crop-breeding programs. The Green Revolution of the 1960s and 1970s was brought about by the deployment of high-yielding cultivars of crops, such as wheat and rice, which were developed at international agricultural centers [2].

I briefly discuss below the various aspects of global plant breeding under five sections. These sections are (i) World Population and Food Needs, (ii) The Need for Global Plant Breeding, (iii) Accomplishments of the Global Plant-Breeding System, (iv) Benefits of Global Plant Breeding, and (v) New Directions in Global Plant Breeding.

World Population and Food Needs

Human population has increased exponentially across centuries, whereas food production has increased only linearly. World population crossed the 7 billion mark in 2011, and it has been predicted to reach 9.73 billion in 2050 (Figure 40.1) [3]. According to FAO estimates, about 815 million people in the world (~10.5%) suffered from chronic undernourishment in 2016 [4]. Almost all the hungry people live in lower-middle-income countries. There are 11 million undernourished people even in developed countries [5].

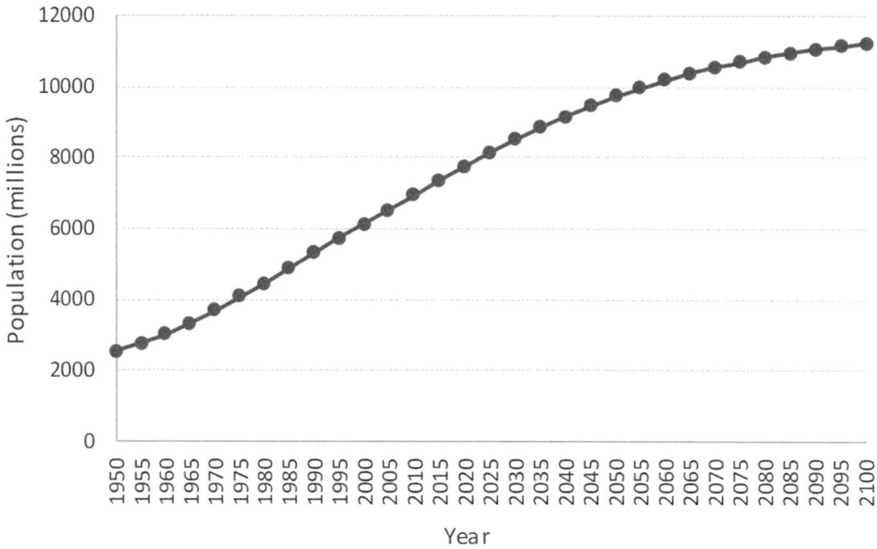

FIGURE 40.1 World population for 1950–2100 (projected beyond 2020). (The graph is based on data from www.populationpyramid.net/world/2050/ [3].)

Agriculture has to provide food for four times as many people per hectare of arable land in less developed countries as it does in developed countries, and by 2050, it might have to support seven times as many people per hectare of arable land in less developed countries [6]. In addition, global warming is expected to have an adverse effect on agricultural production, and the area of potential arable land in the lower latitudes, where most less developed countries are located, might decrease [6]. The agricultural systems at the greatest risk of climate change are the dryland tropics, where the International Crops Research Institute for the Semi-Arid Tropics (ICRISAT), a CGIAR center, has its mandate [7]. Combating climate change will require a set of internationally coordinated policy interventions for reducing greenhouse gas emissions besides addressing regional or global vulnerabilities, development patterns, equity distribution, and technology transfer [8].

Food production must be doubled by 2050 to feed people. Can it be done? There is optimism because cereal yields, like population, doubled during 1960–2000; however, the average yields of root and tuber crops increased by only 36%, and the average yields of pulse crops increased by only 29% during the same interval [6].

The Need for Global Plant Breeding

Someone has appropriately said that from the standpoint of food, the world must be regarded as one civilization, and if there is not enough food for everyone to eat, there will be no peace or prosperity in the world. Thus, it is imperative to produce food for citizens of all nations. Hence, there is the need for global plant breeding.

In industrialized countries, private-sector plant breeding predominates in most of the major field crops [9]. Private industry employed about 70% of the field breeders for agronomic crops in the mid-1990s [10]. Because private companies must make profit, their products are expensive and farmers in poor countries cannot afford them. In developing countries, farmers and farm communities maintain and "develop" their own varieties, especially in regions that are least favorable for commercial crop production [9]. Thus, the CGIAR and other public organizations with international mandates cater to the needs of farmers in less developed countries. Traditionally, CGIAR plant-breeding efforts have used a centralized global research approach, with CGIAR breeders collecting germplasm from various sources, evaluating the germplasm under carefully controlled experimental conditions, and making crosses among superior materials. The best progenies from selected crosses are then distributed as fixed or nearly fixed lines to collaborators in national agricultural research systems for testing.

The universal presence of genotype (or cultivar)-by-environment interaction—that is, differential response of cultivars or genotypes to diverse environments—in crop breeding offers both challenges and opportunities for breeders [11,12]. Multienvironment testing helps characterize genotypes for their range of phenotypic expressions for different agronomic attributes across diverse environments. The CGIAR research centers conduct cultivar/nursery trials across countries, which can provide information on adaptation of test materials [13]. From an efficiency standpoint, most of the CGIAR breeders prefer genotypes/cultivars that exhibit stability of performance across regions/countries. Genotypes with specific adaptation are also identified, which can be used with advantage in specific regions/countries.

Accomplishments of the Global Plant-Breeding System

The CGIAR research centers that are engaged in crop breeding are also known as Future Harvest Centers (FHC), as they provide improved germplasm for future cultivars and breeding methodologies. Their gene banks maintain an extensive collection of genetic resources, including landraces, domesticated species, and sometimes wild relatives of domesticated species that could potentially serve as sources of useful traits [14]. The FHC scientists' strategic research generates information about basic plant biological processes and also provides results leading to the development of novel breeding techniques and selection methods [14]. Germplasm enhancement or prebreeding is a major focus of CGIAR plant genetic improvement work. Prebreeding is essential when exotic (and by definition, unadapted) germplasm is used as a source of needed genetic diversity for elite but narrow-based breeding pools [9]. Some CGIAR centers only do prebreeding; for example, International Maize and Wheat Improvement Center (CIMMYT), International Center for Agricultural Research in the Dry Areas, ICRISAT, and International Rice Research Institute do not develop their own cultivars or hybrids but provide improved germplasm worldwide for further breeding.

There are numerous accomplishments of CGIAR research centers. Modern varieties of wheat, rice, maize, barley, sorghum, pearl millet, potatoes, sweet potatoes, cassava, beans, and many other important crops developed by scientists from FHCs, working in collaboration with colleagues in national breeding programs and, more recently, with farmers, are grown on millions of hectares of land throughout the developing world [15].

CGIAR's innovative program of developing drought-tolerant maize is designed to help smallholder farmers adapt to climate change. According to Bill Gates [2], more than 200 million households in sub-Saharan Africa depend on maize for their livelihoods. Because of climate change, weather patterns have become erratic, because of which farmers are at great risk of maize production losses, sometimes to the extent of having no harvest at all. With funding from Bill and Melinda Gates Foundation, USAID, and the Howard Buffett Foundation, CIMMYT has developed more than 150 new drought-tolerant maize varieties that are adapted to specific regions of Africa [2]. CIMMYT, in partnership with the International Institute of Tropical Agriculture (IITA), has also developed other maize varieties that are adapted not only to drought-prone regions but also to poor soils, and can successfully combat disease, pests, and weeds. Such varieties are expected to enhance farmers' yields by up to 30% and help them fight malnutrition [2].

One-third of the global population suffers from one or more micronutrient deficiencies (hidden hunger), and more than 50% of children and women in 20 states of India are anemic [16]. HarvestPlus, a component of the CGIAR system, in collaboration with ICRISAT, is engaged in developing biofortified crops. They have developed pearl millet biofortified with iron (Fe), which has a relatively higher level of protein, micronutrients, and balanced amino acid profile than other staple cereals. In 2018, Government of India renamed millets, including pearl millet, as nutri-cereals [16].

Benefits of Global Plant Breeding

During the past 50 years, modern varieties developed using improved germplasm from FHCs have achieved significant gains in global food production and generated billions of dollars of benefits for producers and consumers, raising incomes and improving living standards of hundreds of millions of rural poor [17]. Increased crop production has helped reduce the prices of most major food staples, benefiting additional hundreds of millions of poor urban consumers. The FHCs have fostered synergistic activities among public and private national breeding programs and facilitated technology transfer.

As pointed out earlier, HarvestPlus and other CGIAR centers have developed Fe-fortified pearl millet [16]. Their mandate also includes development of crops with high levels of zinc (Zn) and vitamin A. Many biofortified varieties have already been released for production in developing countries.

New Directions in Global Plant Breeding

The centralized plant breeding (CPB) that was responsible for the Green Revolution has been successful in producing uniform, broadly adapted varieties. It is, however, believed that this approach did not benefit resource-poor farmers in marginal areas. The CPB system has a number of weaknesses. The most important among these is its limited success in producing locally adapted cultivars for many specialized environments, particularly for the most difficult environments where poverty is still rampant.

The CPB reduces crop diversity and is characterized by dependence on gene banks for the preservation of genetic resources. The CPB has other limitations as well; it works well in predictable, uniform environments; for example, when farmers can use/afford irrigation and fertilizers. In recent years, researchers in a number of FHCs and national breeding programs have employed a new approach known as participatory plant breeding (PPB) [14,18]. Some non-governmental organizations teach farmers how to use some of the techniques of professional breeding if they suit the farmers' capacities and needs [14]. The PPB involves close farmer–researcher collaboration to bring about genetic improvement within a crop [19]. The PPB is a non-centralized system of breeding that can help broaden the genetic base of crop germplasm and stabilize food production, which results from farmers' developing, identifying, and using locally developed crop varieties that are both acceptable and accessible to them. The PPB exploits genotype-by-environment interactions by identifying locally adapted varieties. The emergence of the PPB movement represents a response to a weakness of the traditional global approach to plant breeding.

A couple of important global initiatives are worth noting. International in scope and founded by Dr. Bikram S. Gill, the Wheat Genetic and Genomic Resources Center at Kansas State University maintains a gene bank of 2,500 wheat accession as well as houses 2,200 cytogenetic stocks developed by wheat scientists [20,21]. Dr. Norman E. Borlaug founded a new organization in 2005, which is now called Borlaug Global Rust Initiative (BGRI) [22]. It contributes toward ensuring sustainable production of crops like wheat by organizing technical workshops on serious biotic threats from rust pathogens. Many countries have come together under the umbrella of BGRI to combat Ug99—a devastating strain of black stem rust.

In 2011, CIMMYT, in collaboration with the Government of India, established Borlaug Institute for South Asia (BISA), in India. The mission of BISA is to boost productivity of wheat and maize in South Asia in the face of climate change and a dwindling supply of natural resources [23].

Conclusion

The world faces a major challenge of feeding and clothing a population of 9.3 billion in 2050. Scientists in the CGIAR system, in collaboration with national breeding programs (centralized breeding) and with local farmers (participatory breeding), have helped develop modern varieties of major food staples (wheat, rice, maize). Increased crop production has helped reduce food prices and achieve a small measure of sustainable food security in developing countries, where farmers cannot afford varieties developed by private companies. Global and local (e.g., PPB) breeding are complementary activities rather than substitutes. PPB helps especially resource-poor farmers who were not the beneficiaries of the Green Revolution of the 1960s and 1970s. Conventional breeding (centralized or PPB) has been highly successful in doubling or even tripling production of some cereals. Gene banks (e.g., CGIAR and Wheat Genetic and Genomic Resources Center at Kansas State University) provide genetic resources for future variety development.

Acknowledgment

Dr. Neeta D. Kang, Betty Cowan Research & Innovation Centre, Christian Medical College & Hospital, Ludhiana, is acknowledged for preparing the population graph using data from www.populationpyramid.net/world/2050/ (Figure 40.1).

References

1. CGIAR. www.cgiar.org/research/research-centers/ (accessed 20 July 2019).
2. Gates, B. You've probably never heard of CGIAR, but they are essential to feeding our future; www.gatesnotes.com/Development/How-CGIAR-is-feeding-our-future (accessed 20 July 2019), 2019.
3. Population pyramids of the world from 1950 to 2100; www.populationpyramid.net/world/2050/ (accessed 20 July 2019).
4. World Hunger. www.worldhunger.org/world-hunger-and-poverty-facts-and-statistics/ (accessed 20 July 2019).
5. Food and Agriculture Organization (FAO), International Fund for Agricultural Development, World Food Program. The state of food insecurity in the world 2015. Strengthening the enabling environment for food security and nutrition. Rome, 2015.
6. Evans, L.T. Is crop improvement still needed? In *Genetic and Productions Innovations in Field Crop Technology*; Kang, M.S., Ed.; Food Products Press—An Imprint of Haworth Press: Binghamton, NY, 2005.
7. Dar, W.D.; Gowda, C.L.L. Declining agricultural productivity and global food security. *Journal of Crop Improvement* 2013, *27*, 242–254.
8. Kang, M.S.; Banga, S.S. Global agriculture and climate change. *Journal of Crop Improvement* 2013, *27*, 667–692.
9. Duvick, D.N. Crop breeding in the twenty-first century. In *Genetic and Productions Innovations in Field Crop Technology*; Kang, M.S., Ed.; Food Products Press—An Imprint of Haworth Press: Binghamton, NY, 2005.
10. Frey, K.J. National plant breeding study I: Human and financial resources devoted to plant breeding research and development in the United States in 1994. Iowa Agriculture and Home Economics Experiment Station, and Cooperative State Research, Education and Extension Service/USDA cooperating, No. 98, 1996.
11. Kang, M.S. Using genotype-by-environment interaction for crop cultivar development. *Advances in Agronomy*. 1998, *62*, 199–252.
12. Yan, W.; Kang, M.S. *GGE Biplot Analysis: A Graphical Tool for Breeders, Geneticists, and Agronomists*; CRC Press: Boca Raton, FL, 2003.
13. Kang, M.S.; Pham, H.N. Simultaneous selection for high yielding and stable crop genotypes. *Agronomy Journal*. 1991, *83*, 161–165.
14. Bellon, M.R.; Morris, M.L. Linking global and local approaches to agricultural technology development: The role of participatory plant breeding research in the CGIAR. *CIMMYT Economics Working Paper 02-03;* CIMMYT: Mexico, DF, 2002.

15. Ceccarelli, S.; Grando, S.; Booth, R.H. International breeding programmes and resource-poor farmers: Crop improvement in difficult environments. In *Participatory Plant Breeding. Proceeding of a Workshop on Participatory Plant Breeding, 26–29 July 1995*, pp. 99–116; Eyzaguirre, P.; Iwanaga, M., Eds.; Wageningen, The Netherlands. IPGRI: Italy, 1996.
16. Govindaraj, M.; Virk, P.S.; Kanatti, A.; Cherian, B.; Rai, K.N.; Anderson, M.S.; Pfeiffer, W.H. Chapter 13: Biofortified pearl millet cultivars offer potential solution to tackle malnutrition in India. In *Quantitative Genetics, Genomics and Plant Breeding* (Second edn.); Kang, M.S., Ed.; CABI: Wallingford, 2019.
17. Evenson, R.; Gollin, D., Eds. *Impact of the CGIAR on International Crop Genetic Improvement*; CABI: Wallingford, 2001.
18. Ceccarelli, S.; Grando, S. Increasing the efficiency of breeding through farmer participation. In Ethics and Equity in Conservation and Use of Genetic Resources for Sustainable Food Security, *pp. 116–121. Proceeding of a Workshop to Develop Guidelines for the CGIAR*, 21–25 April 1997, Foz de Iguacu, Brazil. IPGRI, Rome, Italy, 1997.
19. Weltzien, E.; Smith, M.E.; Meitzner, L.S.; Sperling, L. Technical and institutional issues in participatory plant breeding from the perspective of formal plant breeding. An analysis of issues, results, and current experience. CGIAR Systemwide Program on Participatory Research and Gender Analysis for Technology Development and Institutional Innovation Working Document 3.
20. Wheat Genetic and Genomic Resources Center (WGGRC). www.k-state.edu/wgrc/index.html (accessed 26 July 2019), 2019.
21. Gill, B.S.; Friebe, B.; Raupp, W.J.; Wilson, D.L.; Cox, T.S.; Sears, R.G.; Brown-Guedira, G.L.; Fritz, A.K. Wheat genetics resources center: The first 25 years. *Advances in Agronomy*. 2006, *89*, 73–136.
22. Borlaug Global Rust Initiative. www.globalrust.org/ (accessed 26 July 2019), 2019.
23. Borlaug Institute for South Asia. https://bisa.org/ (accessed 26 July 2019), 2019.

IV

Ecosystem Monitoring

41

Altitudinal Belts: Global Mountains, Patterns, and Mechanisms

Baiping Zhang
Chinese Academy of Sciences

Fang Zhao
Henan University

Introduction 339
Global Altitudinal Belts: Diversity and Complexity 341
Patterns of Altitudinal Belts 341
From High to Low Latitudes • From Coastal to Inland Areas of Continents • From the Outer Margins to the Interior of Mountain Massifs • Difference between Shady and Sunny Slopes and between Windward and Leeward Slopes • The Upward Shifting of Altitudinal Belts with Global Warming
Mechanisms for the Pattern of Altitudinal Belts 343
Latitude • Continentality vs. Isolation • Mass Elevation Effect vs. Summit Syndrome • Physiological Factors • Key Bioclimatic Indexes
Conclusions 345
References 345

Introduction

This entry is to show the diversity and complexity of global altitudinal belts, their basic patterns, and the main mechanisms. The mountains of the world are characterized by numerous and colorful altitudinal vegetation belts and their local-specific combination from mountain base upwards to mountain top (i.e., spectra of altitudinal belts, Figure 41.1), thanks to the extremely high degree of three-dimensional differentiation and heterogeneity of landscapes and the whole environment. On the one hand, these belts follow a general trend of ascending from polar to tropical regions, from island to continental inland, and from the periphery to central parts of large mountain massifs; on the other hand, local factors (geomorphic, human, and soil factors; and even ocean currents, species competition, beetle infestations, invasive pathogens, and even herbivores) could distort the general trend of global-scale distribution and modify the altitudinal limits of these belts to varying degrees. In other words, global and local factors interplay to shape the actual distribution of altitudinal belts. The most prominent and ever taken-for-granted phenomenon is that the latitudinal distribution of treelines and snowlines actually takes on a subtropical/dry tropical double humps with a tropical depression [1,2], and this was ever justified with low seasonal variation of temperature and the consequential lack of "summer heat" during the growing season [3–5]. But the hidden root or geographic factor forming these two humps (at 30°N and 20°S, respectively) has not been revealed. To scientifically illuminate the formation of the two humps can greatly promote the study of global patterns and mechanisms of altitudinal belts and the understanding of extremely high complexity of the ecological world.

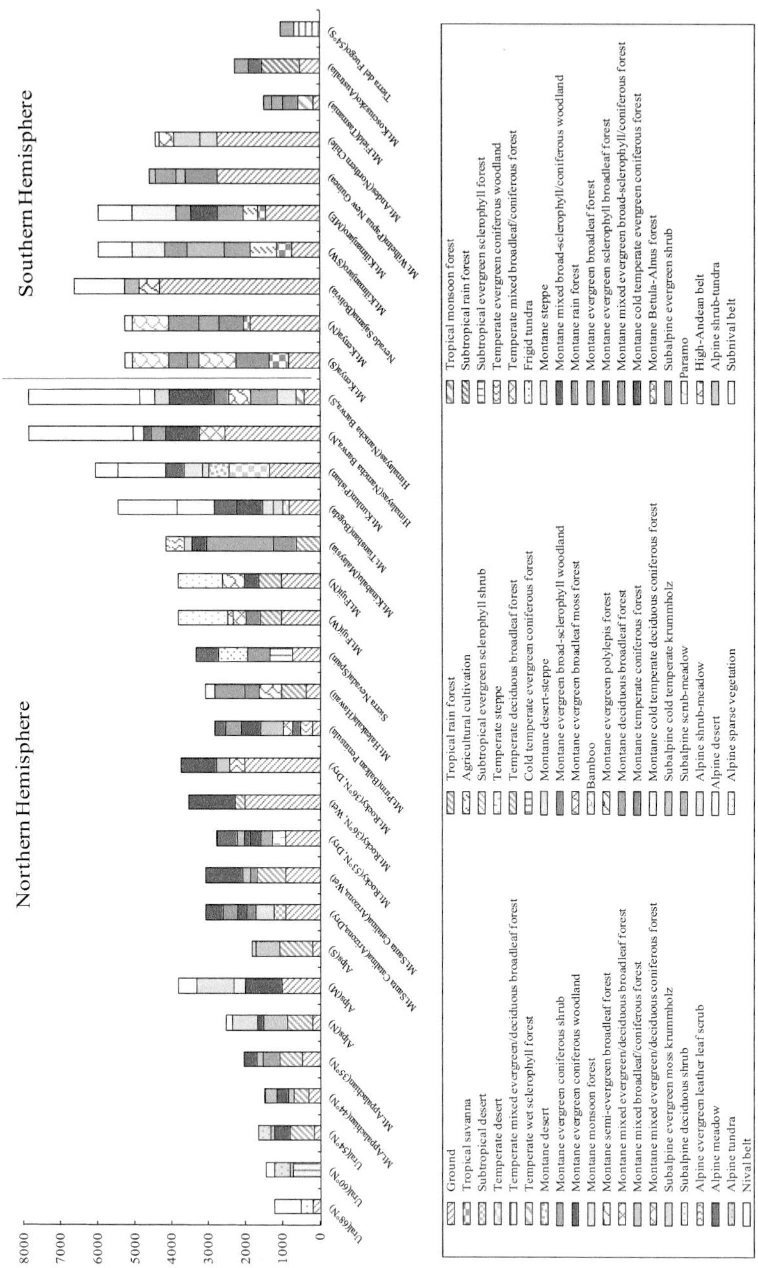

FIGURE 41.1 **(See color insert.)** Representative spectra of altitudinal belts of the global main mountain ranges.

This chapter comprehensively introduces and considers mass elevation effect (MEE) as the main driving factor for the distribution pattern of global altitudinal belts. The conceptual and numerical models of MEE are developed for the main mountain ranges of the world, including the Tibetan Plateau, the Alps, Scandinavian Mts., the Rocky Mts., the Andes, and New Zealand Mts. This can help greatly promote the study of global altitudinal belts and ecological complexity.

Global Altitudinal Belts: Diversity and Complexity

Environmental gradient is about 1,000 times as much in vertical direction as in horizontal, giving rise to the development of distinct altitudinal (vegetation) belts appearing successively with increased elevation in mountains, especially in high mountains. Their types and altitudinal positions are quite different from mountain to mountain and from slope to slope, thanks to the intricate interaction of zonal, non-zonal, and even human factors causing altitudinal belts. As a result, mountain environment displays extremely high degree of three-dimensional differentiation and variety, being characterized by the most diverse landscapes and biological resources in the world. In total, 80 types of altitudinal belts have been identified globally [6] and their types and altitudinal limits can combine in unlimited ways in different mountains and even on varied slopes, making a kaleidoscopic world of altitudinal belts (Figure 41.1). In the southeastern Himalayas, nine altitudinal belts can be identified from piedmont plains to snow-covered peaks, which combine to form an altitudinal spectrum with the largest number of vegetation belts in the world, whereas in the Ural Mountains, only two to three altitudinal belts can be found. The upper timberline (nearly the treeline) can reach up to 4,800 m above sea level in the southeastern Tibetan Plateau and in the central Andes, compared to only about 2,000 m or even lower in similar-latitude mountains. The number, elevation, and combination of altitudinal belts show quite differently even on different flanks/slopes in arid/semiarid temperate mountains where shady slopes are much moister than sunny slopes, but rather moderately in humid tropical and subtropical mountains where shady and sunny slopes have quite similar conditions.

Patterns of Altitudinal Belts

Almost all altitudinal belts have a trend toward increasing in elevation from high to low latitudes, from oceanic to continental climatic regions, from the rim to the interior of immense mountain masses, and from shady slopes to sunny slopes. Especially, the natural high-elevation treeline shows a close association with a physics-driven boundary, the snowline, suggesting a common climatic driver [7]. The upper timberline or forest line (very close to the term of treeline in ecology), as a global phenomenon and the most studied limit of altitudinal belts, is a bioclimatic reference line against which other vertical bioclimatic zones can be defined.

From High to Low Latitudes

Typically, snowline and treeline, just as other altitudinal belt limits, ascend from high to low latitudes, but reach their summit at 30° northern latitude and 20° southern latitude, and between the two summits are relatively low value areas, forming a saddle-shaped global distribution pattern. So, there is no uniform relationship between latitude and treeline elevation, and the pattern is not symmetrical around the equator [5,8].

From Coastal to Inland Areas of Continents

From coastal to inland areas of any continent, the limits of altitudinal belts ascend gradually, usually by 1,000–2,000 m or even higher. The 2,000 m amplitude seen within a narrow latitudinal belt in the temperate zone illustrates that latitude is a very imprecise predictor of treeline elevation [7], or the inland position (roughly continentality) and relief are factors as significant as latitude for the elevation of altitudinal belts. In addition, compared to mountains at similar latitudes in continental areas, treelines on oceanic islands are substantially lower than those on inland mountains [9], generally 1,000–2,000 m lower [10].

From the Outer Margins to the Interior of Mountain Massifs

More than 100 years ago, a tendency was observed for temperature-related parameters, such as treeline and snowline, to occur at higher elevations in the central Alps than on their outer margins [11]. This is more evident in the Tibetan Plateau, where the treeline ascends from 3,500 m in the southeastern rim to about 4,800 m in the interior and the snowline from 5,000 to 6,200 m. In the European Pyrenees, the treeline is at 1,500 m in the rim and 2,300–2,400 m in the interior. In the comparatively high and dry western cordillera of Bolivia, for example, the upper limit of the occurrences of *Polylepis tarapacana* (*P. tomentella*) is about 1,000 m higher than that of the humid and lower eastern cordillera [12]. Polylepis climbs even higher, to 5,200 m, and even to 5,400 m, making the highest treeline in the world [13], if it is regarded as "tree" rather than as krummholz or shrub.

Difference between Shady and Sunny Slopes and between Windward and Leeward Slopes

Usually, a certain type of altitudinal belt appears at a relatively higher position in sunny than in shady slopes. In the Catalina Mts. of Arizona, the fir forest, ponderosa pine forest, oak pine woodland, and dry woodland and Chaparral belts are all at a higher elevation in the south-facing slope than in the north-facing slope [14]. In Mt. Kenya of Africa, the orotropical cloud forest lies between 3,100 and 3,200 m in the humid south slope and between 2,650 and 3,000 m in the arid northern slope [15]. There are some extremes that a certain type of altitudinal belts appears in shady slopes but not in sunny slopes; for example, in northwest arid China, montane forest belt only develops in the northern flank of the Tien Shan Mountains but not in the southern. Some east–west extending ranges often serve as significant geographic demarcation (e.g., the Qinling Mts. in central China, as the transition from subtropical to warm-temperate zone), and different spectra of altitudinal belts develop in the southern and northern flanks. On the other hand, altitudinal belts show differently in windward and leeward slopes, often with a higher position in the leeward. If the sunny slope is the windward slope at the same time, a complicated situation occurs—the altitudinal belt in question may be at a higher position in the shady slope, depending on the degree of aridity resulting from the interplay of so many climatic factors.

The Upward Shifting of Altitudinal Belts with Global Warming

Theoretically, an increase in mean temperature by 6°C would mean an upward shift of treeline or other altitudinal belt limits by 1,200 m [16]. However, the fluctuation of altitudinal belts is related to many other factors (e.g., topography, tree species), also including the effect of site history on the present treeline dynamics [17,18]. With global warming, some altitudinal belts may disappear. In arid and semiarid areas, the dynamics of the present montane forest stands at treeline might be controlled by moisture supply rather than by higher temperatures. In addition, fluctuations of climates and treelines were not synchronous and occurred also at different regional and local intensities. In case of a sudden rise of temperature as has been predicted by many models, mountain treelines would not respond spontaneously, but rather with a time lag of several decades or even centuries. The general trend is being modified by regional, local, and temporal variations and thus is different as to its extent, intensity, and process of change. Consequently, it seems difficult to make generally acceptable statements without restricting them at the same time in view of the many regional and local peculiarities.

Mechanisms for the Pattern of Altitudinal Belts

Many factors together give rise to the great diversity, complexity, and varying elevations of mountain altitudinal belts. Globally, the altitudinal position of all altitudinal belts follows a similar trend to treeline and permanent snowline, suggesting a common physical driver (e.g., temperature). On smaller scales, more factors (topography, microclimate, soil conditions, etc.) participate in the making of altitudinal belts. As a result, altitudinal belts' heterogeneity increases from global to regional, landscape, and local scales. In addition, factors and processes may be not as significant at one scale as they are at another scale. Generally, latitude, continentality, MEE, and physiological factor are the main driving factors. We recognized that the significance of MEE has long been neglected. This chapter tends to highlight its great role in shaping the pattern of global altitudinal belts. In addition, we should seriously consider some key bioclimatic indexes with close relationship to the elevation of altitudinal belts.

Latitude

In the very early stage of altitudinal belt research, it was first recognized that treeline elevation is closely related to the latitude in which it appears, and many regional treeline latitude-elevation models were developed. Some authors related the altitude of climatic treelines to geographical latitude and even calculated so-called "timberline gradients." For example, the gradient is 184 m per degree of latitude between the northern border of Oregon and northern California, 83 m in the Appalachians [19], 71 m in the Ural Mountains, 89 m in Middle Siberia, and 76 m in East Siberia [20]. In the southern Andes between 35°S and 55°S, timberline declines south for about 80 m per degree of latitude [21]. It is clear that the empirically computed gradients are quite different, indicating that latitude is not by far the only determinant of the elevation of treelines and other altitudinal belts.

Continentality vs. Isolation

With increased distance from coastal to continental inland areas, precipitation gradually decreases and continentality increases, leading to ascending of altitudinal belts in the same direction, usually by over 1,000 m. On the contrary, isolation is a limiting factor for the elevation of altitudinal belts [22]; for example, the treeline is found 1,000–2,000 m lower on islands than on immense continents at similar latitudes [10]. Four factors that are thought to contribute to the forest line depression include (i) drought on trade wind exposed island peaks with stable temperature inversions, (ii) the absence of well-adapted high-altitude tree species on isolated islands, (iii) immaturity of volcanic soils, and (iv) an only small mountain mass effect that influences the vertical temperature gradient.

Mass Elevation Effect vs. Summit Syndrome

Larger mountain massifs serve as a heating surface absorbing solar radiation and transforming it to long-wave energy. In other words, climate becomes increasingly warmer and less humid from the rims to the central parts of mountain massifs, making the identical altitudinal belt at a higher position in the interior. This is the so-called "Massenerhebung effect" or "mass elevation effect" [11], sometimes also called "mountain mass effect." It is really a powerful driving factor, with the potential of "pushing" treelines and snowline upward to very high elevation in the interior of extensive and lofty mountain massifs. In fact, some extremely high treelines (at about 4,800 m in the central Andes and southeastern Tibetan Plateau, namely the two "humps" on the global latitudinal distribution curve at 30°N and 20°S, respectively) are actually the result of MEE. Without MEE, treelines could hardly surpass 3,500 m above sea level in any mountains [23]. The magnitude of MEE was ever considered to be closely related to the top and mean elevations of a mountain massif; however, what is really significant is the base elevation of the local basins or valleys in the inner parts of the massif. To a large extent, the local base elevation could

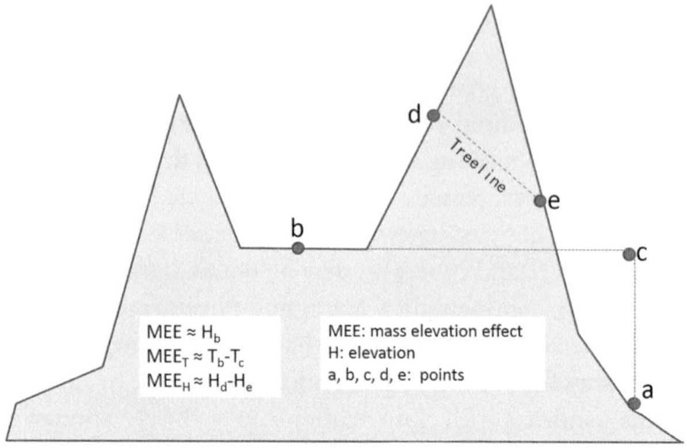

FIGURE 41.2 Conceptual model of MEE.

TABLE 41.1 Numerical MEE Model and the Contribution of MEE Factors for the Main Global Mountain Ranges

MEE Factors	Tibetan Plateau	Alps	Scandinavia	Rocky Mts.	Andes	New Zealand
Latitude	38.41	6.19	17.67	38.52	4.36	28.30
MBE	42.66	56.67	56.10	61.01	82.80	32.84
Hygrometric continentality	18.93	37.14	26.23	0.47	12.84	38.86
Model R2	0.515	0.563	0.476	0.501	0.646	0.544

represent MEE [24], and it has been proved to contribute a lot to the elevation of snowline and timberline [25,26]. Recent studies show that MEE can be quantitatively defined as temperature difference (Δt) at a given elevation between the interior and outer parts of an immense mountain mass [23], as shown in Figure 41.2. Based on multi-source data and calculation, Δt could be as high as 9.0–11.5 K in the Tibetan Plateau [27]. This means that, theoretically, a certain type of altitudinal belt can be about 2,000 m higher (air laps rate ≈0.6 K/100 m) in the interior than in the outers of the plateau. Δt is 7.1 K for the warmest month in the central Andes, with a potential of pushing altitudinal belts 1,180 m higher in the interior, which can well explain the very high treeline at 4,810 in the central Andes [28]. Δt amounts to 2.1–4.1 K in the Colorado Rocky Mountains, meaning an increase of about 680 m in the elevation of altitudinal belts [29] and so could justify the existence of high treeline at 3,800 m there.

Three factors of intra-mountain base elevation (MBE), latitude, and hygrometric continentality are used to model MEE for the main mountain ranges of the world. The result is shown in Table 41.1. We can recognize the great contribution of MBE.

On the contrary, there is another effect that lowers the distribution of altitudinal belts, i.e., summit syndrome [7]. As is known, environment is extremely harsh along the top 200 m vertical range around mountain tops or along the range of mountains, chiefly due to strong wind, steep slope, and poor soil or even non-soil conditions; where normal altitudinal vegetation could not well develop, for example, forests are usually displaced by krummholz or meadow or even bare land. This often gives rise to pseudo-treelines.

Physiological Factors

UV-B (280–315 nm) is damaging and mutagenic to living organisms [30]. It is known to be particularly high in the alpine zone of tropical mountains. This involves the occurrence of dwarf forest at anomalously low altitudes, especially for isolated peaks in tropical mountains. It has also been suggested that temperature could operate via the soil [31], lower temperatures leading to a greater accumulation of organic matter and to changes in nutrient status. The increasingly poor supply of nitrogen and phosphorus depends on decrease in temperature, and the increase in frequency of fog is likely to be important not only in governing the distribution of forest types but also in influencing their biomass and physiognomy. This also leads to relatively lower elevation of forest belts in tropical mountains and islands.

In reality, abovementioned driving factors, latitude, continentality vs. isolation, MEE vs. summit syndrome, and physiological factors, together with other small-scale drivers including slope-facing effect, the existence of well-adapted treeline species, and competition between communities or species, always combine to shape the actual spatial pattern of altitudinal belts.

Key Bioclimatic Indexes

Behind the driving forces mentioned above are some key bioclimatic indexes which can represent or be used to estimate the spatial position of altitudinal belts, for example, the 10°C isotherm for the warmest-month, warmth index of 15°C·month [4,32,33], soil temperature seasonal mean ground temperature of 6.7°C ± 0.8°C or 6.4°C ± 0.7°C [34], aridity, and mean annual precipitation. The first three indexes, in particular, are closely related to treelines. In exploring the "ultimate cause" of altitudinal treelines, it was found that low mean temperatures in the rooting zone during the growing season or all-year-round (at tropical treelines) are the critical factor controlling directly worldwide treeline position [35–38]. The latest research shows that the current natural high-elevation treeline is associated with a growing season that is at least three months long and during which the mean air temperature reaches at least 6.4°C [7].

Conclusions

1. The diversity and complexity of mountain altitudinal belts result from many factors, global and local, zonal and azonal, uplifting (e.g., MEE) and depressing (e.g., summit syndrome and wet climate), etc. Global altitudinal belts have common general patterns and varied local patterns.
2. A difference of 2,000–3,000 m in elevation of treelines exists both in north–south direction and in similar latitudes. The recognition and quantification of MEE greatly deepens our understanding of the complex distribution of altitudinal belts. All treelines above 3,500 m are actually the result of MEE. Without MEE, our world will be much simpler.
3. The magnitude of MEE is closely related to MBE, usually the altitude of intra-mountain basins and valleys, rather with mountain top height or the average elevation of the immense mountain massif.
4. Latitude, MEE, and continentality are the most important geographical factors for the patterns of altitudinal belts. It has been shown that MBE (as a proxy of MEE), latitude, and continentality together could explain as high as 92% of global treeline elevation variation, and that MEE contributes the most (52.2%) [39]. This nearly perfectly explains the global treeline distribution.

References

1. Hermes, K. Die Lage der oberen Waldgrenze in den Gebirgen der Erde und ihr Abstand zur Schneegrenze. Dissertation der Philosophischen Fakultät der Univ. Köln. 1955.
2. Troll, C. The upper timberlines in different climatic zones. *Arctic and Alpine Research*. 1973, 5: 3–18.
3. Ohsawa, M. Altitudinal zonation of forest vegetation on Mount Kerinci, Sumatra: With comparisons to zonation in the temperate region of east Asia. *Journal of Tropical Ecology*. 1985, 1: 193–216.
4. Ohsawa, M. An interpretation of latitudinal patterns of forest limits in South and East Asian mountains. *Journal of Ecology*. 1990, 78: 326–339.

5. Körner, C. A re-assessment of high elevation treeline positions and their explanation. *Oecologia.* 1998, 115 (4): 445–459.
6. Zhang, B., Tan, J. and Yao, Y. *Digital Integration and Patterns of Mountain Altitudinal Belts*; Environmental Science Press: Beijing, 2009 (in Chinese).
7. Körner, C. *Alpine Treelines: Functional Ecology of the Global High Elevation Tree Limits*; Springer Basel: Heidelberg, New York, Dordrecht, London, 2012; 26.
8. Troll, C. *Geoecology of the High Mountain Regions of Eurasia*; Franz Steiner Verlag Gmbh: Wiesbaden, 1972; 1–16.
9. Berdanier, A.B. Global treeline position. *Nature Education Knowledge.* 2010, 1 (11): 11.
10. Leuschner, C. Timberline and alpine vegetation on the tropical and warm-temperate oceanic islands of the world—Elevation, structure and floristics. *Vegetatio.* 1996, 123 (2): 193–206.
11. Quervain, A.D. Die Hebung der atmosphärischen lsothermenin der Schweizer Alpen und ihre Beziehung zu deren Höhengrenzen. *Gerlands Beiträge zur Geophysik.* 1904, 6: 481–533.
12. Kessler, M. *Polylepis-Wälder Boliviens: Taxa, Ökologie, Verbreitung und Geschichte*; Charmer: Berlin, Stuttgart, 1995; 246.
13. Holtmeier, F.K. Mountain timberlines: Ecology, patchiness, and dynamics. In Holtmeier, F.K. eds. *Advances in Global Change Research*; Springer: Berlin, 2009; 36; 1–437.
14. www.geo.arizona.edu/Antevs/biomes/azlifzon.html (accessed August 2012).
15. Bussmann, R.W. Vegetation zonation and nomenclature of African mountains—An overview. *Lyonia.* 2006, 11 (1): 41–66.
16. Zimmermann, N.E., Bolliger, J., Gehrig-Fasel, J., Guisan, A., Kienast, F., Lischke, H., Rickebusch, S. and Wohlgemut, T. Wo wachsen die Bäume in 100 Jahren? *Forum für Wissen*, 2006, 20: 63–71.
17. Payette, S., Filion, L., Delwaide, A. and Begin, C. Reconstruction of the treeline vegetation response to long-term climate change. *Nature.* 1989, 341: 429–432.
18. Lescop, S.K. and Payette, S. Recent advance of the arctic treeline along the eastern coast of Hudson Bay. *Journal of Ecology.* 1995, 83: 929–936.
19. Cogbill, C.V. and White, P.S. The latitude-elevation relationship for spruce-fir forest and treeline along the Appalachian mountain chain. *Vegetatio.* 1991, 94: 153–175.
20. Malyshev, L. Levels of the upper forest boundary in northern Asia. *Vegetatio.* 1993, 109 (2): 175–186.
21. Crawford, R.M.M. *Studies in Plant Survival: Ecological Case Histories of Plant Adaptation to Adversity*; Blackwell Scientific Publications: London, Edinburgh, Boston, MA, Melbourne, Paris, Berlin, Wien, 1989.
22. Severin, D.H., Fabien, A., David, E.V.H., Anke, J., Elisabeth, L., Manuel, J.S. and Carl, B. Patterns of island treeline elevation—A global perspective. *Ecography.* 2015, 38: 1–10.
23. Zhang, B. and Yao, Y. Implications of mass elevation effect for the altitudinal patterns of global ecology. *Journal of Geographical Sciences.* 2016, 26 (7): 871–877.
24. Zhang, S., Zhang, B., Yao, Y., Zhao, F., Qi, W., He, W. and Wang, J. Magnitude and forming factors of mass elevation effect on Qinghai-Tibet plateau. *Chinese Geographical Science.* 2016, 26(6): 745–754.
25. Han, F., Zhang, B., Yao, Y., Zhu, Y. and Pang, Y. Mass elevation effect and its contribution to the altitude of snow- line in the Tibetan plateau and surrounding areas. *Arctic, Antarctic, and Alpine Research.* 2011, 43 (2): 207–212.
26. Han, F., Yao, Y., Dai, S., Wang, C., Sun, R. and Zhang, B. Mass elevation effect and its forcing on timberline altitude. *Journal of Geographical Sciences.* 2012, 22 (4): 609–616.
27. Han, F., Zhang, B., Zhao, F., Wan, L, Tan, J. and Liang, T. Characterizing the mass elevation effect across the Tibetan plateau. *Journal of Mountain Science.* 2018, 15(12): 2651–2665.
28. He, W., Zhang, B., Zhao, F., Zhang, S., Qi, W., Wang, J. and Zhang, W. The mass elevation effect of the Central Andes and its implications for the southern hemisphere's highest treeline. *Mountain Research and Development.* 2016, 36(2): 213–221.

29. Wang, J., Zhang, B., He, W., Yao, Y., Zhang, W. and Zhao, C. A quantitative study on the mass elevation effect of the Rocky Mountains and its significance for treeline distribution. *Physical Geography*. 2017, 38 (3): 231–247.
30. Flenley, J. Ultraviolet insolation and the tropical rainforest: Altitudinal variations, Quaternary and recent change, extinctions, and biodiversity. In Flenley, J. and Bush, M. eds. *Tropical Rainforest Responses to Climatic Change*; Praxis Publishing: Chichester, 2007; 219–235.
31. Grubb, J. Interpretation of the "Massenerhebung" effect on the tropical mountains. *Nature*. 1971, 229 (5279): 44–45.
32. Kira, T, 1945. A new classification of climate in eastern Asia as the basis for agricultural geography. Horticultural Institute. Kyoto: Kyoto University.
33. Fang, J.Y., Oshawa, M. and Kira, T. Vertical vegetation zones along 30°N latitude in humid East Asia. *Vegetatio*. 1996, 126: 135–149.
34. Paulsen, J. and Korner, C. A climate-based model to predict potential treeline position around the globe. *Alpine Botany*. 2014, 124: 1–12.
35. Paulsen, J., Weber, U.M. and Körner, C. Tree growth near treeline: Abrupt or gradual reduction with altitude. *Arctic, Antarctic, and Alpine Research*. 2000, 32 (1): 14–20.
36. Hoch, G. and Körner, C. The carbon charging of pines at the climatic treeline: A global comparison. *Oecologia*. 2003, 135, 10–21.
37. Hoch, G. and Körner, C. A test of treeline theory on a montane permafrost island. *Arctic, Antarctic, and Alpine Research*. 2006, 38 (1): 113–119.
38. Körner, C. and Paulsen, J. A world-wide study of high altitude treeline temperatures. *Journal of Biogeography*. 2004, 31: 713–732.
39. Zhao, F., Zhang, B., Zhang, S., Qi, W., He, W., Wang, J. and Yao, Y. Contribution of mass elevation effect to the altitudinal distribution of global treelines. *Journal of Mountain Science*. 2015, 12 (2): 289–297.

42

Biodegradation and Bioremediation

Introduction 349
Biodegradation Overview 350
Bioremediation Overview 350
Contaminant Biodegradability 351
Contaminant Toxicity 352
Contaminant Bioavailability 353
General Microbial Approaches to Remediation of Contaminants 353
Biostimulation • Bioaugmentation
Bioremediation Technologies 354
Monitored Natural Attenuation or Intrinsic Remediation • In Situ Subsurface Bioremediation • Engineered Biopiles • Soil Composting • Landfarming • Soil Bioreactors • Bioremediation of Shorelines • Deep-Sea Bioremediation
Phytoremediation 356
Air Biofiltration 356
Conclusion 356
References 357

Owen P. Ward
and Ajay Singh
University of Waterloo

Introduction

In their most general sense, biodegradation processes are central to all biological systems. In these processes, cells transform and recycle biological and synthetic chemicals and materials into smaller molecules to generate energy and/or metabolic intermediates needed for cell functioning including maintenance, biosynthesis, and growth. Indeed, these processes also contribute to the unique adaptation properties of organisms, for example, enabling microbes to adjust to changing environments by making the requisite biochemical and physiological changes mediated by biodegradative and biosynthetic metabolic processes. Bioremediation represents an important applied segment among these biodegradation processes in which technologies are generally engineered and optimized to exploit the same biodegradation capabilities especially of microbes, for the removal of organic contaminants and sometimes, metal contaminants from the environment. The principal objective of this entry is to provide an overview of bioremediation processes and systems as well as a perspective on recent developments and future prospects. General principles of biodegradation processes are first considered to provide a context for addressing bioremediation technologies.

Biodegradation Overview

On a macro scale, biodegradation processes often mediate recycling of organic materials, including animal, plant and microbial materials, present in the environment. Enzymes secreted by microbes and other species attack and depolymerize the principal biological polymeric structures, namely, complexes containing carbohydrates, proteins, nucleic acids, lipids, and their derivatives into oligomeric or monomeric components, which may be taken up and metabolized by cells. These processes are promoted in the digestive systems of humans, animals, and other species mediated by endogenous enzymes from these organisms or supported by the microbial populations present or both. These processes also occur everywhere in the environment where microbes can thrive, most notably in soil and in aqueous media. Processes in soil are enhanced around plant roots where nutrients released from the roots support the growth of plant growth-promoting bacteria. Such processes are also actively promoted in the practices of agriculture and horticulture and in myriad solid and liquid waste treatment systems. Some examples of general environmental processes and practices for the biodegradation of natural organic materials include processes for silage making, composting, and wastewater treatment (Table 42.1). The reader is referred to references[1–3] for further discussion of these topics.

Bioremediation Overview

Industrial-scale processes for the recovery of petroleum and the manufacturing, distribution, and use of chemicals invariably lead to the transfer of chemical contaminants into the environment, most notably into soils, groundwater, shorelines, and water bodies. These contaminants range from alkanes and mono- and polycyclic aromatic hydrocarbon compounds, naturally present in crude oil or generated through combustion, to high-volume synthetic chemicals including chlorine-, nitrogen-, and phosphate-containing organic chemicals and many others. In addition, soils become contaminated with inorganic chemicals. Just as microbes have the ability to concentrate and accumulate metals in wastewater treatment processes, specific microbes can accumulate or transform some of these inorganic species or both.

The extent of the threat to the environment posed by petroleum and organic chemicals may be appreciated by reference to the production volumes of these products. Global crude oil production and world refining capacity is approaching 80 million barrels per day. Estimates for global production volumes of synthetic polymers and plastics approximate to 140 million tons per annum.[4] In the United States alone, annual production volumes of each of ethylene, propylene, vinyl chloride, and benzene range from 5 to 20 million tons, with volumes of many other organic chemicals ranging from 1 to 5 million tons.[5]

Some of the most widely distributed organic environmental contaminants are aliphatic and aromatic hydrocarbons, including alkanes, mono- and polycyclic aromatics, present in crude oils or in refined oil fuels or petrochemicals; halogenated aliphatic, especially chlorinated species including chlorinated methanes, ethanes, and ethenes such as trichloroethene (TCE), dichloroethene (DCE), perchloroethene (PCE),

TABLE 42.1 Example Processes for Biodegradation of Natural Organic Materials

- **Silage-making processes**, whereby fibrous plant materials are rendered more digestible for animal feeding by conversion of sugars to organic acids and some enzymatic cellulose/hemicellulose degradation, mediated by microbial anaerobic digestion.
- **Composting processes**, used to aerobically transform plant and other biological and organic materials, most frequently derived from agricultural, horticultural, food, and municipal wastes. A variety of microbial species, most notably fungi and bacteria including mesophiles and thermophiles, participate in these exothermic biodegradative processes to generate beneficial organic fertilizers devoid of pathogens, which are killed as thermophilic conditions are achieved.
- **Biological wastewater treatment processes**, where the objective is to remove contaminants from domestic, municipal, or industrial waste streams. Micro-organisms are primarily responsible for aerobic and anaerobic degradation and assimilation of the organic components together with some removal/concentration of heavy metals. The resulting accumulated biosolids may be further processed and transformed into beneficial organic fertilizers for use in agriculture.

and vinyl chloride (VC); mono- and polyhalogenated phenols and benzoates, including pentachlorophenol (PCP); polychlorinated biphenyls (PCBs); chloroaromatics including 2,4-dichlorophenoxyacetate (2,4-D), dichlorodiphenyltrichloroethane (DDT), and related compounds; nitroaromatics including nitrobenzene and 2,4,6 trinitrotoluene (TNT); organophosphates; nitrate esters (including glycerol trinitrate); nitrogen heterocycles (including the explosive compound hexahydro-1,3,5-trinitrotriazine or RDX); and polyesters, polylactates, polyurethanes, and nylon.[6,7]

The recognition that microbes in soil could degrade, transform, or accumulate many of the chemicals polluting the soil provided momentum for the undertaking of research to exploit these activities to clean up contaminated soil and other media, using processes now termed as bioremediation. Indeed, the observed effectiveness of bioremediation as a process for the degradation of oils contaminating the shorelines at Prince William Sound, Alaska due to the *Exxon Valdez* oil spill led to the development of a variety of strategies for the application of bioremediation processes for soil remediation.[8,9] Bioremediation is generally considered to be the low-cost option in comparison to physical and chemical alternatives.[10]

Bioremediation technologies may be adapted to a variety of different contaminants and environmental conditions, and in particular, conditions can be developed for the selection and enrichment of microbial populations having the requisite and unique degradation abilities to remediate the specific contaminants present at a particular site.[11] Bioremediation technologies are also advantageous where contaminants have become diluted or widely dispersed over time.

A vast array of physical environments exist where biodegradation and bioremediation processes may thrive. Broadly speaking, these environments will all contain water or at least be moist because the presence of water is a prerequisite for microbial growth and biodegradation activity. The nature of the microbial type or population which develops will be influenced by a wide range of factors including the physical properties of the medium; salt concentration; solid content, surface properties and particle size; nature and concentration(s) of contaminant(s); nature and concentration of nutrients present to support microbial growth; relative humidity; presence of oxygen or other electron acceptors; pH; and temperature.

Contaminant Biodegradability

Although many organic chemicals are biodegradable by microbes present in the environment, other natural or synthetic chemicals structures exhibit resistance or recalcitrance to biodegradation and tend to persist in the environment for prolonged periods of time. Microbial transformation of organic pollutants is mediated primarily by enzyme-catalyzed reactions but often with the support of other cell components, associated with cell surfaces, surfactants, chelators, vesicles, and organelles. The biotransforming enzymes may be located within the cell protoplasm or be membrane associated (or a combination of these) or indeed in a few cases may be extracellular. A good example of a combined intracellular/membrane-bound system is the *alk* gene system for alkane degradation, whereas extracellular fungal peroxidases and laccases are non-specific mechanisms for the attack of organic pollutants, mediated by the generation of free radicals.[11] Microbial enzyme transformation of contaminants may lead to complete contaminant mineralization or to the production of breakdown products which may be assimilated into the cells' central pathways for energy or biomass production or both. Alternatively, contaminants may be partially degraded or transformed into metabolites, which the cell cannot further utilize, whereby such metabolites may accumulate in the cell or be excreted. These metabolites may be further transformed by other cell types, among the multitude of microbes present in the ecosystem, which may have different metabolic capacities. Alternatively, the metabolites may be generally recalcitrant to biodegradation and may bioaccumulate. Contaminants in well-aerated environments will predominantly be degraded by aerobic metabolic processes, whereas contaminants in a nonaerated subsurface environment may be anaerobically degraded.

A review of aerobic pathways for the biodegradation of organic contaminants including petroleum hydrocarbons, chlorinated hydrocarbons, nitroaromatics and nitrogen heterocycles, organophosphate derivatives, and plastics reveals the following reaction types: dioxygenases, monooxygenases, hydroxylases, ligninases, and reductases among others. Anaerobic biodegradations typically involve initial activation mechanisms such as carboxylation, methylation, hydroxylation, and dehydrogenation followed by addition reactions, anaerobic reductions, and hydrolytic reactions.[12]

Microbes have the potential to remediate metals in contaminated media by exploiting a combination of mechanisms including precipitation, sorption, biosurfactant complexation, biofilm entrapment, and metabolic uptake. Some microbes can also transform metals including radionuclides, for example, by reducing mercury into a volatile form, which can be removed. In metal bioremediation applications, there is particular interest in exploiting microbes, which have been isolated from mine tailings and heavy metal-contaminated sites, as these strains exhibit superior metal-scavenging capacities, greater resistance to metal toxicity (sometimes mediated by metal efflux mechanisms), and ionizing radiation.[13] Examples of interesting strains for these applications include certain *Pseudomonas* species, including recombinant strains, *Ralstonia eutropha, Thiobacillus ferrooxidans,* and *Deinococcus radiodurans.*

Contaminant Toxicity

The principal concern regarding chemical contamination of the environment relates to contaminant toxicity. Many of the chemical contaminants, but especially the more recalcitrant chemicals, are toxic to higher life-forms or are known carcinogens or both. As such, their presence represents a threat to human and animal health. Many of these chemicals are classified as hazardous by national regulatory agencies, including the United States Environmental Protection Agency (USEPA). For example, 16 polycyclic aromatic hydrocarbons (PAHs) are considered to be priority pollutants by the USEPA and 7 of these are classified as probable human carcinogens.[14]

Toxicity concerns as a result of chemical contamination of the environment do not just relate to the toxicities of the original chemical compounds transferred. There are many examples where the "parent" contaminant may get chemically or biologically transformed into reaction products or catabolites, which are as toxic as or even much more toxic than the parent compound. For example, surfactants present in detergents, such as nonylphenol ethoxylates, are often biotransformed into nonylphenol end products with endocrine-disrupting activity as a result of microbial transformation.[15] The presence of strong chemical oxidizing or reducing agents in the contaminated medium can give rise to non-specific conversions of the contaminants into more reactive and more toxic products.[16]

In heavily contaminated media, the toxicity of the medium may hinder the implementation of bioremediation strategies as a result of the contaminant, at a higher concentration, being toxic to the microbial population which may have the metabolic capability to biotransform the contaminant. In a manner akin to antibiotic resistance, some microbial cells have developed resistance to specific toxic organic compounds such as toluene, by using efflux pumps to extrude toxic contaminants from the cell cytoplasm or periplasmic space.[17] It is also thought that, where toxic contaminants are only partially transformed by one species, efflux pumps may have the potential to excrete the metabolites such that they may be made available to other members of a consortium with the capacity for degradation of the excreted metabolites. In metal bioremediation applications, there is particular interest in exploiting microbes, which have been isolated from mine tailings and heavy metal-contaminated sites, as these strains exhibit greater resistance to metal toxicity (sometimes mediated by metal efflux mechanisms) and ionizing radiation.[13]

The extent of toxicity of a known toxic contaminant cannot be simply related to its concentration in the contaminated medium. Actual toxicity may be diminished by sorption of the contaminant to soil particles or organic matter. In addition, biological species present in or added to the contaminated medium may adsorb the contaminant or secrete biological molecules including chelators or surfactants, which effectively sequester the contaminant, thereby reducing its toxicity. Arguably, the presence of other components in the contaminated medium or indeed produced by the biological organisms present could enhance toxicity. For example, a hydrophobic contaminant, with a tendency to sorb to soil, might become mobilized and solubilized by the presence of lipids or surfactants and possibly become more toxic to the organisms present.

Scientists have developed toxicity-testing methods, whereby contaminant-sensitive organisms, including microbes, algae, plants, fish, macroinvertebrates, and cultured cell lines, are used as biomonitors of pollutant accumulation in soil and aquatic systems and also for following the reduction in the toxicity of the contaminated media as a result of the implemented remediation processes. These biomonitoring tests rely on observing whole-organism biological impacts, related, for example, to growth, behavior, reproduction, or mortality. Alternatively, toxicity tests may involve the measurement of intracellular biomarkers, for example, heat shock proteins; enzymes including antioxidant enzymes, DNA repair enzymes, acetylcholinesterase, cytochrome P450 monooxygenase, and glutathione S transferase; and DNA mutations or chromosomal defects.[18]

Contaminant Bioavailability

Microbes generally thrive and grow in aqueous phases and are most efficient at accessing nutrients that are water soluble. However, petroleum components, as well as most other chemical contaminants, are water insoluble or hydrophobic and hence less bioavailable to microbes. In addition, soil is the most dominant medium for these contaminants, and hydrophobic contaminants tend to bind strongly to soil particles, thereby further reducing their bioavailability to microbes capable of biodegrading the contaminants and rendering them more persistent in the environment. Soil bioremediation processes require adequate amounts of water present in the soil to support microbial activity, and the contaminant-biodegrading organisms have to have the means to access the hydrophobic or soil particle-bound contaminants or both. Two general natural microbial mechanisms are known to facilitate this contaminant access. First, cells may sorb and uptake the contaminant by direct interaction between the microbes and the hydrophobic substance/particulate matter. Alternatively, the microbes may secrete biosurfactants into the surrounding environment, which may pseudosolubilize and mobilize the contaminant through micelle formation, which facilitates contaminant contact with the microbes and promotes cellular uptake of the hydrophobic substrate. As amphipathic molecules, biosurfactants contain both hydrophilic and hydrophobic ends. In the presence of hydrophobic contaminants in an aqueous environment, micelle formation results when the biosurfactant molecules associate with and solubilize the hydrophobic contaminant in the center of the micelle, while the hydrophilic ends of the biosurfactant molecules remain on the outside of the micelle in contact with the aqueous phase.[19,20] These contrasting processes were well illustrated by culturing oil-degrading strains of a *Pseudomonas* sp. and *Rhodococcus* sp. separately and together in aqueous media supplemented with crude oil. The *Pseudomonas* was located predominantly in the aqueous phase and did not associate with oil droplets, whereas the *Rhodococcus* cells were concentrated at the oil–water interface.[21] The greatest degradation occurred where both mechanisms were in play, namely, when a coculture of the strains was used.

In bioremediation processes, the natural mechanisms of contaminant mobilization and pseudosolubilization, promoted by biosurfactants, may be replaced or supplemented by the use of added chemical surfactants. The chemical, nonylphenol ethoxylate surfactant, substantially increased crude oil biodegradation by a consortium of oil-degrading organisms.[22] However, biosurfactants or chemical surfactants may also inhibit degradation of contaminants, whereby isolating the contaminant in the center of the micelles effectively prevent uptake by the degrading microbes. For example, Billingsley et al.[23] observed that at chemical surfactant concentrations above the critical micelle value, the presence of an anionic surfactant promoted, whereas nonionic surfactants inhibited, biotransformation as compared to no surfactant controls. These and related topics are discussed in more detail by Ward.[24]

General Microbial Approaches to Remediation of Contaminants

Two general strategies for the implementation of bioremediation processes were identified around the time of the *Exxon Valdez* oil spill and cleanup, biostimulation, and bioaugmentation. The objective of biostimulation processes is to promote the growth of the segment of microbes present in the contaminated environment, which can degrade the contaminants. If the contaminated environmental medium does not contain microbes, which can degrade the contaminant, bioaugmentation with a selected inoculum already acclimated to grow on the contaminant may be needed.

Biostimulation

Most organic contaminants in the environment are carbon rich and the growth of organisms that can use this carbon source is usually limited by the availability of other nutrients, supplements, and electron acceptors. Nutrients that are most widely used in aerobic bioremediation processes are nitrogen and phosphate, and getting oxygen to the medium is often the most challenging engineering component. In surface soil bioremediation processes, oxygen is supplied from the atmosphere, and atmospheric oxygen is also exploited in reactor-based processes, engineered soil piles, and in situ processes by sparging air to the contaminated soil or slurry.[25] Other strategies involve the supply of pure oxygen gas, ozone, or manganese peroxide. In anaerobic bioremediation, delivery of adequate amounts of suitable electron acceptors to the medium is a priority strategy. Examples of electron acceptors are sulfate, nitrate, and ferric iron.

There are also other more specialized biostimulation approaches. Especially with hydrophobic contaminants, which are less accessible to microbes, surfactants, biosurfactants, or indeed, easily biodegraded solvents and vegetable oils may be added to solubilize these contaminants, releasing them from particulate matter and rendering them more amenable to biodegradation.[19,20,26] Where contaminants are typically degraded by cometabolic processes requiring the presence of primary substrates, remediation of the contaminant may be stimulated by the addition of a primary substrate. A further example of biostimulation, in the case of aerobic bioremediation of clays having low porosities that resist aeration/oxygen delivery systems, is to add low-density bulking agents such as plant materials, sawdust, and wood chips to increase the porosity of the medium and allow easy delivery of air.

Bioaugmentation

Examples where a bioaugmentation strategy has been beneficial include bioremediation of very recalcitrant contaminants such as TNT, phthalate esters, and carbon tetrachloride.[27,28] This strategy is also necessary where the contaminants are not degraded by natural organisms but where recombinant organisms have been engineered, which are capable of degrading the contaminant. Bioaugmentation is also applicable where highly accelerated biodegradation processes are required, for example, for accelerated reactor-based processes for the remediation of hydrocarbon sludges.[29] In many cases, bioaugmentation has been shown not to enhance biodegradation rates beyond what may be achieved through biostimulation, as was most notably illustrated in the case of studies during the *Exxon Valdez* oil spill cleanup.

Bioremediation Technologies

There are a wide range of possible approaches to the bioremediation of contaminated environments. They include soil bioremediation processes involving natural attenuation, in situ subsurface treatments and use of biopiles and composting approaches, and landfarming and bioreactor treatments. Crude oil spills at sea may be addressed through shorelines or deep-sea bioremediation strategies.

Monitored Natural Attenuation or Intrinsic Remediation

This process is the simplest and least-invasive bioremediation approach.[30] The process presumes that the environment contains sufficient nutrients to support the indigenous microbial population in the biodegradation of contaminants. Oxygen availability is often a rate-limiting factor although anaerobic biodegradation may also be important. Intrinsic remediation processes may proceed for years, if not decades, as was shown with remediation of PCBs in the Hudson riverbed.

In Situ Subsurface Bioremediation

This approach involves designing a subsurface configuration to allow growth and contaminant biodegradation by the microbes present through the efficient supply of nutrients, oxygen, or other electron acceptors through injector wells and air-sparging systems into the unsaturated or saturated zone, generally below the contaminated plume.[31] Processes may include a soil vapor extraction or other vapor containment system to prevent the release of volatile contaminants from the site to the atmosphere.

Engineered Biopiles

Engineered soil biopiles are typically fitted with a series of aeration pipes to distribute air to the contaminated soil via positive or negative pressure.[31] With soils contaminated with volatile hydrocarbons, the air pumps generally draw a vacuum such that air enters at the surface of the pile drawing any undegraded volatiles into the pipework system where they may be directed to a biofilter or other volatile organic carbon (VOC) treatment or collection system. Addition of bulking agents to the soil may be required to facilitate airflow through the soil, and care must be taken to control soil humidity such that growth and biodegradation processes are facilitated.

Soil Composting

In the application of composting strategies to soil bioremediation, biomaterials including wood chips or other plant materials are mixed with the contaminated soil in approximately equal amounts. As in general composting, aerobic biodegradation processes result in heat generation. Contaminants are degraded by the bacteria and fungi associated with the compost, and it is known that contaminants may also react irreversibly with plant and humic substance components.[31]

Landfarming

This bioremediation approach involves surface atmospheric aeration processes enhanced by frequent tilling and maintenance of moisture levels at 40–60% saturation via sprinklers or other means. The process is applicable to in situ soils contaminated near the surface or to spreading and treating excavated soils at a designated landfarming location.[31] Weathering, sorption, evaporation or volatilization, leaching, and photo-oxidation processes may cause the removal of certain hydrocarbon compounds during bioremediation, resulting in overestimation of the extent of biodegradation.[7]

Soil Bioreactors

Soil slurry bioreactors have advantages for bioremediation processes similar to the general fermentation and other bioprocesses, namely, those conditions in the mixture can approach homogeneity, and processes can be optimally controlled to promote microbial growth and contaminant biodegradation.[26] Although capital and energy costs are high in these processes, high rates and extents of degradation may be achieved with high levels of process dependability and reliability.[29]

Bioremediation of Shorelines

Marine shoreline bioremediation processes are challenging due to the potential for changing tides to wash away nutrients and microbes from the contaminated shore. The capacities of microbes to remediate contaminated soil, coastlines, and water bodies was first demonstrated on a large scale in the cleanup of the crude oil, which spilled from the Exxon-Valdez tanker which ran aground off the coast of Alaska, near Prince William Sound. Remediation of the hydrocarbons was promoted by the addition of specific oleophilic nitrogen fertilizers, which bound to the oil contaminants to support microbial growth and biodegradation.[8,9] In contrast, addition of microbial inoculants was ineffective.

Deep-Sea Bioremediation

The capacity of microbes to degrade crude oil at sea was illustrated during one of the world's largest environmental disasters, when the explosion at British Petroleum's Deep Water Horizon oil rig resulted in an estimated 4.9 million barrels of crude oil spilling into the Gulf of Mexico. On August 4, 2010, just three weeks after the well was capped, the U.S. government estimated that 50% of the oil was already gone from the Gulf, and the rest was being rapidly degraded. Dr. Kenneth Lee put all of this in a scientific perspective: "The consensus now is that microbial degradation was a major process responsible for cleaning up the Deepwater Horizon spill." (http://www.dfo-mpo.gc.ca/science/publications/article/2012/01-25-12-eng.html).

Phytoremediation

In phytoremediation processes, removal of inorganic or organic contaminants from media such as soil is promoted by plants or associations between plant roots and microbes in soil and aqueous media.[32,33] In these processes, specific contaminants may be extracted, degraded, or immobilized. The mechanisms involved include the following: phytodegradation, whereby root exudates from the plant enhance microbial growth and contaminant biodegradation in the root surroundings, rhizofiltration where soluble contaminants are sorbed onto or into the roots, phytoextraction involving uptake of metals by roots and their translocation in the plants. The plants may be harvested for disposal, and phytovolatilization, where contaminants taken up by the plant, are possibly, but not always, modified and transpired into the atmosphere.

Air Biofiltration

Contaminants in an air stream, for example, VOCs, may be remediated through air biofiltration, whereby the contaminated air passes through a porous medium, which supports growth of microbes as attached biofilms.[34] The contaminants and oxygen are transferred from the air stream to the biofilm where biodegradation occurs, thereby decontaminating the air stream. Biofilter media include wood branches, bark or chips, peat, compost, synthetic media, coated ceramic or other particulates, and various combinations of these constituents. For effective performance, the biofilter environment must support efficient bacterial growth and facilitate effective transfer and sorption of contaminants from the air stream into the biofilm and access to the biodegrading microbes. The air stream or filtration medium must be appropriately humidified to insure that adequate water activity is maintained to support microbial growth and biodegradation while not oversaturating the medium such that air flow rate is retarded. The flow rate of the mobile phase (air in this case) is controlled to achieve a retention time in the biofilter sufficient to remove and biodegrade the contaminants in the air stream to a desired specification.

Conclusion

The purpose of surface or groundwater remediation is not only to enhance the biodegradation, transformation, or detoxification of pollutants, but also to protect the quality of soil to maintain environmental quality and sustain biological productivity within the ecosystem boundaries. Although biological processes are typically implemented at a relatively low cost, implementation of bioremediation technologies requires knowledge of interdisciplinary sciences involving microbiology, chemistry, engineering, ecology, and hydrogeology. Bioremediation methods have been successfully used to treat polluted soils, oily sludges, and groundwater contaminated by industrial chemicals and solvents, petroleum hydrocarbons, pesticides, and heavy metals.

References

1. Bitton, G. *Wastewater Microbiology,* 4th Ed.; Wiley-Blackwell: New Jersey, 2011.
2. Kutzner, H.J. Microbiology of composting. In *Biotechnology,* 2nd Ed.; Rehm, H.-J., Reed, G., Eds.; Wiley-VCH: Weinheim, 2001; Vol. 3C, 35–94.
3. Storm, I.M.; Kristensen, N.B.; Raun B.M.; Smedsgaard J.; Thrane, U. Dynamics in the microbiology of maize silage during whole-season storage. J. Appl. Microbiol. **2010**, *109,* 1017–1026.
4. Shimao, M. Biodegradation of plastics. Curr. Opin. Biotech-nol. **2001**, *12,* 242–247.
5. Singh, A.; Ward, O.P. Soil bioremediation and phytoremediation – an overview. In *Applied Bioremediation and Phytoremediation*; Singh, A., Ward, O.P., Eds.; Soil Biology Series; Springer-Verlag: Germany, 2004; Vol. 1, 1–12.
6. Megharaj, M.; Ramakrishnan, B.; Venkateswarlu, K.; Set-hunathan, N.; Naidu, R. Bioremediation approaches for organic pollutants: A critical perspective. Environ. Intern. **2011**, *37,* 1362–1375.
7. Singh, A.; Ward, O.P., Eds.; *Biodegradation and Bioremediation, Soil Biology Series;* Springer-Verlag: Germany; 2004; Vol. 2.
8. Prince, R.C. Bioremediation of marine oil spills. Trends Biotechnol. **1997**, *15,* 158–160.
9. Head, I.M.; Martin Jones, D.; Röling, W.F.M. Marine microorganisms make a meal of oil. Nature Rev. Microbiol. **2006**, *4,* 173–182.
10. USEPA. Cleaning up the nations waste sites: markets and technology trends. Technology Innovation and Field Services Division, Office of Solid Waste and Emergency Response, EPA 542-R-04-015; 2004.
11. Van Hamme, J.; Singh, A.; Ward, O.P. Recent advances in petroleum microbiology. Microbiol. Mol. Biol. Rev. **2003**, *67,* 503–549.
12. Van Hamme, J.D. Bioavailability and biodegradation of organic pollutants – a microbial perspective. In *Biodegradation and Bioremediation;* Singh, A., Ward, O.P., Eds.; Soil Biology Series; Spring er-Verlag: Germany, 2004; Vol. 2, 37–56.
13. Marques, A.P.G.C.; Rangel, A.O.S.S.; Castro, P.M.L. Remediation of heavy metal contaminated soils: an overview of site remediation techniques. Crit. Rev. Environ. Sci. Tech-nol. **2011**, *41,* 879–914.
14. U.S. Environmental Protection Agency. *Provisional Guidance for Quantitative Risk Assessment of Polycyclic Aromatic Hydrocarbons, EPA/600/R-93/089;* USEPA: Washington, D.C., 1993.
15. Vazquez-Duhalt, R.; Marquez-Rocha, F.; Ponce, E.; Licea, A.F.; Viana, M.T. Nonylphenol, an integrated vision of a pollutant. Appl. Ecol. Environ. Res. **2005**, 4, 1–25.
16. Crawford, R.L.; Hess, T.F.; Paszczynski, A. Combined biological and abiological degradation of xenobiotic compounds. In *Biodegradation and Bioremediation*; Singh, A., Ward, O.P., Eds.; Soil Biology Series, Springer-Verlag: Germany, 2004; Vol. 2, 251–278.
17. Ramos, J.L.; Duque, E.; Godoy, P.; Segura, A. Efflux pumps involved in toluene tolerance in *Pseudomonas putida* DOT-T1E. J. Bacteriol. **1998**, *180,* 3323–3329.
18. Singh, A.; Kuhad, R.C.; Shareefdeen, Z.; Ward, O.P. Methods for monitoring and assessment of bioremediation processes. In *Biodegradation and Bioremediation*; Singh, A., Ward, O.P., Eds.; Soil Biology Series, Springer-Verlag, Germany, 2004; Vol. 2, 279–304.
19. Van Hamme, J.D.; Singh, A.; Ward, O.P. Surfactants in microbiology and biotechnology: Part 1. Physiological aspects. Biotechnol. Adv. **2006**, *24,* 604–620.
20. Singh, A.; Van Hamme, J.D.; Ward, O.P. Surfactants in microbiology and biotechnology: Part 2. Application aspects. Biotechnol. Adv. **2007**, *25,* 99–121.
21. VanHamme, J.D.; Ward, O.P. Physical and metabolic interactions of *Pseudomonas* sp. Strain-B45 and *Rhodococcus* sp. F9-D79 during growth on crude oil and the effect of chemical surfactants thereon. Appl. Environ. Microbiol. **2001**, *67,* 4874–4879.
22. VanHamme, J.; Ward, O.P. Influence of chemical surfactants on the biodegradation of crude oil by a mixed culture. Can. J. Microbiol. **1999**, *45,* 130–137.

23. Billingsley, K.A.; Backus, S.M.; Wilson, S.; Singh, A.; Ward, O.P. Remediation of PCBs in soil by surfactant washing and biodegradation in the wash by *Pseudomonas* sp. LB400. Biotechnol. Lett. **2002**, *24*, 1827–1832.
24. Ward, O.P. Biosurfactants in biodegradation and bioremediation. Adv. Exper. Med. Biol. **2010**, *672*, 65–74.
25. Ward, O.P.; Singh, A.; Van Hamme, J.D.; Voordouw, G. Petroleum microbiology. In *Encyclopedia of Microbiology*, 3rd Edn.; Schaechter, M. Ed.; Elsevier: Amsterdam, 2009; 443–456.
26. Yapa, C.L.; Gana, S.; Ng, H.K. Application of vegetable oils in the treatment of polycyclic aromatic hydrocarbons-contaminated soils. J. Hazard. Mater. **2010**, *177*, 28–41.
27. Tyagi, M.; Fonseca, M.M.R.; de Carvalho, C.C.C.R. Bioaugmentation and biostimulation strategies to improve the effectiveness of bioremediation processes. Biodegradation **2011**, *22*, 231–241.
28. Ye, J.; Singh, A.; Ward, O.P. Biodegradation of nitroaromatics and other nitrogen-containing xenobiotics. World J. Microbiol. Biotechnol. **2004**, *20*, 117–135.
29. Ward, O.P.; Singh, A.; Van Hamme, J. Accelerated biodegradation of petroleum hydrocarbon waste. J. Ind. Microbiol. Biotechnol. **2003**, *30*, 260–270.
30. Delisle, S.; Greer, C.W. Remediation of organic pollutants through natural attenuation. In *Applied Bioremediation and Phytoremediation;* Singh, A., Ward, O.P., Eds.; Soil Biology Series, Springer-Verlag: Germany, 2004; Vol. 1, 159–186.
31. Ward, O.P.; Singh, A. Evaluation of current soil bioremediation technologies. In *Applied Bioremediation and Phytoremediation.* Singh, A., Ward, O.P., Eds.; Soil Biology Series; Springer-Verlag: Germany, 2004; Vol. 1, 187–214.
32. Macek, T.; Kotrba, P., Svatos, A.; Novakova, M.; Demnerova, K.; Mackova, M. Novel roles for genetically modified plants in environmental protection. Trends Biotechnol. **2008**, *26*, 146–152.
33. Gerhardt, K.E.; Xiao-Dong Huang, X.-D.; Glick, B.R.; Bruce M. Greenberg, B.M. Phytoremediation and rhizoremediation of organic soil contaminants: potential and challenges. Plant Sci. **2009**, *176*, 20–30.
34. Shareefdeen, Z.; Singh, A., Eds.; *Biotechnology for Odour and Air Pollution Control;* Springer-Verlag: Germany, 2005.

43

Decision Support: Monitoring, Reporting, and Forecasting Ecological Conditions of Appalachian Trail

Yeqiao Wang
University of Rhode Island

Introduction 359
Methods 360
Conclusion 362
References 362

Introduction

The Appalachian Trail (A.T.) traverses most of the high elevation ridges of the eastern United States, extending 3,676 km across 14 states, from Springer Mountain in Northern Georgia to Mount Katahdin in central Maine. The A.T.'s gradients in elevation, latitude, and moisture and north–south alignment represent a cross-section MEGA-Transect of eastern U.S. forest and alpine areas, and offer a setting for collecting scientific data on the health of ecosystems and species that inhabit them (Dufour and Crisfield, 2008). The high elevation setting of the A.T. provides an ideal landscape for early detection of undesirable changes in the natural resources of the eastern United States, for example, development encroachment, acid precipitation, invasions of exotic species, and climate change impacts.

The A.T. Decision Support System (DSS), or A.T.-DSS, is an Internet-based implementation and dissemination toolset accessible at www.edc.uri.edu/ATMT-DSS. The design of the A.T.-DSS was to facilitate decision support for management of the A.T. with the monitoring, reporting, and forecasting capacity from remote sensing-based Earth observations and geospatial modeling. It was also designed to provide a means to convey meaningful information to the American public. The National Park Service (NPS) Inventory and Monitoring (I&M) program is designed to develop and implement long-term natural resource monitoring and create a targeted decision support system aimed at selecting a suite of reliable and representative metrics, or vital signs, to provide long-term data on changes in ecosystem health (Shriver et al., 2005). Among the vital signs defined by the I&M program and A.T. MEGA-Transect partners, the A.T.-DSS targets phenology and climate change, forest health, and landscape dynamics for system development, data preparation, and modeling, in order to provide time-series and ecological forecasting information for an improved decision support process.

The A.T.-DSS made efforts to create a comprehensive set of seamless indicator data layers consistent with selected A.T. vital signs; to establish a ground monitoring system to complement TOPS and integrate NASA data with *in situ* observations; to assess historical, current, and forecasted ecosystem conditions and trends by coupling TOPS with habitat modeling; and to establish an Internet-based implementation and dissemination system for data visualization, sharing, and management to facilitate collaboration and promote public understanding of the A.T. environment. Benefits of the A.T.-DSS include quick access to useful information to support natural resource management and decision-making; enhanced situational awareness and spatial context of the A.T. lands; improved ability to communicate the status of A.T. environment and natural resources; and improved conservation of biodiversity by habitat assessment modeling, evaluation of forest conditions, and identification of areas of concern.

Methods

The A.T. is an open and complex system. The spatial extent of the A.T.-DSS was adopted from a boundary defined by the NPS and USGS. It was established by selecting all 10-digit Hydrological Unit Code (HUC-10) watersheds within five statute miles of the A.T. land base, termed the A.T. HUC-10 shell (Dieffenbach, 2011). The shell provided an ecologically relevant boundary around the trail to distinguish the data most relevant to the A.T. corridor area.

The A.T.-DSS integrates NASA multi-platform sensor data, NASA Terrestrial Observation and Prediction System (TOPS) models (Nemani et al., 2009; Hashimoto et al., 2011), and *in situ* measurements from A.T. MEGA-Transect partners to support the resource management decision-making process. By integrating NASA's Earth Observation System data and modeling products that link climate models (e.g., through TOPS) and ecological models (e.g., habitat suitability) with *in situ* observations (e.g., USFS Forest Inventory and Analysis data), the A.T.-DSS created a coherent framework for data integration, monitoring, reporting, and forecasting to improve the effectiveness of decision-making in management of the A.T. for natural resource conservation.

The seamless data provided by the A.T.-DSS include MODIS (Moderate-resolution Imaging Spectroradiometer) products, GIMMS (Global Inventory Modeling and Mapping Studies), NDVI (Normalized Difference Vegetation Index), NACP (North American Carbon Program) modeled carbon flux data, and SOGS (Surface Observation and Gridding System) metrological data. The derived data products include soil moisture, ecosystem production—e.g., gross primary production (GPP), net primary production (NPP), and net ecosystem exchange (NEE)—land surface phenology (LSP) metrics, and land cover change. The modeling results out of TOPS also include forecasts of climate change and the impacts to ecosystem functions and services along the A.T. corridor.

The USFS Forest Inventory and Analysis (FIA) data contain detailed forest health information with spatial and temporal records. FIA conducts continuous standardized annual monitoring in all forests within the United States. The data are broad in scope and produce information on forest productivity, biodiversity, vitality, carbon cycling, and soil conservation. The project team obtained FIA data covering 2002–2010 within the A.T. HUC-10 shell area.

The FIA data provide detailed *in situ* measurements which complement remotely sensed observations (e.g., from TOPS). The FIA variables are closely related to key ecological processes like carbon fixation, nutrient cycling, growth, regeneration, mortality, decomposition, etc., and therefore, are designed to detect natural and stress-induced changes in forest condition from climate change, invasive species, air pollution, and other stressors. Areas of degraded forest conditions can be identified from FIA plot level data and evaluated in the broader context of TOPS data for improved understanding of ecosystem processes.

Coupling geospatial data with statistical modeling facilitates assessment of current, historical, and forecasted ecosystem conditions. To demonstrate the capabilities of the A.T.-DSS, a prototype habitat suitability model for the invasive species tree of heaven (*Ailanthus altissima*) was developed. The current and projected distribution of suitable habitats for the species was estimated using geospatial data from the A.T.-DSS (Clark et al., 2014).

Tree of heaven has become a problematic invasive species due to its high tolerance for environmental stress, aggressive dispersal capacity, tenacious establishment, and competitive advantage over native species. It is particularly successful in disturbed landscapes where vast quantities of airborne seeds are able to colonize gaps in forest canopies. Ailanthus colonists rapidly outgrow and allelopathically (chemically) suppress adjacent native tree seedlings, and reproduce via root and stem shoots to form dense stands.

The Internet-based A.T.-DSS (www.edc.uri.edu/ATMT-DSS/) allowed users to build a comprehensive understanding of the current status and trends in terms of driving factors and responsive conditions, and helped characterize habitat condition and primary drivers for simulation and prediction exercises. The visualization and implementation provided effective tools for accessing remote sensing and geospatial data and research conclusions to improve understanding of the changing A.T. environments (Wang et al., 2010). The publically accessible Internet-based A.T.-DSS contained a variety of tools to view and download A.T. data. Toolsets were intuitively organized by their primary functions of Mapping Viewer, viewsheds monitoring, report and forecast, and data gateway (Figure 43.1).

The Mapping Viewer was designed as an interactive mapping tool that allowed users to pan-zoom to areas of interest, toggle the visibility of data layers, and integrate data in combinations that suited their immediate needs. Specific tools were added to the viewer to aid in decision-making, such as the time-series slider for visualizing land cover change, and elevation surface profile mapping. The Mapping Viewer capitalized on the latest technology and was developed using the ArcGIS Flex viewer. The software was fully customizable and allowed system users to easily develop new visualization or analysis tools as priorities change.

The Viewshed Monitoring toolset was designed to depict the A.T. 3-D viewshed using the Google Earth plugin. Among the goals of A.T.-DSS included understanding and preserving the scenic viewsheds for hikers of the trail. Using ArcGIS 10 and Model Builder, A.T.-DSS had viewsheds for 930 of the most noteworthy scenic locations along the A.T. Displaying the results of geoprocessing online, such as viewsheds, could be a formidable task due to the time it took to process geospatial data and the Internet bandwidth available to the end users. Because of time limitations, the A.T.-DSS used pre-generated viewsheds and served them using a common web server. Using client-side JavaScript and the Google Earth web browser Plug-in, A.T.-DSS provided an interactive means to display those viewsheds. By embedding aerial and satellite remote sensing images, the results of this modeling helped users view important characteristics

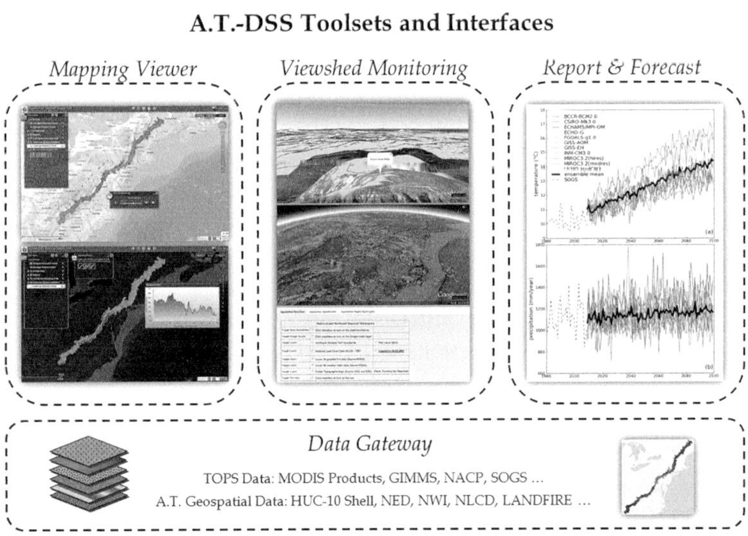

FIGURE 43.1 **(See color insert.)** The toolsets and interfaces were organized by their primary functions, namely, the Mapping Viewer, Viewshed Monitoring, Report & Forecast, and Data Gateway modules (www.edc.uri/edu/ATMT-DSS).

of the A.T. corridor at different spatial scales and from different elevations and view angles. Using this web tool, users were also able to overlay additional geospatial data such as real-time weather data and real-time possible fire locations, in this 3-D context. This functionality enhanced the end user's ability to graphically interpret and analyze A.T.-specific ground-based data for informed decision-making.

The Report & Forecast interfaces provided summaries of the A.T. vital signs phenology and climate, forest health, and landscape dynamics at the landscape scale. The reports contained information about the vital signs for the past, present, and future. The *Data Gateway* provided an avenue for end users to select and download seamless geospatial and remote sensing data for further localized processing as decision support needs.

The Reporting & Forecasting module provided summaries of the pre-calculated vital sign indicators under the categories of phenology and climate, forest health, and landscape dynamics for the past, present, and future. The Data Gateway provided access to seamless geospatial data within the A.T. HUC-10 shell so that users can select data for local processing, analyzes, and mapping.

Conclusion

The A.T.-DSS achieved the designed objectives. The seamless time-series data and modeling outputs provided specific information for understanding the dynamic and changing A.T. environment. The data and information revealed landscape scale patterns and change of environmental vital signs of phenology and climate change, forest health, and landscape dynamics. The developed prototype demonstrated that the geospatial data and modeling outputs can provide information for decision support exercises such as habitat suitability and risk analysis under the impacts such as climate change. The Internet-based decision support interfaces and toolsets represent the state-of-the-art system for visualization, implementation, and dissemination, developed to meet demands of geospatial data at landscape level for the A.T. corridor area. The data and information presented and packaged at different scales can provide information for decision support across different levels of management, as well as convey meaningful information to the American public.

The most significant challenges that the A.T.-DSS encountered came from two perspectives. The first challenge was to develop the data products that would meet the needs of broadly defined user groups in terms of data format, level of details in thematic values, and spatial scales for information presentation and data delivery. In particular, preparation of data for broader scopes of end users, who might have no knowledge and capacity in geospatial data processing, presented challenges. Through a Decision Analysis Workshop and several follow-up project team meetings that engaged user groups in early stage of the project, the A.T.-DSS designed with improved deliverables of data and presentation of information to meet the needs of users. The A.T.-DSS prepared the deliverables in multiple spatial scales, for example, in scales of the entire A.T. corridor, ecoregion province, and subsection levels, so that end users from different perspectives can understand and use directly as references for decision support.

The second significant challenge came from the uncertainty of organizational change from the involved agencies. Any interruption of the commitment from any participating agency at any stage of a decision support system, such as A.T.-DSS, would affect the effectiveness and sustainability of the DSS developed. Further explorations of integrative approach with flexible operational management of the DSS from multiple agencies and user groups for the sustainability and benefit of such a DSS should be continued.

References

Clark, J., Wang, Y., August, P. Assessing current and projected suitable habitats for tree-of-heaven along the Appalachian Trail, *Philosophical Transactions of the Royal Society B*, 2014, 369: 20130192. doi:10.1098/rstb.2013.0192.

Dieffenbach, F. Appalachian National Scenic Trail vital signs monitoring plan. Natural Resource Technical Report NPS/NETN/NRR—2011/389. National Park Service, Northeast Temperate Network, Woodstock, VT, 2011.

Dufour, C., Crisfield, E. The Appalachian Trail MEGA-Transect Harpers Ferry, WV: Appalachian Trail Conservancy, 2008.

Hashimoto, H., Hiatt, S.H., Milesi, C., Melton, F.S., Michaelis, A.R., Votava, P., Wang, W., Nemani, R. Monitoring and forecasting climate impacts on ecosystem dynamics in protected areas using the terrestrial observation and prediction system. In Wang, Y. (Ed.), *Remote Sensing of Protected Lands*, pp. 525–542, 2011, Boca Raton, FL: CRC Press.

Nemani, R., Hashimoto, H., Votava, P., Melton, F., Wang, W., Michaelis, A., Mutch, L., Milesi, C., Hiatt, S., White, M. Monitoring and forecasting ecosystem dynamics using the Terrestrial Observation and Prediction System (TOPS), *Remote Sensing of Environment*, 2009, 113(7): 1497–1509.

Shriver, G., Maniero, T., Schwarzkopf, K., Lambert, D., Dieffenbach, F., Owen, D., Wang, Y., Nugranad-Marzilli, J., Tierney, G., Reese, C., Moore, T. Appalachian Trail vital signs. Boston, MA, 2005.

Wang, Y., Nemani, R., Dieffenbach, F., Stolte, K., Holcomb, G., Robinson, M., Reese, C., McNiff, M., Duhaime, R., Tierney, G., Mitche, B., August, P., Paton, P., LaBash, L. Development of a decision support system for monitoring, reporting and forecasting ecological conditions of the Appalachian Trail, in *2010 IEEE International Geoscience and Remote Sensing Symposium (IGARSS)*, Yokohama, 2010, pp. 2095–2098.

44

Estimation of Net Primary Production and Its Spatial and Temporal Variations in Northeast China: Remote Sensing

Xiaoyi Guo,
Jianjun Zhao, and
Hongyan Zhang
Northeast Normal University

Introduction 365
Study Area and Datasets 366
Northeast China • Dataset and Processing
CASA Model 368
Total Solar Surface Irradiation Algorithm • FPAR Algorithm • Temperature Scalars • Water Scalar • The Maximum Light Use Efficiency • Linear Regression Analysis
Results and Discussion 374
Accuracy Verification and Evaluation • Temporal Dynamics of NPP • Spatial Patterns of NPP
Conclusions 379
References 379

Introduction

Vegetation is the main component of terrestrial ecosystems, and it plays an essential role in mediating the global carbon balance, mitigating the rise of CO_2 concentrations in the atmosphere, and is one of the controlling factors in global climate change [1]. Vegetation production is a fundamental function of terrestrial ecosystems, which includes gross primary production (GPP) and net primary production (NPP). GPP refers to the organic matter produced by green plants absorbing solar energy and assimilating CO_2 through photosynthesis in the ecosystem, whereas NPP represents the GPP remainder after the deduction of the organic matter consumed by the autotrophic respiration of plants. For GPP of vegetation, approximately half of the organic matter is rereleased into the atmosphere by plant respiration, whereas the other part constitutes the NPP. Therefore, NPP can be regarded as the total amount of organic matter produced by photosynthesis when vegetation converts light energy into chemical energy.

NPP is an important biophysical variables for measurement the net carbon content of CO_2 fixed in plants through photosynthesis [2,3]. It is a major determinant of carbon sinks in terrestrial ecosystems and a key regulator of ecological processes [4]. In addition, NPP is also a significant indicator of ecosystem function, which comprehensively reflects the effects of climate change and human activities on terrestrial vegetation [5–7]. Therefore, NPP estimation is important in understanding global change and the terrestrial carbon cycle.

Studies have been conducted to simulate NPP at various scales. Estimation methods of NPP include site observation, and experimental and modeling approaches. Early developed site observations include the sample plot measurement (e.g., direct harvesting method), eddy covariance method, chlorophyll determination, radiolabeling method, and raw material consumption measurement. These methods provide basic parameters and validation data for the construction of estimation models. Site observation and experimental methods could obtain the NPP values with high accuracy at local scale. However, it is difficult to measure NPP at a broad scales. For obtaining NPP values at regional scales, many models of NPP estimation have been developed in recent decades. NPP estimation models can be classified into three categories including statistical models, parameter models, and process models [8]. Statistical models, such as the Miami model, Thornthwaite Memorial model, and Chikugo model, primarily use empirical statistical relationships between meteorological data and the site observation to estimate NPP. Parameter models, also known as light use efficiency models, such as the Carnegie Ames Stanford approach (CASA) model, C-Fix model, and global production efficiency model (GLO-PEM), use solar radiation as the primary variable and temperature and precipitation as the limiting factors to simulate NPP. Process models, such as the Biome-BGC model, integrated terrestrial ecosystem C-budget (InTEC) model and boreal ecosystems productivity simulator (BEPS) model, use meteorological, soil data and plant physiological parameters to estimate NPP.

For tracking the history of broad-scale vegetation dynamics, remote sensing data have been used and regarded as a surrogate measure of primary production [9]. Remote sensing data have promoted an important transformation in simulating vegetation NPP from traditional measurements to simulation. Vegetation index (VI) reflects the characteristics of vegetation spectral curves, which is sensitive to photosynthetic fixed dry matter or leaf area index (LAI). Therefore, the VI may better be used for parameterizing models that may more accurately reflect actual changes in primary production. Paruelo et al. (1997) determined that there was a satisfactory correlation between the observed terrestrial NPP and normalized difference vegetation index (NDVI) [10]. With the accumulation of remotely sensed long time-series data and the establishment of eddy covariance flux sites worldwide, various models of vegetation production rely on remote sensing data. Specifically, models based on the light use efficiency principle have been developed given their simple and easy access to parameters. The CASA fully considers environmental conditions and vegetation characteristics, and it has been widely used in terrestrial ecosystem NPP estimations at various scales [11,12]. Understanding the input parameters for NPP estimate models in specific ecoregions is imperative and urgent [13]. An accurate estimation of fraction of photosynthetically active radiation (FPAR) is one of the key factors in the simulation of productivity using the CASA model. However, the NPP values show significantly different when various parameterization schemes are used to estimation of NPP.

This chapter describes and tests a time series of remote sensing data-driven CASA model for simulating NPP and tracing its dynamics over the recent period for the case of the Northeast China. This study focuses on the impact of different FPAR calculation methods. The objective is to understand the NPP estimation accuracy and provide a basis for the estimation of carbon sources and sinks at a regional scale.

Study Area and Datasets

Northeast China

Northeast China is located between 38°43′N–53°34′N and 115°31′E–133°5′E, covers an area $1.24 \times 10^6 km^2$, and exhibits extensive natural diversity with respect to climate, vegetation, geology, and soil. It covers Heilongjiang Province, Jilin Province, Liaoning Province, and Inner Mongolia Autonomous Regions. This area is sensitive to global change and a regional environmental regulator of northern China. The simulation and monitoring of NPP is of great significance to the regional carbon cycle.

A large portion of the study area is characterized by a temperate monsoon continental climate, with warm summers and cold winters. Northeast China contains cold temperate zone, temperate zone, and warm temperate zone from north to south, and semi-arid, semi-humid, and humid humidity regions from west to east forming a unique pattern of vegetation. The geomorphic units include the Greater Khingan Mountains, Lesser Khingan Mountains, Changbai Mountains, Hulunbuir Plateau, Sanjiang Plain, and Songliao Plain. As shown in Figure 44.1, there are 12 ecoregions in Northeast China [14], i.e., a larch and coniferous forest area in the northern segment of the Greater Khingan Mountains (IA1); wetlands of the Sanjiang Plain (IIA1); a coniferous forest area in the Changbai Mountains in the Lesser Khingan Mountains (IIA2); a coniferous and broadleaf mixed forest area in the eastern piedmont tableland of the Songliao Plain (IIA3); a forest grassland area in the center of the Songliao Plain (IIB1); a mountain grassland forest area in the middle segment of the Greater Khingan Mountains (IIB2); a forest grassland area on the western side of the northern segment of the Greater Khingan Mountains (IIB3); a grassland area in the Xiliaohe Plain (IIC1); a grassland area at the southern end of the Greater Khingan Mountains (IIC2); a plain grassland area in the Hulun Buir Plain (IIC4); low-hill larch and broadleaf forest and artificial vegetation in eastern Liaoning and Shandong (IIIA1); and larch and broadleaf forest areas in the mountains of North China (IIIB2).

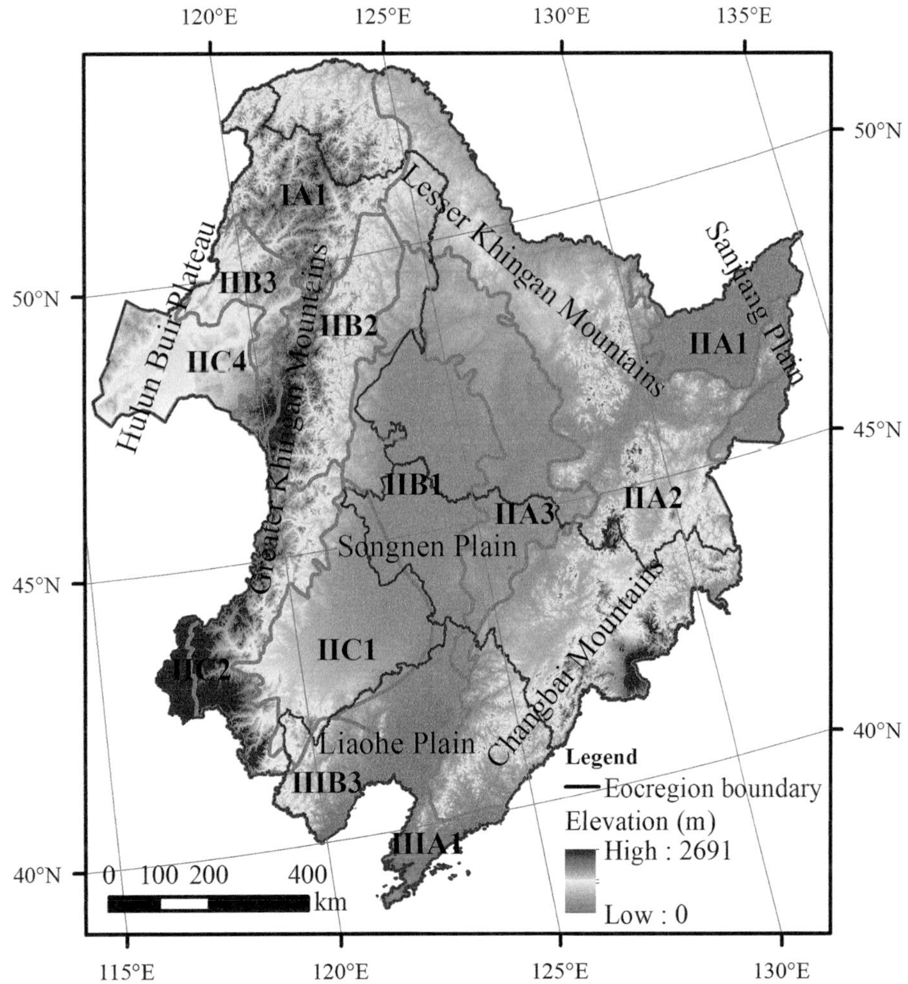

FIGURE 44.1 The location of the study area, Northeast China.

Dataset and Processing

GIMMS NDVI3g

The GIMMS third-generation NDVI version 1 (3g V1) was obtained from the National Aeronautics and Space Administration (NASA) Ames Ecological Forecasting Lab (https://ecocast. arc.nasa.gov/data/pub/gimms/3g.v1/) and was produced by the National Oceanic and Atmospheric Administration (NOAA) advanced very high-resolution radiometers (AVHRR) series satellites. This dataset is the latest version of the GIMMS NDVI product, which has better data quality at high latitudes than previous versions [15]. This dataset has a 15-day interval with a spatial resolution of 1/12° (approximately $8 \times 8\,km^2$) and covers the period from January 1982 to December 2015. Although the GIMMS NDVI3g V1 dataset has been corrected for calibration, solar geometry, heavy aerosols, clouds, and other effects, the original NDVI time series contain residual noise. Therefore, to further improve the quality of the dataset, the Savitzky–Golay method was used to filter the NDVI product by using the TIMESAT program [16]. In this study, the 15-day GIMMS NDVI3g V1 was aggregated to the monthly NDVI using maximum value composition.

Climate Dataset

The meteorological data presented in this chapter include monthly mean temperature, monthly total precipitation, and daylight hour data from 1982 to 2015 that were applied to run the CASA model. All the datasets were acquired from the Chinese Meteorological Science Data Sharing Service Network (http://data.cma.cn/). These data were recorded at 99 meteorological and climatological stations distributed throughout Northeast China. The solar radiation data of stations were obtained by using daylight hour data with an empirical climate model (the method is presented in the "GIMMS NDVI3g" section). ANUSPLIN software was used to interpolate the temperature, precipitation, and daylight hour data from the stations into $8 \times 8\,km^2$ grids to match the GIMMS NDVI3g data. The grid data have the same spatial and temporal resolution and projection system as the GIMMS NDVI3g data.

Land Cover Dataset

The Moderate Resolution Imaging Spectroradiometer (MODIS) land cover product MCD12Q1 was used. The MCD12Q1 product was derived from annual observation data by the Terra and Aqua MODIS sensors to depict the land cover types. This dataset consists of five different land cover classification schemes. Because various studies used different schemes to estimate FPAR based on CASA model, schemes of the University of Maryland (UMD), LAI/FPAR, and plant functional types (PFT) are applied for various cases in this chapter. Due to human activities, land cover in Northeast China changed significantly during the study period [17]. To reflect the change of land cover, MCD12Q1 data in 2000 and 2015 were used to represent the land cover characteristics, respectively.

CASA Model

CASA was developed to estimate crop growth from intercepted photosynthetically active radiation [18,19]. It was expanded to estimate global terrestrial NPP [20]. The NPP is computed as the product of the amount of photosynthetically active radiation (APAR) absorbed by green vegetation and light utilization efficiency (ε). Light utilization efficiency implies that solar radiation converts the APAR into plant biomass increments with which the radiation is converted to plant biomass increment [20]. For each pixel, the calculation of NPP is described as follows:

$$\text{NPP}(x,t) = \text{APAR}(x,t) \times \varepsilon(x,t) \tag{44.1}$$

where NPP(x, t) is the net primary productivity at a pixel x in the month t, APAR(x, t) is the photosynthetically active radiation absorbed by pixel x in month t, and $\varepsilon(x, t)$ is the actual light use efficiency of cell x in month t. The unit for all three is gC/MJ.

The APAR of plants is determined by the total solar surface irradiation and absorbed FPAR of photosynthetically active radiation. The calculation method of APAR is as follows:

$$\text{APAR}(x,t) = S(x,t) \times \alpha \times \text{FPAR}(x,t) \quad (44.2)$$

where $S(x, t)$ is the total solar surface irradiation (MJ/m^2); α is the ratio of solar active radiation that is used by green vegetation in total solar radiation, assuming a constant of 0.5 [21]; FPAR(x, t) is the fraction of photosynthetically active radiation absorbed by the vegetation, which depends on land cover. APAR estimation is further decomposed into two parts, the estimation of $S(x, t)$ and FPAR(x, t). FPAR and VI show strong correlation [22] and they are generally derived from remote sensing data.

Under ideal conditions, plants have the maximum light use efficiency. The light use efficiency is primarily affected by temperature and availability of water under actual conditions. The actual photosynthetic efficiency is expressed as

$$\varepsilon(x,t) = \varepsilon_{\max} \times T_{\varepsilon 1}(x,t) \times T_{\varepsilon 2}(x,t) \times W_{\varepsilon}(x,t) \quad (44.3)$$

where ε_{max} (gC/MJ) is the maximum light use efficiency under ideal conditions, $T_{\varepsilon 1}(x,t)$ and $T_{\varepsilon 2}(x,t)$ are the effects of temperature stress, and W_{ε} is the effect of water stress. The two temperature stress variables serve to depress ε at very high and very low temperatures and to depress ε when the temperature is above or below the optimum temperature.

According to Equations (44.1)–(44.3), NPP is derived from climate variables (i.e., radiation, temperature, and water) and vegetation variables (APAR and ε_{max}) based on the CASA model. Climate variables are obtained from meteorological stations which are interpolated into continuous surfaces. Vegetation variables, FPAR and land cover types, used in the CASA model are derived from remote sensing data. The algorithm of FPAR and classification scheme of land cover are different for various studies. The uncertainty of model simulation generated by FPAR is greater than that of meteorological data at a particular scale [23]. Thus, this study focuses on the impacts of vegetation variables (FPAR and land cover) on the estimation of NPP. A detailed flowchart of data processing and CASA model is presented in Figure 44.2.

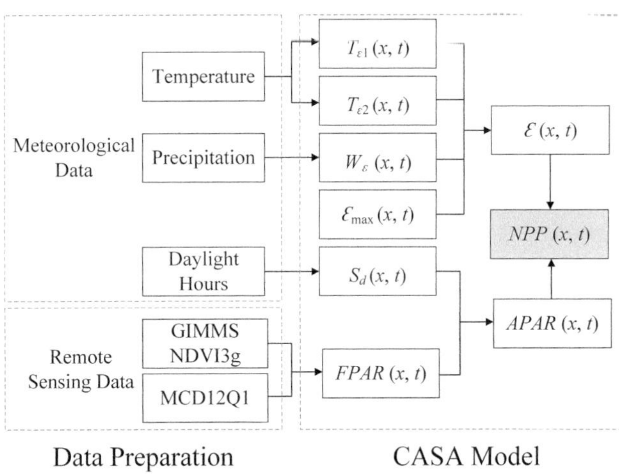

FIGURE 44.2 The flowchart of the proposed method for net primary production estimation using CASA model.

Total Solar Surface Irradiation Algorithm

In the chapter, the solar radiation data for each site were calculated using the empirical model. Daylight hour data of solar radiation from meteorological stations were used to establish the daily total solar radiation model in Northeast China [24]:

$$S_d(x,t)=\left(0.215+0.518\frac{n}{N}\right)S_0(x,t) \quad (44.4)$$

where S_d (x,t) (MJ/m²·day) is the daily total solar radiation; n is the actual daylight hours per day; N is the maximum daylight hours; and $S_0(x,t)$ (MJ/m²·day) is the daily extraterrestrial solar radiation. The monthly total solar radiation should be the sum of the daily S_d. N and S_0 are calculated according to Equations (44.5) and (44.6), respectively [25]:

$$N=\frac{2\omega}{15} \quad (44.5)$$

$$S_0=\frac{24\times 3{,}600\,G}{\pi}\times k\left(\cos\varphi\cos\delta\sin\omega+\frac{\pi\omega}{180}\sin\varphi\sin\delta\right) \quad (44.6)$$

where ω is the solar sunset hour angle; G represents the solar constant, $G = 1{,}367\ \text{W/m}^2$; k is the earth–sun distance correction factor; φ is geographical latitude of the meteorological station; and δ is the solar declination. ω, φ, and δ are calculated by using Equations (44.7), (44.8), and (44.9), respectively:

$$\omega=\arccos(-\tan\varphi\tan\delta) \quad (44.7)$$

$$k=1+0.033\cos\left(\frac{360\,J}{365}\right) \quad (44.8)$$

$$\delta=23.45\sin\left[\frac{360(284+J)}{365}\right] \quad (44.9)$$

where J is the Julian day.

FPAR Algorithm

The VI derived from remote sensing data can effectively reflect the canopy's chlorophyll content. Thus, a linear relationship between FPAR and VI is supported by various studies [26–28]. Five commonly NDVI/SR-FPAR relationships are compared in this chapter.

FPAR Model from NASA-MOD15 (Case A)

The MOD15 product provides a relationship table between the NDVI and FPAR. The algorithm consists of a straightforward direct FPAR retrieval using the pixels biome-class, and the current pixels NDVI measure as the search keys. The relationship between the NDVI and FPAR is divided into three stages for each biome. As shown in Table 44.1, the relationship between NDVI and FPAR in the middle part was polynomially fitted in this study, and then, the estimation results of FPAR were obtained. FPAR is 0 when the NDVI is less than 0.125. In this case, the LAI/FPAR classification scheme (University of Montana biome classes) is selected for land cover types.

TABLE 44.1 NDVI and Corresponding Values of FPAR

Biome	NDVI (x)	Fitting Formula of FRAP (y)	r^2
Grassland/cereal crops	[0.125,0.775)	$y = -87.005x^5 + 180.71x^4 - 137.2x^3 + 47.412x^2 - 6.5335x + 0.4629$	0.9984
	[0.775, 1]	0.9000	
Shrubland	[0.125,0.875)	$y = -4.1234x^4 + 7.6207x^3 - 4.0114x^2 + 1.5x - 0.0006$	0.9998
	[0.875, 1]	0.9074	
Broadleaf crop	[0.125,0.875)	$y = -300.68x^6 + 814.33x^5 - 859.57x^4 + 449.5x^3 - 121.84x^2 + 16.97x - 0.7805$	0.9974
	[0.875, 1]	0.9196	
Savannah	[0.125,0.875)	$y = -64.484x^5 + 148.82x^4 - 126.86x^3 + 49.209x^2 - 7.4574x + 0.4986$	0.9982
	[0.875, 1]	0.9196	
Broadleaf forest	[0.125,0.875)	$y = -54.227x^5 + 122.06x^4 - 102.65x^3 + 40.801x^2 - 6.6992x + 0.447$	0.9987
	[0.875, 1]	0.8853	
Needle leaf forest	[0.125,0.875)	$y = 178.85x^6 - 535.52x^5 + 621.13x^4 - 352.73x^3 + 102.83x^2 - 13.619x + 0.7345$	0.9998
	[0.875, 1]	0.9195	

FPAR Model Proposed by Choudhury [29] (Case B)

There is a linear relationship between the FPAR and NDVI for a certain range of values [30]. The linear relationship is determined by the maximum and minimum FPAR corresponding to the maximum and minimum NDVI for various vegetation types. There is a relationship between FPAR and NDVI as follows:

$$FPAR_{NDVI}(x,t) = \frac{\left[NDVI(x,t) - NDVI_{i,\min}\right] \times \left(FPAR_{\max} - FPAR_{\min}\right)}{NDVI_{i,\max} - NDVI_{i,\min}} + FPAR_{\min} \quad (44.10)$$

where $NDVI_{i,min}$ and $NDVI_{i,max}$ represent the NDVI maximum and minimum values of the i-th vegetation type, respectively; and $FPAR_{min}$ and $FPAR_{max}$ are 0.01 and 0.95, respectively. Zhu et al. discussed the values of $NDVI_{i,min}$ and $NDVI_{i,max}$ for different vegetation types in China [31]. In general, the value of $NDVI_{i,min}$ is 0.023, and the values of $NDVI_{i,max}$ are shown in Table 44.2. The land cover products of the PFT classification system were selected in the scheme.

FPAR Model Proposed by Sellers [22] (Case C)

Other study shows that there is a remarkable linear relationship between FPAR and SR of the AVHRR [32]. The SR is shown in the following equation [33]:

$$SR(x,t) = \frac{1 + NDVI(x,t)}{1 - NDVI(x,t)} \quad (44.11)$$

TABLE 44.2 Maximum NDVI for Various Vegetation Types

Value	Vegetation Types	$NDVI_{max}$
1	Deciduous coniferous forest	0.738
2	Evergreen broadleaved forest	0.647
3	Evergreen broadleaved forest	0.676
4	Deciduous broadleaved forest	0.747
5	Shrubs	0.636
6	Others	0.634

Moreover, Equation (44.10) can be rewritten as [22]

$$\mathrm{FPAR_{SR}}(x,t)=\frac{\left[\mathrm{SR}(x,t)-\mathrm{SR}_{i,\min}\right]\times(\mathrm{FPAR}_{\max}-\mathrm{FPAR}_{\min})}{\mathrm{SR}_{i,\max}-\mathrm{SR}_{i,\min}}+\mathrm{FPAR}_{\min} \tag{44.12}$$

where $\mathrm{SR}_{i,\min}$ represents unvegetated land area, $\mathrm{SR}_{i,\min}$=1.05 [31]; and $\mathrm{SR}_{i,\max}$ is related to vegetation type, the value of which is calculated according to Table 44.3. The values of $\mathrm{FPAR}_{\min}$ and $\mathrm{FPAR}_{\max}$ are the same as those in Choudhury's scheme. The land cover classification scheme adopted is also selected PTF classification.

FPAR Model Proposed by Los [34] (Case D)

A reported study found that the FPAR derived from the NDVI is higher than that from ground observation; meanwhile, the FPAR derived from the SR is lower than that from ground observation [34]. Thus, an improved algorithm of FPAR for CASA model was developed to establish Equation (44.13), and the average is taken as an estimate of the FPAR:

$$\mathrm{FPAR}(x,t)=0.5\mathrm{FPAR_{NDVI}}+0.5\mathrm{FPAR_{SR}} \tag{44.13}$$

where $\mathrm{FPAR_{NDVI}}$ is the result of Equation (44.10); and $\mathrm{FPAR_{SR}}$ is the result of Equation (44.12).

FPAR Model Proposed by Potter [20] (Case E)

Potter developed a method for estimating the FPAR and considered that the value of FPAR cannot be greater than 0.95 [20]. The proposed equation is as:

$$\mathrm{FPAR}(x,t)=\min\left[\frac{\mathrm{SR}(x,t)-\mathrm{SR}_{i,\min}}{\mathrm{SR}_{i,\max}-\mathrm{SR}_{i,\min}},0.95\right] \tag{44.14}$$

where SR(*x, t*) is the result of Equation (44.11), $\mathrm{SR}_{i,\min}$ is the SR value of bare land, and $\mathrm{SR}_{i,\max}$ is related to the vegetation type. $\mathrm{SR}_{i,\min}$ and $\mathrm{SR}_{i,\max}$ are calculated by the maximum and minimum NDVI for various vegetation types in Case B [31]. Therefore, $\mathrm{SR}_{i,\min}$ is 1.08, and $\mathrm{SR}_{i,\max}$ ranges from 4.14 to 6.17.

Temperature Scalars

For Equation (44.3), $T_{\varepsilon 1}(x, t)$ reflects the empirical observation that plants in very cold habitats typically have low maximum growth rates and high root biomass, potentially imposing large respiratory costs. The calculation method is as follows:

$$T_{\varepsilon 1}(x,t)=0.8+0.02T_{\mathrm{opt}}(x)-0.0005T_{\mathrm{opt}}^{2}(x) \tag{44.15}$$

TABLE 44.3 Maximum Light Use Efficiency of Vegetation Types in China

Vegetation Types	Maximum Light Use Efficiency (gC/MJ)
Deciduous coniferous forest	0.485
Evergreen coniferous forest	0.389
Deciduous broadleaved forest	0.692
Evergreen broadleaved forest	0.985
Coniferous and broadleaved mixed forest	0.475
Shrubs	0.429
Grassland	0.542
Cultivated vegetation	0.542
Other	0.542

where $T_{opt}(x)$ is defined as the appropriate temperature for plant growth for pixel x at time t, which is the monthly mean temperature for a pixel when the NDVI reaches the maximum value for the year. If the temperature of a month is less than or equal to 0, $T_{\varepsilon 1}(x, t)$ is equal to 0.

In Equation (44.3), $T_{\varepsilon 2}(x, t)$ reflects the concept that the efficiency of light utilization should be depressed when plants are growing at temperatures displaced from their optimum and has an asymmetric bell shape that declines more quickly at higher than at lower temperatures. The calculation method is as follows [20,22]:

$$T_{\varepsilon 2}(x,t) = \frac{\dfrac{1.1814}{1+e^{0.2(T_{opt}(x)-10-T(x,t))}}}{1+e^{0.3(-T_{opt}(x)-10+T(x,t))}} \tag{44.16}$$

where $T(x, t)$ is the average temperature for pixel x at time t.

Water Scalar

The water scalar is calculated on a monthly time step as a function of the ratio of estimated evapotranspiration to potential evapotranspiration:

$$W_{\varepsilon}(x,t) = 0.5 + \frac{0.5E(x,t)}{E_p(x,t)} \tag{44.17}$$

where $E(x, t)$ [mm] refers to the regional actual evapotranspiration, and $E_p(x, t)$ [mm] is the regional potential evapotranspiration.

$E(x, t)$ is calculated according to the regional real evapotranspiration model as follows [35]:

$$E(x,t) = \frac{P(x,t)\times R(x,t)\times\left[P^2(x,t)+R^2(x,t)+P(x,t)\times R(x,t)\right]}{\left[P(x,t)+R(x,t)\right]\times\left[P^2(x,t)+R^2(x,t)\right]} \tag{44.18}$$

where $P(x, t)$ [mm] is the monthly total precipitation at pixel x, and $R(x, t)$ [MJ/m^2] is the net solar radiation for the pixel x at month t. $R(x, t)$ is calculated by the empirical formula as follows:

$$R(x,t) = \left[E_{p0}(x,t)\times P(x,t)\right]^{0.5} \times \left\{0.369 + 0.598\times\left[\frac{E_{p0}(x,t)}{P(x,t)}\right]^{0.5}\right\} \tag{44.19}$$

where $E_{p0}(x,t)$ [mm] is the local potential evapotranspiration, which is related to the temperature for pixel x at time t. $E_{p0}(x, t)$ is calculated using the vegetation-climate model proposed by Thornthwaite [36]:

$$E_{p0}(x,t) = 16\times\left[\frac{10T(x,t)}{I(x)}\right]^{a(x)} \tag{44.20}$$

where $I(x)$ is the heat index from the sum of 12 months, and $a(x)$ is a constant that differs depending on the location. $I(x)$ and $a(x)$ are calculated as follows, respectively:

$$I(x) = \sum_{1}^{12}\left[\frac{T(x,t)}{5}\right]^{1.514} \tag{44.21}$$

$$a(x) = \left[0.6751\times I^3(x) - 77.1I^2(x) + 17{,}920I(x) + 492{,}390\right]\times 10^{-6} \tag{44.22}$$

$E_p(x, t)$ is calculated according to Boucher's complementary as follows:

$$E_p(x,t) = \frac{E(x,t) + E_{p0}(x,t)}{2} \tag{44.23}$$

The Maximum Light Use Efficiency

An important parameter of the CASA model is the maximum light use efficiency of vegetation which refers to the utilization efficiency of photosynthetically active radiation by vegetation under ideal conditions without any restrictions. The maximum light use efficiency can be assigned a fixed value when processing data corresponding to a small area [2]. Different types of vegetation have different ecological and physiological structures. Therefore, it directly affects the value calculated for the light use efficiency. Zhu et al. (2006) used remote sensing data and meteorological data combined with observed NPP data to simulate the maximum light use efficiency of typical vegetation types in China (Table 44.3) [31].

Linear Regression Analysis

The spatio-temporal variations in NPP were analyzed using a least-squares linear regression analysis method at the pixel and regional scales. In the method, the year was defined as the independent variable, and the NPP value was the dependent variable. The slope of the linear regression represents the magnitude of the trend in vegetation dynamics from 1982 to 2015. The slope of the trend is defined as follows:

$$\text{Slope} = \frac{n \times \sum_{i=1}^{n} i \times \text{NPP}_i - \left(\sum_{i=1}^{n} i\right)\left[\sum_{i=1}^{n} \text{NPP}_i\right]}{n\sum_{i=1}^{n} i^2 - \left(\sum_{i=1}^{n} i\right)^2} \tag{44.24}$$

where slope represents the trend in the NPP time series, n represents the cumulative number of years in the study period, i represents the number of the year, and NPP_i represents the NPP value in year i. The vegetation dynamics exhibit a positive trend if slope is greater than 0 and a negative trend if slope is less than 0.

Results and Discussion

Accuracy Verification and Evaluation

It is inevitable that uncertainty exists in the NPP estimation. Therefore, it is particularly important to verify the accuracy of NPP estimation results. Validation method directly compares the ground-measured NPP data with the model simulation values. Indirect verification compares the NPP estimates with previous studies, which can be used to confirm the reliability and stability of the model. Due to the lack of field observation data of NPP in this chapter, the indirect method is employed to verify the evaluation of NPP results of this study. The MOD17A3 product of MODIS has a good correlation with the measured NPP.

First, the calculated NPP was compared with the MOD17A3 product [37]. Figure 44.3 shows that the mean NPP values of all cases of FPAR in this chapter are significantly and positively correlated with that of the MOD17A3 data at pixel level, although the correlation coefficients are relatively low for all schemes of estimation FPAR. This indicated that the simulation results of all cases are reliable. Cases A and E exhibit the strongest correlations with the MOD17A3 data ($r^2 = 0.29$), whereas Case C apparently exhibits the weakest correlation with the MOD17A3 data ($r^2 = 0.18$). The FPAR calculation scheme

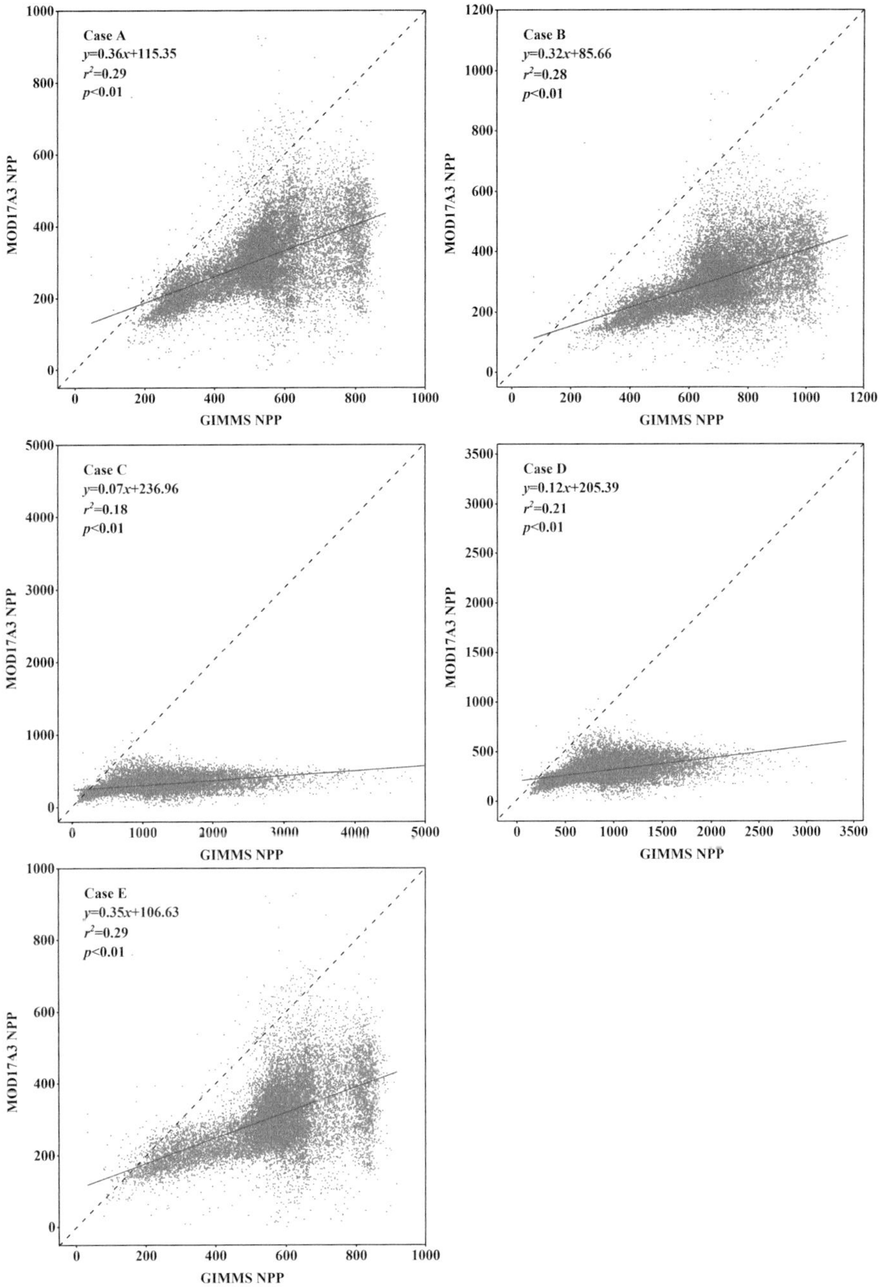

FIGURE 44.3 The corrections between NPP derived from GIMMS NDVI3g and MOD17A3 NPP product.

of MODIS was adopted in Case A. Therefore, Case A is the closest to MOD17A3 dataset. In addition, it is found that the NPP results derived from CASA model with direct involvement of the NDVI (such as Case A, Case B, and Case D) are similar to MOD17A3 in this study, because the FPAR of MOD17A3 data also depends on NDVI.

Change in FPAR leads to influence on the results of NPP estimation. The NPP values derived from various cases in this study are higher than those of MOD17A3. This finding is consistent with reported study in Northeast China, for which the NPP estimations in farmland, forest, grassland, shrubs, and desert were 19.1%, 8.0%, 1.3%, 7.9%, and 7.4% higher than those of MOD17A3, respectively [38]. Study compared the mean of the MOD17A3 NPP product data for the period of 2000–2006 with the ground-measured data at 128 forest sites in the study area, which found that there was a close correlation between MOD17A3 and observation; however, the ground-measured NPP was also higher than the MOD17A3 NPP [39].

Table 44.4 shows that the NPP estimation results derived from remote sensing data in Northeast China for different studies based on various models. Generally, the estimated results in this chapter are consistent with the mean NPP value range. The results of Cases A, B, and E are more similar to other studies, whereas the results of Case C evidently are higher. The NPP values derived from GIMMS NDVI data are significantly higher than those from MODIS data. Based on the above two methods of accuracy evaluation for the Northeast China, the FPAR model has a great impact on the estimation NPP.

Temporal Dynamics of NPP

The FPAR has a greater impact on the estimation of NPP values (Figure 44.4). The average annual NPP of all cases is 751.92 gC/m^2·a during 1982 and 2015 in Northeast China. The annual NPP values based on various FPAR schemes exhibit similar fluctuation trajectories at regional scale. For example, the NPP values in 1992 and 2003 are the smallest for all the cases. The lowest NPP value occurred in 1992 because of the Mount Pinatubo volcanic eruption, which reduced the sunshine by 10%, resulting in a 2-year volcanic winter. The dip in 2003 was due to the continuous decline in precipitation leading to successive years of drought starting in 2002 [1]. The NPP values of Cases A and E have the smallest interannual fluctuation with 15.50 and 15.62 of standard deviation, respectively. The mean interannual NPP values of Case A are the smallest (535.08), whereas those of Case C are the largest (1,074.74). The NPP results calculated from above two FPAR estimation methods are approximately twice as different. In addition, there is a non-significant trend among all cases for the NPP values in Northeast China during 1982 and 2015. Various estimations of FPAR schemes have little influence on the overall trend of NPP.

Spatial Patterns of NPP

This study calculated the average values of NPP for each case over the past 34 years (1982–2015). NPP values for various cases show a similar spatial pattern (Figure 44.5). The spatial distribution of NPP exhibits strong spatial variations, which is comprehensively affected by vegetation type, climate, and topography.

TABLE 44.4 Comparison of Other Results Based on Various Remote Sensing Dataset

Researcher	Remote Sensing data	Model	Study Area	Period	NPP (g C/m^2·a)
Case A	AVHRR NDVI3g	CASA	Northeast China	1982–2015	535.0821
Case B	AVHRR NDVI3g	CASA	Northeast China	1982–2015	696.4969
Case C	AVHRR NDVI3g	CASA	Northeast China	1982–2015	1,074.7405
Case D	AVHRR NDVI3g	CASA	Northeast China	1982–2015	885.6187
Case E	AVHRR NDVI3g	CASA	Northeast China	1982–2015	567.6671
Mao et al.	AVHRR NDVIg and MODIS	CASA	Northeast China	1982–2010	600–800
Zhang et al.	GIMMS	CASA	Northeast China Transect	1982–1999	58–811
Cheng et al.	MOD13Q1	CASA	Heilongjiang, Jilin, Liaoning	2001–2010	305
Zhao et al.	MOD15A2	GLOPEM-CEVSA	Northeast China	2000–2008	445
Li et al.	MODIS LAI	InTEC	Forest in Northeast China	2008	378
Mao et al.	MODIS LAI	BEPS	Forest in Northeast China	2012	369.92

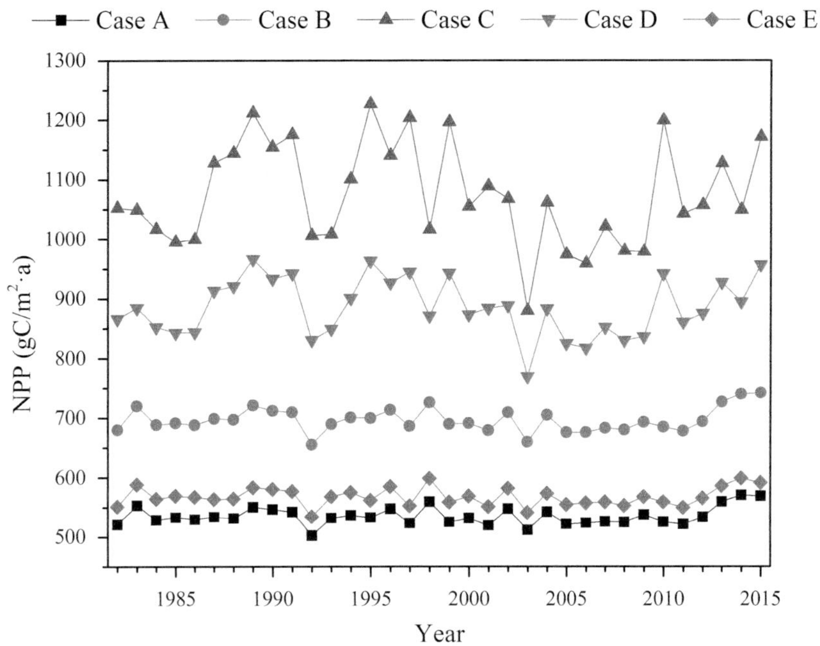

FIGURE 44.4 **(See color insert.)** Annual NPP value for various cases from 1982 to 2015 in Northeast China.

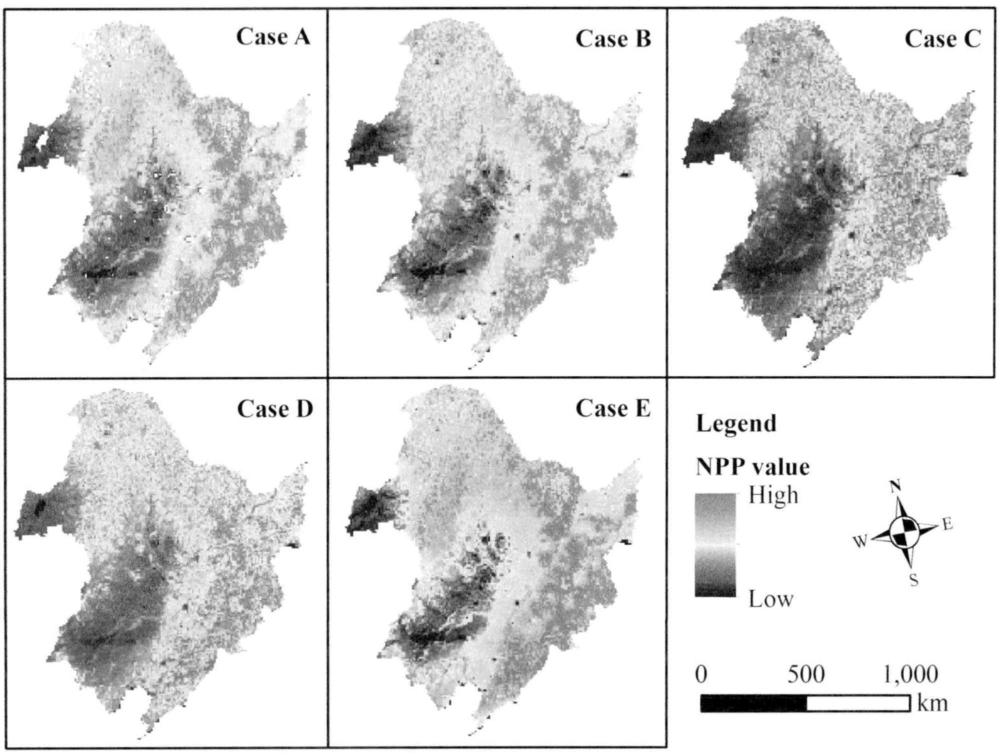

FIGURE 44.5 **(See color insert.)** Spatial patterns of NPP estimations based on various FPAR cases in Northeast China.

The NPP values of the Changbai Mountains, the Lesser Khingan Mountains, and the middle section of the Greater Khingan Mountains are relatively high at pixel level during the study period. These regions are dominated by deciduous broadleaved forest and temperate evergreen coniferous forest with adequate hydrothermal conditions. The northern part of Greater Khingan Mountains is dominated by deciduous coniferous forest; however, its NPP value is lower than that of the Changbai Mountains and Lesser Khingan Mountains due to limitation of temperature. Moreover, the regions with relatively low values of NPP include the Songnen Plain, Sanjiang Plain, and Liaohe Plain which are predominantly covered by cropland. The Hulun Buir Plateau are the primary grassland areas in Northeast China, characterized by an arid climate, less precipitation, long-term overgrazing and reclamation and a grassland desertification and salinization phenomenon; thus, the NPP values of this region are the lowest.

This study analyzed the average NPP value in 12 ecoregions in the study area for all cases (Table 44.5). According to the order of average NPP values, various cases exhibit similar ability of simulation in ecoregion IIA2, IIC1, and IIC4. The ecoregion IIA2 always has the highest NPP value for various schemes of FPAR. Meanwhile, ecoregion IIC1 and IIC4 generally have the lowest NPP values. However, the orders of average NPP values of different cases show a diversity for most of ecoregions. Obviously, the scheme of FPAR impacts on the result of NPP values in various ecoregions, which should be seriously discussed for particular ecoregion.

The linear trends of the NPP were calculated at the 95% confidence level in Northeast China at pixel level. Most of pixels show non-significant changes in study area. There were more areas with significant increasing pixels in NPP value than those with a significant decrease for all cases. In all cases, the pixel with a significant increasing trend in NPP value accounted for 7%–16% of the study area, and the pixel with a significant decreasing trend in NPP value accounted for 1%–4% of the study area (Table 44.6). The pixels exhibiting an increase in NPP value are predominantly occurred in the western Liaohe Plain. For Cases C and D, some pixels located on the Lesser Khingan Mountains also exhibited a significant increasing trend in NPP values. In contrast, the pixels exhibiting a significant decrease in NPP values

TABLE 44.5 Comparison of Average NPP Value for Various Cases in the Different Ecoregions

Ecoregions	Case A	Case B	Case C	Case D	Case E
IA1	566.26	670.30	1,077.44	873.87	580.63
IIA1	527.40	691.45	1,040.67	866.06	561.98
IIA2	670.40	844.17	1,572.65	1,208.41	697.92
IIA3	522.37	732.15	1,265.86	999.00	578.28
IIB1	427.10	593.34	666.52	629.93	474.82
IIB2	608.30	766.26	1,191.28	978.77	644.18
IIB2	443.76	483.38	460.65	472.01	401.46
IIB3	504.55	670.29	1,035.89	853.09	550.72
IIC1	318.24	460.24	339.24	399.74	332.02
IIC2	411.02	576.85	556.88	566.86	467.46
IIC2	362.50	521.45	413.79	467.62	403.88
IIC4	300.03	413.16	315.52	364.34	302.41
IIIA1	527.09	670.58	783.89	727.23	558.64
IIIB3	485.12	661.35	707.76	684.56	555.75

TABLE 44.6 Percentage of Pixels in Northeast China That Exhibited Positive and Negative Trends in NPP

Trend	Case A (%)	Case B (%)	Case C (%)	Case D (%)	Case E (%)
Positive	7	7	16	14	7
Negative	4	2	1	1	4

are mainly occurred in the Changbai Mountains. Furthermore, the amount of pixels exhibiting a significant increasing trend for Cases C and D are more than that of other cases. Possible reasons for the trend include the conversion of grasslands into cropland, agricultural management technology improvements, grassland protection, and reasonable grazing and other measures which provide favorable conditions for the growth of vegetation in this region. The Changbai Mountains are severely influenced by human activities. Land cover/use changes may be one of the primary contributing factors in this area. A decreasing trend in the NPP values is caused by the conversion of forest to cropland. Additionally, urbanization can also lead to a reduced NPP value.

Conclusions

The FPAR has a significant impact on the estimation results. The NPP values derived from various cases distinctly present different. The results in Cases A and E are the smallest NPP values among all cases, whereas those in Case C are the greatest. The NPP value derived from the FPAR estimated using the NDVI is less than that using the SR. The average NPP values of all cases in this study are higher than those of MOD17A3 at regional scale. Similar temporal and spatial variations in the NPP in Northeast China are presented for all cases in the study area. The overall annual NPP in Northeast China shows non-significant trend during the 34-year study period. The NPP in the western Liaohe Plain significantly increased for all case, whereas the NPP in the Changbai Mountains significantly decreased. The NPP in other areas remained unchanged. Previous studies have emphasized the effects of temperature- and water-induced stress on the NPP. However, the key information provided by remote sensing models is the radiance and reflectance of plants. Therefore, better FPAR models must be developed to improve estimation accuracy. When applying the CASA model in particular regions, other aspects should be considered. The ε_{max} is influenced more by the carbon capacity of plant species than by the overall type of vegetation. In addition, field observation values are needed from those areas corresponding to remote sensing images to obtain higher-accuracy results.

References

1. Zhu, Q.; Zhao, J.; Zhu, Z.; Zhang, H.; Zhang, Z.; Guo, X.; Bi, Y.; Sun, L. Remotely sensed estimation of net primary productivity (NPP) and its spatial and temporal variations in the greater Khingan mountain region, China. *Sustainability* 2017, 9, 1213.
2. Field, C.B.; Randerson, J.T.; Malmström, C.M. Global net primary production: Combining ecology and remote sensing. *Remote Sensing of Environment* 1995, 51, 74–88.
3. Jia, L.I.; Cui, Y.; Liu, J.; Shi, W.; Qin, Y. Estimation and analysis of net primary productivity by integrating MODIS remote sensing data with a light use efficiency model. *Ecological Modelling* 2013, 252, 3–10.
4. Field, C.B. Primary production of the biosphere: Integrating terrestrial and oceanic components. *Science* 1998, 281, 237–240.
5. Melillo, J.M.; McGuire, A.D.; Kicklighter, D.W.; Moore, B.; Vorosmarty, C.J.; Schloss, A.L. Global climate change and terrestrial net primary production. *Nature* 1993, 363, 234–240.
6. Running, S.W.; Nemani, R.R.; Heinsch, F.A.; Zhao, M; Reeves M.; Hashimoto, H. A continuous satellite-derived measure of global terrestrial primary production. *BioScience* 2004, 54, 547–560.
7. Crabtree, R.; Potter, C.; Mullen, R.; Sheldon, J.; Huang, S.; Harmsen, J.; Rodman, A.; Jean, C. A modeling and spatio-temporal analysis framework for monitoring environmental change using NPP as an ecosystem indicator. *Remote Sensing of Environment* 2009, 113, 1486–1496.
8. Ruimy, A.; Saugier, B.; Dedieu, G. Methodology for the estimation of terrestrial net primary production from remotely sensed data. *Journal of Geophysical Research: Atmospheres* 1994, 99, 5263–5283.

9. Box, E.O.; Holben, B.N.; Kalb, V. Accuracy of the AVHRR vegetation index as a predictor of biomass, primary productivity and net CO_2 flux. *Vegetatio* 1989, 80, 71–89.
10. Paruelo, J.M.; Epstein, H.E.; Lauenroth, W.K.; Burke, I.C. ANPP estimates from NDVI for the central grassland region of the United States. *Ecology* 1997, 78, 953–958.
11. Donmez, C.; Berberoglu, S.; Curran, P.J. Modelling the current and future spatial distribution of NPP in a Mediterranean watershed. *International Journal of Applied Earth Observation and Geoinformation* 2011, 13, 336–345.
12. Pachavo, G.; Murwira, A. Remote sensing net primary productivity (NPP) estimation with the aid of GIS modelled shortwave radiation (SWR) in a Southern African Savanna. *International Journal of Applied Earth Observation and Geoinformation* 2014, 30, 217–226.
13. Yu, D.; Shi, P.; Shao, H.; Zhu, W.; Pan, Y. Modelling net primary productivity of terrestrial ecosystems in East Asia based on an improved CASA ecosystem model. *International Journal of Remote Sensing* 2009, 30, 4851–4866.
14. Wu, S.; Yang, Q.; Zheng, D. Delineation of eco-geographic regional system of China. *Journal of Geographical Sciences* 2003, 13, 309–315.
15. Zhao, J.; Wang, Y.; Zhang, Z.; Zhang, H.; Guo, X.; Yu, S.; Du, W.; Huang, F. The variations of land surface phenology in Northeast China and its responses to climate change from 1982 to 2013. *Remote Sensing* 2016, 8, 400.
16. Jönsson, P.; Eklundh, L. TIMESAT—a program for analyzing time-series of satellite sensor data. *Computers & Geosciences* 2004, 30, 833–845.
17. Liu, J.; Kuang, W.; Zhang, Z.; Xu, X.; Qin, Y.; Ning, J.; Zhou, W.; Zhang, S.; Li, R.; Yan, C.; Wu, S. Spatiotemporal characteristics, patterns, and causes of land-use changes in China since the late 1980s. *Journal of Geographical Sciences* 2014, 24, 195–210.
18. Monteith, J.L. Solar radiation and productivity in tropical ecosystems. *Journal of Applied Ecology* 1972, 9, 747–766.
19. Monteith, J.L.; Moss, C.J. Climate and the efficiency of crop production in Britain. *Philosophical Transactions of the Royal Society of London. Series B, Biological Sciences* 1977, 281, 277–294.
20. Potter, C.S.; Randerson, J.T.; Field, C.B.; Matson, P.A.; Vitousek, P.M.; Mooney, H.A.; Klooster, S.A. Terrestrial ecosystem production: A process model based on global satellite and surface data. *Global Biogeochemical Cycles* 1993, 7, 811–841.
21. Bishop, J.K.B.; Rossow, W.B. Spatial and temporal variability of global surface solar irradiance. *Journal of Geophysical Research: Oceans* 1991, 96, 16839–16858.
22. Sellers, P.J.; Tucker, C.J.; Collatz, G.J.; Los, S.O.; Justice, C.O.; Dazlich, D.A.; Randall, D.A. A revised land surface parameterization (SiB2) for atmospheric GCMS. Part II: The generation of global fields of terrestrial biophysical parameters from satellite data. *Journal of Climate* 1996, 9, 706–737.
23. Sjöström, M.; Zhao, M.; Archibald, S.; Arneth, A.; Cappelaere, B.; Falk, U.; de Grandcourt, A.; Hanan, N.; Kergoat, L.; Kutsch, W.; Merbold, L. Evaluation of MODIS gross primary productivity for Africa using eddy covariance data. *Remote Sensing of Environment* 2013, 131, 275–286.
24. He, L. Study of the Spatio-temporal Variation Characteristics of the Solar Radiation in Growing Season over Northeast China. Master's Thesis, Northeast Normal University: Changchun, 2013.
25. Duffie, J.A.; Beckman, W.A. *Solar Engineering of Thermal Processes*; Wiley: New York, 1991.
26. Myneni, R.B.; Williams, D.L. On the relationship between FAPAR and NDVI. *Remote Sensing of Environment* 1994, 49, 200–211.
27. Myneni, R..; Hoffman, S.; Knyazikhin, Y.; Privette, J.; Glassy, J.; Tian, Y.; Wang, Y.; Song, X.; Zhang, Y.; Smith, G.; Lotsch, A. Global products of vegetation leaf area and fraction absorbed PAR from year one of MODIS data. *Remote Sensing of Environment* 2002, 83, 214–231.
28. Fensholt, R.; Sandholt, I.; Rasmussen, M.S. Evaluation of MODIS LAI, fAPAR and the relation between fAPAR and NDVI in a semi-arid environment using in situ measurements. *Remote Sensing of Environment* 2004, 91, 490–507.

29. Choudhury, B.J. Relationships between vegetation indices, radiation absorption, and net photosynthesis evaluated by a sensitivity analysis. *Remote Sensing of Environment* 1987, 22, 209–233.
30. Hatfield, J.; Asrar, G.; Kanemasu, E. Intercepted photosynthetically active radiation estimated by spectral reflectance. *Remote Sensing of Environment* 1984, 14, 65–75.
31. Zhu, W.; Pan, Y.; He, H.; Yu, D.; Hu, H. Simulation of maximum light use efficiency for some typical vegetation types in China. *Chinese Science Bulletin* 2006, 51, 457–463.
32. Sellers, P.J. Canopy reflectance, photosynthesis, and transpiration. II. The role of biophysics in the linearity of their interdependence. *Remote Sensing of Environment* 1987, 21, 143–183.
33. Smith, H. *Plants and the Daylight Spectrum*; Academic: San Diego, Calif, 1981.
34. Los, S.O.; Collatz, G.J.; Sellers, P.J.; Malmström, C.M.; Pollack, N.H.; Defries, R.S.; Bounoua, L.; Parris, M.T.; Tucker, C.J.; Dazlich, D.A. A global 9-yr biophysical land surface dataset from NOAA AVHRR data. *Journal of Hydrometeorology* 2000, 1, 183–199.
35. Lieth, H. Modeling the Primary Productivity of the World. In *Primary Productivity of the Biosphere*; Lieth, H., Whittaker, R.H., Eds.; Springer Berlin Heidelberg: Berlin, Heidelberg, 1975; pp. 237–263. ISBN 978-3-642-80913-2.
36. Thornthwaite, C.W. An approach toward a rational classification of climate. *Geographical Review* 1948, 38, 55–94.
37. Zhao, M.; Heinsch, F.A.; Nemani, R.R.; Running, S.W. Improvements of the MODIS terrestrial gross and net primary production global data set. *Remote Sensing of Environment* 2005, 95, 164–176.
38. Zhao, G.S.; Wang, J.B.; Fan, W.Y.; Ying, T.Y. Vegetation net primary productivity in Northeast China in 2000–2008: Simulation and seasonal change. *Chinese Journal of Applied Ecology* 2011, 22, 621–630.
39. Guo, Z.X.; Wang, Z.M.; Zhang, B.; Liu, D.W.; Yang, G.; Song, K.S.; Fang, L.I. Analysis of temporal-spatial characteristics and factors influencing vegetation NPP in Northeast China from 2000 to 2006. *Resources Science* 2008, 30, 1226–1235.

45

Optical Properties of Reflected Light from Leaves

Introduction 383
Theory 384
Samples and Measurements 385
Results 386
Comparison between BRF and IpRF of Leaves • Optical Properties of Light Reflected from Leaf Samples • Comparison with Model Results
Conclusions 390
References 390

Zhongqiu Sun
Northeast Normal University

Introduction

Optical properties of leaves are the essential for vegetation remote sensing applications, which allow us to investigate natural resources using remote sensing at different time and scales [1–3]. These applications are mainly based on the intensity signals obtained by the sensors. Light reflected from leaves should be described not only in terms of its intensity but also in terms of its polarization properties [4,5]; however, remote sensing has paid a little attention to the measured polarization results of land surfaces, such as leaves or vegetation covers [4]. Several researchers have pointed out that using polarization measurement results the reflected light of a leaf can be separated by two parts: polarized and non-polarized, the former derived from the first surface of leaf and the later primarily controlled by the internal tissue of leaf [6–8]. These investigations indicated that polarization measurements of leaves provided useful signals which cannot be measured directly from spectral or intensity measurements and showed that polarization had the potential to characterize vegetation covers [9–11].

Based on the polarization measurement results taken from airborne and spaceborne polarimeters [12,13], some models were proposed to calculate the polarization of vegetated land [14–17]. However, the focus of these models was on the contribution of land surfaces to the polarization of atmosphere. This is because the polarization of atmosphere is larger than that of natural surfaces. Even under this background, it is also necessary for measuring the polarization signals of vegetated surfaces, because polarization is possible to characterize the properties of leaves and vegetation covers [4,11,18–20]. Moreover, understanding the polarization of low boundary is also useful to investigate the impact on the retrieval of aerosol properties [21–25]. Though some airborne- and spaceborne-based measurements have been performed for vegetation, the development process from single leaf to the canopy from the ground-based study is necessary for polarized remote sensing [26–29]. The physical polarization properties of leaves can be measured in laboratory and field, and they are also important for characterizing the radiative transfer within the vegetation covers [30,31].

In this study, the focus is on the measurements of spectral intensity and polarization of leaves from different plant species at different viewing directions over the 2π space. The relationship between total intensity and *I* parameter, which is derived from the Stokes parameters, will be discussed, and we also investigate whether the combination of the BRF (bidirectional reflectance factor) model and BPRF (bidirectional polarized reflectance factor) model is suitable to simulate the Dolp (degree of linear polarization) of these leaves. These results can not only deepen our understanding of optical properties of leaves, but also confirm the significance of polarized signals in optical remote sensing.

Theory

The polarization properties of light reflected from leaves can be represented using the Stokes parameters (*I*, *Q*, *U*, and *V*); *V* is usually negligible for the reflection of natural surfaces [4]:

$$I = (L_{0^\circ} + L_{45^\circ} + L_{90^\circ} + L_{135^\circ})/2 \tag{45.1}$$

$$Q = L_{0^\circ} - L_{90^\circ} \tag{45.2}$$

$$U = L_{45^\circ} - L_{135^\circ} \tag{45.3}$$

$$I_p = \sqrt{Q^2 + U^2} \tag{45.4}$$

$$\text{Dolp} = \frac{\sqrt{Q^2 + U^2}}{I} \tag{45.5}$$

where L_x is the polarized reflected radiance from a surface at different polarizer directions, the reference plane for four polarizer directions (0°, 45°, 90°, and 135°) is taken to be the meridional plane of the instrument, I_p is the reflected polarized radiance from a sample, and Dolp is the degree of linear polarization. Theoretically, *I* parameter is equal to the total intensity of reflected radiance from leaf surface, and the total reflection can be defined by BRF [32]:

$$\text{BRF}(\lambda,\theta_s,\theta_v,\phi_s,\phi_v) = \frac{dL_{\text{sample}}(\lambda,\theta_s,\theta_v;\phi_s,\phi_v)}{dL(\lambda,\theta_s,\theta_v;\phi_s,\phi_v)}\rho_\lambda \tag{45.6}$$

where λ is the wavelength, θ_s is the solar zenith angle, θ_v is the viewing zenith angle, ϕ_s is the solar azimuth angle, ϕ_v is the viewing azimuth angle, dL_{sample} is the reflected radiance from leaf surface, and dL is measured from an ideal and diffuse surface. The Spectralon panel, whose hemispherical spectral reflectance (ρ_λ) was provided by the manufacturer, was used as a perfect Lambertian panel in this study. *I* parameter reflectance factor (IpRF) is similar to BRF, and it is defined as follows:

$$\text{IpRF}(\lambda,\theta_s,\theta_v;\phi_s,\phi_v) = \frac{dI_{sample}(\lambda,\theta_s,\theta_v;\phi_s,\phi_v)}{dL(\lambda,\theta_s,\theta_v;\phi_s,\phi_v)}\rho_\lambda \tag{45.7}$$

where dI_{sample} is the *I* parameter radiance of leaf. In the polarization measurements, a polarizer is in front of the fiber-optic cable of spectrometer, and its extinction ratio will reduce the measured radiance from leaf surface [33]. When the extinction ratio of polarizer is considered, the IpRF can be redefined as follows:

$$I = \left(\frac{L_{0^\circ,\text{measured}}}{1-\text{ext}_p} + \frac{L_{45^\circ,\text{measured}}}{1-\text{ext}_p} + \frac{L_{90^\circ,\text{measured}}}{1-\text{ext}_p} + \frac{L_{135^\circ,\text{measured}}}{1-\text{ext}_p}\right)\Big/2 \tag{45.8}$$

where ext_p is the extinction ratio of the polarizer, and $L_{x,\mathrm{measured}}$ is the directly measured polarized reflected radiance at different polarizer directions. The extinction of our polarizer has been calculated by the average difference between the polarized radiance at several polarization states and the reflected radiance of the Spectralon panel [33].

The polarization of leaf can be defined using BPRF [29,30]:

$$\mathrm{BPRF}(\lambda,\theta_s,\theta_v,\phi_s,\phi_v)=\frac{dI_p(\lambda,\theta_s,\theta_v;\phi_s,\phi_v)}{dL(\lambda,\theta_s,\theta_v;\phi_s,\phi_v)}\rho_\lambda \tag{45.9}$$

where dI_p is the polarized radiance from leaf surface. Based on the BRF that can be replaced by IpRF, the Dolp is described as follows:

$$\mathrm{Dolp}=\frac{\sqrt{Q^2+U^2}}{I}=\frac{\mathrm{BPRF}}{\mathrm{IpRF}}=\frac{\mathrm{BPRF}}{\mathrm{BRF}} \tag{45.10}$$

The detailed BRF model [34] and BPRF model [24] of leaf, which is used in this study, can be found in previous studies.

Samples and Measurements

The leaf samples, as shown in Figure 45.1, were taken from Balata (Ficus elastica Roxb. ex Hornem., see Figure 45.1a), Populus ussuriensis (Figure 45.1b), Apricot tree (Armeniaca vulgaris Lam. Figure 45.1c), Read twig dogwood (Cornus alba L., Figure 45.1d), and Grape (Vitis vinifera L., Figure 45.1e). The characteristics of Balata leaf are described as follows: shiny oval leaf, thick leathery leaf, a smooth and flat leaf surface, and a glabrous surface. The dimensions of this single leaf are 19 cm long and 11 cm wide. This sample has a special feature, i.e., a glabrous leaf, which avoids the effect of hair on the reflection property. The characteristics of Populus ussuriensis leaf are described as follows: shiny oval leaf, thin leathery leaf, a relative smooth and flat leaf surface with veins, and sparse pubescence along the leaf veins. The characteristics of Apricot leaf are described as follows: oval leaf, thin leathery leaf, and a

FIGURE 45.1 **(See color insert.)** (a) Balata leaf, (b) Populus ussuriensis leaf, (c) Apricot leaf, (d) Red twig dogwood leaf, and (e) Grape leaf; the main vein was parallel to the incident direction in the measurements.

relative smooth and flat leaf surface with minor veins. The characteristics of Red twig dogwood leaf are described as follows: ovate leaf, a relative flat leaf with veins and white pubescence on the leaf surface. The characteristics of Grape leaf are described as follows: ovate leaf and shallowly lobed surface, with sparse pubescence and veins.

In the measurement of optical properties of leaves, we used Northeast Normal University Laboratory Goniospectrometer System (NENULGS) to measure the reflection and polarization of leaves. Our measurement system is comprised of a goniometer, an artificial illumination, and an analytical Spectral Devices FieldSpec 3 (ASD FS3, Colorado, USA) spectroradiometer [35]. A calcite Glan–Thompson prism is in front of the fiber-optic cable when the polarization measurements are performed. The photometric and polarimetric measurements were obtained on the hemisphere above leaf samples using NENULGS. The interval of viewing zenith angle was 10°, the interval of azimuth angle was 15°, and the incident angle was 40°. Under these measurement geometries, the total measurement directions were 79.

Results

Comparison between BRF and IpRF of Leaves

In our previous study [33], the BRFs of some natural surfaces have been confirmed to be similar to IpRFs when the extinction ratio of polarizer was considered. The similarity between BRF and IpRF of leaves is also illustrated in this study. In Figure 45.2a, the BRF was larger than IpRF of Balata leaf in the wavelength range from 400 to 1000 nm when the extinction ratio of polarizer was not considered. However, the good match between BRF and IpRF was found when the extinction ratio of polarizer was considered, as shown in Figure 45.2b. Table 45.1 also shows the average relative differences between BRF and IpRF of leaves at selected wavelengths over all the measurement directions. The small difference values indicated that BRF can be replaced by IpRF of our leaf samples if the extinction ratio of polarizer was considered in the measurements. This finding also confirmed that both intensity and polarization properties of leaves can be derived from polarimetric measurements.

Optical Properties of Light Reflected from Leaf Samples

Figure 45.3 shows the Dolp of all the leaf samples at different viewing zenith angles in the principal plane, as computed using the I and Q, and it is found that the maximum Dolp of the Balata leaf (in Figure 45.3a) does not occur in the specular plane ($\theta s = \theta v = 40°$) and is instead observed at a viewing zenith angle of 50°. This is because Dolp of the sample is not only derived by the I parameter but also

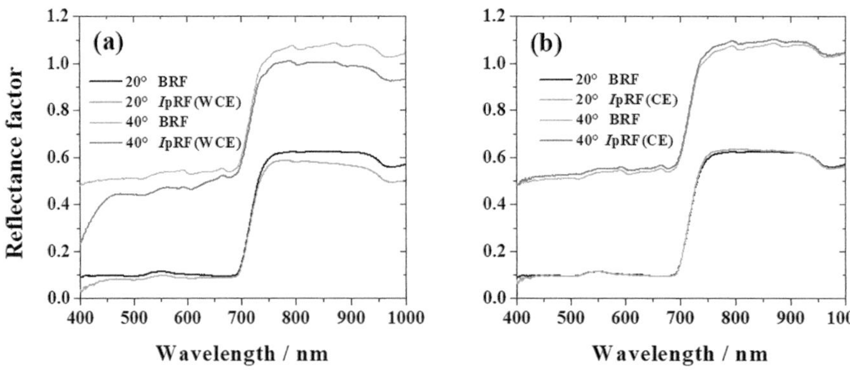

FIGURE 45.2 **(See color insert.)** Comparison between the BRF and IpRF of the Balata leaf in the forward scattering directions in the principal plane: (a) without considering the extinction (WCE) of the polarizer, and (b) considering the extinction (CE) of the polarizer, the illumination zenith angle was 40°.

TABLE 45.1 Absolute Value of Relative Difference (BRF−IpRF)/BRF between BRF and IpRF of Our Leaf Samples at Selected Wavelengths over All the Measurement Directions

Incident Angle (°)	Samples	560 nm	670 nm	865 nm
40	Balata	0.018	0.019	0.017
40	Populus ussuriensis	0.022	0.03	0.029
40	Apricot	0.03	0.037	0.011
40	Red twig dogwood	0.018	0.04	0.016
40	Grape	0.03	0.041	0.022

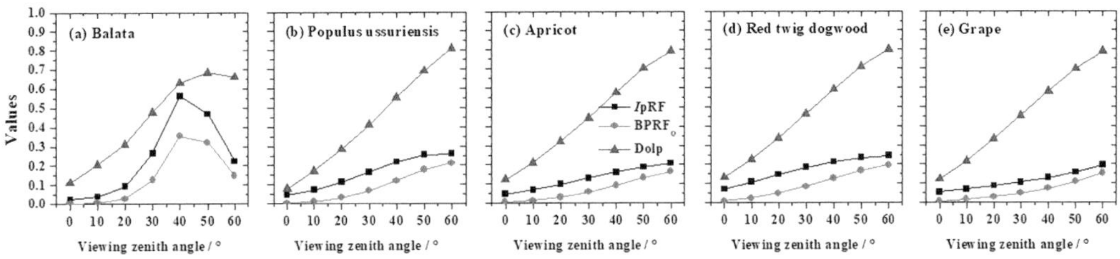

FIGURE 45.3 $BPRF_Q$ (circles) (which is the same as Equation (15.9), but the numerator is the *Q* parameter); IpRF (squares); and Dolp (triangles, $-Q/I$) of all the leaf samples at 670 nm at different viewing zenith angles in the forward scattering direction in the principal plane when the incident zenith angle was 40°. The smooth surface generates a strong specular reflection, as shown in (a); the relative smooth leaf surface generates a weak specular reflection, as shown in (b); and the rough leaf surfaces broaden the specular lobe and the increase in diffuse scattering, as shown in (c)–(e).

related to the *Q* parameter. Although the maximum values of *I* and *Q* parameters of Balata leaf were observed in the specular direction ($\theta s = \theta v = 40°$), their maximum derivative was not located in the same viewing direction. This result means that the BPRF has a more direct relationship with the polarization property of the sample than does the Dolp. In Figure 45.3b, the maximum Dolp and the maximum IpRF do not occur in the same location, such as the former at 60° and the later at 50°. In Figure 45.3c and d, the same phenomenon is clear, i.e., Dolp, IpRF, and $BPRF_Q$ increase with the increase in viewing zenith angle. The different reflection characteristics of leaves attributed to the physical properties of leaves are numerous [30,34]. These results indicate that the combination of IpRF (or BRF) and BPRF is essential for characterizing the optical properties of reflected light from a leaf.

Comparison with Model Results

Our results have confirmed the BRF can be replaced by IpRF, which means that Dolp is the ratio of BPRF to IpRF of leaves. If the BPRF and IpRF can be modeled using appropriate models, the Dolp is also calculated by the combination of polarimetric and photometric models [30,36]. To further ensure that our measured BRF is effective for replacing by IpRF of Balata leaf, both of them are used to invert the BRF model parameters. It is clear that the parameters inverted by the BRFs and *I*pRFs are similar, as shown in Table 45.2, which lists the best fitted parameters of the BRF model for the BRF and IpRF at selected wavelengths. Because the BRF has been proposed to be replaced by IpRF, our main focus is on the comparison between the modeled and measured IpRF of these leaves. The values of model parameters and RSME of other leaf samples are shown in Table 45.3. Except for the results at 670 nm, the average relative difference (ARD) at other selected wavelengths is small. This is because the BRF values at 670 nm are smaller than other wavelengths, and a small change in BRF will lead to a relatively large change in ARD. An interesting phenomenon is that the inverted refractive index of the leaf, *n*,

TABLE 45.2 Best Fitted Parameters of the BRF Model [34], Which Are Inverted by the BRF and IpRF, for Balata Leaf at Selected Wavelengths

IZA (°)	Reflectance Factor	Wavelength (nm)	n	$k_l(\lambda)$	δ	RMSE	ARD
40	BRF	560	1.459	0.0352	0.1676	0.02	0.13
		670	1.469	0.0181	0.1655	0.019	0.22
		865	1.410	0.5721	0.1429	0.02	0.02
	IpRF	560	1.461	0.0349	0.1657	0.017	0.13
		670	1.472	0.0175	0.1645	0.018	0.21
		865	1.411	0.5718	0.1431	0.02	0.021

IZA corresponds to the incident zenith angle, and ARD represents the absolute values of average relative difference between the model and measured results over all the viewing directions.

TABLE 45.3 Best Fitted Parameters of the BRF Model, Which Are Inverted by the IpRF, for Other Leaves at Selected Wavelengths

IZA (°)	Samples	Wavelength (nm)	n	$k_l(\lambda)$	δ	RMSE	ARD
40	Populus ussuriensis	560	1.6166	0.0777	0.3222	0.006	0.037
		670	1.5775	0.0166	0.3176	0.006	0.079
		865	1.6013	0.4418	0.3269	0.010	0.016
	Apricot	560	1.4915	0.1305	0.2643	0.014	0.054
		670	1.4807	0.0219	0.2647	0.014	0.155
		865	1.4685	0.4420	0.2601	0.014	0.022
	Red twig dogwood	560	1.4450	0.1281	0.3134	0.011	0.044
		670	1.4420	0.0210	0.3233	0.010	0.115
		865	1.3610	0.4666	0.2971	0.011	0.019
	Grape	560	1.4368	0.1366	0.2799	0.010	0.042
		670	1.4413	0.0212	0.2863	0.011	0.134
		865	1.4450	0.4565	0.2964	0.011	0.018

IZA corresponds to the incident zenith angle, and ARD represents the absolute values of average relative difference between the model and measured results over all the viewing directions.

as shown in Tables 45.2 and 45.3, which varied from 1.368 to 1.6166 with the change in wavelength, is in the range of refractive indexes of leaves used and retrieved by previous studies [34,37,38], and then, the inverted refractive index of leaves at the selected wavelength is used as the input parameter in the BPRF model.

The best fitted parameters of the BPRF model are shown in Table 45.4. The small value of RMSE and the comparison between the modeled and measured BPRFs in Figure 45.4 appear to indicate that the model results match well with our measurements. However, the values of ARD at all the selected wavelengths exceed 0.3, because the BPRF is small (10–3) in the backward scattering direction; a slight change causes a great difference. It is also found that the inverted roughness parameter (σ) described in the BPRF model (in Table 45.4) has the similar change trend to the roughness parameter (δ) which is given in Tables 45.2 and 45.3. The parameter value increases with the samples from Balata, Apricot, Grape, Red twig dogwood to Populus ussuriensis, which indicates that the BRF and BPRF models have the potential to describe the structure of the leaf surface.

Based on the comparison between modeled and measured results, we find that the photometric and polarimetric models are effective for simulating the IpRFs and BPRFs of the leaves, respectively. Then, the modeled Dolp, which is the combination of the modeled IpRF and BPRF, is compared with the measured Dolp of leaves, as shown in Figure 45.5. It is found that there is a relatively good match with the modeled Dolp when the measured Dolp of leaf is large. This is because Dolp is derived from the IpRF

TABLE 45.4 Best Fitted Parameters of the BPRF Model for the Single Leaf at Selected Wavelengths

Samples	Wavelength (nm)	$a\lambda$	b	k	Σ	ζ	RMSE	ARD
Balata	560	0.0012	6.254	−6.009	0.1234	5.87	0.009	0.53
	670	0.0019	6.107	−5.229	0.1246	6.74	0.009	0.57
	865	0.0740	0.120	−0.449	0.1087	6.77	0.010	0.55
Populus ussuriensis	560	0.9750	4.084	2.805	0.1972	−3.35	0.003	0.37
	670	0.8751	4.171	2.947	0.1976	−3.58	0.003	0.31
	865	0.1532	0.695	0.509	0.1776	−3.24	0.004	0.43
Apricot	560	2.4300	6.584	3.691	0.1344	−1.51	0.003	0.47
	670	2.1810	6.660	3.678	0.1366	−1.71	0.002	0.36
	865	0.0785	0.393	−0.274	0.1589	−4.50	0.003	0.47
Red twig dogwood	560	0.2841	2.696	1.477	0.1909	−4.27	0.003	0.33
	670	0.2715	2.882	1.582	0.1906	−3.81	0.002	0.40
	865	0.1029	1.049	0.686	0.1858	−4.32	0.002	0.45
Grape	560	0.4405	3.355	1.852	0.1724	−3.62	0.002	0.36
	670	0.5163	3.863	2.128	0.173	−3.43	0.002	0.43
	865	0.0986	0.746	0.227	0.1691	−3.74	0.002	0.34

The incident zenith angle is 40°, and ARD is the average relative difference between measured and modeled results in all viewing directions. The input refractive index of the leaves at selected wavelength is retrieved by the BRF model.

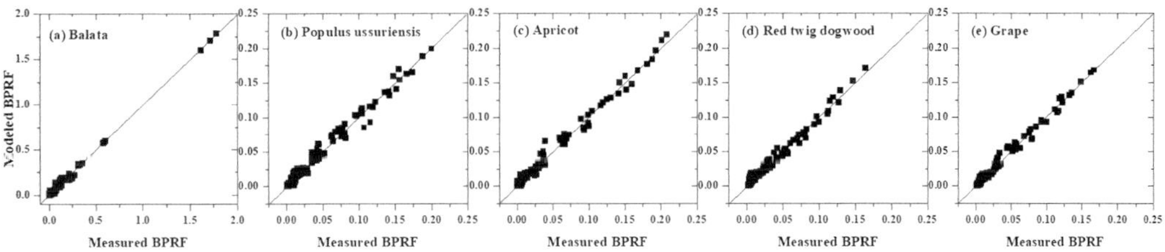

FIGURE 45.4 Comparison between the measured and modeled BPRFs of the leaves at all wavelengths, and the illumination angle was 40° for other leaf samples.

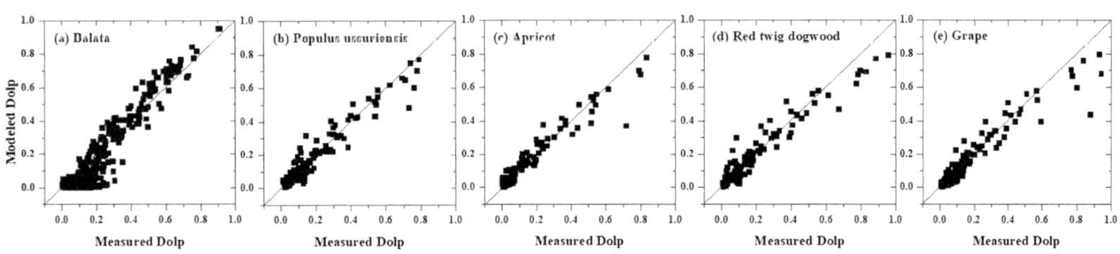

FIGURE 45.5 Comparison between the measured and modeled Dolp of all the leaves at all the wavelengths, the illumination angle was 40° for leaf samples.

and BPRF, and the good match between the model results and measurements locates in the forward scattering directions for the two parameters. The results indicate that the combination of the polarimetric and photometric models is effective in simulating Dolp of our single leaves in the forward scattering directions.

Conclusions

This study describes the photometric and polarimetric properties of leaves from different plant species under laboratory condition. We confirmed that BRF can be replaced by IpRF of leaves if the extinction ratio of polarizer was considered. Moreover, the photometric and polarimetric measurements can be simulated using the BRF and BPRF models, and the combination of modeled results is suitable for calculating Dolp of leaves. Based on these findings, we concluded that both the intensity and polarization can be derived from polarimetric measurements (Stokes parameters), the IpRF (similar to BRF), BPRF, and Dolp, which deepens the understanding of the optical properties of leaves. Moreover, polarimetric measurements should be considered in the remote sensing applications in the future, because non-polarized part of reflected light of leaves derives from the interior of the leaves, which may have stronger relationship with biochemical properties of leaves than with total reflection. Further studies of polarization of leaves should be performed to confirm this idea.

References

1. Schweiger, A.K.; Cavender-Bares, J.; Townsend, P.A.; Hobbie, S.E.; Madritch, M.D.; Wang, R.; et al. Plant spectral diversity integrates functional and phylogenetic components of biodiversity and predicts ecosystem function. *Nature Ecol. Evol.* 2018, 2, 976–982.
2. Asner, G.P.; Scurlock, J.M.O.; Hicke, J.A. Global synthesis of leaf area index observations: implications for ecological and remote sensing studies. *Global Ecol. Biogeogr.* 2003, 12, 191–205.
3. Tucker, C.J.; Sellers, P.J. Satellite remote sensing of primary production. *Int. J. Remote Sens.* 1986, 7, 1395–1416.
4. Talmage, D.A.; Curran, P.J. Remote sensing using partially polarized light. *Int. J. Remote Sens.* 1986, 7, 47–64.
5. Tyo, J.S.; Goldstein, D.L.; Chenault, D.B.; Shaw, J.A. Review of passive imaging polarimetry for remote sensing applications. *Appl. Opt.* 2006, 45, 5453–5469.
6. Grant, L. Diffuse and specular characteristics of leaf reflectance. *Remote Sens. Environ.* 1987, 22, 309–322.
7. Grant, L.; Daughtry, C.S.T.; Vanderbilt, V.C. Polarized and non-polarized leaf reflectances of coleus blumei. *Environ. Exp. Bot.* 1987, 24, 139–145.
8. Vanderbilt, V.C.; Grant, L.; Daughtry, C.S.T. Polarization of light scattered by vegetation. *Proc. IEEE* 1985, 73, 1012–1024.
9. Woessner, P.; Hapke, B. Polarization of light scattered by clover. *Remote Sens. Environ.* 1987, 21, 243–261.
10. Rondeaux, G.; Herman, M. Polarization of light reflected by crop canopies. *Remote Sens. Environ.* 1991, 38, 63–75.
11. Vanderbilt, V.C.; Grant, L.; Biehl, L.L.; Robinson, B.F. Specular, diffuse, and polarized light scattered by two wheat canopies. *Appl. Optics* 1985, 24, 2408–2418.
12. Cairns, B.; Russell, E.E.; Travis, L.D. The Research Scanning Polarimeter: calibraton and ground-based measurements. In *Proceedings of SPIE*, 1999, 3754.
13. Deschamps, P-Y.; Bréon, F.M.; Leroy, M.; Podaire, A.; Bricaud, A.; Burize, J-C.; et al. The POLDER mission: instrument characteristics and scientific objectives. *IEEE Trans. Geosci. Remote Sens.* 1994, 32, 598–615.
14. Bréon, F.M.; Tanré, D.; Lecomte, P.; Herman, M. Polarized reflectance of bare soils and vegetation: measurements and models. *IEEE Trans. Geosci. Remote Sens.* 1995, 33, 487–499.
15. Litvinov, P.; Hasekamp, O.; Cairns, B. Models for surface reflection of radiance and polarized radiance: comparison with airborne multi-angle photopolarimetric measurements and implications for modeling top-of-atmosphere measurements. *Remote Sens. Environ.* 2011, 115, 781–792.

16. Nadal, F.; Bréon, F.M. Parameterization of surface polarized reflectance derived from POLDER spaceborne measurements. *IEEE Trans. Geosci. Remote Sens.* 1999, 37, 1709–1718.
17. Maignan, F.; Bréon, F-M.; Fédèle, E.; Bouvier, M. Polarized reflectances of natural surfaces: spaceborne measurements and analytical modeling. *Remote Sens. Environ.* 2009, 113, 2642–2650.
18. Vanderbilt, V.C.; Grant, L. Plant canopy specular reflectance model. *IEEE Trans. Geosci. Remote Sens.* 1985, 23, 722–730.
19. Peltoniemi, J.I.; Gritsevich, M.; Puttonen, E. Reflectance and polarization characteristics of various vegetation types. In *Light Scattering Reviews 9: Light Scattering and Radiative Transfer Berlin*, Heidelberg: Springer: 2015, 257–294.
20. Suomalainen, J.; Hakala, T.; Puttonen, E.; Peltoniemi, J. Polarised bidirectional reflectance factor measurements from vegetated land surfaces. *J. Quant. Spectrosc. Radiat. Transfer* 2009, 110, 1044–1056.
21. Waquet, F.; Leon, J-F.; Cairns, B.; Goloub, P.; Deuzé, J.L.; Auriol, F. Analysis of the spectral and angular response of the vegetated surface polarization for the purpose of aerosol remote sensing over land. *Appl. Optics* 2009, 48, 1228–1236.
22. Litvinov, P.; Hasekamp, O.; Cairns, B.; Mishchenko, M. Reflection models for soil and vegetation surfaces from multiple-viewing angle photopolarimetric measurements. *J. Quant. Spectrosc. Radiat. Transfer* 2010, 111, 529–539.
23. Litvinov, P.; Hasekamp, O.; Dubovik, O.; Cairns, B. Model for land surface reflectance treatment: physical derivation, application for bare soil and evaluation on airborne and satellite measurements. *J. Quant. Spectrosc. Radiat. Transfer* 2012, 113, 2023–2039.
24. Diner, D.; Xu, F.; Martonchik, J.; Rheingans, B.; Geier, S.; Jovanovic, V.; et al. Exploration of a polarized surface bidirectional reflectance model using the ground-based multiangle spectropolarimetric imager. *Atmos.* 2012, 3, 591–619.
25. Breon, F-M.; Maignan, F. A BRDF–BPDF database for the analysis of Earth target reflectances. *Earth Syst. Sci. Data.* 2017, 9, 31–45.
26. Kallel, A. Leaf polarized BRDF simulation based on Monte Carlo 3-D vector RT modeling. *J. Quant. Spectrosc. Radiat. Transfer* 2018, 221, 202–224.
27. Suomalainen, J.; Hakala, T.; Peltoniemi, J.; Puttonen, E. Polarised multiangular reflectance measurements using the finnish geodetic institute field goniospectrometer. *Sensors* 2009, 9, 3891–3907.
28. Sun, Z.; Wu, D.; Lv, Y.; Zhao, Y. Bidirectional polarized reflectance factors of vegetation covers: influence on the BRF models results. *IEEE Trans. Geosci. Remote Sens.* 2017, 55, 5687–5701.
29. Sun, Z.; Wu, D.; Lv, Y.; Zhao, Y. Polarized reflectance factors of vegetation covers from laboratory and field: a comparison with modeled results. *J. Geophys. Res.: Atmos.* 2017, 122, 1042–1065.
30. Sun, Z.; Wu, D.; Lv, Y.; Lu, S. Optical properties of reflected light from leaves: a case study from one species. *IEEE Trans. Geosci. Remote Sens.* 2019, 57, 4388–4406.
31. Sun Z.; Peng Z.; Wu, D.; Lv, Y. Photopolarimetric properties of leaf and vegetation covers over a wide range of measurement directions. *J. Quant. Spectrosc. Radiat. Transfer* 2018, 206, 273–285.
32. Schaepman-Strub, G.; Schaepman, M.E.; Painter, T.H.; Dangel, S.; Martonchik, J.V. Reflectance quantities in optical remote sensing—definitions and case studies. *Remote Sens. Environ.* 2006, 103, 27–42.
33. Sun, Z.; Huang, Y.; Bao, Y.; Wu, D. Polarized remote sensing: a note on the Stokes parameters measurements from natural and manmade targets using a spectrometer. *IEEE Trans. Geosci. Remote Sens.* 2017, 55, 4008–4021.
34. Bousquet, L.; Lachérade, S.; Jacquemoud, S.; Moya, I. Leaf BRDF measurements and model for specular and diffuse components differentiation. *Remote Sens. Environ.* 2005, 98, 201–211.
35. Sun, Z.; Wu, Z.; Zhao, Y. Semi-automatic laboratory goniospectrometer system for performing multi-angular reflectance and polarization measurements for natural surfaces. *Rev. Sci. Instrum.* 2014, 85, 014503.

36. Sun, Z.; Wu, D.; Lv, Y. Photopolarimetric properties of a manmade target over a wide range of measurement directions. *Opt. Express* 2017, 25, A85–A100.
37. Woolley, J.T. Reflectance and transmittance of light by leaves. *Plant Physiol.* 1971, 47, 656–662.
38. Comar, A.; Baret, F.; Obein, G.; Simonot, L.; Meneveaux, D.; Viénot, F.; et al. ACT: a leaf BRDF model taking into account the azimuthal anisotropy of monocotyledonous leaf surface. *Remote Sens. Environ.* 2014, 143, 112–121.

46

Remote Sensing of CO_2 Emissions from Wildfires

Introduction 393
GOSAT Data 394
Method 394
Case Study 394
Wildfire Smoke Plume Areas 395
ΔX_{CO2} Retrieval and Simulation 397
Mean Height of Wildfire Smoke Plumes 398
CO_2 Released from Wildfires 398
Compared with BBM and ERs Methods 398
Conclusions 400
References 400

Meng Guo
Northeast Normal University

Introduction

Wildfires have been considered an important reason of the interannual variability of CO_2, but an accurate assessment of global CO_2 emissions from wildfires is necessary to confirm this hypothesis (Simmonds et al., 2005). At present, the amount of CO_2 released from burning of biomass, especially from wildfires, is still uncertain because CO_2 emissions depend not only on the fuel type and combustion phase but also on the season, the wind speed, and other factors. These uncertainties make it a challenge to accurately quantify the CO_2 released from forest fires.

Seiler and Crutzen (1980) proposed the biomass burning model (BBM) to calculate greenhouse gas emissions from forest fires for the first time. Reported studies improved the model and obtained satisfactory results at different sites (Amiro et al., 2010; Dixon and Krankina, 1993; Goto and Suzuki, 2013; Robinson, 1989). The aboveground fuel load, combustion efficiency, and emission factors are the major parameters in the BBM. All of these parameters change with species and vegetation type, which makes it difficult to accurately estimate and finally determine CO_2 emissions (Rosa et al., 2011). Because of the advantages of remote sensing technology in obtaining data and repeating observations in large geographical regions, remote sensing can be widely used in estimating CO_2 emissions from wildfires. Greenhouse gas-monitoring satellites such as GOSAT (Greenhouse Gases Observing Satellite), OCO-2 (Orbiting Carbon Observatory-2), and TanSat (CarbonSat) can play a key role in tracking biomass burning emissions, which will help better understand the contributions of biomass burning to regional atmospheric composition and climate change. This study tried to use GOSAT data to calculated CO_2 emissions from the 2010 Russia wildfires.

GOSAT Data

GOSAT was launched on January 23, 2009, from Tanegashima Island, Japan, carrying two main instruments, including Thermal and Near-infrared Sensor for Carbon Observation Fourier Transform Spectrometer (TANSO-FTS) and Cloud and Aerosol Imager (CAI). FTS has four bands: three of short-wave infrared (SWIR) (approximately 0.76, 1.6, and 2.0 μm) with a ground pixel size of about 80 km^2 that are used to estimate the columnar concentrations of CO_2; and one TIR band (5.5 ~ 14.3 μm), which is used to describe the vertical profile of CO_2 concentration in the upper troposphere. CAI imagery has four bands (0.380, 0.675, 0.870, and 1.600 μm) with a spatial resolution of 500 m and has been mainly used to map the characteristics of clouds and aerosols from space.

GOSAT is the world's first satellite that was designed to estimate CO_2 and CH_4 concentrations (X_{CO2} and X_{CH4}, in ppm) from SWIR bands, with a return cycle of 3 days (Guo et al., 2012; Oguma et al., 2011).

Method

As noted above, the GOSAT X_{CO2} data have units of concentration in ppm, and it was necessary to convert these numbers to density values (mg•m^{-3}) using Equation (46.1):

$$\mathrm{mg} \bullet \mathrm{m}^{-3} = M / 22.4 \times \mathrm{ppm} \times \left[273/(273+T)\right] \times (\mathrm{Ba}/1{,}013.25), \quad (46.1)$$

where M is the molecular weight of CO_2, and ppm is the concentration of CO_2, T is the temperature (°C), and Ba is the atmospheric pressure (hPa).

This study calculated the height of a smoke plume (H) using the isotherm pressure equation proposed by Wu (1990):

$$H = \left(R \cdot T / g \cdot M\right) \cdot \ln\left(P_0 / P\right) - h, \quad (46.2)$$

where R is the universal gas constant (8.31 J mole^{-1} K^{-1}), T is the thermodynamic temperature (K), g is the gravitational acceleration (9.8 m s^{-2}), P_0 is the standard atmospheric pressure (1,013.25 hPa), P is the desired atmospheric pressure (hPa), and h (km) is the height of land surface.

The CO_2 *emissions* (E_{CO2}) for a smoke plume were calculated by

$$E_{CO2} = \rho \cdot V = \mathrm{mg} \cdot \mathrm{m}^{-3} \cdot S \cdot H, \quad (46.3)$$

where S is the area of a wildfire smoke plume.

Case Study

In summer 2010, intense wildfires occurred in western Russia near Moscow due to high temperatures and the absence of precipitation (Chubarova et al., 2012; Witte et al., 2011). Based on the identification of smoke plumes from forest fires using GOSAT CAI and other previously investigated regions, this study was restricted to the region between 45°N–63°N and 32°E–70°E. Further, this study focused only on the time that serious fires occurred, from July 22 to August 18, 2010, which was the same as Witte et al. (2011).

Wildfire Smoke Plume Areas

GOSAT CAI L1B+ data were used to identify the smoke plumes that were released from Russian forest fires in 2010. Figure 46.1 shows the example of smoke plumes on August 8, 2010, which could be identified efficiently from the image. This study defined a threshold for band 1 to obtain the approximate smoke plume areas (Figure 46.1c). Based on a visual inspection, a region of interest (ROI) was built to restrict the smoke plumes to obtain the final forest fire smoke plumes (Figure 46.1d). Using the same method, smoke plumes for each wildfire day were identified. Elevated radiance values of the CAI band 1 indicated a higher concentration of smoke plumes and more CO_2 emissions. Finally, because of the three-day revisit interval, this study had to examine a mosaic 3 days of smoke plumes to obtain the plume areas of the entire region.

Figure 46.2 shows the three-day summed smoke plume distribution and areas. The smoke plume areas increased from July 23 to August 9 because of the dramatic growth of the fire points and the burned areas, and then as the wildfires came under control, the smoke plume areas decreased. In total, the smoke plume areas reached about 1,781 million hectares for the study region in 2010.

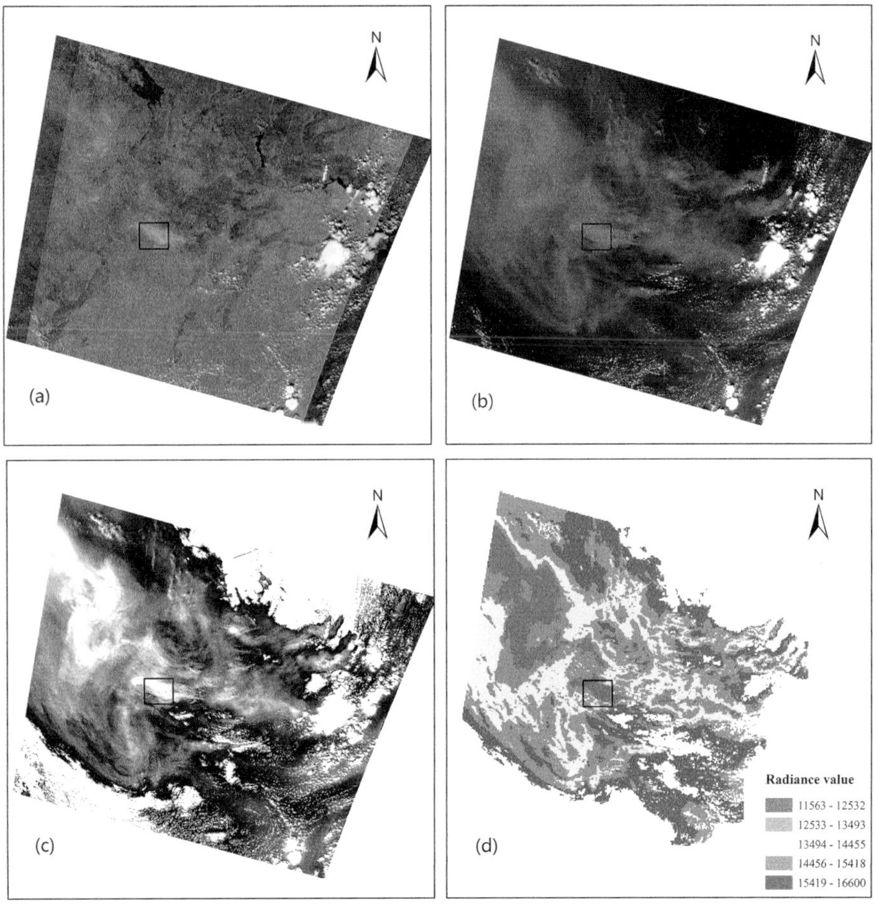

FIGURE 46.1 **(See color insert.)** Example to identify smoke plumes on August 8, 2010. CAI image with band 431 (a), CAI image with band 1 (b), smoke plume areas identified by using thresholds of band 1 (c), and a subset of smoke plumes using ROI (d) (Guo et al. 2017).

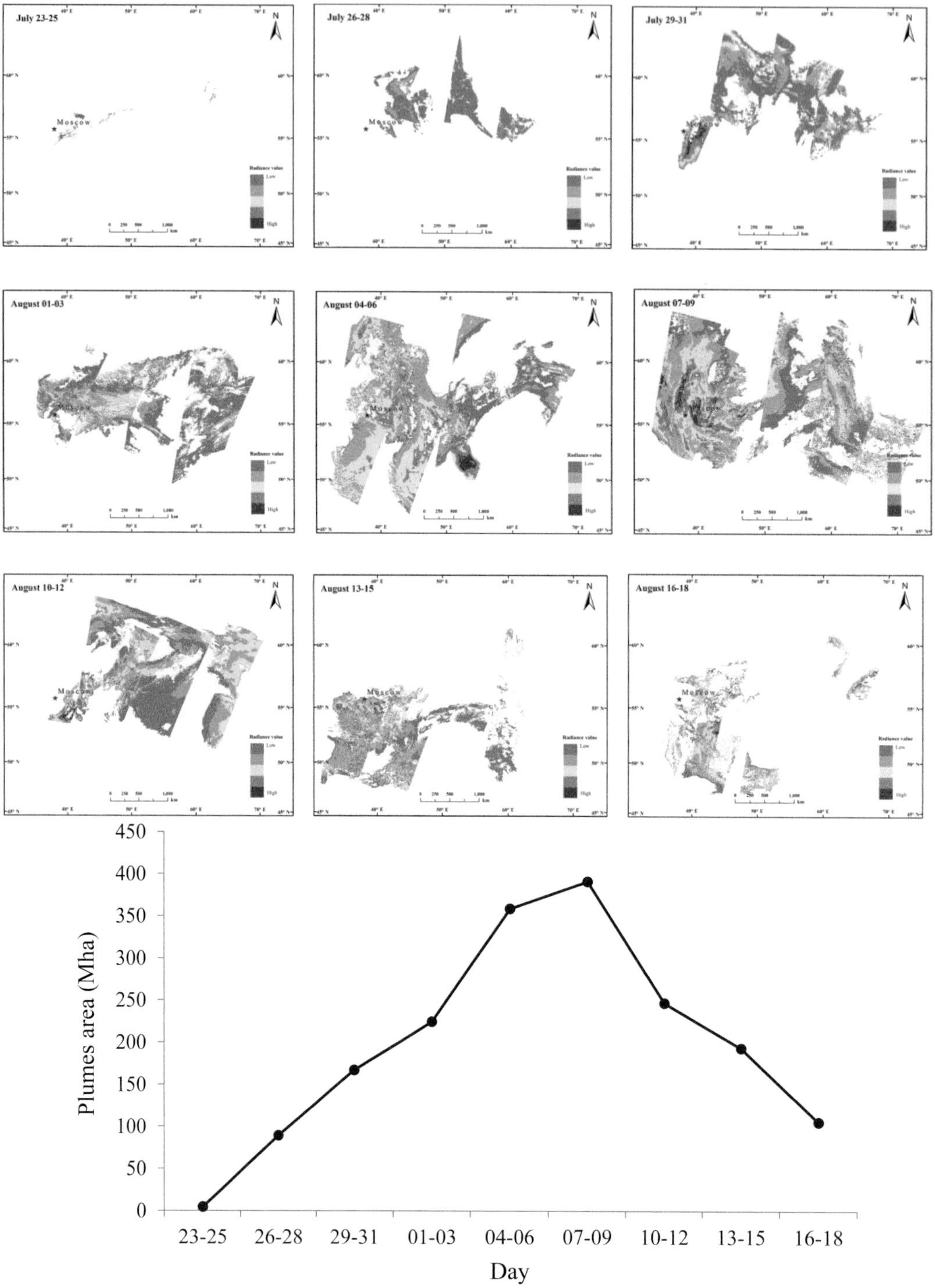

FIGURE 46.2 (See color insert.) Three-day smoke plume areas in the study region. The preceding nine figures show the distribution of three-day smoke plumes, and the last figure shows the smoke plume area in million hectares (23–25 of *X*-axis means July 23–25 and 1–3 means August 1–3). Here, the three-day description means the sum of 3 days but without overlapping regions (Guo et al. 2017).

ΔX_{CO2} Retrieval and Simulation

Referring to the intermediate results of Ross et al. (2013), this study found the FTS L1B points that contained wildfire smoke plumes as shown in Table 46.1. SX_{CO2} is the X_{CO2} values that were exposed to smoke plumes, and BX_{CO2} is the background values. The background points must be on the same day and not far from the points exposed to smoke plumes, and sometimes, BX_{CO2} is the mean value of more than one point. This study assumed that an increase in the CO_2 concentration was the result of wildfire emissions alone. In total, 25 FTS L1B point data were observed during the study duration. Based on artificial identification, this study classified the smoke plumes as light smoke, smoke, and thick smoke, and we found that thick smoke usually corresponded to higher ΔX_{CO2} values.

Table 46.1 shows the different smoke classifications that corresponded to different ΔX_{CO2} values. This study used a linear regression model to calculate the X_{CO2} value of the smoke plumes that were not covered by the FTS L1B points. A scatter plot of ΔX_{CO2} against the CAI band 1 radiance values is shown in Figure 46.3. This graph illustrates that higher radiance values usually corresponded to higher CO_2 emissions ($R^2 = 0.44$, $P < 0.001$). We could calculate the CO_2 emissions from wildfires for each smoke plume pixel with this regression equation.

TABLE 46.1 FTS L1B Point Data for Exposures to Smoke from Russian Forest Fires in 2010

ID	Time	Lat	Lon	Smoke Classification	SX_{CO2} (ppm)	BX_{CO2} (ppm)	ΔX_{CO2} (ppm)
1	20100801101735	56.2002	45.1421	Smoke	389.112	389.055	0.056
2	20100805105011	55.4093	41.0231	Thick smoke	402.024	386.006	16.017
3	20100805105015	55.4095	41.0207	Thick smoke	400.112	386.006	14.105
4	20100805105020	55.4088	41.0185	Thick smoke	401.468	386.006	15.462
5	20100807101741	55.9293	45.0298	Smoke	388.500	383.000	5.500
6	20100807101746	55.9326	45.0297	Smoke	389.946	383.000	6.946
7	20100807101751	55.9358	45.0304	Smoke	388.348	383.000	5.348
8	20100808104944	55.7508	37.6287	Smoke	388.736	385.655	3.081
9	20100809104948	55.7134	37.6118	Smoke	389.709	385.655	4.054
10	20100810104953	55.4384	37.4946	Smoke	398.088	385.655	12.433
11	20100811104958	55.1632	37.3787	Smoke	391.350	385.655	5.695
12	20100812105012	55.4476	41.0439	Light smoke	386.878	385.655	1.224
13	20100813105016	55.4473	41.0413	Light smoke	385.720	385.655	0.065
14	20100814105021	55.4472	41.0396	Light smoke	387.079	385.655	1.425
15	20100815105025	55.9688	36.8649	Smoke	388.874	385.655	3.219
16	20100816105044	53.4402	35.8906	Light smoke	386.710	385.655	1.055
17	20100817105048	53.4415	35.8874	Light smoke	388.309	385.655	2.655
18	20100813101743	55.9331	45.0332	Thick smoke	398.309	383.730	14.579
19	20100813101748	55.9356	45.0328	Thick smoke	401.248	383.730	17.518
20	20100813101753	55.9396	45.0338	Thick smoke	400.359	383.730	16.629
21	2010081410509	55.9958	41.3305	Light smoke	383.685	378.960	4.725
22	20100814105013	55.9964	41.329	Light smoke	383.031	378.960	4.070
23	20100814105032	53.9962	36.1053	Light smoke	388.255	378.960	9.295
24	20100814105036	53.9956	36.1008	Light smoke	387.056	378.960	8.096
25	20100814105041	53.9973	36.0976	Light smoke	386.535	378.960	7.574

Source: Data from Guo et al. (2017).

Note: SX_{CO2} means X_{CO2} of smoke plumes; BX_{CO2} means X_{CO2} in clear sky; ΔX_{CO2} was the difference value between SX_{CO2} and BX_{CO2}.

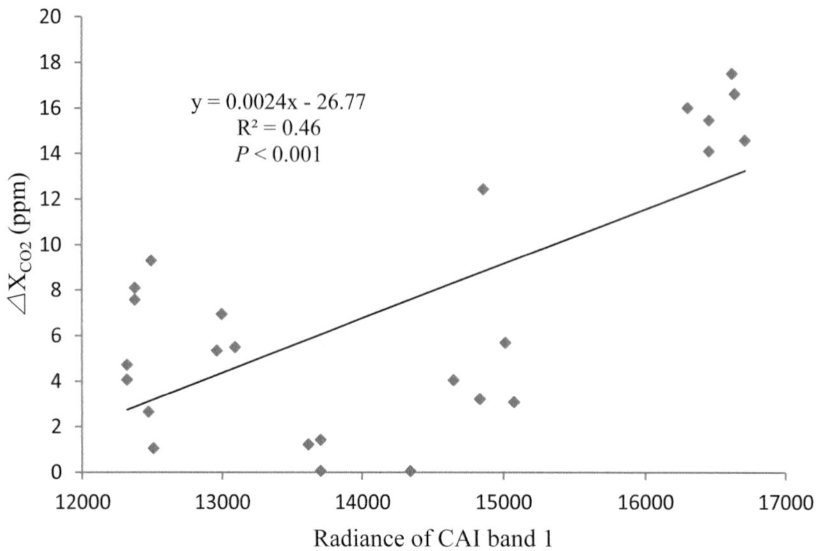

FIGURE 46.3 Scatter plots of ΔX_{CO2} against CAI band 1 radiance values (Guo et al. 2017).

Mean Height of Wildfire Smoke Plumes

The GOSAT-FTS TIR L2 data provided CO_2 profiles at 28 levels of atmospheric pressure. This study selected 10 GOSAT TIR L2 data points from July 29 to August 18 that covered the smoke plumes and 10 data points in clear sky conditions, and then, we analyzed the CO_2 concentration changes against atmospheric pressure (Figure 46.4). Compared with the profile of CO_2 concentrations when the skies were clear, it showed that the elevation of 800 hPa was the break point when the FTS TIR points were exposed to smoke plumes. Based on Equation (46.2), 1.58 km was obtained as the height of smoke plume. Figure 46.4a also indicates that on August 13 and 16, the break point of CO_2 concentration appeared at approximately 700 hPa, and in Figure 46.4b, the effect on August 9 was quite different from that of other days at 800 hPa.

CO_2 Released from Wildfires

The previously determined smoke plume areas and height were put into Equation (46.3) to calculate the weight of the CO_2 emissions for each smoke plume pixel, and later, the values were summed to obtain the amount of CO_2 released in 3 days (Figure 46.5). In total, 255.76 Tg CO_2 was released from Russian forest fires in 2010. From August 4 to 6, wildfires in Russia emitted the highest amount of CO_2 at 54.51 Tg. At the beginning of the study time, from July 23 to 25, wildfires released only 0.44 Tg CO_2, and the CO_2 emissions subsequently increased with the increase of fire points and burned areas. After August 6, the CO_2 release decreased to 14.72 Tg by August 16–18.

Compared with BBM and ERs Methods

Using the BBM (Seiler and Crutzen, 1980), this study multiplies the burned area, biomass fuel load, combustion factor, and emission factor of CO_2 to obtain CO_2 emissions from the 2010 Russia wildfires. The burned area of 5 million hectares was used in BBM because it was confirmed by two Russian Institutes (EMERCOM and the Irkutsk Institute of Solar and Terrestrial Physics) for the same time (GFMC State DUMA Report on the Fires in Western Russia 2010). With regard to the other parameters, the results of Akagi et al. (2010) were used. At last, CO_2 emissions of 261.82–302.48 Tg were obtained, which were slightly

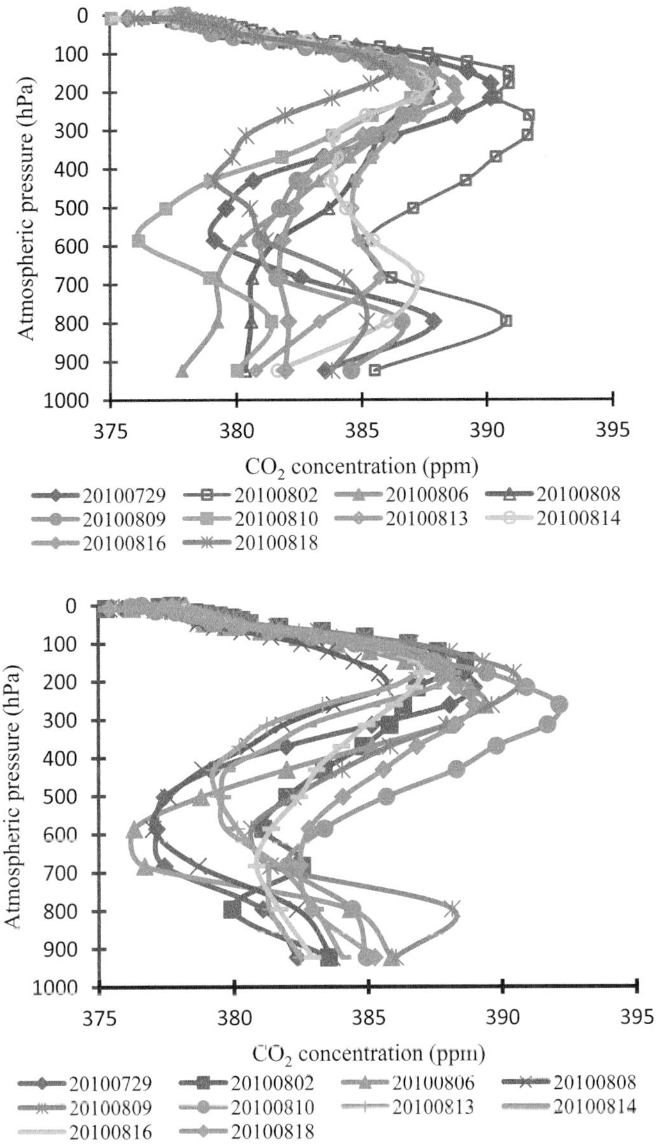

FIGURE 46.4 **(See color insert.)** CO_2 concentration changes against atmospheric pressure. FTS TIR points were covered by smoke plumes (a) and in clear sky (b) (Guo et al. 2017).

higher than those calculated from the GOSAT data but were within the tolerance range. Differences in the burned areas, the time scale and the study regions may be the reasons for the different results. In any case, this was an excellent result for us, which means that it was achieved by using remote sensing data to estimate the CO_2 emissions without considering the wildfire process and parameters.

Reported studies estimated CO emissions from the 2010 Russia wildfires and obtained quite different results. Yurganov et al. (2011) estimated that the total CO released from Russian wildfires was 34–40 Tg during July and August 2010. Konovalov et al. (2011) estimated that the 2010 western Russia forest fires emitted ~9.7 Tg CO, and R'Honi et al. (2013) argued that the total CO emitted from July and August 2010 in the west of Russia was between 19 and 33 Tg. Using the emission ratio (ER) of CO (0.11±0.07) that was found by Simpson et al. (2011), who analyzed smoke plume samples from Canadian boreal forest fires,

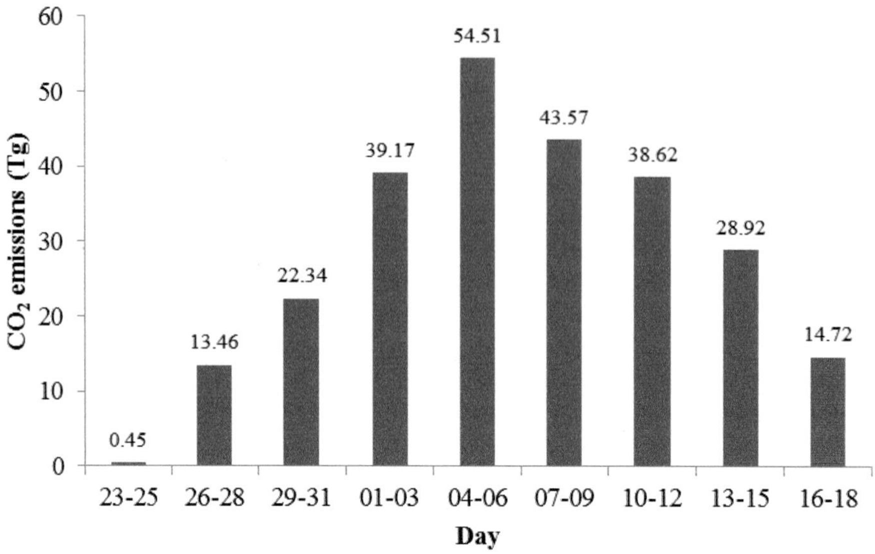

FIGURE 46.5 Three-day CO_2 emissions from Russian forest fires in 2010. 23–25 means the sum of values on July 23, 24, and 25 (Guo et al. 2017).

we obtained CO_2 emissions: 485.71–571.43 Tg (by Yurganov et al.), ~138.57 Tg (by Konovalov et al.), and 271.43–471.42 Tg (by R'Honi, et al.). One could see that CO_2 emissions determined from the GOSAT data (255.76 Tg) were closest to those calculated by using R'Honi's CO emissions, though somewhat lower. The different results of various researchers might have been caused by different data sources or dissimilar study regions.

Conclusions

Following are the main conclusions drawn from this exercise:

1. GOSAT CAI provided data well suited for distinguishing smoke plumes from clear or cloudy skies.
2. GOSAT-FTS L1B data could be used to calculate the ΔXCO_2 by comparing the XCO_2 values that were exposed to smoke plumes from those that represented clear sky.
3. GOSAT-FTS TIR data, which provided the CO_2 profiles, could be used to monitor the height of the smoke plumes. The smoke plume height was approximately 800 hPa, or 1.58 km above the ground level, in the Russian 2010 wildfires.
4. GOSAT could be used to calculate CO_2 emissions from wildfires. With the method that proposed in this study, we determined that the 2010 Russian wildfires emitted 255.76 Tg CO_2, and this result was verified by the BBM and ER method.

References

Akagi, S.K.; Yokelson, R.J.; Wiedinmyer, C.; Alvarado, M.J. Emission factors for open and domestic biomass burning for use in atmospheric models. *Atmospheric Chemistry & Physics* **2010**, *11*, 27523–27602.

Amiro, B.D.; Barr, A.G.; Barr, J.G.; Black, T.A.; Bracho, R.; Brown, M.; Chen, J.; Clark, K.L.; Davis, K.J.; Desai, A.R.; Dore, S.; Engel, V.; Fuentes, J.D.; Goldstein, A.H.; Goulden, M.L.; Kolb, T.E.; Lavigne, M.B.; Law, B.E.; Margolis, H.A.; Martin, T.; McCaughey, J.H.; Misson, L.; Montes-Helu, M.;

Noormets, A.; Randerson, J.T.; Starr, G.; Xiao, J. Ecosystem carbon dioxide fluxes after disturbance in forests of North America. *Journal of Geophysical Research-Biogeosciences* **2010**, *115*(G00K02), 4869–4890.

Chubarova, N.; Nezval', Y.; Sviridenkov, I.; Smirnov, A.; Slutsker, I. Smoke aerosol and its radiative effects during extreme fire event over Central Russia in summer 2010. *Atmospheric Measurement Techniques* **2012**, *5*, 557–568.

Dixon, R.K.; Krankina, O.N. Forest fires in Russia: carbon dioxide emissions to the atmosphere. *Canadian Journal of Forest Research* **1993**, *23*, 700–705.

Goto, Y.; Suzuki, S. Estimates of carbon emissions from forest fires in Japan, 1979–2008. *International Journal of Wildland Fire* **2013**, *22*, 721–729.

Guo, M.; Li, J.; Xu, J.; Wang, X.; He, H.; Wu, L. CO_2 emissions from the 2010 Russian wildfires using GOSAT data. *Environmental Pollution* **2017**, *226*, 60–68.

Guo, M.; Wang, X.; Liu, Y.; Li, J.; Wang, H.; Matsuoka, N.; Tani, H. The effects of sand dust storms on greenhouse gases. *International Journal of Remote Sensing* **2012**, *33*, 6838–6853.

Konovalov, I.B.; Beekmann, M.; Kuznetsova, I.N.; Yurova, A.; Zvyagintsev, A.M. Atmospheric impacts of the 2010 Russian wildfires: integrating modelling and measurements of the extreme air pollution episode in the Moscow megacity region. *Atmospheric Chemistry and Physics* **2011**, *11*, 12141–12205.

Oguma, H.; Morino, I.; Suto, H.; Yoshida, Y.; Eguchi, N.; Kuze, A.; Yokota, T. First observations of CO_2 absorption spectra recorded in 2005 using an airship-borne FTS (GOSAT TANSO-FTS BBM) in the SWIR spectral region. *International Journal of Remote Sensing* **2011**, *32*, 9033–9049.

R'Honi, Y.; Clarisse, L.; Clerbaux, C.; Hurtmans, D. Exceptional emissions of NH3 and HCOOH in the 2010 Russian wildfires. *Atmospheric Chemistry & Physics* **2013**, *13*, 4171–4181.

Robinson, J.M. On Uncertainty in the Computation of Global Emissions from Biomass Burning. *Climatic Change* **1989**, *14*, 243–262.

Rosa, I.M.D.; Pereira, J.M.C.; Tarantola, S. Atmospheric emissions from vegetation fires in Portugal (1990–2008): estimates, uncertainty analysis, and sensitivity analysis. *Atmospheric Chemistry & Physics* **2011**, *11*, 2625–2640.

Ross, A.N.; Wooster, M.J.; Boesch, H.; Parker, R. First satellite measurements of carbon dioxide and methane emission ratios in wildfire plumes. *Geophysical Research Letters* **2013**, *40*, 4098–4102.

Seiler, W.; Crutzen, P.J. Estimates of gross and net fluxes of carbon between the biosphere and the atmosphere from biomass burning. *Climatic Change* **1980**, *2*, 207–247.

Simmonds, P.G.; Manning, A.J.; Derwent, R.G.; Ciais, P.; Ramonet, M.; Kazan, V.; Ryall, D. A burning question. Can recent growth rate anomalies in the greenhouse gases be attributed to large-scale biomass burning events? *Atmospheric Environment* **2005**, *39*, 2513–2517.

Simpson, I.J.; Akagi, S.K.; Barletta, B.; Blake, N.J.; Choi, Y.; Diskin, G.S.; Fried, A.; Fuelberg, H.E.; Meinardi, S.; Rowland, F.S.; Vay, S.A.; Weinheimer, A.J.; Wennberg, P.O.; Wiebring, P.; Wisthaler, A.; Yang, M.; Yokelson, R.J.; Blake, D.R. Boreal forest fire emissions in fresh Canadian smoke plumes: C-1-C-10 volatile organic compounds (VOCs), CO_2, CO, NO_2, NO, HCN and CH_3CN. *Atmospheric Chemistry & Physics* **2011**, *11*, 6445–6463.

Witte, J.C.; Douglass, A.R.; da Silva, A.; Torres, O.; Levy, R.; Duncan, B.N. NASA A-Train and Terra observations of the 2010 Russian wildfires. *Atmospheric Chemistry & Physics* **2011**, *11*, 9287–9301.

Wu, J. The change of the atmospheric pressure with elevation, Tianjin Education, **1990**, 46.

Yurganov, L.N.; Rakitin, V.; Dzhola, A.; August, T.; Fokeeva, E.; George, M.; Gorchakov, G.; Grechko, E.; Hannon, S.; Karpov, A.; Ott, L.; Semutnikova, E.; Shumsky, R.; Strow, L. Satellite- and ground-based CO total column observations over 2010 Russian fires: accuracy of top-down estimates based on thermal IR satellite data. *Atmospheric Chemistry Physics* **2011**, *11*, 7925–7942.

Index

A

Agriculture, 17
  biodiversity, 17
    agricultural species, 19
    agroecosystems, 18
    global impact, 17–18
    loss of diversity, 19
    offsite impacts, agricultural activities, 19
    principle of agroecosystem management, 19
  organic, 251
    agricultural practices, 252
    certification, 253
    history and extent of, 253
    sustainability, 254
Agroforestry, 55
  biodiversity, 55–61
  definition, 73
  preservation, 76–77
  tropical, 55
  value of, 74
  in world forests, 75–76
Alpine tundra, nitrogen deposition, 315
  Changbai Mountain, 316
  D. angustifolia, 316
  Deyeuxia angustifolia, 316
  elevated temperature, 317
  encroachment, 315
  responses to environmental change, 317
  vegetation, 315
Altitudinal belts, 339
  bioclimatic indexes, 345
  continentality *vs.* isolation, 343
  diversity and complexity, 341
  ecological factors, 343
  geographical factors
    continentality, 343
    latitude, 343
    slope-facing effect, 343
  mass elevation effect *vs.* summit syndrome, 343–344
  patterns of, 341
    coastal to inland areas, 341
    high to low latitudes, 342
    outer margins to interior of mountain massifs, 342
    shady *vs.* sunny slopes, 342
    upward shifting, global warming, 342
    windward *vs.* leeward slopes, 341
  physiological factors, 344
Appalachian trail, 359
  decision support, 359–362
  ecological conditions, 359–362
  forecasting, 359–362
  monitoring, 359–362
  reporting, 359–362
Assessment
  attribute ratings, 11
  community-based, 3
  community forests (CFs), 4
  human-elephant conflict (HEC), 3
  major problems, 7
  Nepal, 3
  reduce conflict, 9
  site selection, 4
  statistical analysis, 7

B

Biodegradation, 349
  air biofiltration, 356
  bioremediation technologies, 354
  contaminant bioavailability, 351
  contaminant biodegradability, 353
  contaminant toxicity, 352
  general microbial approaches, 353
  phytoremediation, 356
  remediation of contaminants, 353
Biodiversity
  agriculture, 17
    agricultural species, 19
    agroecosystems, 18

Biodiversity (*cont.*)
global impact, 17–18, 47–48
loss of diversity, 19
offsite impacts, agricultural activities, 19
principle of agroecosystem management, 19
offsite impacts, agricultural activities, 48
principle of agroecosystem management, 49–50
climate change, 23
ecosystems and ecosystem services, 25
experimental approaches, 26
impacts, 24
individuals, populations, and habitats impacts, 24
interactions between species and community, 25
mitigation and adaptation, 28–29
modeling approaches and scenario development, 27
non-knowledge management, 27
politics, 29
relative assessment, 27
species and community impacts, 25
system resilience, 25
vulnerability indices, 27
conservation, 35–40
biology, 41
dimensions, 35
discovery, 61–63
extinction rates, 42
how much do we know, 35
genetic diversity, 35
global action, 41
phyletic diversity, 35
phylogenetic tree of life, 36
protected areas, 40
species diversity, 60–61
threats and environment, 40
habitat suitability, 45
biological assumptions, 47
data and approach limitations, 48
data requirements, 47
endangered species management, 50
habitat change, 49–50
invasive species management, 50
limitations, 48–49
due to biological assumptions, 49
due to data approach, 48
mechanistic approaches, 49
modeling, 47
n-dimensional hypervolume, 46
niche theory, 46
species fitness and habitat quality, 45
suitability, 46
important biodiversity areas (IBAs), 120
International Union for the Conservation of Nature (IUCN), 120
prioritizing conservation sites, 119
alliance for zero extinction (AZE) sites, 120
biodiversity prioritization, 120
conservation and development, 123
game conservation in Uganda, 124
identification and delineation criteria, 119
key biodiversity areas (KBA), 120
Uganda, 120
district forest services, 119
National Forest Authority (NFA), 119
protected area, 120
Wildlife Authority (UWA), 119
tropical agroforestry, 55
certifications, 60
definitions, 55, 57
planned, 60
preservation, agroforestry and biodiversity, 59
studies, 60
trends in the world forests, 58
value of, 56, 58
values, 65
conflict resolution, 71
instrumental, 68
intrinsic, 66
value battleground, 70
Biological, nitrogen fixation, 321
associative, 324
biochemistry, 322
cell-free, 323
control of nitrogenase, 323
free-living, 321
genetics, 324
purification of nitrogenase, 323
requirement for adenosine triphosphate, 323
tertiary structure, 324
symbiotic and free-living fixers, 321
Biome, 75
Afrotropical highlands, 119
boreal Forest, 77
definition, 75
desert, 79
Guinea–Congo Forests, 119
Lake Victoria Basin, 119
montane, 75
polar, 77
soil of, 77
Somali–Masai, 119
Sudan–Guinea Savanna, 119
temperate grassland and forest, 77
tropics, 79
Bioremediation, biodegradation, 349
air biofiltration, 356
bioremediation technologies, 354
contaminant bioavailability, 351
contaminant biodegradability, 353
contaminant toxicity, 352
general microbial approaches, 353
phytoremediation, 356
remediation of contaminants, 353

C

Climate change, 23–30
biodiversity, 17, 23, 35, 45, 55, 65, 119, 359
ecosystems and ecosystem services, 52–53
experimental approaches, 54
individuals, populations, and habitats impacts, 52
mitigation and adaptation, 55–56
modeling approaches and scenario development, 54
non-knowledge management, 55
politics, 56
species and community impacts, 52
system resilience, 53
vulnerability indices, 54–55
CO_2
emission, 393
GOSAT data, 394
release, 398
smoke plume, 395
wildfire, 395
Community forests (CFs), 136
attributes of, 138
equity, 137
National Forestry Policies, 138
operation, 135
sustainability of silviculture practices, 136
Conservation
biodiversity, 3, 17, 23, 35, 45, 55, 65, 119, 359
biology, 41
dimensions, 35
discovery, 61–63
extinction rates, 42
how much do we know, 35
genetic diversity, 35
global action, 41
phyletic diversity, 35
phylogenetic tree of life, 36
protected areas, 40
species diversity, 60–61
threats and environment, 40
decision-making, 81
monitoring strategies, 83
sampling design, 83–84
structured, 82
Crop and atmosphere, 261
controlling factors, 262
impacts of trace gas, 261
internal transfer resistances, 264
plant physiological controls, 263
trace gas exchange, 262
Crop yield, ozone effects, 325
crop loss assessment, 326
critical levels, 326
effects on crops, 326
open top chamber, 325
protein contents, 326
quality effects, 328
uptake, 326

D

Decision-making
monitoring strategies, 83
sampling design, 83–84
structured, 82
Decision support, 359–362
Appalachian trail, 359–362
ecological conditions, 359–362
forecasting, 359–362
monitoring, 359–362
reporting, 359–362
Desertification, 267
climate change, 275–276
concern, 273–275
definition, 267–268
land degradation and desertification LDD), 268
causes, 269–271
Desertification, land degradation, and drought (DLDD), 275–276
Diversity, ecosystems, 166
agroecosystems, 172
biodiversity at ecosystem level, 166
deciduous forests, 169
deserts, 169
ecosystem management, 173
ecosystem types, 166
forest, 168
grasslands, 169
polar ecosystems, 171
species, 87
benefits of, 90–91
definition, 134
maintenance of, 89
patterns of, 88–89
quantification of, 87
threats to, 89–90
tropical, 168
rainforests, 168
seasonal forests, 168
urban ecosystems, 172
wetlands, 171
world ecosystems, 167

E

Ecological and evolutionary process, 281
climate change, 283
evolutionary changes, 282
evolutionary conservation strategies, 284
jeopardized ecosystem health, 282
maintenance of genetic biodiversity, 284
natural resource overexploitation, 282

Ecological and evolutionary process (*cont.*)
protect evolutionary process, 283
sustainable natural resource, 283–284
Ecology, 289
comparative, 290
evolutionary, 291
experimental, 290
functional, 289
general patterns, 290
physiology to ecophysiology, 289
Economic valuation, environmental, 205
economic valuation, 206
ecosystem services, 151–153
modern origin of perspectives, 208
market value and emergy value, 208
non-economic valuation, 207
nonmarket value and emergy value, 209
Ecosystem production, 305
definition, 305
measurement of, 306
net ecosystem production (NEP), 305
temporal and spatial patterns, 307
terrestrial *vs.* aquatic ecosystems, 307
Ecosystems diversity, 166
agroecosystems, 172
biodiversity at ecosystem level, 166
deciduous forests, 169
deserts, 169
ecosystem management, 173
ecosystem types, 166
forest, 168
grasslands, 169
polar ecosystems, 171
tropical, 168
rainforests, 168
seasonal forests, 168
wetlands, 171
world ecosystems, 167
urban ecosystems, 172
Ecosystems, forest nutrient cycling, 177
climate influence, 178
biotic factors, 179
abiotic factors, 179
role of disturbance, 180
Ecosystems, 143 165, 177, 183, 191
diversity, 166
functions and services, 183
accounting and assessment, 187
adaptive management, 187
bundling, 187
distribution issues, 188
early evaluation syntheses, 185
ecosystem services, 183
information dissemination, 188
integrated measurement, 186
major world report, 186
modeling, 187
participation, 188
property rights, 187
scale matching, 187
scaling, 187
science–policy interface, 188
trade-offs, 186
valuation, 184
soil animal functioning, 191
categories, 183
decomposition and nutrient cycling, 193
earthworms, 185
impacts, 194
earthworms, 194–195
protozoa, 195
termites, 195
influence of, 193
pest control, 194
protozoa, 194
soil biodiversity, 195
soil food web, 192
spatial distribution, 193–194
water cycling, 194
Ecosystem services (ESs), 143, 151, 157
evaluation, 143
affects the value, 147
biophysical, 144
coordinated biophysical and social, 147
economic, 145
spatial analysis, 145
exotic species, 93
land systems approach, 151
coupled human-environment system, 151
definition, 151
ecosystem goods and services, 152
land systems change, 152
pollinators and pollination, 157
cultural services, 161
definitions, 158
maintaining and supporting, 162
natural resource systems, 157
pollination services, 158
provisional systems, 159–161
regulating services, 161
supportive services, 158–159
spatial analysis, 145
Ecosystem services' partnership (ESP), 186
accounting and assessment, 187
adaptive management and new institutions, 187
distribution issues, 188
information dissemination, 188
integrated measurement, modeling, valuation, and decision science, 186

participation, 188
property rights, 187
scale matching, 187
science-policy interface, 188
trade-offs, 186
valuation, 184
Ecotone, 197
directional transitions, 200
emergent properties, 201–202
hierarchies, 198
landscape-scale, 200
patch-scale, 198–199
plant-scale, 198
shifting transitions, 200–201
spatial and temporal dynamics, 200–201
stationary transitions, 200–201
Elevated CO_2 levels, leaves, 299
canopy and ecosystem scale, 301
molecular and cellular scale, 299
whole leaf scale, 301
whole plant scale, 301
Environmental goods and services, 205
economic valuation, 206
modern origin of perspectives, 208
market value and emergy value, 208
non-economic valuation, 207
nonmarket value and emergy value, 209
Evaluation, ecosystem services, 143
affects the value, 147
biophysical, 144
coordinated biophysical and social, 147
economic, 145–146
spatial analysis, 145
Evolution, origin of land plants, 293
adaptations, 293
algae, 294
earliest, 294
embryophytes, 294–295
evolutionary relationship, 295–296
evolve, when, 294
Evolutionary process, ecological, 281
climate change, 283
evolutionary changes, 282
evolutionary conservation strategies, 284
jeopardized ecosystem health, 282
maintenance of genetic biodiversity, 284
natural resource overexploitation, 282
protect evolutionary process, 283
sustainable natural resource, 283–284
Exotic species, 93
impact of, 96
invasive, 94
prevention, control, management, and risk assessment, 98–99
species lists and data sources, 100
who is who, 93–94

F

Forests, temperate, 213
climate, 214–215
land formation history, 215
structure and composition, 216
temperate deciduous, 217
temperature evergreen, 218
threats, 220
Forests, tropical rain, 227
conservation, 232
definition, what is, 227
importance, 230
threats, 231
where occur? 227
Africa, 229
Asia, 229
Island Rain forest, 229
Madagascar, 231–232
New Guinea and Australia, 229–230
South and Central America, 228
Forests, community (CFs), 136
attributes of, 138
community, 136
equity, 137
National Forestry Policies, 138
operation, 135
sustainability of silviculture practices, 136
temperate, 205
tropical rain, 227
Forest nutrient cycling, ecosystems, 177
abiotic factors, 179
climate influence, 178
biotic factors, 179
role of disturbance, 180
Functional ecology, 289
comparative, 290
definition, 147
evolutionary, 291
experimental, 290
functional, 289
general patterns, 290
physiology to ecophysiology, 289
Functions and services, 183
accounting and assessment, 187
adaptive management, 187
bundling, 187
distribution issues, 188
early evaluation syntheses, 185
ecosystem services, 183
information dissemination, 188
integrated measurement, 186
major world report, 186
modeling, 187
participation, 188

Functions and services (*cont.*)
property rights, 187
scale matching, 187
scaling, 187
science–policy interface, 188
trade-offs, 186
valuation, 184

G

Genetics, population, 113
coalescent, 116
DNA sequence change, 113
evolutionary forces, 113
genetic novelty, 114
migration, spatial dimension, 117
mutation drift, 115
natural selection, 115
neutrality hypothesis, 115
random drift, 114
recombination, 116
Global plant breeding, 331
accomplishments, 333
benefits, 334
need for, 332
new directions, 334
world population and food needs, 331
Goods and services, environmental, 205
economic valuation, 206
modern origin of perspectives, 208
market value and emergy value, 208
non-economic valuation, 207
nonmarket value and emergy value, 209
Grasslands and savanna, 235
biodiversity and productivity, 242
determinants of vegetation structure, 237
fire, 239
herbivores, 239–241
plant available moisture/nutrients, 237–239
ecological concerns, 242–243
geographic distribution, 236–237
paradigms/models, 241–242
structural properties, 236–237
sustainable use, 242–243

H

Habitat suitability, 45
biological assumptions, 47
data and approach limitations, 48
data requirements, 47
endangered species management, 50
habitat change, 49–50
invasive species management, 50
limitations, 48–49
due to data approach, 48
due to biological assumptions, 49
mechanistic approaches, 49
modeling, 47
n-dimensional hypervolume, 46
niche theory, 46
species fitness and habitat quality, 45
suitability, 46
Human-elephant conflict, 3
Human-wildlife conflict, 148

I

Important biodiversity areas (IBAs), 120
International Union for the Conservation of Nature (IUCN), 120
Indicator species, 106
expected values, 108
benchmarks, 108

K

Keystone species, 105
expected values and benchmark, 108
indicator, 106
keystone, 107

L

Land plants, 293
adaptations, 293
algae, 294
earliest, 294
embryophytes, 294–295
evolutionary relationship, 295–296
evolve, when, 294
origin and evolution, 293
Land systems approach, ecosystem services, 151
coupled human-environment system, 151
definition, 151
ecosystem goods and services, 152
land systems change, 152
Leaves, elevated CO_2 levels, 299
canopy and ecosystem scale, 301
molecular and cellular scale, 299
whole leaf scale, 301
whole plant scale, 301

M

Mountain mass effect, 339
altitudinal belts, 339
bioclimatic indexes, 345
continentality *vs.* isolation, 343
diversity and complexity, 341
ecological factors, 343

geographical factors
continentality, 343
latitude, 343
slope-facing effect, 343
mass elevation effect *vs.* summit syndrome, 343–344
patterns of, 341
coastal to inland areas, 341
high to low latitudes, 342
outer margins to interior of mountain massifs, 342
shady *vs.* sunny slopes, 342
upward shifting, global warming, 342
windward *vs.* leeward slopes, 341
physiological factors, 344

N

Net ecosystem production (NEP), 305
definition, 305
measurement of, 306
temporal and spatial patterns, 307
terrestrial *vs.* aquatic ecosystems, 307
Net primary production (NPP), 365
CASA model, 368–369
estimation, 365
FPAR Algorithm, 370–373
maximum light use efficiency, 374
Northeast China, 366
temperature scalars, 368
total solar surface irradiation algorithm, 370
water scalar, 373
Nitrogen deposition, 315
alpine tundra vegetation, 315
Changbai Mountain, 316
D. angustifolia, 316
Deyeuxia angustifolia, 316
elevated temperature, 317
encroachment, 315
responses to environmental change, 317
Nitrogen fixation, 321
associative, 324
biochemistry, 322
cell-free, 323
purification of nitrogenase, 323
requirement for adenosine triphosphate, 323
control of nitrogenase, 323
free-living, 321
genetics, 324
tertiary structure, 324
symbiotic and free-living fixers, 321
Non-economic valuation, environmental goods and services, 205
economic valuation, 206
ecosystem services, 151
modern origin of perspectives, 208
market value and emergy value, 208
non-economic valuation, 207
nonmarket value and emergy value, 209

O

Optical properties, 383
leaves, 383
light, 386
reflected light, 386
Organic, agriculture, 251
agricultural practices, 252
certification, 253
history and extent of, 253
sustainability, 254
Organic farming systems, 252–255
Origin of Species, 36
Origin and evolution, land plants, 293
adaptations, 293
algae, 294
earliest, 294
embryophytes, 294–295
evolutionary relationship, 295–296
evolve, when, 294
Ozone effects, 325
crop loss assessment, 326
critical levels, 326
effects on crops, 326
open top chamber, 325
protein contents, 326
quality effects, 328
uptake, 326

P

Partnership, ecosystem services' (ESP), 186
accounting and assessment, 187
adaptive management and new institutions, 187
distribution issues, 188
information dissemination, 188
integrated measurement, modeling, valuation, and decision science, 186
participation, 188
property rights, 187
scale matching, 187
science-policy interface, 188
trade-offs, 186
valuation, 184
Plant breeding, global, 331
accomplishments, 333
benefits, 334
need for, 332
new directions, 334
world population and food needs, 331

Pollinators and pollination, 157
cultural services, 161
definitions, 158
maintaining and supporting, 162
natural resource systems, 157
pollination services, 158
pollinators, 157–162
provisional systems, 159–161
regulating services, 161
supportive services, 158–159
Population genetics, 113
coalescent, 116
DNA sequence change, 113
evolutionary forces, 113
genetic novelty, 114
migration, spatial dimension, 117
mutation drift, 115
natural selection, 115
neutrality hypothesis, 115
random drift, 114
recombination, 116
Primary production, 365
CASA model, 368–369
estimation, 365
FPAR Algorithm, 370–373
maximum light use efficiency, 374
Northeast China, 366
temperature scalars, 368
total solar surface irradiation algorithm, 370
water scalar, 373
Prioritizing conservation sites, 119
alliance for zero extinction (AZE) sites, 120
biodiversity prioritization, 120
conservation and development, 123
game conservation in Uganda, 124
identification and delineation criteria, 119
key biodiversity areas (KBA), 120
Uganda, 120
district forest services, 119
National Forest Authority (NFA), 119
protected area, 120
Wildlife Authority (UWA), 119
Production, ecosystem, 305
definition, 305
measurement of, 306
net ecosystem production (NEP), 305
temporal and spatial patterns, 307
terrestrial *vs.* aquatic ecosystems, 307
Production, net primary (NPP), 365
CASA model, 368–369
estimation, 365
FPAR Algorithm, 370–373
maximum light use efficiency, 374
Northeast China, 366
temperature scalars, 368
total solar surface irradiation algorithm, 370
water scalar, 373

R

Reflected light, 386
optical properties, 383
leaves, 383
light, 386
Remote sensing, 359, 393
decision support, 359–362
Appalachian trail, 359–362
ecological conditions, 359–362
forecasting, 359–362
monitoring, 359–362
reporting, 359–362
CO_2 emission, 393
CO_2 release, 398
GOSAT data, 394
smoke plume, 395
wildfire, 395

S

Savannas and grasslands, 235
biodiversity and productivity, 242
determinants of vegetation structure, 237
fire, 239
herbivores, 239–241
plant available moisture/nutrients, 237–239
ecological concerns, 242–243
geographic distribution, 236–237
paradigms/models, 241–242
structural properties, 236–237
sustainable use, 242–243
Seed banks and seed dormancy, 129
characteristics, 129
depth of burial, 130
environmental factors, 131
periodicity and intermittent germination, 130
persistence, 130
seed banks, 129
seed dormancy, 129
soil seed bank, 129
weeds, 129
Soil animal functioning, 191
decomposition and nutrient cycling, 193
impacts, 194
earthworms, 194–195
protozoa, 195
termites, 195
influence of, 193
pest control, 194
protozoa, 194
soil biodiversity, 195
soil food web, 192
spatial distribution, 193–194
water cycling, 194

Species, 87, 93, 105,
expected values and benchmark, 108
indicator, 106
keystone, 107
Species, diversity, 87
benefits of, 90–91
definition, 87
maintenance of, 89
patterns of, 88–89
quantification of, 87
threats to, 89–90
Species, exotic 93
impact of, 96
invasive, 94
prevention, control, management, and risk assessment, 98–99
species lists and data sources, 100
who is who, 93–94
Sustainability, 136
silviculture, community forests, 136

T

Temperate forests, 213
climate, 214–215
land formation history, 215
structure and composition, 216
temperate deciduous, 217
temperature evergreen, 218
threats, 220
Trace gas exchange, crops and atmosphere, 261
controlling factors, 262
impacts of trace gas, 261
internal transfer resistances, 264
plant physiological controls, 263
Tropical agroforestry, 55
certifications, 60
definitions, 55, 57
planned, 60
preservation, agroforestry and biodiversity, 59
studies, 60
trends in the world forests, 58
value of, 56, 58
Tropical rain forests, 227
conservation, 232
definition, what is, 227
importance, 230
threats, 231
where occur? 227
Africa, 229
Asia, 229
Island Rain forest, 229
Madagascar, 231–232
New Guinea and Australia, 229–230
South and Central America, 228
valuing of biodiversity
conflict resolution, 71
instrumental, 68
intrinsic, 66
value battleground, 70

U

Uganda, 119

V

Valuation, environmental economics, 205
economic valuation, 206
ecosystem services, 151–153
modern origin of perspectives, 208
market value and emergy value, 208
non-economic valuation, 207
nonmarket value and emergy value, 209

W

Weeds, 129
characteristics, 129
depth of burial, 130
environmental factors, 131
periodicity and intermittent germination, 130
persistence, 130
seed banks, 129
seed dormancy, 129
soil seed bank, 129
Wildfire, 395
CO_2 emission, 393
CO_2 release, 398
GOSAT data, 394
smoke plume, 395
Wildlife conservation, 4–8, 71, 81, 119–126

Y

Yield, crop ozone effects, 325
crop loss assessment, 326
critical levels, 326
effects on crops, 326
open top chamber, 325
protein contents, 326
quality effects, 328
uptake, 326

For Product Safety Concerns and Information please contact our
EU representative GPSR@taylorandfrancis.com Taylor & Francis
Verlag GmbH, Kaufingerstraße 24, 80331 München, Germany